AF333743

Interferometry and Synthesis in Radio Astronomy

Interferometry and Synthesis in Radio Astronomy

A. RICHARD THOMPSON
National Radio Astronomy Observatory

JAMES M. MORAN
Harvard-Smithsonian Center for Astrophysics

GEORGE W. SWENSON, JR.
*Department of Electrical and Computer Engineering
and Department of Astronomy
University of Illinois at Urbana-Champaign*

KRIEGER PUBLISHING COMPANY
MALABAR, FLORIDA
1998

Original Edition 1986
Reprint Edition 1991 with corrections
Reissue 1994 with minor correction
Reissue 1998 with minor corrections

Printed and Published by
KRIEGER PUBLISHING COMPANY
KRIEGER DRIVE
MALABAR, FLORIDA 32950

> **FROM A DECLARATION OF PRINCIPLES JOINTLY ADOPTED BY**
> **A COMMITTEE OF THE AMERICAN BAR ASSOCIATION AND A**
> **COMMITTEE OF PUBLISHERS:**
> This publication is designed to provide accurate and authoritative information in regard to the
> subject matter covered. It is sold with the understanding that the publisher is not engaged in
> rendering legal, accounting, or other professional service. If legal advice or other expert
> assistance is required, the services of a competent professional person should be sought.

Library of Congress Cataloging-In-Publication Data

Thompson, A.R. (Anthony Richard), 1931-
  Interferometry and synthesis in radio astronomy / A. Richard
Thompson, James M. Moran, George W. Swenson, Jr.
      p. cm.
  Reprint. Originally published: New York: Wiley, 1986.
  Includes indexes.
  ISBN 1-57524-087-4 (alk. paper)
  1. Radio interferometers. 2. Radio astronomy. I. Moran, James
M. II. Swenson, George W. (George Warner), 1922-. III. Title.
  [QB479.3.T47 1991]
  522'.682---dc20
                                                    90-48717
                                                         CIP

10   9   8   7   6   5   4   3   2

To
Sheila, Barbara, Janice,
Sarah, Susan, and Michael

...truste wel that alle the conclusiouns that han ben founde, or elles possibly mighten be founde in so noble an instrument as an Astrolabie, ben un-knowe perfitly to any mortal man...

GEOFFREY CHAUCER
A Treatise on the Astrolabe
circa 1391

PREFACE

The techniques of radio interferometry as applied to astronomy and astrometry have developed enormously in the past four decades, and the attainable angular resolution has advanced from degrees to milliarcseconds, a range of over six orders of magnitude. As arrays for synthesis mapping[†] have developed, techniques in the radio domain have overtaken those in optics in providing the finest angular detail in astronomical images. The same general developments have introduced new capabilities in astrometry and in the measurement of the earth's polar and crustal motions. The theories and techniques that underlie these advances continue to evolve, but have reached by now a sufficient state of maturity that it is appropriate to offer a detailed exposition.

The book is intended primarily for graduate students and professionals in astronomy, electrical engineering, physics, or related fields who wish to use interferometric or synthesis-mapping techniques in astronomy, astrometry, or geodesy. It is also written with radio systems engineers in mind and includes discussions of important parameters and tolerances for the types of instruments involved. Our aim is to explain the underlying principles of the relevant interferometric techniques but to limit the discussion of details of implementation. Such details of the hardware and the software are largely specific to particular instruments and are subject to change with developments in elec-

---

[†] We define synthesis mapping as the reconstruction of images from measurements of the Fourier transforms of their brightness distributions. In this book the terms map, image, and brightness distribution are largely interchangeable.

tronic engineering and computing techniques. With an understanding of the principles involved, the reader should be able to comprehend the instructions and instrumental details that are encountered in the user-oriented literature of most observatories.

The book does not stem from any course of lectures, but the material included is suitable for a graduate level course. A teacher with experience in the techniques described should be able to interject easily any necessary guidance to emphasize astronomy, engineering, or other aspects as required.

The first two chapters contain a brief review of radio astronomy basics, a short history of the development of radio interferometry, and a basic discussion of the operation of an interferometer. Chapter 3 discusses the underlying relationships of interferometry from the viewpoint of the theory of partial coherence and may be omitted from a first reading. Chapter 4 introduces coordinate systems and parameters that are required to describe synthesis mapping. It is appropriate then to examine configurations of antennas for multielement synthesis arrays in Chapter 5. Chapters 6–8 deal with various aspects of the design and response of receiving systems, including the effects of quantization in digital correlators. The special requirements of very-long-baseline interferometry (VLBI) are discussed in Chapter 9. The foregoing material covers in detail the measurement of complex visibility and leads to the derivation of radio maps discussed in Chapters 10 and 11. The former presents the basic Fourier transformation method, and the latter the more powerful algorithms that incorporate both calibration and transformation. Precision observations in astrometry and geodesy are the subject of Chapter 12. There follow discussions of factors that can degrade the overall performance, namely, effects of propagation in the atmosphere, the interplanetary medium and the interstellar medium in Chapter 13, and radio interference in Chapter 14. Propagation effects are discussed at some length since they involve a wide range of complicated phenomena that place fundamental limits on the measurement accuracy. The final chapter describes related techniques including intensity interferometry, speckle interferometry, and lunar occultation observations.

References are included to seminal papers and to many other publications and reviews that are relevant to the topics of the book. Numerous descriptions of instruments and observations are also referenced for purposes of illustration. Details of early procedures are given wherever they are of help in elucidating the principles or origin of current techniques, or because they are of interest in their own right. Because of the diversity of the phenomena described, it has been necessary, in some cases, to use the same mathematical symbol for different quantities. A glossary of principal symbols and usage follows Chapter 15.

The material in this book comes only in part from the published literature, and much of it has been accumulated over many years from discussions, seminars, and the unpublished reports and memoranda of various observatories. Thus we acknowledge our debt to colleagues too numerous to mention individually. Our special thanks are due to a number of people for critical reviews of portions of the book, or other support. These include D. C. Backer, D. S. Bagri, R. H. T. Bates, M. Birkinshaw, R. N. Bracewell, B. G. Clark, J. M. Cordes, T. M. Cornwell, L. R. D'Addario, J. L. Davis, R. D. Ekers, J. V. Evans, M. Faucherre, S. J. Franke, J. Granlund, L. J. Greenhill, C. R. Gwinn, T. A. Herring, R. J. Hill, W. A. Jeffrey, K. I. Kellermann, J. A. Klobuchar, R. S. Lawrence, J. M. Marcaide, N. C. Mathur, L. A. Molnar, P. C. Myers, P. J. Napier, P. Nisenson, H. V. Poor, M. J. Reid, J. T. Roberts, L. F. Rodriguez, A. E. E. Rogers, A. H. Rots, J. E. Salah, F. R. Schwab, I. I. Shapiro, R. A. Sramek, R. Stachnik, J. L. Turner, R. F. C. Vessot, N. Wax, and W. J. Welch. The reproduction of diagrams from other publications is acknowledged in the captions, and we thank the authors and the publishers concerned for permission to use this material. For major contributions to the preparation of the manuscript we wish to thank C. C. Barrett, C. F. Burgess, N. J. Dimond, J. M. Gillberg, J. G. Hamwey, E. L. Haynes, G. L. Kessler, K. I. Maldonis, A. Patrick, V. J. Peterson, S. K. Rosenthal, A. W. Shepherd, J. F. Singarella, M. B. Weems, and C. H. Williams. We are grateful to M. S. Roberts and P. A. Vanden Bout, former Director and present Director of the National Radio Astronomy Observatory, and to G. B. Field and I. I. Shapiro, former Director and present Director of the Harvard–Smithsonian Center for Astrophysics, for encouragement and support. Much of the contribution by J. M. Moran was written while on sabbatical leave at the Radio Astronomy Laboratory of the University of California, Berkeley, and he is grateful to W. J. Welch for hospitality during that period. G. W. Swenson, Jr. thanks the Guggenheim Foundation for a fellowship during 1984–1985. Finally, we acknowledge the support of our home institutions: the National Radio Astronomy Observatory which is operated by Associated Universities, Inc. under contract with the National Science Foundation; the Harvard–Smithsonian Center for Astrophysics which is operated by Harvard University and the Smithsonian Institution; and the University of Illinois.

A. RICHARD THOMPSON
JAMES M. MORAN
GEORGE W. SWENSON, JR.

Charlottesville, Virginia
Cambridge, Massachusetts
Urbana, Illinois

CONTENTS

1. INTRODUCTION AND HISTORICAL REVIEW   **1**

  1.1 Applications of Radio Interferometry   1

  1.2 Some Basic Terms and Definitions   3
      Cosmic Signals, 3
      Source Positions and Nomenclature, 8
      Reception of Cosmic Signals, 9

  1.3 Development of Radio Interferometry   11
      Michelson Interferometer, 11
      Principles of Radio Interferometry and
      Early Instruments, 14
      Phase Switching and the Correlator Interferometer, 19
      Survey Instruments and High-Resolution Measurements, 21
      Detailed Mapping and Special Uses of Interferometers, 23
      Earth Rotation in Synthesis Mapping, 26
      Development of Synthesis Arrays, 28
      Very-Long-Baseline Interferometry, 29

2. INTRODUCTORY THEORY OF INTERFEROMETERS AND CORRELATOR ARRAYS   **40**

  2.1 Simple Radio Interferometer   40

  2.2 Broadband Interferometer   44

2.3 Cosmic Source Synthesis   48
       Convolution and Extended Sources, 48
       Source Synthesis, 51
       Two-Dimensional Mapping, 52

2.4 Generalized, Two-Dimensional Relationship Between
       Visibility and Brightness   54

3. FURTHER THEORY OF THE INTERFEROMETER RESPONSE   57

3.1 Properties of the Incident Radiation   57
       Incident Field, 57
       Source Coherence Function, 60

3.2 Properties of the Receiving System   63
       Components of the Receiving System, 63
       *Response of the Receiving System to the
       Incident Radiation, 64*

3.3 Response to a Completely Incoherent Source   67

3.4 Response to a Completely Coherent Source   70

3.5 Space–Frequency Equivalence   71

Appendix 3.1   Van Cittert–Zernike Theorem   73

Appendix 3.2   Correlation, Convolution, and the
                        Wiener–Khinchin Relation   76

**4. GEOMETRICAL RELATIONSHIPS AND OTHER
       PRACTICAL CONSIDERATIONS   78**

4.1 Coordinate Systems for the Fourier
       Transform Relationship   78

4.2 Antenna Spacing Coordinates and (u, v) Loci   86

4.3 (u', v') Plane   90

4.4 Fringe Frequencies   91

4.5 Visibility Frequencies   92

4.6 Calibration of the Baseline   93

4.7 Antenna Mounts   94

4.8 Beamwidth and Beamshape Effects   96

4.9 Polarimetry   97
       Parameters Defining Polarization, 97
       Interferometer Response in Terms of Stokes Parameters, 99
       Instrumental Polarization, 105

4.10 Discrete Fourier Transformation   106

4.11 Sampling Theorem   110

Appendix 4.1   Conversion between Hour Angle–Declination
and Azimuth–Elevation Coordinates   112

5. DESIGN OF ARRAYS   114

5.1   Spectral Sensitivity Function of a
       Nontracking Array   115
5.2   Nontracking Interferometers and Arrays   118
5.3   Phased Arrays and Correlator Arrays   121
5.4   Spectral Sensitivity and Transfer
       Functions of a Tracking Array   123
5.5   One-Dimensional Tracking Arrays   125
5.6   Two-Dimensional Tracking Arrays   131
5.7   VLBI Arrays   135
5.8   VLBI Antennas in Space   138
5.9   Theoretical Relationships and
       Aperture Synthesis   139

6. RESPONSE OF THE RECEIVING SYSTEM   143

6.1   Frequency Conversion, Fringe Rotation,
       and Complex Correlators   143
       Frequency Conversion, 143
       Response of a Single-Sideband System, 145
       Upper-Sideband Reception, 146
       Lower-Sideband Reception, 148
       Multiple Frequency Conversion, 149
       Delay Tracking and Fringe Rotation, 149
       Simple and Complex Correlators, 150
       Response of a Double-Sideband System, 150
       *Relative Advantages of Double- and Single-Sideband
       Systems, 153*
       Sideband Separation, 154
6.2   Response to the Noise   155
       Signal and Noise Processing in the Correlator, 155
       Noise in the Measurement of Complex Visibility, 160
       Signal-to-Noise Ratio in a Synthesized Map, 162
       Noise in Visibility Amplitude and Phase, 165
       Discussion, 167

xvi   Contents

6.3	Effect of Bandwidth	169
	Mapping in the Continuum Mode, 169	
	Wide Field Mapping with a Multichannel System, 174	
6.4	Effect of Visibility Averaging	174

7.	**DESIGN OF THE ANALOG RECEIVING SYSTEM**	**180**
7.1	Principal Subsystems of the Receiving Electronics	180
	Low-Noise Input Stages, 181	
	Local Oscillator, 182	
	IF and Signal Transmission Subsystems, 182	
	Delay and Correlator Subsystems, 183	
	VLBI Systems, 184	
7.2	Local Oscillator and General Considerations of Phase Stability	184
	Round-Trip Phase Measuring Schemes, 184	
	Swarup and Yang System, 185	
	Frequency-Offset, Round-Trip System, 186	
	Automatically Correcting System, 191	
	Phase-Locked Loops and Reference Frequencies, 192	
	Phase Stability of Filters, 194	
7.3	Frequency Responses of the Signal Channels	195
	Optimum Response, 195	
	Tolerances on Variation of the Frequency Response: Degradation of Sensitivity, 196	
	Tolerances on Variation of the Frequency Response: Gain Errors, 198	
	Delay-Setting Tolerances, 200	
	Implementation of Bandpass Tolerances, 201	
7.4	Polarization-Mismatch Tolerances	201
7.5	Phase Switching	202
	Reduction of Responses to Spurious Signals, 202	
	Implementation of Phase Switching Using Walsh Functions, 203	
	Interaction of Phase Switching with Fringe Rotation and Delay Adjustment, 205	
7.6	Automatic Level Control and Gain Calibration	206
	Appendix 7.1 Single-Sideband Mixer	207

8. DIGITAL SIGNAL PROCESSING   **210**

8.1   Bivariate Gaussian Probability Distribution   211
8.2   Periodic Sampling   212
        Nyquist Rate, 212
        Correlation of Sampled but Unquantized Waveforms, 213
8.3   Sampling with Quantization   216
        Two-Level Quantization, 217
        Four-Level Quantization, 221
        Three-Level Quantization, 227
8.4   Comparison of Quantization Schemes   228
8.5   Accuracy in Digital Sampling   230
        Principal Causes of Errors, 230
        Tolerances in Three-Level Sampling, 231
8.6   Digital Delay Circuits   234
8.7   Quadrature Phase Shift of a Digital Signal   235
8.8   Digital Correlators for Continuum Observations   236
8.9   Digital Correlators for Spectral-Line Observations   236
        Principles of Digital Spectral Measurements, 236
        Spectral Correlator Systems, 241
Appendix 8.1   Evaluation of $\sum_{q=1}^{\infty} R_{\infty}^2(q\tau_s)$ 244
Appendix 8.2   Evaluation of the Probability Integral
                for Two-Level Quantization   245

9. VERY-LONG-BASELINE INTERFEROMETRY   **247**

9.1   Early Development   247
9.2   Differences Between VLBI and
        Conventional Interferometry   250
9.3   Basic Performance of a VLBI System   251
        Time and Frequency Errors, 251
        Retarded Baselines, 258
        Noise in VLBI Observations, 259
        Probability of Error in the Signal Search, 262
        Coherent and Incoherent Averaging, 266
9.4   Phase Stability and Atomic Frequency Standards   269
        Analysis of Phase Fluctuations, 269
        Oscillator Coherence Time, 277
        Precise Frequency Standards, 280

Rubidium Gas-Cell Resonator, 283
Cesium-Beam Resonator, 285
Hydrogen-Maser Frequency Standard, 285
Local Oscillator Stability, 290
Phase Calibration System, 290

9.5   Time Synchronization   291

9.6   Recording Systems   292

9.7   Processing Systems and Algorithms   295
Fringe Rotation Loss (η_R), 296
Fringe Sideband Rejection Loss (η_S), 299
Discrete Delay Step Loss (η_D), 301
Summary of Processing Losses, 304

9.8   Bandwidth Synthesis   305

9.9   Phased Arrays as VLBI Elements   308

10.   CALIBRATION AND FOURIER TRANSFORMATION OF VISIBILITY DATA   314

10.1   Calibration of the Visibility   314
Corrections for Predictable or Directly Measurable Effects, 315
Use of Calibration Sources, 316
Calibration of Spectral Line Data, 317
Calibration of Polarization Data, 318

10.2   Derivation of Brightness from Visibility   320
Model Fitting, 320
Mapping by Direct Fourier Transformation, 321
Implementation of the Direct Fourier Transform, 325
Mapping by Discrete Fourier Transformation, 327
Convolving Functions and Aliasing, 328
Aliasing and the Signal-to-Noise Ratio, 332
Interpretation of the Brightness Variable in the Map, 334

10.3   Practical Considerations   335
Possible Causes of Errors in Maps, 335
Visibility at Low Spatial Frequencies, 337
Miscellaneous Hints on Planning and Reduction of Observations, 338

11. IMAGE PROCESSING AND ENHANCEMENT   **342**

11.1   Limitation of (u, v) Coverage and
Methods of Compensation   342
CLEAN Algorithm, 343
*Implementation and Performance of the CLEAN
Algorithm, 347*
Constrained Optimization Techniques, 349

11.2   Mapping with Incomplete Phase Data   352
Mapping with the Visibility Modulus Only, 352
Mapping with Uncalibrated Phase Data, 356

11.3   Miscellaneous Application Notes   365

**12. INTERFEROMETER TECHNIQUES FOR ASTROMETRY
AND GEODESY**   **369**

12.1   Requirements for Astrometry   370

12.2   Solution for Baseline and
Source-Position Vectors   372
Connected-Element Systems, 372
Measurements with VLBI Systems, 374

12.3   Time and the Motion of the Earth   378
Precession and Nutation, 378
Polar Motion, 379
Universal Time, 380
Measurement of Polar Motion and UT1, 381

12.4   Geodetic Measurements   383

12.5   Mapping Astronomical Masers   384

Appendix 12.1   Least-Mean-Squares Analysis   388

13. PROPAGATION EFFECTS   **405**

13.1   Neutral Atmosphere   406
Basic Physics, 406
Refraction and Propagation Delay, 411
Absorption, 417
Origin of Refraction, 421
Smith–Weintraub Equation, 426
Phase Fluctuations, 428
Water Vapor Radiometry, 436

13.2   Ionosphere   439
Basic Physics, 440
Refraction and Propagation Delay, 444
Calibration of Ionospheric Delay, 446
Absorption, 447
Small- and Large-Scale Irregularities, 447

13.3   Scintillation Caused by Plasma Irregularities   449
Gaussian Screen Model, 449
Power-Law Model, 454

13.4   Interplanetary Medium   455
Refraction, 455
Interplanetary Scintillation, 458

13.5   Interstellar Medium   460
Dispersion and Faraday Rotation, 460
Interstellar Scintillation, 462

14.   RADIO INTERFERENCE   470

14.1   General Considerations   470
14.2   Connected-Element Arrays   473
Fringe-Frequency Averaging, 473
Decorrelation of Broadband Signals, 477
14.3   Very-Long-Baseline Systems   478
14.4   Interference from Airborne and Space Transmitters   480

15.   RELATED TECHNIQUES   482

15.1   Intensity Interferometer   482
15.2   Lunar Occultation Observations   487
15.3   Measurements on Antennas   490
15.4   Optical Interferometry   493
Modern Michelson Interferometer, 494
Optical Intensity Interferometer, 497
Speckle Interferometry, 497

PRINCIPAL SYMBOLS   503

AUTHOR INDEX   513

SUBJECT INDEX   521

Interferometry and Synthesis in Radio Astronomy

1

INTRODUCTION AND HISTORICAL REVIEW

The subject of this book can be broadly described as the principles of radio interferometry applied to the measurement of natural radio signals from cosmic sources. The uses of such measurements lie mainly within the domains of astronomy, astrometry, and geodesy. As an introduction we consider in this chapter the applications of the technique, some basic terms and concepts, and the historical development of the instruments and their uses.

1.1 APPLICATIONS OF RADIO INTERFEROMETRY

Radio interferometers and synthesis arrays, which are basically ensembles of two-element interferometers, are used to make measurements of the fine angular detail in the radio emission from the sky. The angular resolution of single radio antennas is insufficient for many astronomical purposes. Practical considerations limit the resolution to a few tens of arcsec. For example, the beamwidth of a 100-m-diameter antenna at 7-mm wavelength is approximately 17 arcsec. In the optical range the diffraction limit of large telescopes (diameter ~ 6 m) is about 0.025 arcsec, but the angular resolution achievable from the ground by conventional techniques is limited to about 1 arcsec by turbulence in the troposphere. For progress in astronomy it is particularly important to measure the positions of radio sources with sufficient accuracy to allow identification with objects detected in the optical and other parts of the electromagnetic spectrum. It is also very important to be able to compare parameters such as brightness, polarization, and frequency spectrum with similar angular detail in the radio and optical domains. Radio interferometry enables such studies to be made. The angular resolution desired is progressing from the arcsecond to the submilliarcsecond range as techniques improve, and more accurate information may be obtainable from telescopes above the earth's atmosphere.

Precise measurement of the angular positions of stars and other cosmic objects is the concern of astrometry. This includes the study of the small changes in celestial positions attributable to the parallax introduced by the earth's orbital motion, as well as those resulting from the intrinsic motions of the objects. Such measurements are an essential step in the establishment of the distance scale of the universe. Astrometric measurements have also provided a means to test the general theory of relativity and to establish the dynamical parameters of the solar system which are important, for example, in spacecraft navigation. In making astrometric measurements it is essential to establish a system of position references. The inertial mass of the universe provides the ultimate fixed reference, and radio measurements of the distant quasars presently offer the best prospects for the establishment of such a coordinate system. Radio techniques provide an accuracy of the order of 10^{-3} arcsec for absolute positions and 10^{-5} arcsec or less for the relative positions of objects closely spaced in angle. Optical measurements of stellar images, as seen through the earth's atmosphere, allow the positions to be determined with a precision of about 0.05 arcsec. However, future measurements of the relative positions of large numbers of stars from an observatory in earth orbit can be expected to provide a more precise system of optical positions.

As part of the measurement process, astrometric observations include a determination of the orientation of the instrument relative to the celestial reference frame. Ground-based observations therefore provide a measure of the variation of the kinematic parameters of the earth. In addition to the well-known precession and nutation of the direction of the axis of rotation, there are irregular shifts of the earth's axis relative to the surface. These shifts are referred to as polar motion and are attributed to the gravitational effects of the sun and moon on the equatorial bulge of the earth and to dynamic effects in the earth's crust and atmosphere. The same causes give rise to changes in the angular rotation velocity of the earth, which are manifest as corrections that must be applied to the system of universal time. During the 1970s it became clear that radio techniques could provide an accurate measure of these effects, and in the late 1970s the first radio programs devoted to the monitoring of universal time and polar motion were set up jointly by the U.S. Naval Observatory and the U.S. Naval Research Laboratory, and also by the National Aeronautics and Space Administration (NASA) and the National Geodetic Survey.

In addition to revealing angular changes in the motion and orientation of the earth, precise interferometer measurements entail an astronomical determination of the vector spacing between the antennas, which is usually more precise than can be obtained by conventional surveying techniques. Very-long-baseline interferometry (VLBI) involves antenna spacings of hundreds or thousands of kilometers, and the uncertainty with which these spacings can be determined decreased from a few meters in 1967, when such measurements were first made, to less than 10 cm in little more than a decade. This improvement in precision provides further evidence for our changing perception of the earth from a rigid body to one in which dynamic crustal processes are measurable. Average relative motions of widely spaced sites, on separate tectonic plates, are believed to lie in the range 1–10 cm per year and are therefore detectable.

Interferometric techniques have also been applied to the tracking of vehicles on the lunar surface and the determination of the positions of spacecraft. In this book, however, we limit our concern to measurements of natural signals from astronomical objects. The attainment of the highest angular resolution in the radio domain of the spectrum results in part from the ease with which radiofrequency signals can be processed electronically. This capability allows much more complex signal processing than is possible at shorter wavelengths. The phase variations induced by the earth's neutral atmosphere are also less severe than at shorter wavelengths. Future technology will provide even higher resolution in the infrared and optical domains from observatories above the earth's atmosphere. However, radio waves will remain of vital importance in astronomy since they reveal objects that do not radiate in other parts of the spectrum, and they are able to pass through galactic dust clouds that obscure the view in the optical range.

1.2 SOME BASIC TERMS AND DEFINITIONS

This section is written for the reader who is unfamiliar with radio astronomy. It presents a brief review of some background information useful when approaching the subject of radio interferometry.

Cosmic Signals

The voltages induced by cosmic-source radiation are generally referred to as signals, although they do not contain information in the usual engineering sense. Such signals are generated by natural processes and have the form of Gaussian random noise. That is to say, the voltage as a function of time at the terminals of a receiving antenna can be described as a series of very short pulses of random occurrence which combine as a waveform with Gaussian-distributed amplitude. In a bandwidth $\Delta\nu$ the envelope of the radiofrequency waveform has the appearance of random pulses with duration of order $\Delta\nu^{-1}$. For most radio sources the characteristics of the signals are stationary with time, at least on the scale of minutes or hours typical of the duration of an observation such as we are considering. Gaussian waveforms of this type are assumed to be identical in character to the noise voltages generated in resistors and amplifiers. Such waveforms are ergodic, that is, ensemble averages and time averages are equivalent.

Most of the power is in the form of *continuum radiation*, the power spectrum of which shows slow variation with frequency and may be regarded as constant over the receiving bandwidth of most instruments. Figure 1.1 shows continuum spectra of three radio sources. Radio emission from the radio galaxy Cygnus A and from the quasar 3C48 is generated by the synchrotron mechanism (see, e.g., Rybicki and Lightman 1979) in which high-energy electrons in magnetic fields radiate as a result of their orbital motion. The radiating electrons are generally highly relativistic, and under these conditions the radiation emitted by each one is concentrated in the direction of its instantaneous motion. An observer therefore sees pulses of radiation from those electrons whose orbital motion lies in, or close to, a plane containing the observer. The observed polarization of the radiation is mainly linear, and any circularly polarized component is generally very small. The overall linear polarization from a source, however, is usually not large, being randomized by the variation of the direction of the magnetic field within the source and by Faraday rotation. The power in the electromagnetic pulses from the electrons

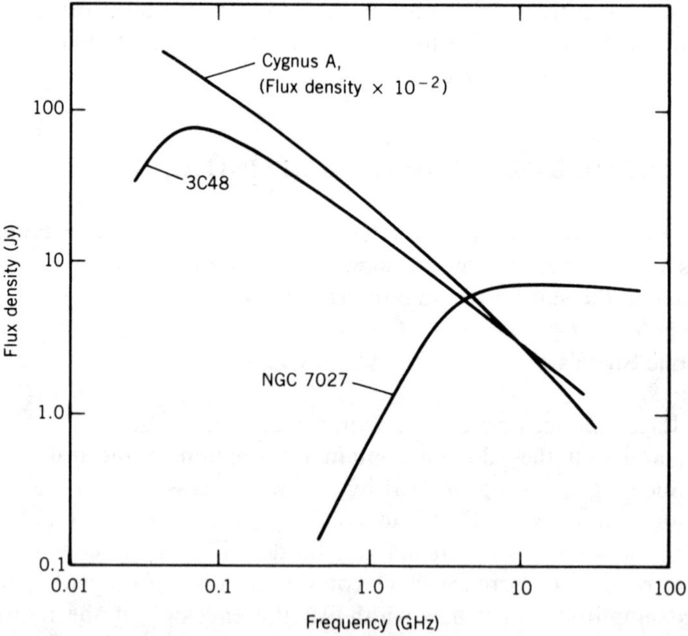

Figure 1.1   Continuum spectra of three discrete sources: Cygnus A, a radio galaxy; 3C48, a quasar; and NGC 7027, an ionized nebula within our galaxy. Data are from Conway, Kellermann, and Long (1963), Kellermann and Pauliny-Toth (1969), and Thompson (1974). One jansky (Jy) = 10^{-26} W m^{-2} Hz^{-1}.

is concentrated at harmonics of the orbital frequency, and a continuous distribution of electron energies results in a continuum radio spectrum. The individual pulses from the electrons are too numerous to be separable, and the electric field appears as a continuous random process with zero mean. The variation of the spectrum as a function of frequency is related to the slope of the energy distribution of the electrons. In the quasar in Fig. 1.1, which is a very much more compact object than the radio galaxy, the electron density and magnetic fields are high enough to produce self-absorption of the radiation at low frequencies.

NGC 7027, whose spectrum is shown in Fig. 1.1, is a planetary nebula within our galaxy in which the gas is ionized by radiation from a central star. The radio emission is a thermal process and results from free–free collisions between unbound electrons and ions within the plasma. At the low-frequency end of the spectral curve the nebula is opaque to its own radiation and emits like a blackbody following the Rayleigh–Jeans law. As the frequency increases, the absorptivity, and hence the emissivity, decrease approximately as ν^{-2} (Oster 1961; Rybicki and Lightman 1979) where ν is the frequency. This behavior counteracts the ν^2 dependence of the Rayleigh–Jeans law, and thus the spectrum becomes flat when the nebula is no longer opaque to the radiation. Radiation of this type is unpolarized.

In contrast to continuum radiation *spectral line radiation* is generated at specific frequencies by atomic and molecular processes. A fundamentally important line is that of neutral atomic hydrogen at 1420.405 MHz, which results from the transition between two energy levels of the atom, the separation of which is related to the spin vector of the electron in the magnetic field of the nucleus. The natural width of the hydrogen line is negligibly small, but Doppler shifts caused by thermal motion of the atoms and large-scale motion of gas clouds spread the line radiation. The overall Doppler spread within our galaxy covers several hundred kilohertz. Information on galactic structure is obtained by comparison of these velocities with those of models describing galactic rotation.

Our galaxy and others like it also contain large molecular clouds at temperatures of 10–100 K in which new stars are continually forming. These clouds give rise to many atomic and molecular transitions in the radio and far-infrared ranges. Over 500 such radio lines have been detected (Lovas, Snyder, and Johnson 1979). Those considered to be most important are listed by the International Astronomical Union (IAU 1982), and a number of these are given in Table 1.1. Figure 1.2 shows the spectrum of radiation of many molecular lines from the Orion Nebula in the band from 241 to 242 GHz. A small number of lines, notably those of OH, H_2O, and SiO, show very intense emission from sources of very small apparent angular diameter. This emission is believed to be generated by a maser process (see, e.g., Reid and Moran 1981).

The strength of the radio signal received from a discrete source is expressed as the *spectral flux density* or *spectral power flux density* and is measured in

TABLE 1.1   Important Radio Lines

Chemical Name	Chemical Formula	Transition	Frequency (GHz)
Hydrogen	H	$1^2S_{1/2}, F = 1\text{–}0$	1.420
Hydroxyl radical	OH	$^2\Pi_{3/2}, J = 3/2, F = 1\text{–}2$	1.612
	OH	$^2\Pi_{3/2}, J = 3/2, F = 1\text{–}1$	1.665
	OH	$^2\Pi_{3/2}, J = 3/2, F = 2\text{–}2$	1.667
	OH	$^2\Pi_{3/2}, J = 3/2, F = 2\text{–}1$	1.721
Formaldehyde	H_2CO	$1_{10}\text{–}1_{11}$, six F-transitions	4.830
Water	H_2O	$6_{16}\text{–}5_{23}$, five F-transitions	22.235
Ammonia	NH_3	$1,1\text{–}1,1$, eighteen F-transitions	23.694
	NH_3	$2,2\text{–}2,2$, seven F-transitions	23.723
	NH_3	$3,3\text{–}3,3$, seven F-transitions	23.870
Silicon monoxide	SiO	$v = 2, J = 1\text{–}0$	42.820
	SiO	$v = 1, J = 1\text{–}0$	43.122
	SiO	$v = 1, J = 2\text{–}1$	86.243
Hydrogen cyanide	HCN	$J = 1\text{–}0$, three F-transitions	88.631
Formyl ion	HCO^+	$J = 1\text{–}0$	89.189
Carbon monoxide	$^{12}C^{18}O$	$J = 1\text{–}0$	109.782
Carbon monoxide	$^{13}C^{16}O$	$J = 1\text{–}0$	110.201
Carbon monoxide	$^{12}C^{16}O$	$J = 1\text{–}0$	115.271
Carbon monoxide	$^{12}C^{16}O$	$J = 2\text{–}1$	230.538
Carbon	C	$^3P_1\text{–}^3P_0$	492.162

watts per square meter per hertz (W m^{-2} Hz^{-1}). Usage varies, but for brevity this quantity is often referred to as *flux density*. The unit of flux density is the jansky (Jy); 1 Jy = 10^{-26} W m^{-2} Hz^{-1}. It is used for both spectral line and continuum radiation. The measure of radiation integrated in frequency over a spectral band has units of W m^{-2} and is referred to as power flux density, or simply flux density. In the standard definition of the IEEE (1977), power flux density is equal to the time average of the Poynting vector of the wave. In mapping a radio source the desired quantity is the *radio brightness B*, the units of which are W m^{-2} Hz^{-1} sr^{-1}, that is, spectral flux density per steradian. Brightness refers to a particular direction, and hence only brightness averaged over a finite solid angle can be measured.

In radiation theory the quantity *intensity*, or *specific intensity*, usually represented by I_ν, is the measure of radiated energy flow per unit area, per unit time, per unit frequency bandwidth, and per unit solid angle (see, e.g., Rybicki and Lightman 1979). Thus in Fig. 1.3 the power flowing in direction s within solid angle $d\Omega$, frequency band $d\nu$, and area dA is $I_\nu(s)\,d\Omega\,d\nu\,dA$. This can be applied to emission from the surface of a radiating object, to propagation

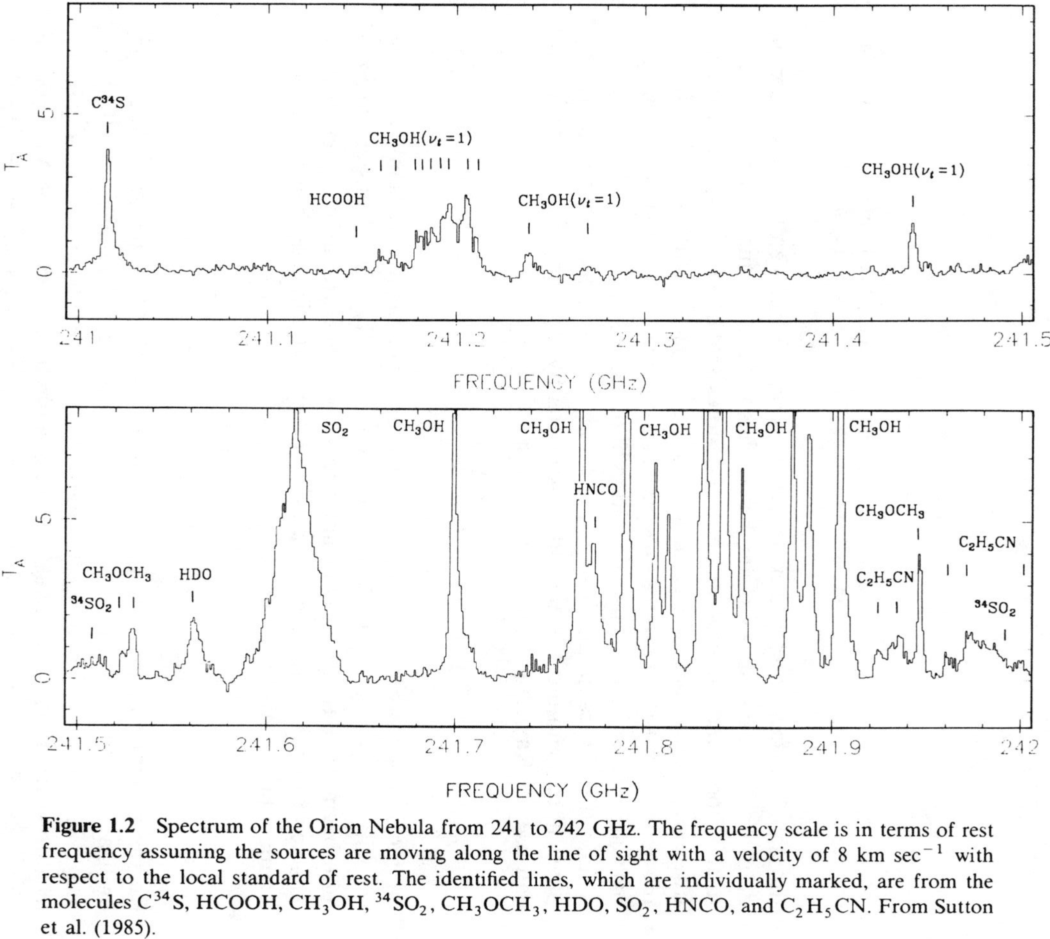

Figure 1.2 Spectrum of the Orion Nebula from 241 to 242 GHz. The frequency scale is in terms of rest frequency assuming the sources are moving along the line of sight with a velocity of 8 km sec^{-1} with respect to the local standard of rest. The identified lines, which are individually marked, are from the molecules $C^{34}S$, HCOOH, CH_3OH, $^{34}SO_2$, CH_3OCH_3, HDO, SO_2, HNCO, and C_2H_5CN. From Sutton et al. (1985).

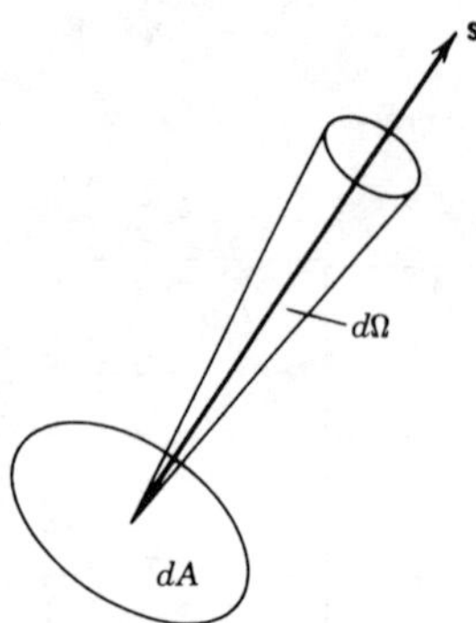

Figure 1.3   Elements of solid angle and surface area illustrating the definition of intensity. dA is normal to s.

through a surface in space, or to reception on the surface of a transducer or detector. The last case is the one that applies to the receiving systems discussed in this book, and the solid angle then denotes the area of the celestial sphere from which the radiation emanates. In using the term brightness, B, we follow the general custom in radio astronomy and emphasize here that B is simply a particular case of the classical definition of intensity.

For thermal radiation from a blackbody the brightness is related to the physical temperature T of the radiating matter by the Planck law, for which we can use the Rayleigh–Jeans approximation

$$B = \frac{2kT\nu^2}{c^2},\tag{1.1}$$

in the domain where $h\nu \ll kT$. Here k is Boltzmann's constant, c is the velocity of light, and h is Planck's constant. The Rayleigh–Jeans approximation begins to introduce significant errors for frequencies above a few tens of gigahertz and for low temperatures. For any radiation mechanism, one can define a brightness temperature T_B given by

$$T_B = \frac{c^2 B}{2k\nu^2}.\tag{1.2}$$

In the examples in Fig. 1.1, T_B is of the order of 10^4 K for NGC 7027 and indicates the thermal temperature of the electrons. For Cygnus A and 3C48, T_B is of the order of 10^8 K or greater and is a measure of the energy density of the electrons and the magnetic fields, and not a physical temperature. As a spectral line example, T_B for the carbon monoxide lines from molecular clouds is typically 10–100 K. In this case T_B is proportional to the excitation temperature associated with the energy levels of the transition, and is related to the temperature and density of the gas as well as to the temperature of the radiation field.

Source Positions and Nomenclature

The positions of radio sources are measured in the celestial coordinates *right ascension* and *declination*. On the celestial sphere these quantities are analogous, respectively, to longitude and latitude on the earth. The zero of right ascension is arbitrarily chosen as the point at which the sun crosses the celestial equator in the constellation of Aries at a given epoch. In the international system of nomenclature (IAU 1974) radio sources are designated as follows. The first four characters give the hour and minutes of right ascension, the fifth the sign of the declination, and the remaining three the degrees and truncated tenths of a degree of declination for the mean equator and equinox of 1950. For example, the source at R.A. $01^h34^m49.83^s$, Dec. $32°54'20.5''$ is designated $0134 + 329$. Early nomenclature still persists for some sources. Thus in Fig. 1.1 Cygnus A is the strongest radio source in the constellation of Cygnus, and 3C48 indicates source number 48 in the third Cambridge survey (Edge et al. 1959). NGC 7027 is an optical designation and refers to the New General Catalog of nonstellar objects by Dreyer (1888).

Positions of objects in celestial coordinates vary as a result of precession and nutation of the earth's axis of rotation, aberration, and proper motion. Positions of radio sources are usually listed for the standard epochs 1950 or 2000. Procedures for the reduction of these positions to those for specific dates and times are given in the *Explanatory Supplement to the Astronomical Ephemeris*. Several tens of thousands of radio sources have been cataloged (Kesteven and Bridle 1977). Most of the radio sources that have been detected are believed to be radio galaxies or quasars that lie far beyond our own galaxy.

Reception of Cosmic Signals

The antennas used most commonly in radio astronomy, and in particular in the instruments discussed in this book, are of the reflector type mounted to allow tracking over most of the sky. The exceptions are mainly instruments designed for meter or longer wavelengths where very large structures are required to obtain the desired angular resolution. The collecting area A of an antenna for radiation incident in the center of the main beam is equal to the geometrical area multiplied by an aperture efficiency factor which is typically within the range 0.3–0.7. The received power P_r delivered to a matched load in a bandwidth $\Delta\nu$ from a randomly polarized source of flux density S is given by

$$P_r = \tfrac{1}{2}AS\Delta\nu. \tag{1.3}$$

The factor $\tfrac{1}{2}$ takes account of the fact that the antenna responds to only one-half the power in the randomly polarized wave.

It is often useful to specify the noise power in a receiving system[†] in terms of the temperature T_S of a matched resistive load that would produce an equal

---

[†] In radio astronomy the terms *receiver* and *receiving system* are broadly used and generally denote the electronic system following the output of the antenna(s) and may, or may not, include one or more detectors or correlators (which we define later) and subsequent processing and recording equipment.

power level in an equivalent noise-free receiver. The available noise power from such a resistor, in the classical limit, is (Nyquist 1928)

$$P_n = kT_S \Delta\nu. \tag{1.4}$$

T_S is called the *system temperature* and consists of two parts—T_R, the *receiver temperature* which represents the internal noise from the receiving amplifier, and T_A', the *antenna temperature* which represents the unwanted noise from the antenna produced by ground radiation, atmospheric attenuation, and other sources. The term antenna temperature is also used to express the power received from a radio source,

$$P_r = kT_A \Delta\nu, \tag{1.5}$$

and T_A is then related to the flux density by Eqs. (1.3) and (1.5). In this book antenna temperature usually has the latter meaning.

The ratio of the signal power from the source to the noise power in the receiving amplifier is T_A/T_S. Because of the random nature of the signal and noise, measurements of the power levels made at time intervals separated by $(2\Delta\nu)^{-1}$ can be considered essentially independent. A measurement in which the signal level is averaged for a time τ contains approximately $2\Delta\nu\,\tau$ independent samples. The signal-to-noise ratio $\mathcal{R}_{sn}$ at the output of a power-measuring device attached to the receiver is increased in proportion to the square root of the number of independent samples and is of the form

$$\mathcal{R}_{sn} = C\frac{T_A}{T_S}\sqrt{\Delta\nu\tau}, \tag{1.6}$$

where C is a constant. This simple analysis appears to have been first used by Dicke (1946). More detailed examination of the behavior of radio astronomy receivers (see, e.g., Tiuri 1964) shows that Eq. (1.6) is generally applicable and that the constant is of order unity but depends on the details of the system. Typical values of $\Delta\nu$ and τ are 50 MHz and 5 hr, which result in a value of 10^6 for the factor $(\Delta\nu\,\tau)^{1/2}$. As a result, it is possible to detect a signal for which the power level is little more than 10^{-6} times the system noise. The following calculation may help to illustrate the low energies involved. Consider a large radio telescope with a total collecting area of 10^4 m^2 pointed toward a radio source of flux density 1 mJy ($= 10^{-3}$ Jy) and accepting signals over a bandwidth of 50 MHz. In 10^3 years the total energy accepted is about 10^{-7} J (1 erg), which is comparable to a few percent of the kinetic energy in a single falling snowflake. To detect the source with the same telescope, and a system temperature of 50 K, would require an observing time of about 5 min during which time the energy received would be about 10^{-15} J.

1.3 DEVELOPMENT OF RADIO INTERFEROMETRY

Michelson Interferometer

Interferometric techniques in astronomy date back to the optical work of Michelson (1890, 1920) and Michelson and Pease (1921) who were able to obtain sufficiently fine angular resolution to measure the diameters of some of the nearer and larger stars such as Arcturus and Betelgeuse. The principle of their optical interferometer is illustrated in Fig. 1.4. Beams of light from a star

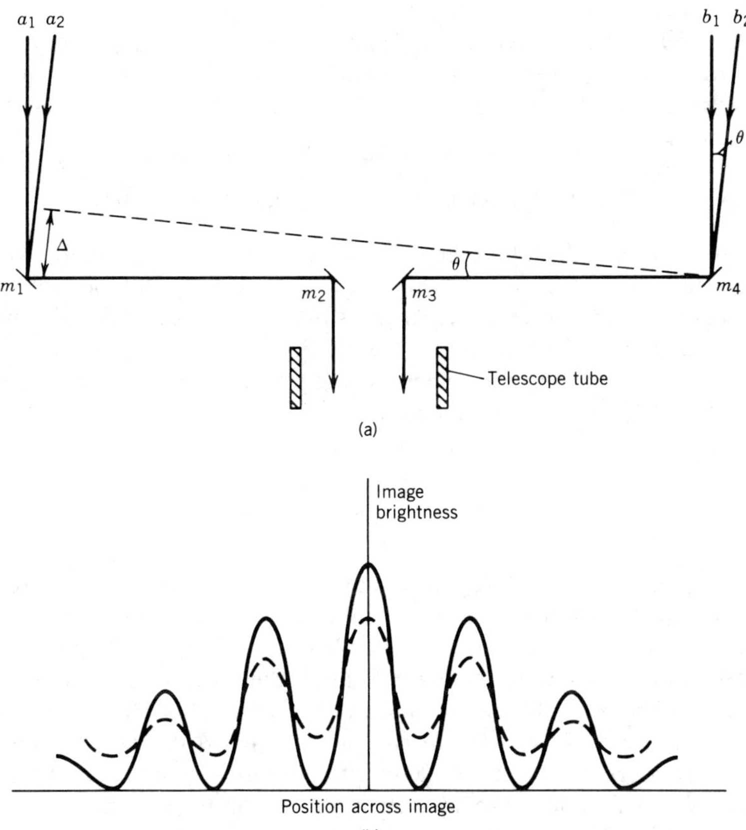

Figure 1.4 (a) Schematic diagram of Michelson's stellar interferometer. The incoming rays are guided into the telescope aperture by mirrors m_1 to m_4, of which the outer pair define the two apertures of the interferometer. Rays a_1 and b_1 traverse equal paths to the eyepiece at which the image is formed, but rays a_2 and b_2, which approach at an angle θ to the instrumental axis, traverse paths that differ by a distance Δ. (b) The brightness of the image as a function of position angle in a direction parallel to the spacing of the interferometer apertures. The solid line shows the fringe profiles for an unresolved star ($\mathcal{V}_M = 1.0$) and the broken line for a partially resolved star for which $\mathcal{V}_M = 0.5$.

fall upon two apertures and are combined in a telescope. The resulting stellar image has a finite width and is shaped by effects that include atmospheric turbulence, diffraction at the mirrors, and the bandwidth of the radiation. Maxima in the light intensity resulting from interference occur at angles θ for which the difference Δ in the path lengths from the star to the point at which the light waves are combined is an integral number of wavelengths at the center of the optical passband (Anderson 1922). If the angular width of the star is small compared with the spacing in θ between adjacent maxima, the image of the star is crossed by alternate dark and light bands, known as interference fringes. If, however, the width of the star is comparable to the spacing between maxima, one can visualize the resulting image as being formed by the superposition of images from a series of points across the star. The maxima and minima of the fringes from different points do not coincide, and the fringes tend to merge into a smooth continuum. As a measure of the relative amplitude of the fringes, Michelson defined the *fringe visibility* $\mathscr{V}_M$ as

$$\mathscr{V}_M = \frac{\text{brightness of maxima} - \text{brightness of minima}}{\text{brightness of maxima} + \text{brightness of minima}}. \tag{1.7}$$

Note that with this definition the visibility is normalized and is unity when the source width is small compared with the fringe width. As will be evident from the discussions in Chapter 2, the fringe visibility is proportional to the modulus of the Fourier transform of $B_1(\xi)$, the brightness profile of the star projected onto a plane containing the centers of the star and the interferometer apertures. Thus if (ξ, η) are coordinates on the sky with ξ measured parallel to the aperture spacing vector and η normal to it,

$$B_1(\xi) = \int B(\xi, \eta)\, d\eta. \tag{1.8}$$

For example, the profile $B_1(\xi)$ for a uniform circular disk is a semicircle. The fringe visibility as a function of u, the aperture spacing measured in wavelengths, is shown in Fig. 1.5 for several different brightness models. If the fringe visibility is measurably less than unity the object is said to be *resolved* by the interferometer. Michelson and Pease used principally the circular disk model to interpret their observations and determined the stellar diameter by varying the aperture spacing of the interferometer to locate the first minimum in the visibility function. The adjustment of such an instrument and the visual estimation of $\mathscr{V}_M$ required great skill and care, since the fringes were not stable but vibrated across the image in a random manner as a result of atmospheric fluctuations and mechanical instability. The published results on stellar diameters measured with this method were never extended beyond seven bright stars in Pease's (1931) list; for a detailed review see Hanbury Brown (1968). However, the use of electro-optical techniques now offers much greater instrumental capabilities based on Michelson's system as is discussed in Chapter 15.

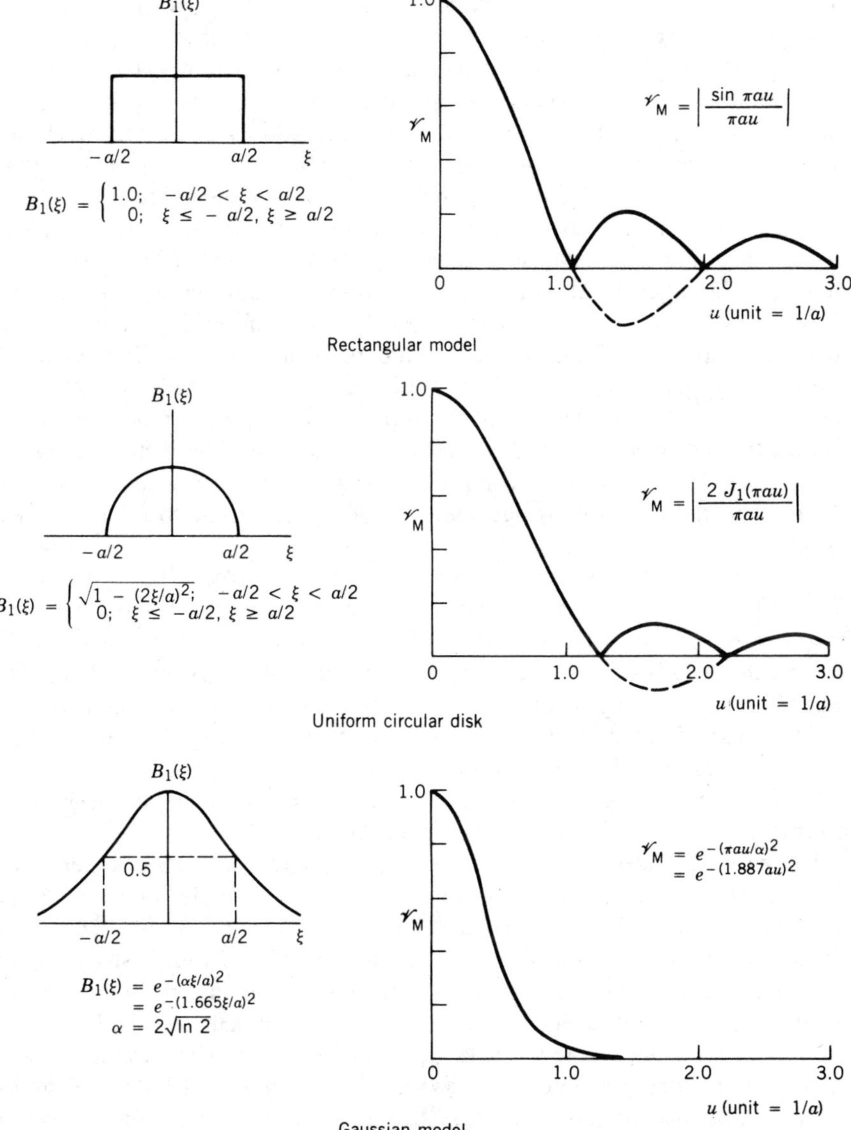

Figure 1.5 Three models of one-dimensional brightness distribution B_1 and the corresponding Michelson visibility functions $\mathcal{V}_M$: ξ is the angular variable on the sky, a is the characteristic angular width of the model, and u is the spacing of the receiving apertures measured in wavelengths. The solid lines in the curves of $\mathcal{V}_M$ indicate the modulus of the Fourier transform of $B_1(\xi)$, and the broken lines indicate negative values of the transform.

Principles of Radio Interferometry and Early Instruments

The principle of the resolution of a source by a fringe pattern, which we reemphasize here, is equally applicable to the radio and optical cases. As is shown in Chapter 2, the fringe pattern is closely approximated by a sinusoidal function of angle. Consider, for example, the response of an interferometer to a source consisting of two point components with angular spacing equal to one-half the fringe period. The fringe-pattern responses in the interferometer output are then in antiphase and cancel one another. In the case of an arbitrary brightness profile the responses from different elements of the source correspond to different fringe phases and combine vectorially. Thus, in general, the response to a source depends upon both the flux density of the source and the form and angular width of the brightness distribution relative to the width of the fringes. Observations with an appropriate range of spacings provide information on both the flux density and the source profile. The theory of synthesis mapping is a development of this basic idea. We shall show later that the output of the interferometer is proportional to one component of the Fourier transform of the source brightness distribution. The reciprocal nature of this transform provides a convenient means of recovering the brightness profile from the measured fringe values, which we continue to refer to as the visibility.

In general, the brightness distribution of a radio source is not symmetrical about any central point, and, consequently, its Fourier transform is complex. This means that we must measure both the amplitude and the phase of the fringes; the details of how the phase is measured will be explained later. The important point to note here is that the measurement of the *complex visibility* as a function of antenna spacing is the basis of mapping by Fourier synthesis in radio astronomy. In the early observations the instrumental stability often did not permit measurement of the visibility phase, and angular widths were interpreted in terms of symmetrical models as in Fig. 1.5.

The early discoveries of cosmic radio emission (Jansky 1933, Reber 1940) were made by measuring the power received in a single antenna. Radio interferometry in astronomy was developed following World War II, during which further attention was drawn to the possibilities of radio astronomy by the effects of solar activity on radar receivers (Appleton 1945, Southworth 1945). In 1946 Ryle and Vonberg constructed a radio analog of the Michelson interferometer using dipole antenna arrays at 175 MHz. The baseline, that is, the spacing between the antennas, was varied between 10 and 140 wavelengths. A diagram of such an instrument and the type of record obtained are shown in Fig. 1.6. In this and most other meter-wavelength interferometers of the 1950s and 1960s the antenna beams were pointed in the meridian and scanned in right ascension by the motion of the earth.

A somewhat different instrument with which a number of important results prior to 1950 were obtained was the sea interferometer, a radio analog of the Lloyd's mirror optical interferometer. It was used mostly in Australia where

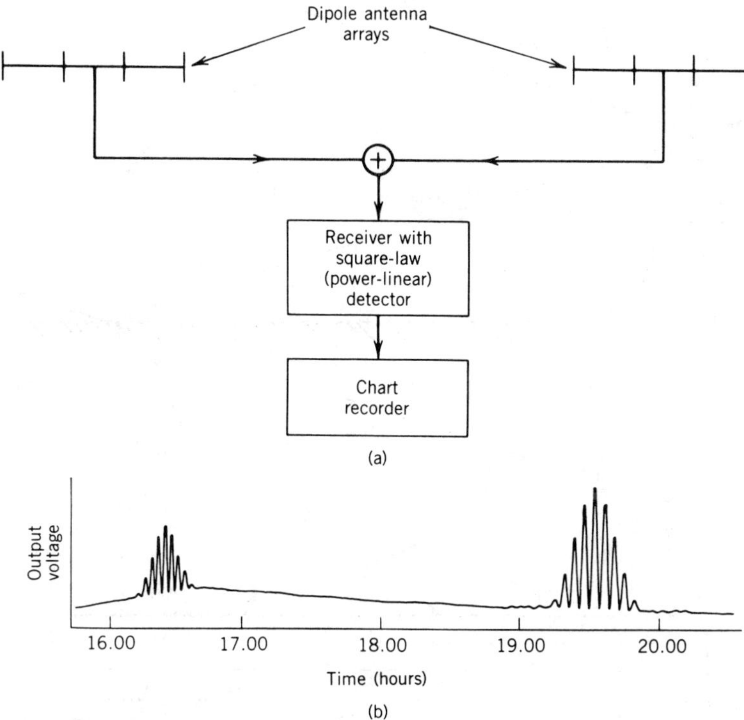

Figure 1.6   (a) Simple interferometer in which the signals are combined additively. (b) Record from such an interferometer with east–west antenna spacing. The ordinate is the total power received and the abscissa is time. The source at the left is Cygnus A and the one at the right Cassiopeia A. The increase in level near Cygnus A results from the galactic background radiation which is concentrated toward the plane of our galaxy but is completely resolved by the interferometer fringes. The record is from Ryle (1952).

horizon-looking radar antennas had been installed at several coastal locations near Sydney at elevations of 60–120 m above the sea. Radiation from sources rising over the eastern horizon was received both directly and by reflection from the sea, as shown in Fig. 1.7. The frequencies used were in the range 40–400 MHz, the middle part of the range being the most satisfactory because of ionospheric effects at the low-frequency end and sea roughness at the high end. Although the low elevation angle of the observations maximized the unwanted ionospheric fluctuation and refraction effects, there were compensating advantages. The sudden appearance of a rising source was useful in separating individual sources, and the sensitivity at the peak of a fringe corresponded to four times that for direct reception with the single antenna because of the reflected wave.

McCready, Pawsey, and Payne-Scott (1947) used the sea interferometer for a study of radiation from the sun. Their observations included the first records of interferometer fringes in radio astronomy, made in January–February 1946.

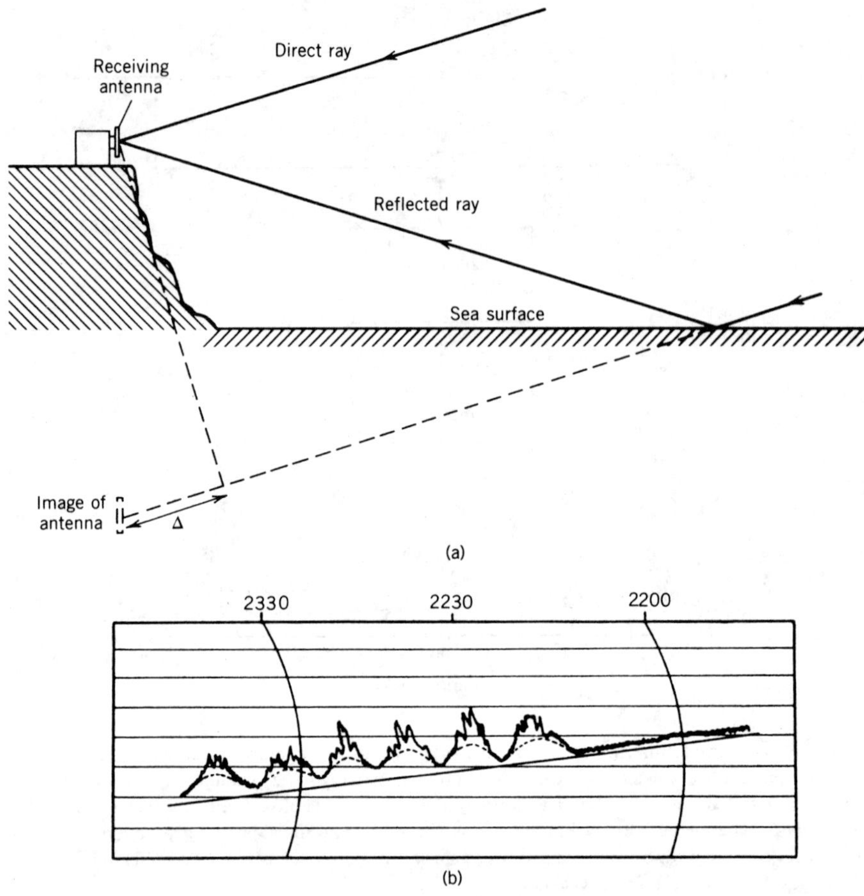

Figure 1.7   (a) Schematic diagram of a sea interferometer. The fringe pattern is similar to that which would be obtained with the actual receiving antenna and one at the position of its image in the sea. The reflected ray undergoes a phase change of 180° on reflection and travels an extra distance Δ in reaching the receiving antenna. (b) Sea interferometer record of the source Cygnus A at 100 MHz by Bolton and Stanley (1948). The source rose above the horizon at approximately 22.17. The broken line was inserted to show that the record could be interpreted in terms of a steady component and a fluctuating component of the source: the fluctuations were later shown to be of ionospheric origin. The fringe width was approximately 1.0° and the source was unresolved. Part (b) is reprinted by permission from *Nature*, Vol. 161, No. 4087, p. 313; copyright © 1948 Macmillan Journals Limited.

The fringe spacing of approximately 0.5° substantially resolved the quiet sun, that is, the largely constant radiation from the solar atmosphere as distinct from active regions. However, the positions and widths of active regions could be measured and correlated with sunspots. This study was the first occasion that the Fourier transform relationship between the source profile and the visibility function was mentioned with respect to radio observations. The sea interferometer also provided the first positive evidence of the occurrence of discrete nonsolar sources. Hey, Parsons, and Phillips (1946) had mapped the galactic background radiation at 5-m wavelength using a low-resolution antenna and found fluctuations in the signal strength from the Cygnus region. They surmised that the fluctuations must arise from a source of small angular width. This idea was confirmed by Bolton and Stanley (1948) who used the sea interferometer to show that the source, Cygnus A, was narrower than 8 arcmin and to determine its position with comparable accuracy. An even stronger source, Cassiopeia A, was discovered by Ryle and Smith (1948) using a two-element interferometer operating at 85.5 MHz at Cambridge, England. A total of 50 sources were cataloged with this instrument (Ryle, Smith, and Elsmore 1950). Several other strong sources, Taurus A, Virgo A, and Centaurus A (Bolton, Stanley, and Slee 1949), were also discovered, and these were fairly rapidly identified with prominent optical objects; the Crab Nebula and galaxies NGC 4486 and NGC 5128, respectively. The two galaxies were known to possess anomalous optical features: NGC 4486 contains a bright jet and NGC 5128 a dark band. No prominent objects were found in the positions of Cygnus A or Cassiopeia A, however, and as more sources were listed it soon became apparent that the identification of radio sources with well-known optical objects was the exception rather than the rule. This situation led to efforts to measure positions and angular widths with greater precision. Mills (1952), using a two-element interferometer at 101 MHz, obtained positional accuracy of ± 50 arcsec in right ascension and ± 1.5 arcmin in declination for Cygnus A. Smith (1951) obtained similar results for four sources with rms errors as small as ± 20 arcsec in right ascension and ± 40 arcsec in declination. In these sets of measurements the accuracy depended on the absolute calibration of the positions of the interferometer fringes. Calibration techniques, which were described in detail by Smith (1952a), included observing circumpolar sources at both upper and lower culminations (meridian crossings), measuring the transit time with different azimuths of the antenna baseline, interchanging antennas, and making careful comparisons of the electrical lengths of cables and preamplifiers. The optical identification of the Cygnus and Cassiopeia radio sources by Baade and Minkowski (1954a, b) was a direct result of the improved radio positions. Cygnus A proved to be a distant galaxy and Cassiopeia A a supernova remnant, but the interpretation of the optical observations was not fully understood at the time.

Further confirmation of the identification of radio sources was obtained through measurements of angular sizes at radio wavelengths. Mills (1953) extended his interferometer baseline by using a small transportable array of Yagi elements which could be located at distances up to 10 km from the main antenna. The signal from this remote antenna was transmitted back over a radio link, and fringes were formed. The variation of the fringe amplitude with the antenna spacing was measured for different azimuths and the resulting angular widths were interpreted by Mills in terms of models with elliptical brightness contours. The results indicated that in the case of the Crab Nebula the whole optical nebula is a radio source, but for NGC 4486 and NGC 5128 the radio emission is concentrated around the jet and band structures in these galaxies. Smith (1952b) also measured the variation of fringe amplitude with antenna spacing, but he used smaller antenna spacings than Mills and concentrated upon precise measurements of small changes in the fringe amplitude by using three antennas and connecting the receiver between different pairs. These results gave a diameter for Cassiopeia A that agreed closely with the extent of the optical nebulosity (Smith 1952c).

A third group working on the angular widths of radio sources, Jennison and Das Gupta (1953, 1956) at the Jodrell Bank Experimental Station of the University of Manchester, England, used a different technique, *intensity interferometry*. It had been shown by Hanbury Brown and Twiss (1954) that if the signals from the two antennas are passed through square-law detectors, the fluctuations in the intensity that result from the Gaussian fluctuations in the received field strength are correlated. The degree of correlation varies in proportion to the square of the fringe amplitude that would be obtained in a conventional interferometer in which the signals are combined before detection. Such a scheme has the advantage that it is not necessary to preserve the radiofrequency phase of the signals in bringing them to the location at which they are combined. Thus the local oscillators in the receivers at the antennas do not have to be synchronized, and the system is relatively insensitive to atmospheric fluctuations. It had been supposed that angular dimensions in the stellar range might be encountered. Thus Jennison and Das Gupta's interferometer was designed so that the detected signals fell within the audiofrequency range and could be transmitted over telephone channels or recorded on magnetic tape if very large antenna separations were required. In fact, spacings up to approximately 10 km only were sufficient and the signals were transmitted using a VHF radio link. The disadvantage of the intensity interferometer is that it requires a high signal-to-noise ratio, and even for Cygnus A and Cassiopeia A, the two highest flux density sources in the sky, it was necessary to construct large arrays of dipoles. These operated at 125 MHz.

The most important result of the Jennison and Das Gupta measurements was the discovery that for Cygnus A the modulus of the Fourier transform of the east–west brightness profile falls close to zero and then increases to a secondary maximum as the antenna spacing is increased. The amplitude of the secondary maximum was too large to be explained by a single-component source model, and hence a multiple-component model was required. Two symmetric models were possible: in one the phase of Fourier transform changed by 180° in going through the minimum and in the other it remained at 0°. The intensity interferometer gives no information on the phase, so a subsequent program of measurements was made with a conventional interferometer by Jennison and Latham (1959). Because the instrumental phase of the equipment was not stable enough to permit calibration, three antennas were used and three sets of fringes for the three pair combinations were recorded simultaneously. If ϕ_{mn} is the phase of the fringe pattern for antennas

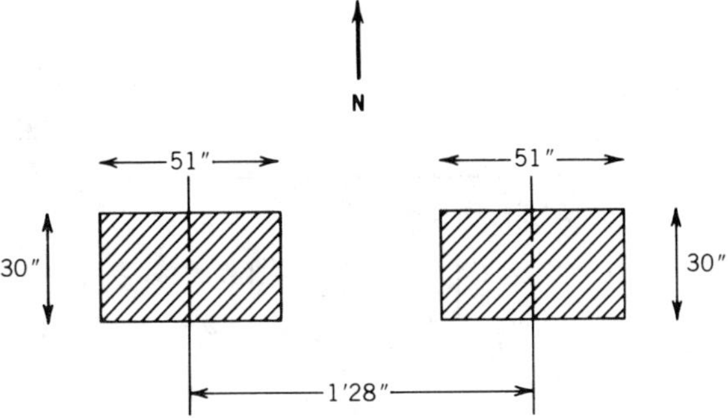

Figure 1.8 Two-component model of Cygnus A derived by Jennison and Das Gupta (1953) using the intensity interferometer. Reprinted by permission from *Nature*, Vol. 172, No. 4387, p. 996; copyright © 1953 Macmillan Journals Limited.

m and n, it is easy to show that at any instant the combination

$$\phi_{123} = \phi_{12} + \phi_{23} + \phi_{31} \tag{1.9}$$

is independent of instrumental and atmospheric phase effects and is a measure of the corresponding combination of visibility phases (Jennison 1958). By moving one antenna at a time it was found that the phase does indeed change by approximately 180° at the visibility minimum; therefore, the two-component model in Fig. 1.8 is the appropriate one. The intensity interferometer will be discussed further in Chapter 15, but we should remark here that it has not proved to be an important instrument in radio astronomy because of its lack of sensitivity. It has been used very effectively, however, in the optical range. The use of combinations of simultaneous visibility measurements typified by Eq. (1.9), sometimes referred to as closure relationships, was virtually forgotten for almost 20 years. It later became an important technique in interpreting visibility measurements when direct calibration is impractical or insufficiently accurate, and is discussed in Chapter 11.

Phase Switching and the Correlator Interferometer

The most important technical improvement in early radio interferometry was the introduction of *phase switching* by Ryle (1952). In the earliest interferometers the signals from the two antennas were added and applied to the input of a receiver with a square-law (power-linear) detector. If V_1 and V_2 represent the signal voltages as functions of time from the two antennas, the output from the simple adding interferometer is proportional to $(V_1 + V_2)^2$. In the phase-switching system, shown in Fig. 1.9, the phase of one of the signals is periodically reversed, so the output of the detector can be thought of as alternating between $(V_1 + V_2)^2$ and $(V_1 - V_2)^2$. A synchronous detector takes the difference between these two terms which is equal to $4V_1V_2$. Thus the output of a phase-switching interferometer is the time average of the product of the signal voltages, that is, it is proportional to the cross correlation of the

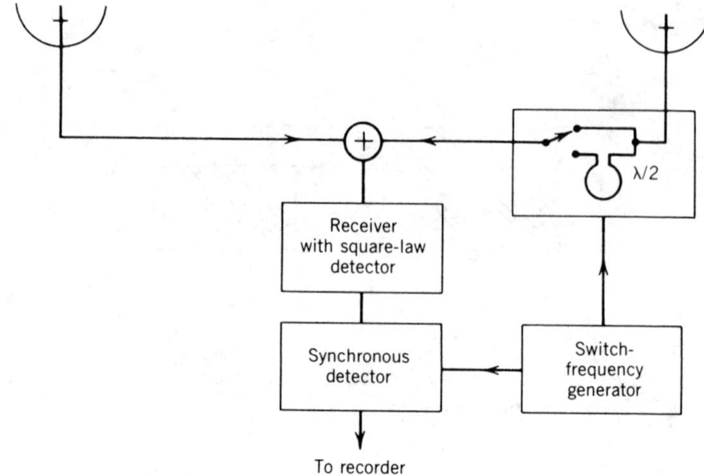

Figure 1.9   Phase-switching interferometer. The signal from one antenna is periodically reversed in phase, indicated here by switching an additional half wavelength of path into the transmission line.

two signals. The circuitry that performs the multiplication and averaging of the signals in a modern interferometer is known as a *correlator*. In a correlator interferometer the terms V_1^2 and V_2^2 in the output of an adding interferometer do not occur, and the response to a source is shown in Fig. 1.10. If amplifiers are installed at the antennas, the time averages of the terms V_1^2 and V_2^2 largely represent the amplifier noise. However, the time average of V_1V_2 represents only the product of the correlated components of the waveforms that result from the radio source, since the noise contributions of the two amplifiers are uncorrelated. Thus for a weak source the terms V_1^2 and V_2^2 may be greater than V_1V_2 by several orders of magnitude. The output of the adding interferometer is therefore highly sensitive to variations in the gain or system noise level since these produce proportional changes in V_1^2 and V_2^2. In the correlator interferometer the output is proportional only to the correlated component of the

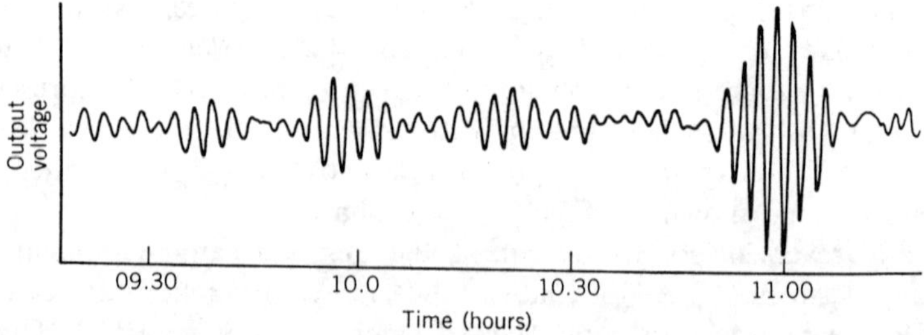

Figure 1.10   Output of a phase-switching interferometer as a function of time showing the response to a number of sources. From Ryle (1952).

input signals, and small gain changes result only in proportional variations in the amplitude of the fringe visibility. With this reduction in the sensitivity to instrumental gain it became practicable to install amplifiers at the antennas to overcome attenuation in the transmission lines. This advance resulted in the use of longer antenna spacings and larger arrays. Most interferometers from about 1950 onward incorporated phase switching, and it provided the first means of implementing the action of a correlator. It is now no longer necessary to use phase switching to obtain the voltage-multiplying action, but it is often included to help eliminate various sources of error as is described in Chapters 7 and 8.

Unlike the two-element interferometer, the sea interferometer could not be phase switched. Although a number of improvements that could be made to it were described by Bolton and Slee (1953), its obvious limitations resulted in a decrease in its use after the early 1950s.

Survey Instruments and High-Resolution Measurements

By the middle 1950s the number of identified sources was not yet sufficient for significant statistical studies. The thrust of much of the work at that time was therefore toward surveying larger numbers of sources and obtaining accurate positions and angular widths to facilitate identifications.

Determination of accurate positions using interferometers requires, of course, that the angular position of the fringes, that is, the phase response, be calibrated. Instruments that were constructed for this purpose generally used spacings of no more that 1 km and a symmetrical arrangement of antennas. A large interferometer at Cambridge used four antennas located at the corners of a rectangle 580 m east-west by 49 m north-south (Ryle and Hewish 1955). This arrangement provided both east-west and north-south fringe patterns for measurement of right ascension and declination. A different type of survey instrument was developed by Mills et al. (1958) at Fleurs, near Sydney, consisting of two long, narrow antenna arrays in the form of a cross, as shown in Fig. 1.11. Each array produced a *fan beam*, that is, a beam that is narrow in a plane containing the long axis of the array and wide in the orthogonal direction. The outputs of these two arrays were combined in a phase-switching

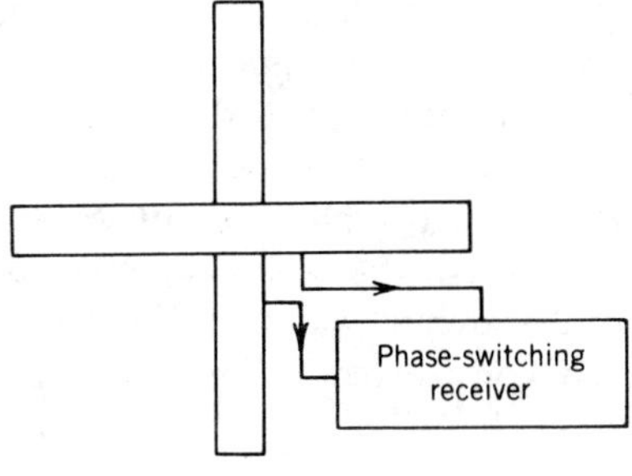

Figure 1.11 Simplified diagram of the Mills cross radio telescope. The cross-shaped area represents the apertures of the two antennas.

receiver and the voltage-multiplying action provided a power response pattern that was equal to the product of the voltage responses of the two arrays. The arrays were 457 m long and the cross produced a *pencil beam* (i.e., a narrow beam of approximately circular cross section) of width 49 arcmin at 85.5 MHz. The beam was directed toward the meridian and could be steered in elevation by adjusting the phase of the dipoles in the north-south array. The sky survey made with this instrument provided a list of over 2200 sources (Mills, Slee, and Hill 1958, 1960, 1961; Hill and Mills 1962). Early in this work a comparison of results with those from the Cambridge interferometer, which was initially operated at 81.5 MHz (Shakeshaft et al. 1955), showed poor agreement between the source lists for a common area of sky (Mills and Slee 1957). The discrepancy was found to result principally from the occurrence of *source confusion* in the Cambridge observations. When two or more sources are simultaneously present within the antenna beams they produce fringe patterns with slightly different frequencies, resulting from the differences in their declinations. The combined fringe patterns contain amplitude variations analogous to beat tones in acoustics, and the maxima in the resulting output do not correspond to individual sources. Evidently narrower beamwidths were required for the interferometer antennas, a problem that did not arise in the Mills cross which was designed to provide the required resolution for accurate positions in the single pencil beam. This incident established the practical criterion that the density of sources cataloged in a survey should not, on average, exceed one in about twenty times the solid angle of the antenna beams (Pawsey 1958, Hazard and Walsh 1959). The frequency of the Cambridge interferometer was increased to 159 MHz, thereby reducing the solid angles of the beams by a factor of 4, and a new list of 471 sources was rapidly compiled (Edge et al. 1959). This was the famous 3C survey (source numbers are preceded by 3C indicating the third Cambridge catalog) a revised version of which (Bennett 1962) became a cornerstone of radio astronomy for the following decade.

In the 1960s a new and larger generation of survey instruments began to appear. At Cambridge these included a large interferometer with one antenna elongated in the east–west direction and the other north–south, and also a large T-shaped array which had characteristics similar to those of a cross. In each of these instruments the north–south-elongated element was not constructed in full, but the response with such an aperture was synthesized by using a small antenna that was moved in steps along the required aperture, a different position being used for each 24-hour scan in right ascension (Ryle, Hewish, and Shakeshaft 1959). These instruments are discussed further in Chapter 5. The large interferometer produced the 4C catalog containing over 4800 sources (Pilkington and Scott 1965; Gower, Scott, and Wills 1967). At Molonglo in Australia a larger Mills cross (Mills et al. 1963) was constructed with arrays 1 mile long producing a beam of 2.8 arcmin width at 408 MHz. Crosses of comparable dimensions were also constructed in the northern hemisphere at Bologna in Italy (Braccesi et al. 1969) and at Serpukhov in the Soviet Union (Vitkevich and Kalachev 1966).

Measurements of the angular widths of most of the sources found in the surveys called for resolutions of less than 1 arcmin necessitating antenna spacings of tens of kilometers at the meter wavelengths at which most instruments were operated at that time. An early interferometer aimed at achieving the angular width requirement was developed by Hanbury Brown, Palmer, and Thompson (1955) at the Jodrell Bank Experimental Station, England, (later to be known as the Nuffield Radio Astronomy Laboratories). Three sources were found to have diameters less than 12 arcsec using spacings up to 20 km at 158 MHz observing frequency (Morris, Palmer, and Thompson 1957). This interferometer used an offset local oscillator technique which took the place of a phase switch and also enabled the frequency of the fringe pattern to be slowed down to within the response time of the chart recorders commonly used at that time. A radio link was used to bring the signal from the distant antenna. During the 1960s this instrument was extended to achieve resolution of less than 1 arcsec and greater sensitivity (Elgaroy, Morris, and Rowson 1962). This program later led to the development of a multielement, radio-linked interferometer known as the MERLIN array (Davies, Anderson, and Morison 1980).

Detailed Mapping and Special Uses of Interferometers

After completion of the 4C and other large catalogs, the emphasis in the late 1960s and the 1970s turned toward more detailed examination of individual sources. This called for high-resolution mapping in two dimensions. Interferometers constructed for this purpose generally operated at centimeter wavelengths and used tracking antennas. For a given antenna spacing the resolution was one to two orders of magnitude greater than that of the meter-wavelength arrays that were typical of the large survey instruments. The technique required for mapping brightness distributions had been demonstrated in part in early work on the sun by Stanier (1950) and O'Brien (1953), who used variable-spacing interferometers to study the solar brightness distribution, but assumed symmetrical models. To measure the two-dimensional brightness distribution of a radio source requires measurement of the amplitude and phase of the fringe patterns in two dimensions, which can be accomplished by using different lengths and azimuths of the antenna baseline. The interferometer at the Owens Valley Radio Observatory, California (Read 1961), provides a good example of one of the earliest instruments used extensively for determining radio structures. It consists of two 27.5-m diameter paraboloid antennas on equatorial mounts with a rail track system that allows the spacing between them to be varied up to 490 m in both the east–west and north–south directions. Studies by Maltby and Moffet (1962) and Fomalont (1968) il-

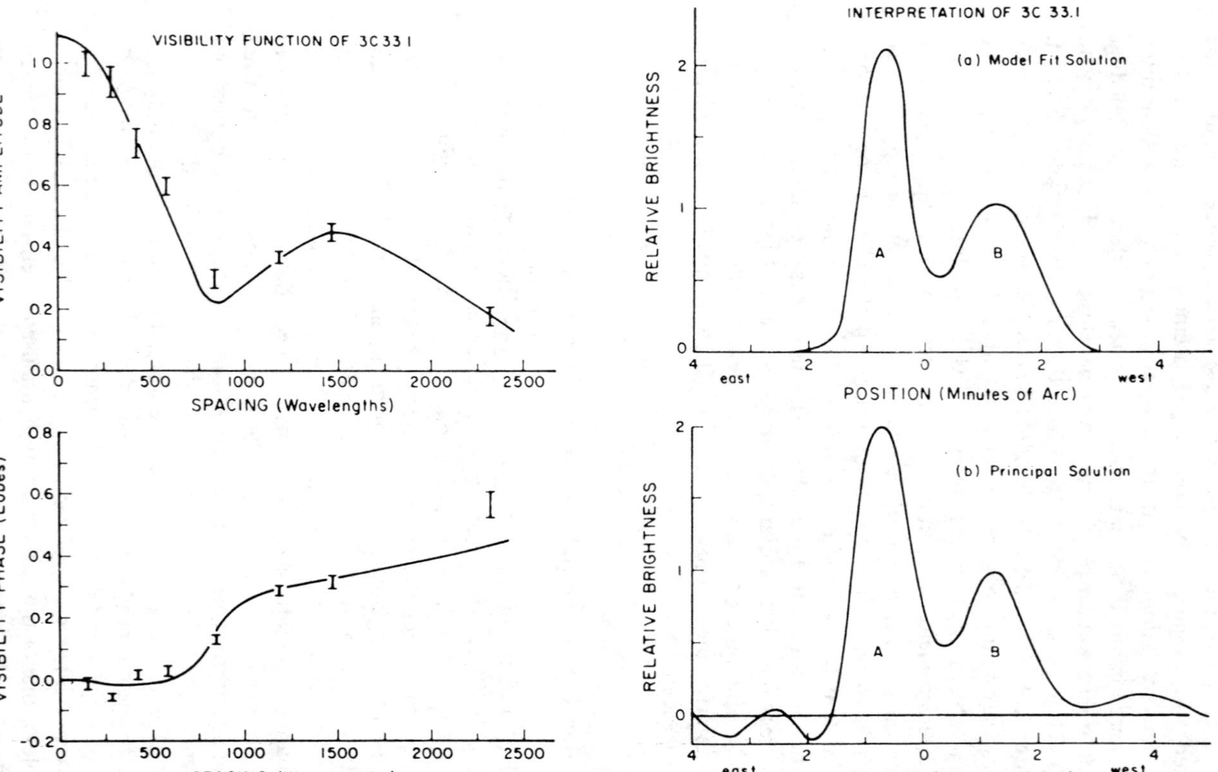

Figure 1.12   Example of interferometer measurements of one-dimensional brightness: the east–west profile of the source 3C33.1 as determined by Fomalont (1968) using the interferometer at the Owens Valley Radio Observatory at 1425 MHz. The points at the left show the measured amplitude and phase of the visibility. The profile at the top right was obtained by fitting Gaussian components to the visibility data, as shown by the curves through the measured visibility points. The profile at the lower right was obtained by Fourier transformation of the observed visibility values. The unit of visibility phase is 2π radians.

lustrate the use of this instrument for measurement of brightness distributions, an example of which is shown in Fig. 1.12.

Calibration of the phase of the interference fringes is required for such mapping and various advances facilitated this process. First, by the early 1960s, the positions of a number of small-diameter extragalactic sources were known to about 1 arcsec and observation of these could therefore be used to calibrate the fringes. The positions were obtained either from measurements of the optical objects for those sources that were identified, of from observations of the occultations of sources by the moon's limb, as discussed in Chapter 15. Second, the introduction of electronic computers allowed the expected phase as a function of hour angle to be calculated rapidly for arbitrary baselines and source positions. The observed and predicted fringe patterns for numbers of sources could be compared, and best-fit values obtained for the baseline vector and for the instrumental phase. Finally, continued improvements in the design of electronic receiving equipment resulted in improved stability of the instrumental gain and phase.

Interferometers were also widely used for more general applications. For example, it became clear that an interferometer offers important advantages in searching for very-low-level radio emission from small-diameter objects such as stars. With a single antenna it is necessary to average a large number of records taken either as drift scans through the position of the object, or with the antenna tracking and the beam alternately on and off the source position. In such observations the receiver records the total power from the antenna, and the sensitivity that can be achieved is often limited by effects such as the variation of the thermal radiation from the ground that is picked up in the sidelobes as the antenna follows the source. With an interferometer of the correlator type only signals that are coherent at the two antennas produce an output. A source can be tracked for many hours and the output continuously analyzed for the component at the expected fringe frequency. The fundamental limit set by the receiver noise is more easily achieved in this manner. The technique was particularly successful for the measurement of radio emission from stars (Hjellming 1974). The ability of an interferometer to reject sidelobe pickup is also helpful in measuring small changes in the flux density as the plane of polarization of the antennas is rotated.

By the early 1960s some interferometers had been fitted with spectral-line receiving systems. One purpose for which spectral-line interferometry proved to be particularly useful was the observation of absorption spectra in the 21-cm hydrogen line. Emission in this line extends over wide regions near the galactic plane and is easily mapped with a single antenna. At the line frequency the gas also absorbs the continuum radiation from any more distant source that is observed through it. Comparison of the emission and absorption spectra of the gas yields information on its temperature and density. Measurement of absorption spectra of sources with a single antenna is very difficult because the antenna responds also to the broadly distributed emitting gas within the antenna beam. The emission spectrum must be subtracted, and this can usually be done with sufficient accuracy only for strong sources. With an interferometer the broad emission features on the sky are almost entirely resolved and the absorption spectrum can be observed directly (e.g. Clark, Radhakrishnan, and Wilson 1962; Hughes, Thompson, and Colvin 1971).

Earth Rotation in Synthesis Mapping

The next important step in the development of interferometer techniques was the use of the variation of the antenna baseline provided by the rotation of the earth. For a source at a high declination the position angle of the baseline projected onto a plane normal to the direction of the source rotates through 180° in 12 hr. Thus if the source is tracked across the sky for a series of 12-hr periods, each one with a different antenna spacing, the required two-dimensional visibility data can be collected while the antenna spacing is varied in one dimension only. Figure 1.13 illustrates this principle as described by Ryle (1962). The Cambridge One-Mile radio telescope was the first instrument

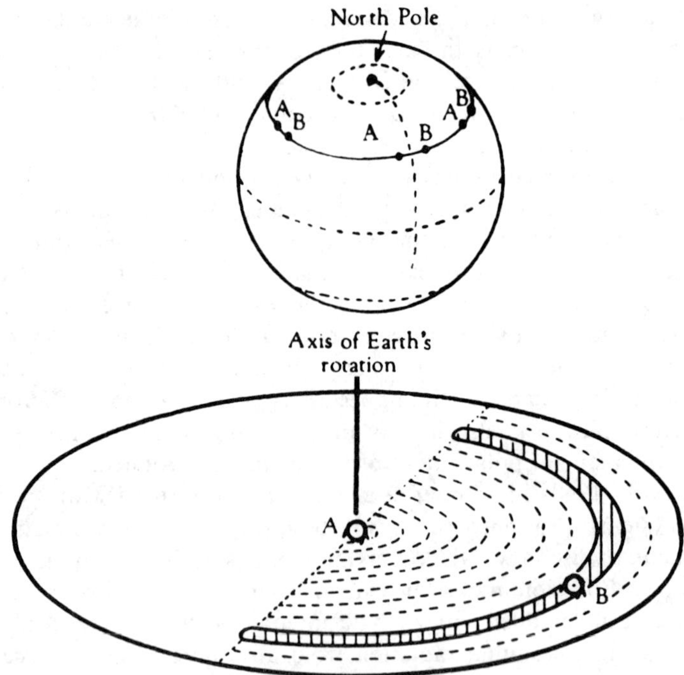

Figure 1.13   Use of earth rotation in synthesis mapping as explained by Ryle (1962). The antennas A and B are spaced on an east–west line. By varying the distance between the antennas from one day to another, and observing for 12 hr with each configuration, it is possible to encompass all the spacings from the origin to the elliptical outer boundary of the lower diagram. Only 12 hr of observing are required, since during the other 12 hr the spacings covered are identical but the positions of the antennas are effectively interchanged. Reprinted by permission from *Nature*, Vol. 194, No. 4828, p. 517; copyright © 1962 Macmillan Journals Limited.

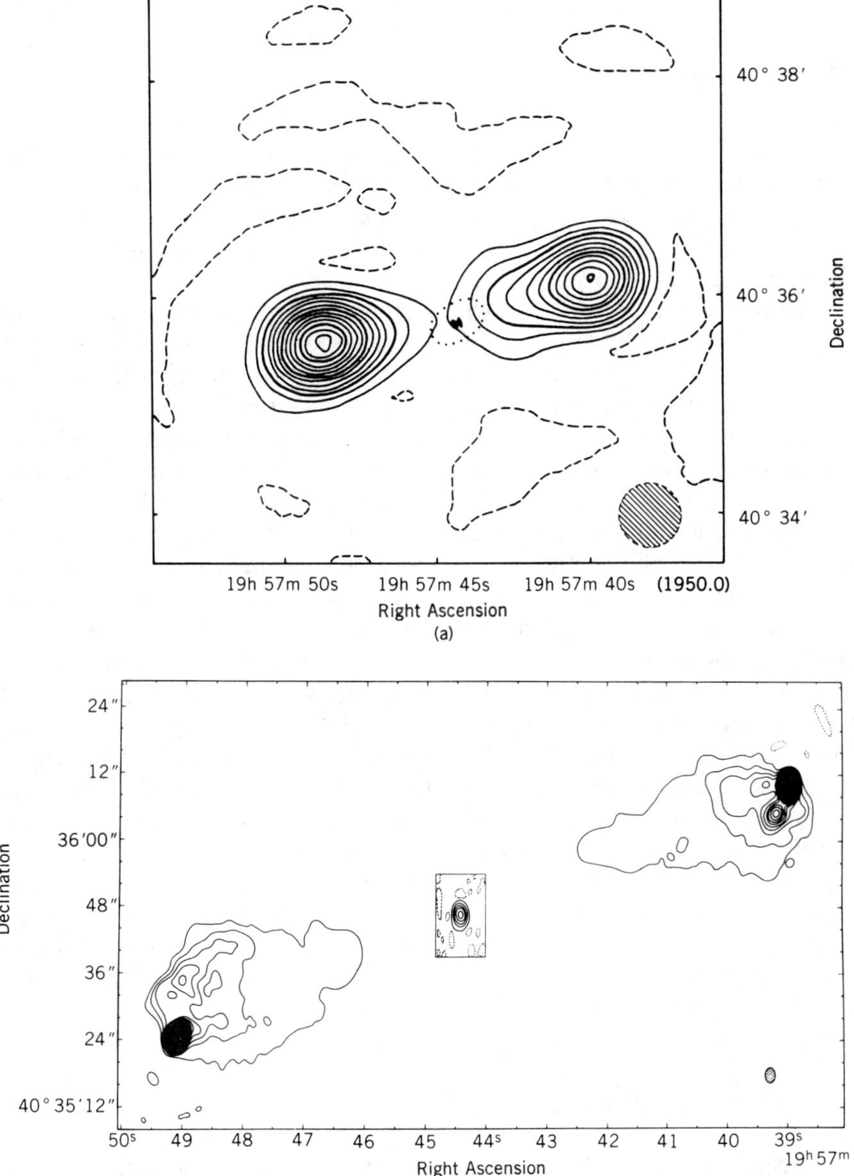

Figure 1.14 Two maps of the source Cygnus A. (a) One of the first results (Ryle, Elsmore, and Neville 1965) from the Cambridge One-Mile telescope using the earth-rotation principle shown in Fig. 1.13. The frequency is 1.4 GHz. The map has been scaled in declination so that the half-power beam contour is circular, as shown by the shaded area in the lower right corner. The dotted ellipse shows the outer boundary of the optical source, and its central structure is also indicated. (b) Map by Hargrave and Ryle (1974) using the Cambridge Five-Kilometer telescope at 5 GHz. This map showed for the first time the radio nucleus associated with the central galaxy and the high brightness at the outer edges of the radio lobes. Part (a) is reprinted by permission from *Nature*, Vol. 205, No. 4978, p. 1260; copyright © 1965 Macmillan Journals Limited. Part (b) is reprinted by permission of the Royal Astronomical Society.

designed to exploit fully the earth rotation technique and to apply it to large numbers of radio sources. The first maps published from this instrument, those of the strong sources Cassiopeia A and Cygnus A (Ryle, Elsmore, and Neville 1965), exhibited a degree of structural detail unprecedented in earlier studies and heralded the development of synthesis mapping. The map of Cygnus A is shown in Fig. 1.14a.

The use of earth rotation was not a sudden development in radio astronomy, however, and it had been used in solar studies for a number of years. In 1955 Christiansen and Warburton had obtained a two-dimensional map of the quiet sun at 21-cm wavelength using both east–west and north–south grating arrays. These arrays consisted of 32 (east–west) and 16 (north–south) uniformly spaced paraboloid antennas. Each array produced a series of fan beams in a manner analogous to the responses of an optical diffraction grating. (Thus, uniformly spaced arrays are sometimes referred to as *grating arrays*.) As the sun moved through the sky it was scanned at different angles by the different beams, and a two-dimensional map could be synthesized by Fourier analysis of the scan profiles. Later instruments for solar mapping used grating arrays, crossed in the manner of a Mills cross, to produce a rectangular matrix pattern of beams on the sky. The rotation of the earth then enabled sufficient scans to be obtained to provide daily maps of active regions and other features. Instruments of this type included crosses at 21-cm wavelength at Fleurs, Australia (Christiansen and Mullaly 1963), at 10-cm wavelengh at Stanford, California (Bracewell and Swarup 1961), and a T-shaped array at 1.9-m wavelength at Nançay, France (Blum, Boischot, and Ginat 1957; Blum 1961). Grating arrays at 3- and 7.5-cm wavelength were also constructed at Toyokawa, Japan (Hatanaka 1963). Early nonsolar uses of earth rotation in interferometry included a two-element tracking interferometer at Jodrell Bank, England (Rowson 1963). Also a survey of the north polar region at 178 MHZ was made by Ryle and Neville (1962) using earth rotation to demonstrate the feasibility of the technique.

Development of Synthesis Arrays

Following the success of the Cambridge One-Mile telescope, interferometers such as the NRAO instrument at Green Bank, West Virginia (Hogg et al. 1969), were rapidly adapted for synthesis mapping. Several large arrays designed to provide increased mapping speed, sensitivity, and angular resolution were brought into operation during the 1970s. Prominent among these were the Five-Kilometer radio telescope at Cambridge, England (Ryle 1972), the Westerbork synthesis radio telescope in the Netherlands (Baars et al. 1973), and the Very Large Array (VLA) in New Mexico (Thompson et al. 1980). These instruments permit mapping of radio sources with a resolution of less than 1 arcsec at centimeter wavelengths. By using n_a antennas, where n_a varies up to 27 in the arrays mentioned, as many as $n_a(n_a - 1)/2$ simultaneous baselines are obtained. If the array is designed to minimize redundancy, as

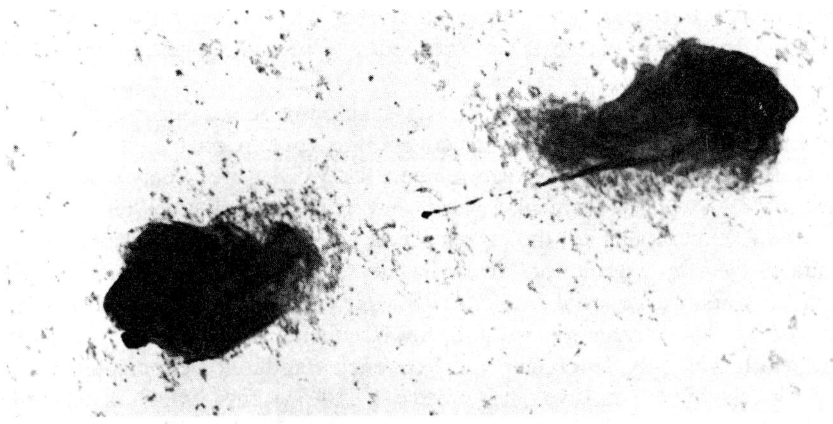

Figure 1.15   Radio map of Cygnus A made with the VLA at 4.9 GHz by Perley, Dreher, and Cowan (1984). Observations with four configurations of the array were combined and the resolution is 0.4 arcsec. The display of the image shown here involves a nonlinear process to enhance the contrast of the fine structure. This emphasizes the jet from the central galaxy to the northwestern lobe (top right) and the filamentary structure in the main lobes. Comparison with other records of Cygnus A in this chapter illustrates the technical advances made during three decades. Reproduced by permission of NRAO/AUI.

discussed in Chapter 5, the speed with which the visibility function is measured is approximately proportional to n_a^2. Maps of Cygnus A obtained with two of the arrays mentioned above are shown in Figs. 1.14b and 1.15. A review of the development of synthesis instruments at Cambridge is given in the Nobel lecture by Ryle (1975).

Also during the early 1980s several synthesis instruments for operation at millimeter wavelengths, largely for spectral line observations, were in various stages of operation or development. These included arrays at Hat Creek, California; Owens Valley, California; Nobeyama, Japan; and the Plateau de Bure, France.

Very-Long-Baseline Interferometry

Investigation of the angular diameters of quasars and other objects that appear nearly pointlike in structure has presented an important challenge throughout the development of radio astronomy. During the early 1960s the long-baseline interferometer at Jodrell Bank, mentioned above, was the foremost instrument in this field. An advance that led to an immediate increase of an order of magnitude in resolution, and subsequently to several orders more, was the use of independent local oscillators and signal recorders. By using local oscillators at each antenna that are controlled by high precision frequency standards, it is possible to preserve the coherence of the signals for time intervals long enough to measure interference fringes. The received signals are converted to an intermediate frequency low enough that they can be recorded directly on magnetic tape, and the tapes are subsequently brought together and played into a correlator.

This technique became known as very-long baseline interferometry (VLBI), and the early history of its development is discussed by Cohen et al. (1968) and Burke (1969a). The technical requirements for VLBI were widely discussed in the early 1960s (e.g., Matveenko, Kardashev, and Sholomitskii 1965). An early successful experiment in the development of the technique was done in January 1967 by a group at the University of Florida that detected fringes from the burst radiation of Jupiter at 18 MHz (Brown, Carr, and Block 1968). Because of the strong signals and low frequency, the required recording bandwidth was only 2 kHz and the frequency standards were crystal oscillators. Much more sensitive and precise VLBI systems, which used wider bandwidths and atomic frequency standards, were developed by three groups. A Canadian group built an analog recording system with a bandwidth of 1 MHz based on television tape recorders (Broten et al. 1967). They obtained fringes at a frequency of 448 MHz on baselines of 183 and 3074 km on several quasars in April 1967. In the United States, another group from the National Radio Astronomy Observatory and Cornell University developed a computer-compatible digital system with a bandwidth of 360 kHz (Bare et al. 1967). They obtained fringes at 610 MHz on a baseline of 220 km on several quasars in May 1967. A third group from MIT joined in the development of the NRAO-Cornell system in early 1967 and obtained fringes at a frequency of 1665 MHz on a baseline of 845 km on several OH masers in June 1967 (Moran et al. 1967). The exciting scientific results from these early successes fueled the rapid development of VLBI.

The initial experiments used signal bandwidths of less than a megahertz, but by the 1980s systems capable of recording signals with bandwidths greater than 100 MHz were available, with corresponding improvements in sensitivity. Real-time linking of the signals from remote telescopes to the correlator via a geostationary satellite has been demonstrated (Yen et al. 1977). Also, experiments have been performed in which the local oscillator signal was distributed over a satellite link (Knowles et al. 1982). These developments have lessened the distinction between VLBI and more conventional forms of interferometry. However, VLBI has many peculiar technical considerations, which are discussed in Chapter 9.

The extremely high angular resolution that can be achieved by the VLBI technique is well illustrated by a measurement by Burke et al. (1972) who obtained a resolution of 2×10^{-4} arcsec using antennas in Westford, Massachusetts, and Simeiz in the Crimea, operating at a wavelength of 1.3 cm. Early results, obtained using a few baselines only, were generally interpreted in terms of the simple models in Fig. 1.5. During the mid-1970s, however, several groups of astronomers began to combine their facilities to obtain measurements over ten or more baselines simultaneously. These observations led to more complex models (see, e.g., Cohen et al. 1975). Important results were the

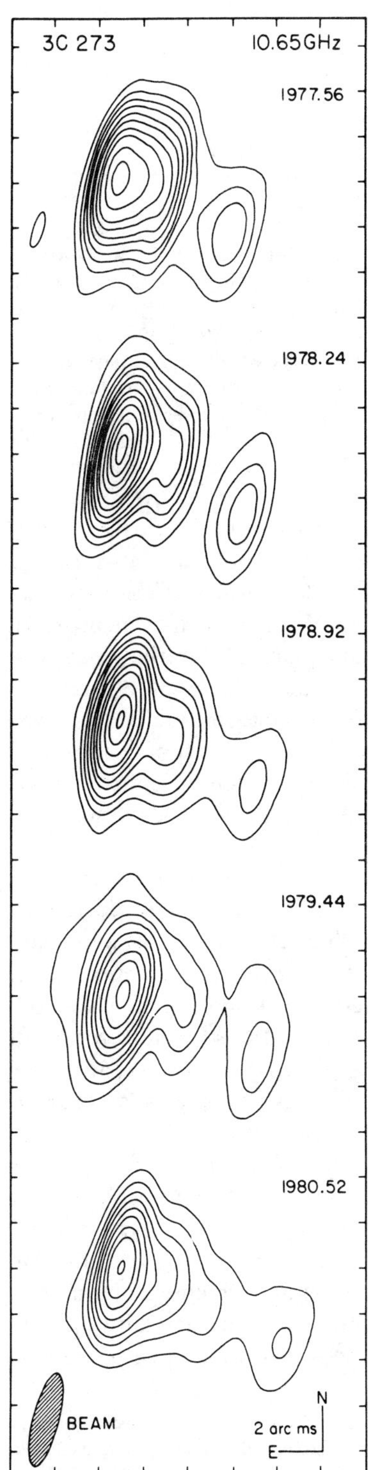

Figure 1.16 VLBI maps of the quasar 3C273 at five epochs, showing the relative positions of two components. From the distance of the object, deduced from the optical red shift, the apparent relative velocity of the components exceeds the velocity of light, but can be explained by relativistic and geometrical effects. The observing frequency is 10.65 GHz. From Pearson et al. (1981). Reprinted from *Nature*, Vol. 290, No. 5805, p. 366; copyright © 1981 Macmillan Journals Limited.

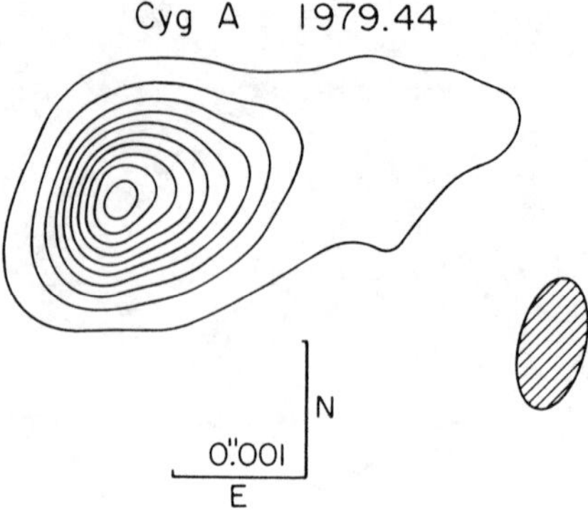

Figure 1.17 VLBI map of the small central component of Cygnus A by Linfield (1981), made using four antennas at 10.65 GHz. The half-power contour of the synthesized beam, shown by the shaded ellipse, has dimensions of 0.5×1 milliarcsec. The contour interval for the source is 8×10^8 K in brightness temperature. The major axis of the source is in the same direction as the jet in Fig. 1.15.

discovery and investigation of superluminal (apparently faster than light) motions in quasars (Whitney et al. 1971, Pearson et al. 1981), as shown in Fig. 1.16, and the measurement of proper motion in H_2O-line masers (Genzel et al. 1981). A problem in VLBI observations is that the use of nonsynchronized local ocillators usually precludes calibration of the phase of the fringe patterns. To overcome this problem the phase closure relationship of Eq. (1.9) was first applied to VLBI data by Rogers et al. (1974). The technique rapidly developed into a method to obtain brightness maps known as hybrid mapping. This and related procedures are also used in mapping with connected-element[†] arrays and are discussed in Chapter 11. For examples of hybrid mapping see Figs. 1.16 and 1.17

The great potential of VLBI in astrometry and geodesy was immediately recognized (e.g. Gold 1967). Its use in these applications developed rapidly during the 1970s and 1980s; see, for example, Whitney et al. (1976) and Clark et al. (1985). In the United States, NASA and several other federal agencies set up a cooperative program of geodetic measurements (Flinn 1981). This work evolved in part from the use of deep-space communications facilities for VLBI observations. The program includes the use of transportable antennas for periodic monitoring of the positions of many different sites. Similar geodetic studies are being undertaken or planned on several continents.

The development of VLBI introduced a new facet into the apparent paradox in the quantum mechanical description of interferometry (Burke 1969b). The radio interferometer is the analog of Young's two-slit interference experiment. It is well known (Louden 1973) that a single photon creates an interference pattern, but that any attempt to determine which slit the photon entered will destroy the interference pattern, otherwise the uncertainty principle would be violated. Discussion of VLBI suggests that it might be possible to

---

[†] The term *connected-element*, or *linked-element*, is used to describe arrays of the conventional type in which the signals are brought to the correlators in real time, usually by transmission lines or radio links, in contrast to VLBI systems. This classification is useful although the distinction is not always clear, for example, in the case of measurements with spacing typical of VLBI but using direct signal transmission via satellite to the correlators.

determine at which antenna a particular photon arrived, since its signature is captured on the tape as well as in the fringe pattern generated during correlation. However, it is necessary to have linear amplifiers in the receiving system. These amplifiers preserve the signal phase and hence do not disturb the interference pattern in the classical limit. Such an amplifier must obey the uncertainty principle, $\Delta E \Delta t \simeq h/2\pi$, where ΔE and Δt are the uncertainties in energy and time. This principle can be written in terms of uncertainty in photon number, Δn, and phase, $\Delta\phi$, as

$$\Delta n \, \Delta\phi \simeq 1, \tag{1.10}$$

where $\Delta E = h\nu\Delta n$ and $\Delta\phi = 2\pi\nu\Delta t$. Hence to preserve phase there must be an uncertainty of at least one photon per unit bandwidth per unit time in the output of the amplifier. An alternative viewpoint is that every amplifier must have spontaneous noise emission so that the minimum system temperature is approximately $h\nu/k$. Hence the signal-to-noise ratio is less than unity in the single-photon limit, and it is impossible to determine at which antenna a single photon entered. Oliver (1965) discusses in detail the limitations imposed by quantum noise.

The bibliography that follows lists books and special issues of journals that are useful as sources, particularly on instrumental development in radio astronomy.

BIBLIOGRAPHY

Alder, B., S. Fernbach, and M. Rotenberg, Eds., *Methods in Computational Physics*, Vol. 14, Academic Press, New York, 1975.

Berkner, L. V., Ed., *IRE Trans. Antennas Propag.*, Special Issue on Radio Astronomy, Vol. AP-9, No. 1, 1961.

Biraud, F., Ed., *Very Long Baseline Interferometry Techniques*, Cepadues, Toulouse, France, 1983.

Bracewell, R. N., Ed., Paris Symposium on Radio Astronomy, *IAU Symposium No. 9*, Stanford University Press, Stanford, California, 1959.

Bracewell, R. N., Radio Astronomy Techniques, in *Handbuch Der Physik*, Vol. 54, S. Flugge, Ed., Springer-Verlag, Berlin, 1962.

Christiansen, W. N. and J. A. Högbom, *Radiotelescopes*, Cambridge University Press, Cambridge, England, 1969 (2nd ed. 1985).

Findlay, J. W., Ed., *Proc. IEEE*, Special Issue on Radio and Radar Astronomy, Vol. 61, No. 9, 1973.

Haddock, F. T., Ed., *Proc. IRE*, Special Issue on Radio Astronomy, Vol. 46, No. 1, 1958.

Kraus, J. D., Ed., *IEEE Trans. Mil. Electron.*, Special Issue on Radio and Radar Astronomy, Vol. Mil-8, Nos. 3 and 4, 1964, also issued by *IEEE Trans. Antennas Propag.*, Vol. AP-12, No. 7, 1964.

Kraus, J. D., *Radio Astronomy*, McGraw-Hill, New York, 1966.

Meeks, M. L., Ed., *Methods of Experimental Physics*, Vol. 12, parts B and C, Academic Press, New York, 1976.

Pawsey, J. L., Ed., *Proc. IRE, Aust.*, Special Issue on Radio Astronomy, Vol. 24, No. 2, 1963.

Pawsey, J. L. and R. N. Bracewell, *Radio Astronomy*, Oxford University Press, Oxford, England, 1955.

Shklovsky, I. S., *Cosmic Radio Waves*, translated by R. B. Rodman and C. M. Varsavsky, Harvard University Press, Cambridge, Mass., 1960.

Sullivan, W. T., III, Ed., *The Early Years of Radio Astronomy*, Cambridge University Press, Cambridge, England, 1984.

Wild, J. P., Ed., *Proc. IREE Aust.*, Special Issue on The Culgoora Radioheliograph, Vol. 28, No. 9, 1967.

Yen, J. L., Image Reconstruction in Synthesis Radio Telescope Arrays, in *Array Signal Processing*, S. Haykin Ed., Prentice Hall, Englewood Cliffs, New Jersey, 1985, pp. 293–350.

REFERENCES

Anderson, J. A., The Wave-Length in Astronomical Interferometer Measurements, *Astrophys, J.*, **55**, 48–70, 1922.

Appleton, E V., Departure of Long-Wave Solar Radiation from Black-Body Intensity, *Nature*, **156**, 534–535, 1945.

Baade, W. and R. Minkowski, Identification of the Radio Sources in Cassiopeia, Cygnus A, and Puppis A, *Astrophys. J.*, **119**, 206–214, 1954a.

Baade, W. and R. Minkowski, On the Identification of Radio Sources, *Astrophys, J.*, **119**, 215–231, 1954b.

Baars, J. W. M., J. F. van der Brugge, J. L. Casse, J. P. Hamaker, L. H. Sondaar, J. J. Visser, and K. J. Wellington, The Synthesis Radio Telescope at Westerbork, *Proc. IEEE*, **61**, 1258–1266, 1973.

Bare, C., B. G. Clark, K. I. Kellermann, M. H. Cohen, and D. L. Jauncey, Interferometer Experiment with Independent Local Oscillators, *Science*, **157**, 189–191, 1967.

Bennett, A. S., The Revised 3C Catalog of Radio Sources, *Mem. R. Astron. Soc.*, **68**, 163–172, 1962.

Blum, E. J., Le Réseau Nord–Sud à Lobes Multiples, *Ann. Astrophys.* **24**, 359–366, 1961.

Blum, E. J., A. Boischot, and M. Ginat, Le Grand Interféromètre de Nançay, *Ann. Astrophys*, **20**, 155–164, 1957.

Bolton, J. G. and O. B. Slee, Galactic Radiation at Radio Frequencies, V. The Sea Interferometer, *Aust. J. Phys.*, **6**, 420–433, 1953.

Bolton, J. G. and G. J. Stanley, Variable Source of Radio Frequency Radiation in the Constellation of Cygnus, *Nature*, **161**, 312–313, 1948.

Bolton, J. G., G. J. Stanley, and O. B. Slee, Positions of Three Discrete Sources of Galactic Radio Frequency Radiation, *Nature*, **164**, 101–102, 1949.

Bracessi, A., M. Ceccarelli, G. Colla, R. Fanti, A. Ficarra, G. Gelato, G. Greuff, and G. Sinigaglia, The Italian Cross Radio Telescope, III. Operation of the Telescope, *Nuovo Cimento B*, **62**, 13–19, 1969.

Bracewell, R. N. and G. Swarup, The Stanford Microwave Spectroheliograph Antenna, a Microsteradian Pencil Beam Interferometer, *IRE Trans. Antennas Propag.*, **AP-9**, 22–30, 1961.

Broten, N. W., T. H. Legg, J. L. Locke, C. W. McLeish, R. S. Richards, R. M. Chisholm, H. P. Gush, J. L. Yen, and J. A. Galt, Observations of Quasars Using Interferometer Baselines up to 3074 km, *Nature*, **215**, 38, 1967.

Brown, G. W., T. D. Carr, and W. F. Block, Long Baseline Interferometry of S-Bursts from Jupiter, *Astrophys. Lett.*, **1**, 89–94, 1968.

Burke, B. F., Long Baseline Interferometry, *Physics Today*, **22**, No. 7, 54–63, 1969a.

Burke, B. F., Quantum Interference Paradox, *Nature*, **223**, 389–390, 1969b.

Burke, B. F., K. J. Johnston, V. A. Efanov, B. G. Clark, L. R. Kogan, V. I. Kostenko, K. Y. Lo, L. I. Matveenko, I. G. Moiseev, J. M. Moran, S. H. Knowles, D. C. Papa, G. D. Papadopoulos, A. E. E. Rogers, and P. R. Schwartz, Observations of Maser Radio Source with an Angular Resolution of 0".0002, *Soviet Astron.-AJ*, **16**, 379–382, 1972.

Christiansen, W. N. and R. F. Mullaly, Solar Observations at a Wavelength of 20 cm with a Crossed-Grating Interferometer, *Proc. IRE Aust.*, **24**, 165–173, 1963.

Christiansen, W. N. and J. A. Warburton, The Distribution of Radio Brightness over the Solar Disk at a Wavelength of 21 cm, III. The Quiet Sun—Two Dimensional Observations, *Aust. J. Phys.*, **8**, 474–486, 1955.

Clark, B. G., V. Radhakrishnan, and R. W. Wilson, The Hydrogen Line in Absorption, *Astrophys. J.*, **135**, 151–174, 1962.

Clark, T. A., B. E. Corey, J. L. Davis, G. Elgered, T. A. Herring, H. F. Hinteregger, C. A. Knight, J. I. Levine, G. Lundqvist, C. Ma, E. F. Nesman, R. B. Phillips, A. E. E. Rogers, B. O. Rönnäng, J. W. Ryan, B. R. Schupler, D. B. Shaffer, I. I. Shapiro, N. R. Vandenberg, J. C. Webber, and A. R. Whitney, Precision Geodesy Using the Mark-III Very-Long-Baseline Interferometer System, *IEEE Trans. Geosci. Remote Sens.*, **GE-23**, 438–449, 1985.

Cohen, M. H., D. L. Jauncey, K. I. Kellermann, and B. G. Clark, Radio Interferometry at One-Thousandth Second of Arc, *Science*, **162**, 88–94, 1968.

Cohen, M. H., A. T. Moffet, J. D. Romney, R. T. Schilizzi, D. B. Shaffer, K. I. Kellermann, G. H. Purcell, G. Grove, G. W. Swenson Jr., J. L. Yen, I. I. K. Pauliny-Toth, E. Preuss, A. Witzel, and D. Graham, Observations with a VLBI Array. I. Introduction and Procedures, *Astrophys. J.*, **201**, 249–255, 1975.

Conway, R. G., K. I. Kellermann, and R. J. Long, The Radio Frequency Spectra of Discrete Radio Sources, *Mon. Not. R. Astron. Soc.*, **125**, 261–284, 1963.

Davies, J. G., B. Anderson, and I. Morison, The Jodrell Bank Radio-Linked Interferometer Network, *Nature*, **288**, 64–66, 1980.

Dicke, R. H., The Measurement of Thermal Radiation at Microwave Frequencies, *Rev. Sci. Instrum.*, **17**, 268–275, 1946.

Dreyer, J. L. E., New General Catalog of Nebulae and Clusters of Stars, *Mem. R. Astron. Soc.*, **49**, part 1, 1888 (reprinted, *R. Astron. Soc. London*, 1962).

Edge, D. O., J. R. Shakeshaft, W. B. McAdam, J. E. Baldwin, and S. Archer, A Survey of Radio Sources at a Frequency of 159 Mc/s, *Mem. R. Astron. Soc.,*, **68**, 37–60, 1959.

Elgaroy, O., D. Morris, and B. Rowson, A Radio Interferometer for Use with Very Long Baselines, *Mon. Not. R. Astron. Soc.*, **124**, 395–403, 1962.

Explanatory Supplement to the Astronomical Ephemeris, H. M. Stationary Office, London, 1961.

Flinn, A. E., Applications of Space Technology to Geodynamics, *Science*, **213**, 89–96, 1981.

Fomalont, E. B., The East–West Structure of Radio Sources at 1425 MHz, *Astrophys. J. Suppl.*, **15**, 203–274, 1968.

Genzel, R., M. J. Reid, J. M. Moran, and D. Downes, Proper Motions and Distances of H_2O Maser Sources. I. The Outflow in Orion-KL, *Astrophys. J.*, **244**, 884–902, 1981.

Gold, T., Radio Method for the Precise Measurement of the Rotation Period of the Earth, *Science*, **157**, 302–304, 1967.

Gower, J. F. R., P. F. Scott, and D. Wills, A Survey of Radio Sources in the Declination Ranges −07° to 20° and 40° to 80°, *Mem. R. Astron. Soc.*, **71**, 49–144, 1967.

Hanbury Brown, R., Measurement of Stellar Diameters, *Ann. Rev. Astron. Astrophys.*, **6**, 13–38, 1968.

Hanbury Brown, R., H. P. Palmer, and A. R. Thompson, A Rotating-Lobe Interferometer and its

Application to Radio Astronomy, *Philos. Mag.*, ser.7, **46**, 857–866, 1955.

Hanbury Brown, R. and R. Q. Twiss, A New Type of Interferometer for Use in Radio Astronomy, *Philos. Mag.*, ser.7, **45**, 663–682, 1954.

Hargrave, P. J. and M. Ryle, Observations of Cygnus A with the 5-km Radio Telescope, *Mon. Not. R. Astron. Soc.*, **166**, 305–327, 1974.

Hatanaka, T., Radio Astronomy in Japan, *Proc. IRE Aust.*, **24**, 243–251, 1963.

Hazard, C. and D. Walsh, A Comparison of an Interferometer and Total-Power Survey of Discrete Sources of Radio Frequency Radiation, in *The Paris Symposium on Radio Astronomy*, R. N. Bracewell, Ed., Stanford University Press, Stanford, California, 1959, pp. 477–486.

Hey, J. S., S. J. Parsons, and J. W. Phillips, Fluctuations in Cosmic Radiation at Radio Frequencies, *Nature*, **158**, 234, 1946.

Hill, E. R. and B. Y. Mills, Source Corrections to the Sydney Radio Source Survey Catalogue, *Aust. J. Phys.*, **15**, 437–440, 1962.

Hjellming, R. M., Radio Stars, in *Galactic and Extragalactic Radio Astronomy*, G. L. Verschuur and K. I. Kellermann, Eds., Springer-Verlag, New York, 1974, pp. 159–178.

Hogg, D. E., G. H. Macdonald, R. G. Conway, and C. M. Wade, Synthesis of Brightness Distribution in Radio Sources, *Astron. J.*, **74**, 1206–1213, 1969.

Hughes, M. P., A. R. Thompson, and R. S. Colvin, An Absorption-Line Study of Galactic Neutral Hydrogen at 21 cm Wavelength, *Astrophys. J. Suppl.*, **23**, 232–367, 1971.

IAU, *Transactions of the International Astronomical Union*, **15B**, 142, 1974.

IAU, *Transactions of the International Astronomical Union*, **18B**, 273 and 278, 1982.

IEEE, Standard Definitions of Terms for Radio Wave Propagation, Std 211–1977, The Institute of Electrical and Electronics Engineers, New York, 1977.

Jansky, K. G., Electrical Disturbances Apparently of Extraterrestrial Origin, *Proc. IRE*, **21**, 1387–1398, 1933.

Jennsion, R. C., A Phase Sensitive Interferometer Technique for the Measurement of the Fourier Transforms of Spatial Brightness Distributions of Small Angular Extent, *Mon. Not. R. Astron. Soc.*, **118**, 276–284, 1958.

Jennison, R. C. and M. K. Das Gupta, Fine Structure in the Extra-terrestrial Radio Source Cygnus 1, *Nature*, **172**, 996–997, 1953.

Jennison R. C. and M. K. Das Gupta, The Measurement of the Angular Diameter of Two Intense Radio Sources, *Philos. Mag.*, ser.8, **1**, 55–75, 1956.

Jennsion, R. C. and V. Latham, The Brightness Distribution Within the Radio Sources Cygnus A (19N4A) and Cassiopeia A (23N5A), *Mon. Not. R. Astron. Soc.*, **119**, 174–183, 1959.

Kellermann, K. I. and I. I. K. Pauliny-Toth, The Spectra of Opaque Radio Sources, *Astrophys. J.*, **155**, L71–L78, 1969.

Kesteven, M. J. L. and A. H. Bridle, Index of Extragalactic Radio Source Catalogs, *J. R. Astron. Soc. Can.*, **71**, 21–39, 1977.

Knowles, S. H., W. B. Waltman, J. L. Yen, J. Galt, D. N. Fort, W. H. Cannon, D. Davidson, W. Petrachenko, and J. Popelar, A Phase-Coherent Link via Synchronous Satellite Developed for Very Long Baseline Radio Interferometry, *Radio Sci.*, **17**, 1661–1670, 1982.

Linfield, R., VLBI Observations of Jets in Double Radio Galaxies, *Astrophys. J.*, **244**, 436–446, 1981.

Loudon, R., *The Quantum Theory of Light*, Oxford University Press, London, 1973, p. 229.

Lovas, F. J., L. E. Snyder, and D. R. Johnson, Recommended Rest Frequencies for Observed Interstellar Molecular Transitions, *Astrophys. J. Suppl.*, **41**, 451–480, 1979.

McCready, L. L., J. L. Pawsey, and R. Payne-Scott, Solar Radiation at Radio Frequencies and its Relation to Sunspots, *Proc. R. Soc. A*, **190**, 357–375, 1947.

Maltby, P. and A. T. Moffet, Brightness Distribution in Discrete Radio Sources, *Astrophys. J.*

Suppl., **7**, 93–163, 1962.

Matveenko, L. I., N. S. Kardashev, and G. B. Sholomitskii, Large Base-Line Radio Interferometers, Radiofizika, **8**, 651–654, 1965; Eng. Trans. in *Soviet Radiophys.*, **8**, 461–463, 1965.

Michelson, A. A., On the Application of Interference Methods to Astronomical Measurements, *Philos. Mag.*, ser. 5, **30**, 1–21, 1890.

Michelson, A. A., On the Application of Interference Methods to Astronomical Measurements, *Astrophys. J.*, **51**, 257–262, 1920.

Michelson, A. A. and F. G. Pease, Measurement of the Diameter of α Orionis with the Interferometer, *Astrophys. J.*, **53**, 249–259, 1921.

Mills, B. Y., The Positions of Six Discrete Sources of Cosmic Radio Radiation, *Aust. J. Sci. Res.* A5, 456–463, 1952.

Mills, B. Y., The Radio Brightness Distribution Over Four Discrete Sources of Cosmic Noise, *Aust. J. Phys.*, **6**, 452–470, 1953.

Mills, B. Y., R. E. Aitchison, A. G. Little, and W. B. McAdam, The Sydney University Cross-Type Radio Telescope, *Proc. IRE Aust.*, **24**, 156–165, 1963.

Mills, B. Y., A. G. Little, K. V. Sheridan, and O. B. Slee, A High-Resolution Radio Telescope for Use at 3.5 m, *Proc. IRE*, **46**, 67–84, 1958.

Mills, B. Y. and O. B. Slee, A Preliminary Survey of Radio Sources in a Limited Region of the Sky at a Wavelength of 3.5 m, *Aust. J. Phys.*, **10**, 162–194, 1957.

Mills, B. Y., O. B. Slee, and E. R. Hill, A Catalogue of Radio Sources Between Declinations +10° and −20°, *Aust. J. Phys.*, **11**, 360–387, 1958. See also: **13**, 676–699, 1960; and **14**, 497–507, 1961.

Moran, J. M., P. P. Crowther, B. F. Burke, A. H. Barrett, A. E. E. Rogers, J. A. Ball, J. C. Carter, and C. C. Bare, Spectral Line Interferometry with Independent Time Standards at Stations Separated by 845 Kilometers, *Science*, **157**, 676–677, 1967.

Morris, D., H. P. Palmer, and A. R. Thompson,, Five Radio Sources of Small Angular Diameter, *Observatory*, **77**, 103–106, 1957.

Nyquist, H., Thermal Agitation of Electric Charge in Conductors, *Phys. Rev.*, **32**, 110–113, 1928.

O'Brien, P. A., The Distribution of Radiation Across the Solar Disk at Metre Wave-Lengths, *Mon. Not. R. Astron. Soc.*, **113**, 597–612, 1953.

Oliver, B. M., Thermal and Quantum Noise, *Proc. IEEE*, **53**, 436–454, 1965.

Oster, L., Emission and Absorption of Thermal Radio Radiation, *Astrophys. J.*, **134**, 1010–1012, 1961.

Pawsey, J. L., Sydney Investigations and Very Distant Radio Sources, *Pub. Astron. Soc. Pac.*, **70**, 133–140, 1958.

Pearson, T. J., S. C. Unwin, M. H. Cohen, R. P. Linfield, A. C. S. Readhead, G. A. Seielstad, R. S. Simon, and R. C. Walker, Superluminal Expansion of Quasar 3C273, *Nature*, **290**, 365–368, 1981.

Pease, F. G., Interferometer Methods in Astronomy, *Ergeb. Exakten Naturwiss.*, **10**, 84–96, 1931.

Perley, R. A., J. W. Dreher, and J. J. Cowan, The Jet and Filaments in Cygnus, *Astrophys. J.*, **285**, L35–L38, 1984.

Pilkington, J. D. H. and P. F. Scott, A Survey of Radio Sources Between Declinations 20° and 40°, *Mem. R. Astron. Soc.*, **69**, 183–224, 1965.

Read, R. B., Two-Element Interferometer for Accurate Position Determinations at 960 Mc, *IRE Trans. Antennas Propag.*, **AP-9**, 31–35, 1961.

Reber, G., Cosmic Static, *Astrophys. J.*, **91**, 621–624, 1940.

Reid, M. J. and J. M. Moran, Masers, *Ann. Rev. Astron. Astrophys.*, **19**, 231–276, 1981.

Rogers, A. E. E., H. F. Hinteregger, A. R. Whitney, C. C. Counselman, I. I. Shapiro, J. J. Wittels, W. K. Klemperer, W. W. Warnock, T. A. Clark, L. K. Hutton, G. E. Marandino, B. O.

Rönnäng, O. E. H. Rydbeck, and A. E. Niell, The Structure of Radio Sources 3C273B and 3C84 Deduced from the Closure Phases and Visibility Amplitudes Observed with Three-Element Interferometers, *Astrophys. J.*, **193**, 293–301, 1974.

Rowson, B., High Resolution Observations with a Tracking Radio Interferometer, *Mon. Not. R. Astron. Soc.*, **125**, 177–188, 1963.

Rybicki, G. B. and A. P. Lightman, *Radiative Processes in Astrophysics*, Wiley-Interscience, New York, 1979.

Ryle, M., A New Radio Interferometer and its Application to the Observation of Weak Radio Stars, *Proc. R. Soc. A*, **211**, 351–375, 1952.

Ryle, M., The New Cambridge Radio Telescope, *Nature*, **194**, 517–518, 1962.

Ryle, M., The 5-km Radio Telescope at Cambridge, *Nature*, **239**, 435–438, 1972.

Ryle, M., Radio Telescopes of Large Resolving Power, *Science*, **188**, 1071–1079, 1975.

Ryle, M., B. Elsmore, and A. C. Neville, High Resolution Observations of Radio Sources in Cygnus and Cassiopeia, *Nature*, **205**, 1259–1262, 1965.

Ryle, M. and A. Hewish, The Cambridge Radio Telescope, *Mem. R. Astron. Soc.*, **67**, 97–105, 1955.

Ryle, M., A. Hewish, and J. R. Shakeshaft, The Synthesis of Large Radio Telescopes by the Use of Radio Interferometers, *IRE Trans. Antennas Propag.*, **7**, S120–S124, 1959.

Ryle, M. and A. C. Neville, A Radio Survey of the North Polar Region with a 4.5 Minute of Arc Pencil-Beam System, *Mon. Not. R. Astron. Soc.*, **125**, 39–56, 1962.

Ryle, M. and F. G. Smith, A New Intense Source of Radio Frequency Radiation in the Constellation of Cassiopeia, *Nature*, **162**, 462–463, 1948.

Ryle, M., F. G. Smith, and B. Elsmore, A Preliminary Survey of the Radio Stars in the Northern Hemisphere, *Mon. Not. R. Astron. Soc.*, **110**, 508–523, 1950.

Ryle, M. and D. D. Vonberg, Solar Radiation at 175 Mc/s, *Nature*, **158**, 339–340, 1946.

Shakeshaft, J. R., M. Ryle, J. E. Baldwin, B. Elsmore, and J. H. Thomson, A Survey of Radio Sources Between Declinations $-38°$ and $+83°$, *Mem. R. Astron. Soc.*, **67**, 106–154, 1955.

Smith, F. G., An Accurate Determination of the Positions of Four Radio Stars, *Nature*, **168**, 555, 1951.

Smith F. G. The Determination of the Position of a Radio Star, *Mon. Not. R. Astron. Soc.*, **112**, 497–513, 1952a.

Smith, F. G., The Measurement of the Angular Diameter of Radio Stars, *Proc. Phys. Soc. B.*, **65**, 971–980, 1952b.

Smith, F. G., Apparent Angular Sizes of Discrete Radio Sources—Observations at Cambridge, *Nature*, **170**, 1065, 1952c.

Southworth, G. C., Microwave Radiation from the Sun, *J. Franklin Inst.*, **239**, 285–297, 1945.

Stanier, H. M., Distribution of Radiation from the Undisturbed Sun at a Wave-Length of 60 cm, *Nature*, **165**, 354–355, 1950.

Sutton, E. C., G. A. Blake, C. R. Masson, and T. G. Phillips, Molecular Line Survey of Orion A from 215–247 GHz, *Astrophys. J. Suppl.*, **58**, 341–378, 1985.

Thompson, A. R., The Planetary Nebulae as Radio Sources, in *Vistas in Astronomy*, Vol. 16, A. Beer, Ed., Pergamon Press, Oxford, 1974.

Thompson, A. R., B. G. Clark, C. M. Wade, and P. J. Napier, The Very Large Array, *Astrophys. J. Suppl.*, **44**, 151–167, 1980.

Tiuri, M. E., Radio Astronomy Receivers, *IEEE Trans. Antennas Propag.*, **AP-12**, 930–938, 1964.

Vitkevich, V. V. and P. D. Kalachev, Design Principles of the FIAN Cross-Type Wide Range Telescope, in *Radio Telescopes*, D. V. Skobel'tsyn, Ed., *Proc. P. N. Lebedev Phys. Inst.*, [*Acad. Sci. USSR*] Vol. 28, translated by Consultants Bureau, New York, 1966.

Whitney, A. R., A. E. E. Rogers, H. F. Hinteregger, C. A. Knight, J. I. Levine, S. Lippincott, T. A. Clark, I. I. Shapiro, and D. S. Robertson, A Very Long Baseline Interferometer System for Geodetic Applications, *Radio Sci.*, **11**, 421–432, 1976.

Whitney, A. R., I. I. Shapiro, A. E. E. Rogers, D. S. Robertson, C. A. Knight, T. A. Clark, R. M. Goldstein, G. E. Marandino, and N. R. Vandenberg, Quasars Revisited: Rapid Time Variations Observed via Very Long Baseline Interferometry, *Science*, **173**, 225–230, 1971.

Yen, J. L., K. I. Kellermann, B. Rayhrer, N. W. Broten, D. N. Fort, S. H. Knowles, W. B. Waltman, and G. W. Swenson, Jr., Real-Time, Very-Long-Baseline Interferometry Based on the Use of a Communications Satellite, *Science*, **198**, 289–291, 1977.

2

INTRODUCTORY THEORY OF INTERFEROMETERS AND CORRELATOR ARRAYS

In order to expedite the study of interferometers and synthesis arrays it will be helpful to begin with a streamlined mathematical analysis based on simplified physical assumptions. Having thus introduced some of the properties of the devices and having shown their potentialities, the more detailed theory will be developed in later chapters. The history of interferometry in radio astronomy was discussed in Chapter 1. The basic similarity of the theory to that of optical interferometry was recognized in the early days of the science, so that prior studies on the nature of optical radiation fields and their interactions with optical devices have been valuable precedents to the radio theory.

2.1 SIMPLE RADIO INTERFEROMETER

Consider a pair of antennas as shown in Fig. 2.1, situated on the surface of the earth and pointed at a distant cosmic radio source whose angular dimensions are very small. Suppose that each antenna has a radio receiving system that admits the same narrow band of frequencies. The source is so distant compared with the spacing between antennas that the incoming rays to the two antennas are essentially parallel. Now let the voltages from the two antennas be multiplied together and the resulting voltage passed through a lowpass filter

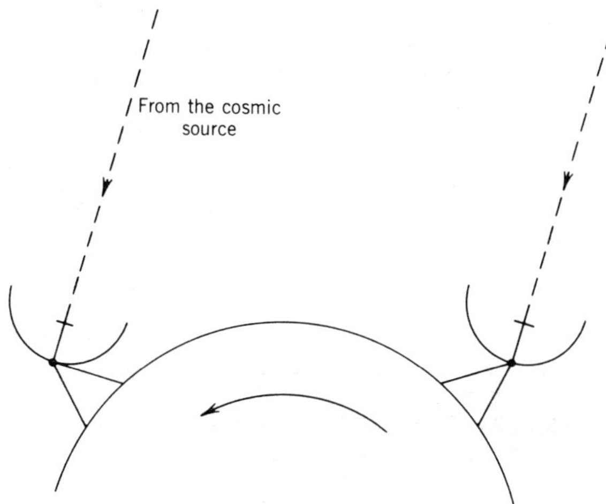

Figure 2.1   Elementary interferometer. The arrow indicates the rotation of the earth.

whose output is recorded on a graph against time. As noted in Chapter 1 the combination of the multiplying circuit and the lowpass filter or time-averaging circuit is referred to as a correlator.

As the earth rotates, the two antennas generally experience differing radial velocities with respect to the source. If the source were emitting a monochromatic wave, that is, a single-frequency signal, this signal would have a different Doppler shift in frequency at the two antennas. Then the output of the lowpass filter would be a sine wave whose frequency equalled the difference or beat between the two Doppler-shifted frequencies. Although the source does not, in fact, emit a discrete frequency (unless it is a manmade spacecraft) this reasoning yields the correct frequency of the beats (or fringes) which represent the output of the interferometer. The fringe frequency varies with the hour angle of the source and is also a function of other parameters of the interferometer and the source, as discussed in Chapter 4. The principal astronomical information from an interferometer observation of a cosmic source is contained in the amplitudes and phases of the recorded interference fringes. It must be emphasized that these amplitudes and phases are those of the modulation resulting from the interference between the outputs of the two antennas, and not those of the radiofrequency radiation itself.

Another simple line of reasoning again involves the assumption that the receivers pass bands of noise narrow enough to give the general appearance of sinusoids at the outputs, but broad enough to accommodate the Doppler shifts. The geometrical situation is shown in Fig. 2.2, where the antenna spacing is east–west. The two antennas are separated by a distance D, the baseline, and observe the same distant cosmic source. The source will be assumed for the moment to have infinitesimal angular dimensions. The outputs are multiplied and filtered as before and the result recorded. The wavefront from the source

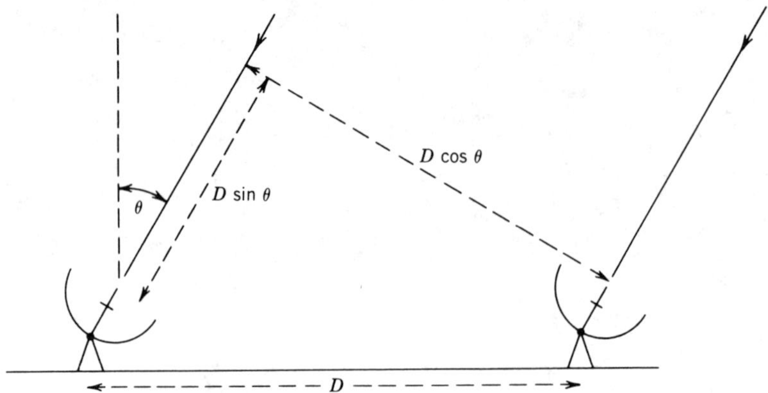

Figure 2.2   Geometry of an elementary interferometer.

in direction θ is essentially plane because of the great distance traveled, and it reaches the right-hand antenna at a time $\tau_g = (D/c)\sin\theta$ before it reaches the left-hand one. τ_g is called the geometrical delay and c is the velocity of light. Thus, in terms of the frequency v, the output of the multiplier is proportional to

$$F = 2\sin(2\pi vt)\sin 2\pi v(t - \tau_g) = \cos 2\pi v\tau_g - \cos(4\pi vt)\cos(2\pi v\tau_g)$$

$$- \sin(4\pi vt)\sin(2\pi v\tau_g). \qquad (2.1)$$

The center frequency of the receivers is generally in the range of tens of megahertz to hundreds of gigahertz. As the earth rotates the most rapid rate of variation of θ is equal to the earth's rotational velocity which is of the order of 10^{-4} rad sec^{-1}. As D/λ, where $\lambda = c/v$ is the wavelength, cannot be more than, say, 10^{10} for terrestrial baselines, the rate of variation of $v\tau_g$ is smaller than that of vt by at least five orders of magnitude. The more rapidly varying terms in Eq. (2.1) are easily filtered out leaving

$$F = \cos\left(\frac{2\pi D}{\lambda}\sin\theta\right) = \cos\left(\frac{2\pi D\xi}{\lambda}\right), \qquad (2.2)$$

where $\xi = \sin\theta$ is the direction cosine measured with respect to the baseline. For sidereal sources the variation of θ with time as the earth rotates generates quasisinusoidal fringes at the correlator which are the output of the interferometer. Figure 2.3 shows an example of this function, in effect the directional power reception pattern of the interferometer. We have assumed that the antennas are either isotropic or that they track the source as the earth rotates, so that the reception patterns of the individual antennas do not affect the result.

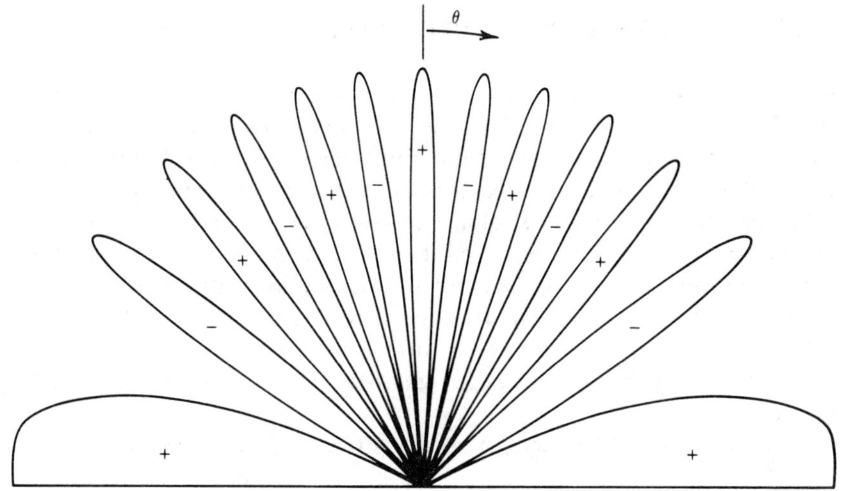

Figure 2.3 Polar plot of the fringe function $F = \cos[(2\pi D/\lambda)\sin\theta]$ for $-90° \le \theta \le 90°$, with $D/\lambda = 3$. The $+$ and $-$ signs indicate the polarity of the function.

A development of this analysis can be made if we consider two Fourier components of the received signal at frequencies ν_1 and ν_2. At some angle θ the signals at frequency ν_1 from the two antennas will interfere constructively in the multiplier and at other angles the interference will be destructive, giving rise to the quasisinusoidal fringe function F_1 as in Eq. (2.2). For frequency ν_2 the coefficient $2\pi D/\lambda = 2\pi D\nu_2/c$ will be different, so F_2 will have a different period from F_1 at any given angle θ. This difference in period gives rise to interference between F_1 and F_2, so that the fringe maxima have superimposed on them a modulation function which also depends upon θ. Similar statements can be made about other combinations of frequencies and, in particular, about continuous bands of frequencies. For example, if the signals at the correlator are of uniform power spectral density over a band of width $\Delta\nu$ and center frequency ν_0, the output becomes

$$F(\nu_0) = \frac{1}{\Delta\nu} \int_{\nu_0-\Delta\nu/2}^{\nu_0+\Delta\nu/2} \cos\left(\frac{2\pi D\xi\nu}{c}\right) d\nu$$

$$= \cos\left(\frac{2\pi D\xi\nu_0}{c}\right) \frac{\sin(\pi D\xi\,\Delta\nu/c)}{\pi D\xi\,\Delta\nu/c}. \tag{2.3}$$

Thus the fringe pattern has a sinc-function envelope. This is an example of the general result, to be discussed in the following section, that in the case of uniform power spectral density at the antennas the envelope of the fringe pattern is the Fourier transform of the instrumental frequency response.

2.2 BROADBAND INTERFEROMETER

Figure 2.4 shows an interferometer of the same general type as in Fig. 2.2, but with the amplifiers H_1 and H_2, the multiplier, and an integrator shown explicitly and with an instrumental time delay τ_i inserted into one arm. Assume that for a point source each antenna delivers the same signal voltage V to the correlator, and that one voltage lags the other by a time delay $\tau = \tau_g - \tau_i$ as determined by the baseline D and the source direction θ. The integrator has a time constant $2T$; that is, it sums the output from the multiplier for $2T$ seconds and then resets to zero after the sum is suitably recorded. The output of the integrator may be a voltage, a current, or a coded set of logic levels, but in any case it represents a physical quantity with the dimensions of voltage squared.

The output from the integrator resulting from a point source is

$$r = \frac{1}{2T} \int_{-T}^{T} V(t)V(t - \tau)\, dt. \tag{2.4}$$

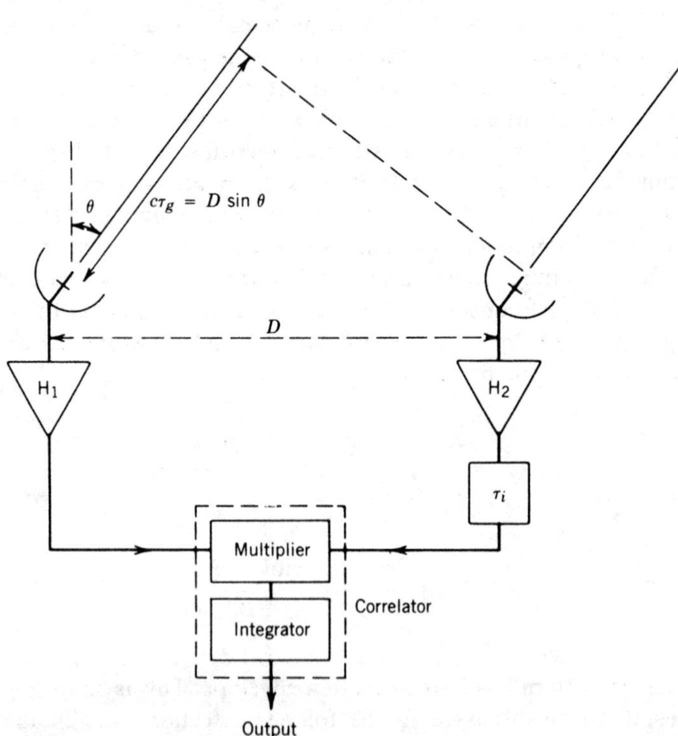

Figure 2.4   Elementary interferometer showing bandpass amplifiers H_1 and H_2, geometrical time delay τ_g, instrumental time delay τ_i, and the correlator consisting of a multiplier and integrator.

The two amplifiers are assumed to have identical bandpass characteristics, including finite bandwidths $\Delta\nu$ outside which no frequency components are admitted. The integration time $2T$ is chosen to be very much larger than $\Delta\nu^{-1}$. Thus, Eq. (2.4) can be written approximately as

$$r(\tau) = \lim_{T\to\infty} \frac{1}{2T}\int_{-T}^{T} V(t)V(t-\tau)\,dt, \qquad (2.5)$$

which is an (unnormalized) autocorrelation function.[†] The condition $T \to \infty$ is a physical impossibility, of course, as the time constant must clearly be finite and much less than the fringe period. We use it here for mathematical convenience, and note that it is a reasonable approximation since it is only necessary that a very large number of variations of the signal amplitude, which have a duration $\sim \Delta\nu^{-1}$, occur in time $2T$.

As described in Chapter 1, the signal from a natural cosmic source can be considered as a continuous random process that results in a broad spectrum of which the phases are a random function of frequency. It will be assumed for our immediate purpose that the time-averaged amplitude of the cosmic signal in any finite band is constant with frequency over the passband of the receiver.

The squared amplitude of a frequency spectrum is known as the power density spectrum or power spectrum. The power spectrum of a signal is the Fourier transform of the autocorrelation function of that signal. This statement is known as the Wiener–Khinchin relation (see Appendix 3.2 of Chapter 3), which applies to signals that are either deterministic or statistical in nature. It can be written

$$\int_{-\infty}^{\infty} r(\tau)e^{-j2\pi\nu\tau}\,d\tau = |H(\nu)|^2 \qquad (2.6)$$

and

$$\int_{-\infty}^{\infty} |H(\nu)|^2 e^{j2\pi\nu\tau}\,d\nu = r(\tau), \qquad (2.7)$$

where $H(\nu)$ is the amplitude (voltage) response, and hence $|H(\nu)|^2$ is the power spectrum of the signal input to the multiplier. In this case, as the cosmic signal is assumed to have a constant amplitude spectrum, the amplitude spectrum $H(\nu)$ is determined solely by the passband characteristics (frequency response) of the amplifiers. Thus the output of the interferometer as a function of the time delay τ is the Fourier transform of the power spectrum of the cosmic signal as bandlimited by the amplifiers. Assume, as a simple example, a

[†] For simplicity we consider only the signals from a point source, which are identical except for a time delay. In practical systems the input waveforms at the correlator contain system noise and other nonidentical components which can be taken into account by considering the *cross-correlation* function.

Gaussian passband centered at ν_0:

$$|H(\nu)|^2 = \frac{1}{2\sigma\sqrt{2\pi}}\left\{\exp\left[-\frac{(\nu - \nu_0)^2}{2\sigma^2}\right] + \exp\left[-\frac{(\nu + \nu_0)^2}{2\sigma^2}\right]\right\}. \quad (2.8)$$

Note that to perform the Fourier transforms in Eqs. (2.6) and (2.7) we must include a negative frequency response centered on $-\nu_0$. The negative frequencies have no physical meaning but arise mathematically from the use of the exponential kernel of the transform. The function used to represent the spectrum must be symmetrical with respect to zero frequency since the autocorrelation function is real. Thus the interferometer response is

$$r(\tau) = e^{-2\pi^2\tau^2\sigma^2}\cos(2\pi\nu_0\tau), \quad (2.9)$$

which is illustrated in Fig. 2.5a. Note that $r(\tau)$ is a cosinusoidal function multiplied by an envelope function, in this case a Gaussian, whose shape and width depend on the amplifier passband. This envelope function is often referred to as the *delay pattern, bandwidth pattern,* or *fringe washing function.*

By setting the instrumental delay τ_i to zero and substituting for the geometrical delay $\tau_g = (D/c)\sin\theta$ in Eq. (2.9) we obtain

$$r(\tau_g) = \exp\left[-2\left(\frac{\pi D\sigma}{c}\sin\theta\right)^2\right]\cos\left(\frac{2\pi\nu_0 D}{c}\sin\theta\right). \quad (2.10)$$

The period of the fringes varies inversely as the quantity $\nu_0 D/c = D/\lambda_0$ and does not depend on the bandwidth parameter σ. The width of the bandwidth pattern, however, is a function of both σ and D: wide bandwidths and long baselines result in narrow fringe envelopes. This result is quite general. For example, a rectangular amplifier passband of width $\Delta\nu$, as considered in Eq. (2.3), results in an envelope pattern of the form $[\sin(\pi\Delta\nu\,\tau)]/(\pi\Delta\nu\,\tau)$, as shown in Fig. 2.5b.

In mapping applications the fringe envelope is usually considered a nuisance, although there are some applications, particularly in very-long-baseline interferometry, in which the envelope is useful. In most cases it is desirable to observe the fringes in the vicinity of the maximum of the pattern, where the fringe amplitudes are greatest. This condition can be achieved by changing the instrumental delay τ_i periodically so as to keep $\tau = \tau_g - \tau_i$ suitably small. If τ_i is adjusted in steps of the reciprocal of the center frequency[†] ν_0, the interferometer response remains cosinusoidal with τ_g.

The instrumental delay is calculated for a nominal source position θ_0 which becomes the reference position. Consider the response to a narrow source at

---

[†] Other adjustment methods are more commonly used and are described in Chapter 7.

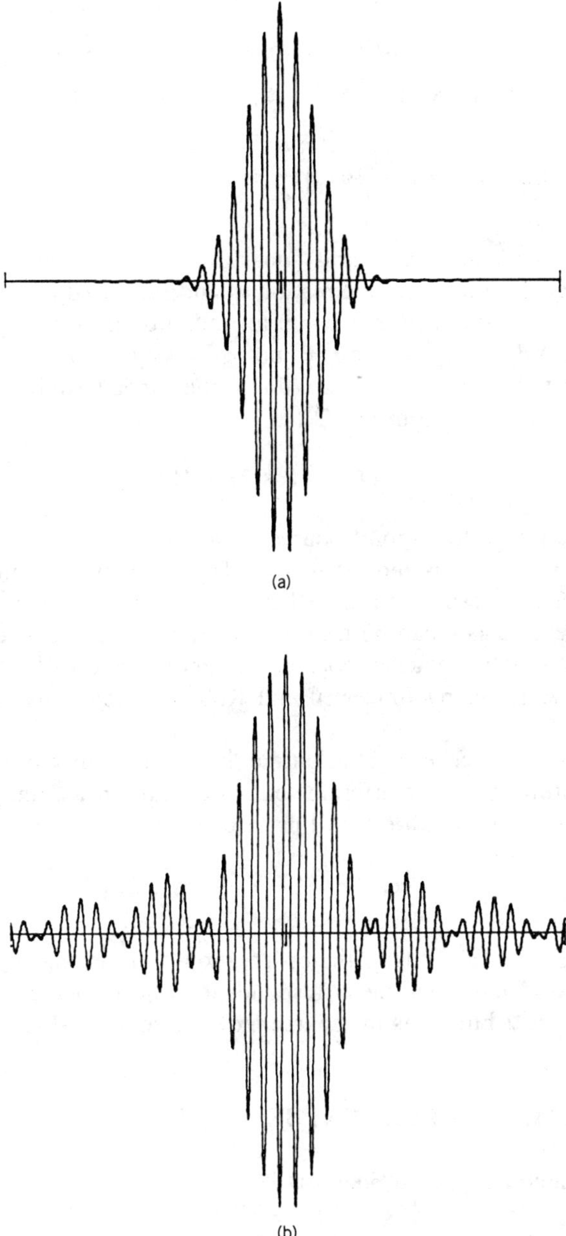

(a)

(b)

Figure 2.5 Point-source response of an interferometer with (a) Gaussian and (b) rectangular passbands. The abscissa is the geometrical delay τ_g. The bandwidth pattern determines the envelope of the sinusoidal fringe term.

position θ in the vicinity of θ_0. Let $\theta = \theta_0 - \theta'$, where θ' is very small. Then

$$\cos(2\pi\nu_0\tau) \simeq \cos\left\{2\pi\nu_0\left[\frac{D}{c}(\sin\theta_0 - \theta'\cos\theta_0) - \tau_i\right]\right\}. \qquad (2.11)$$

We introduce a quantity u that is equal to the component of the antenna spacing projected normal to the direction θ_0 and measured in wavelengths. Let $\tau_i = \tau_g(\theta_0) = (D/c)\sin\theta_0$ so that $\tau = 0$ for radiation from direction θ_0. Define $u = (\nu_0 D/c)\cos\theta_0 = (D/\lambda_0)\cos\theta_0$, and $\xi' = \sin\theta' \simeq \theta'$. Then since the bandwidth pattern is near maximum (unity) the fringe term in the response of the correlator output is, from Eq. (2.11),

$$F(\xi') = \cos(2\pi u\xi'). \qquad (2.12)$$

This is the response to a point source at $\theta = \theta_0 - \theta'$ of an interferometer whose net delay $\tau_g - \tau_i$ is zero at $\theta = \theta_0$. The quantity u is interpreted as a *spatial frequency*. It can be measured in cycles per radian, since the spatial variable ξ', being small, can be measured in radians. The concepts of spatial frequency and spatial-frequency spectra are fundamental to the topics covered in this book. A paper by Bracewell and Roberts (1954) was seminal in this subject.

If we shift the reference angle θ_0 on the sky by one-fourth of a fringe period without adjusting τ_i, the interferometer power pattern assumes a sinusoidal rather than a cosinusoidal form. In this case

$$F(\xi_1') = \cos\left(2\pi u\xi_1' - \tfrac{1}{2}\pi\right) = \sin(2\pi u\xi_1'), \qquad (2.13)$$

where $\xi_1' = \xi' + 1/4u$. Thus in the observation of an extended source, which we discuss later, the interferometer responds to both the odd and even parts of the brightness distribution as the source moves across the sky.

2.3 COSMIC SOURCE SYNTHESIS

Convolution and Extended Sources

Consider a single antenna receiving radiation of frequency ν from a point source, as in Fig. 2.6. The antenna reception pattern is a plot of the effective area of the antenna as a function of angle from the optimum direction of pointing. Suppose that the pattern maximum is in direction θ and the source direction is θ''; then the effective area for the direction θ'' is $A_1(\theta'' - \theta, \nu)$.[†] The flux density received from the source is $S(\nu)\Delta\nu$ in an effective receiving bandwidth $\Delta\nu$, where $S(\nu)$ is the (spectral) flux density. Thus the output power

---

[†] The subscript 1 indicates the one-dimensional forms of functions A and B.

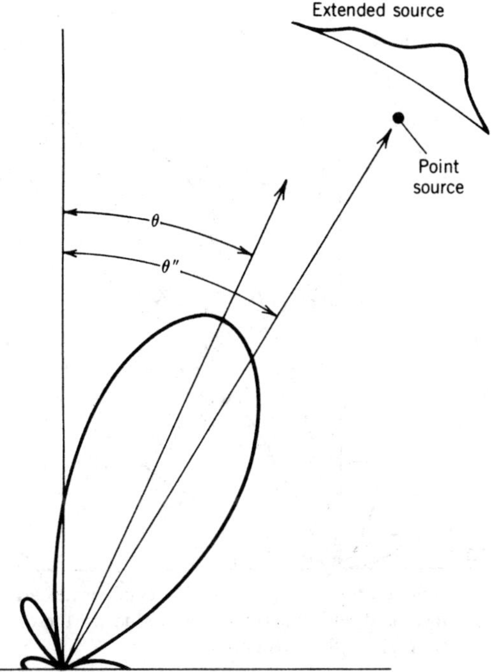

Figure 2.6 Power reception pattern of a typical antenna, which is a polar plot of effective area versus angle of arrival of plane-wave radiation.

of the antenna is $\frac{1}{2}S(\nu)\Delta\nu A_1(\theta'' - \theta, \nu)$, where the factor $\frac{1}{2}$ takes account of the ability of the antenna to respond to one component only of a randomly polarized source. Figure 2.6 also shows the same antenna receiving radiation from an extended source of brightness $B_1(\theta'', \nu)$. The component of the output power at frequency ν in bandwidth $\Delta\nu$ contributed by each element $d\theta''$ of the source is $\frac{1}{2}B_1(\theta'', \nu)\Delta\nu A_1(\theta'' - \theta, \nu)\,d\theta''$ and the total output power from the antenna at pointing direction θ is

$$\frac{\Delta\nu}{2}\int_{\text{source}} B_1(\theta'', \nu)A_1(\theta'' - \theta, \nu)\,d\theta''. \tag{2.14}$$

The integral is the cross correlation of the antenna reception pattern and the brightness distribution of the source. It is convenient to define $\mathscr{A}_1(\theta - \theta'') = A_1(\theta'' - \theta)$, in which case Expression (2.14) becomes

$$\frac{\Delta\nu}{2}\int_{\text{source}} B_1(\theta'', \nu)\mathscr{A}_1(\theta - \theta'', \nu)\,d\theta''. \tag{2.15}$$

The integral in Expression (2.15) is the *convolution* integral; see, for example, Bracewell (1965) or Champeney (1973). We can say that the output power of

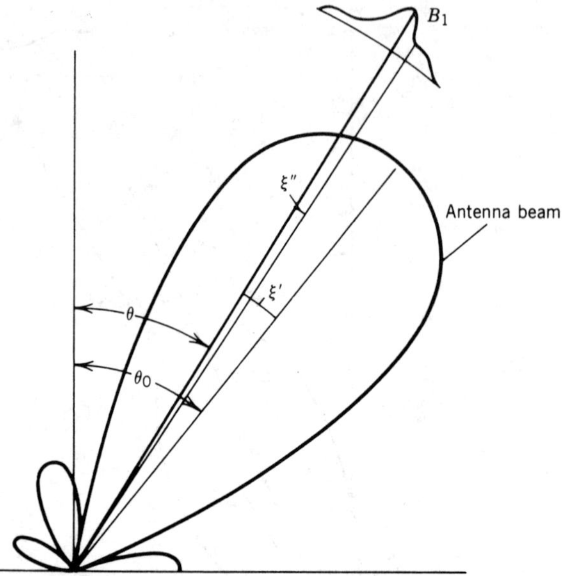

Figure 2.7   Angular relationships in Eq. (2.16). The small angles are indicated by their direction cosines ξ' and ξ'' measured with respect to the normal to the direction θ. The mirror-image antenna beam is reversed about the direction θ_0.

the antenna is given by the convolution of the source with the mirror image of the antenna's power reception pattern. In the case of a point source, $B_1(\theta'', \nu)$ is a delta function of θ''. Thus the mirror-image reception pattern can be described as the response of the antenna to a point source.

The response of an interferometer to an extended source can be found by convolving the brightness distribution of the source with the mirror image of the interferometer power reception pattern. The interferometer output, being the product of two voltages, has the dimensions of power. Thus, using Eq. (2.12) multiplied by the antenna reception pattern as the power reception pattern of the interferometer, and $B_1(\xi')$ as the brightness of the source, noting that $\cos(2\pi u\xi') = \cos(-2\pi u\xi')$, and omitting constant gain factors, we obtain the interferometer output as a function of ξ':

$$R(\xi') = \Delta\nu \int_{\text{source}} \cos[2\pi u(\xi' - \xi'')] \mathcal{A}_1(\xi' - \xi'') B_1(\xi'') \, d\xi''.$$

$$(2.16)$$

This expression is the convolution of $\mathcal{A}_1(\xi')\cos 2\pi u\xi'$ with $B_1(\xi')$. Figure 2.7 illustrates the angular relationships involved.

The *convolution theorem* states that the Fourier transform of the convolution of two functions is the product of their Fourier transforms:

$$f * g \rightleftharpoons \mathscr{F}\mathscr{G},$$

$$(2.17)$$

where $f \rightleftharpoons \mathscr{F}$, $g \rightleftharpoons \mathscr{G}$, $*$ indicates convolution, and $\rightleftharpoons$ indicates Fourier transformation. A proof of the convolution theorem can be found in almost any text on Fourier transforms. Now for a particular value $u = u_0$, the Fourier transform of the fringe term is given by

$$\cos(2\pi u_0 \xi') \rightleftharpoons \tfrac{1}{2}[\delta(u - u_0) + \delta(u + u_0)], \qquad (2.18)$$

where δ is the delta function. We define $b_1(u) \rightleftharpoons B_1(\xi')$ and $r(u) \rightleftharpoons R(\xi')$, and assume that the antenna beams are wide enough that the effective area A_1 has a constant value over the source. Then

$$\begin{aligned}
r(u) &= \tfrac{1}{2}[\delta(u - u_0) + \delta(u + u_0)]b_1(u)A_1\,\Delta\nu \\
&= \tfrac{1}{2}A_1\,\Delta\nu[b_1(u_0)\delta(u - u_0) + b_1(-u_0)\delta(u + u_0)]. \qquad (2.19)
\end{aligned}$$

Thus $r(u)$, the Fourier transform with respect to ξ' of the output of the interferometer $R(\xi')$, consists of two delta functions situated at plus and minus u_0 on the u axis. This constitutes the angular (spatial) frequency spectrum of the source brightness distribution as filtered by the interferometer, which passes only the spatial frequencies $\pm u_0$. We assume here that the antennas are of negligible dimensions compared with D and can therefore be represented by delta functions in u space.

The negative spatial frequency $-u_0$, like the negative frequencies in Eq. (2.8), has no physical meaning. It arises from our use, for mathematical convenience, of the exponential Fourier transform rather than the sine and cosine transforms which correspond more directly to the physical situation. As a result, our spatial frequency spectra will be symmetrical about the origin in the Hermitian sense; that is, with even real parts and odd imaginary parts. They will contain no more information than would unilateral spectra with positive frequencies only.

Source Synthesis

Consider an interferometer with an east–west baseline and an extended cosmic source that passes through the interferometer reception pattern from east to west as the earth rotates. Each hour angle of the source corresponds to a different value of the angle between the baseline and the direction to the source, and thus to a different spatial frequency u. The variable ξ' also increases with time, so the output of the interferometer varies cosinusoidally (or sinusoidally) with time over small intervals. Equation (2.19) shows that at any instant the amplitude of the fringe pattern from the correlator $r(u)$, is equal to the even part of the Fourier transform of $B_1(\xi')$, that is, to the cosine transform of the brightness with respect to the point on the source at position angle θ_0 at that time. With respect to this same point on the source at an instant one-quarter of a fringe cycle later in time the output fringe pattern is equal to the sine transform, which is the imaginary part of the brightness

Fourier transform. Thus the output fringe pattern contains both the real and imaginary parts of $b_1(u)$, the Fourier transform of $B_1(\xi')$. In practice, a phase reference position on the sky is designated, central to the region of interest, and the fringe pattern is then computed for a hypothetical point source of unit flux density at this position. The amplitude and phase of the observed fringe pattern are measured with respect to the hypothetical fringes. The resulting values, when expressed as a complex quantity, are known as the *complex visibility*, $\mathscr{V}(u)$, which is equal to $b_1(u)$ in Eq. (2.19). The complex visibility differs from Michelson's fringe visibility $\mathscr{V}_M$ discussed in Chapter 1. in that the latter is normalized to unity and represents only the amplitude of the fringes and not the phase. The complex visibility, as derived from the measured output of an interferometer or array, includes the effects of the antenna beams, the finite receiving bandwidth, and any instrumental imperfections. Usually these effects are not large, and methods of correcting for them are discussed in later chapters. In practice the antennas often track the source, so B_1 is multiplied by A_1 rather than convolved with $\mathscr{A}_1$. Then Eq. (2.16) becomes

$$R(\xi') = \Delta\nu \int_{\text{source}} \cos[2\pi u(\xi' - \xi'')] A_1(\xi'') B_1(\xi'') \, d\xi''. \qquad (2.20)$$

If the phases and amplitudes of the fringes are observed at frequent enough intervals in u, the complex spatial spectrum represented by $\mathscr{V}(u)$ is obtained. Then by Fourier transforming this spectrum and correcting for A_1 the source brightness distribution can be reconstructed.

While observations of a source for many hours with a single baseline can yield, in principle, a wide range of spatial frequencies, in practice they usually cannot produce a complete enough spatial spectrum to permit satisfactory reconstruction of a source profile. Several different baselines are thus required. These may be obtained from the $n_a(n_a - 1)/2$ pairs in a set of n_a antennas that form an array. Alternatively, the same result may be obtained from a single pair of antennas by moving one of them to successive points and repeating the observation. In either case it must be assumed that the source does not change significantly during the time of the observations.

Two-Dimensional Mapping

In the particular case of an east–west baseline and an *equatorial* source, the source can be mapped in the east–west dimension only, as the spatial-frequency spectrum so derived has only one frequency dimension, u. In effect, we are scanning the source with a fan-shaped beam whose plane is north–south. To synthesize a map in two dimensions requires a two-dimensional spatial-frequency spectrum in the (u, v) frequency domain, v being the north–south frequency component as shown in Fig. 2.8. This may be accomplished for an equatorial source by observing with a two-dimensional array of interferom-

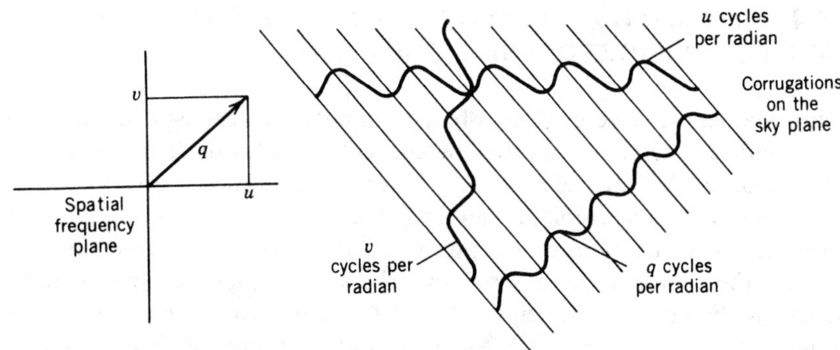

Figure 2.8 The brightness distribution of a radio source is composed of a spatial spectrum of two-dimensional Fourier components that can be visualized as a set of sinusoidal corrugations, each of which extends over the entire sky. Four parameters characterize each component. The spatial frequency q and the angle of the corrugations on the sky are usually specified in terms of the frequencies u and v measured along the east–west and north–south directions. The amplitude of the corrugations, and their phase at some specified reference point, are referred to as the amplitude and phase of the complex visibility $\mathscr{V}(u, v)$ at the point (u, v) in the spatial-frequency plane.

eters. Although we have considered only east–west baselines, the results derived in terms of θ and ξ hold for any baseline direction provided that θ is measured with respect to a plane that is normal to the baseline. However, θ is a linear function of hour angle only for an east–west baseline and an equatorial source.

A source at a high declination (near the celestial pole) can be mapped in two dimensions with either one- or two-dimensional arrays, as will be explained in Chapter 4. At intermediate declinations the requirement for antenna spacings depends on the declination and structure of the source and on the geographical latitudes of the antennas. From Fig. 2.2, the spatial frequency u is proportional to the projection of the terrestrial baseline onto the celestial sphere at the position of the source under observation. The two-dimensional situation, similar but geometrically more complicated, is developed in detail in Chapter 4. As the earth rotates the baseline projection on the celestial sphere rotates and foreshortens. The plot of the length and direction of the baseline projection is an ellipse in the (u, v) plane. The parameters of the ellipse depend upon the declination of the source, the length and orientation of the baseline, and the latitude of the center of the baseline. Each baseline has a different ellipse, and the ensemble of ellipses indicates the spatial frequencies that can be measured by the array. The design of a synthesis array consists in achieving a distribution of measurements in u and v consistent with the angular resolution, field of view, declination range, and sidelobe level specified for the array, and is discussed in Chapter 5.

2.4 GENERALIZED, TWO-DIMENSIONAL RELATIONSHIP BETWEEN VISIBILITY AND BRIGHTNESS

In this chapter we have explained how an interferometer responds to a radio source and introduced a number of concepts to provide a broad basis of understanding. We now give a more direct mathematical derivation of the relationship between visibility and brightness that is not restricted to the one-dimensional functions or to east–west baselines. Suppose that the antennas track the source under observation, which is the most common situation, and let the direction of the center of the field to be mapped be designated by the unit vector $\mathbf{s}_0$, as in Fig. 2.9. An element of the source of solid angle $d\Omega$ at position $\mathbf{s} = \mathbf{s}_0 + \boldsymbol{\sigma}$ contributes a component of power $\frac{1}{2}A(\boldsymbol{\sigma})B(\boldsymbol{\sigma})\Delta\nu\,d\Omega$ at each of the two antennas, which we assume to be identical. It is easily seen that this component has the dimensions of power since the units of B are $\mathrm{W\,m^{-2}\,Hz^{-1}\,sr^{-1}}$. From the considerations outlined in the derivation of Eqs. (2.1) and (2.2) the resulting component of the correlator output is proportional to the received power and to the fringe term, that is, to $A(\boldsymbol{\sigma})B(\boldsymbol{\sigma})\Delta\nu\cos(2\pi\nu\tau_g)d\Omega$. If the vector $\mathbf{D}_\lambda$ specifies the baseline in wavelengths, then $\nu\tau_g = \mathbf{D}_\lambda \cdot \mathbf{s} = \mathbf{D}_\lambda \cdot (\mathbf{s}_0 + \boldsymbol{\sigma})$. Thus the output from the correlator is represented by

$$r(\mathbf{D}_\lambda, \mathbf{s}_0) = \Delta\nu \int_{4\pi} A(\boldsymbol{\sigma})B(\boldsymbol{\sigma})\cos\left[2\pi\mathbf{D}_\lambda \cdot (\mathbf{s}_0 + \boldsymbol{\sigma})\right] d\Omega$$

$$= \Delta\nu \cos(2\pi\mathbf{D}_\lambda \cdot \mathbf{s}_0) \int_{4\pi} A(\boldsymbol{\sigma})B(\boldsymbol{\sigma})\cos(2\pi\mathbf{D}_\lambda \cdot \boldsymbol{\sigma}) d\Omega$$

$$- \Delta\nu \sin(2\pi\mathbf{D}_\lambda \cdot \mathbf{s}_0) \int_{4\pi} A(\boldsymbol{\sigma})B(\boldsymbol{\sigma})\sin(2\pi\mathbf{D}_\lambda \cdot \boldsymbol{\sigma}) d\Omega.$$

$$(2.21)$$

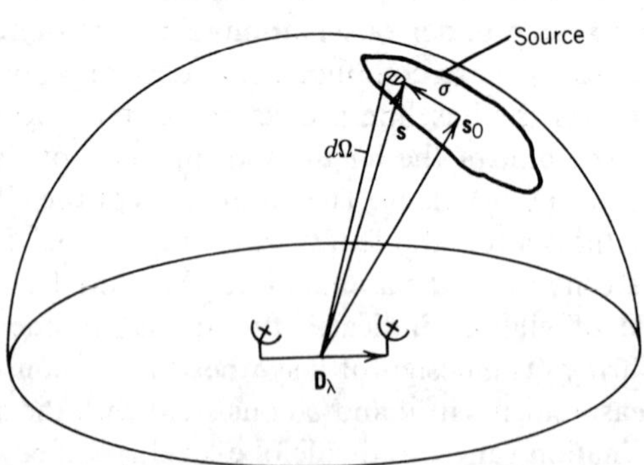

Figure 2.9 Baseline and position vectors that specify the interferometer and the source. The source is represented by the outline on the celestial sphere.

Let A_0 be the antenna collecting area in direction s_0 in which the beam is pointed. We introduce a normalized reception pattern $A_N(\sigma) = A(\sigma)/A_0$ and consider the modified brightness distribution $A_N(\sigma)B(\sigma)$. Now we define the complex visibility[†] as

$$\mathscr{V} = |\mathscr{V}|e^{j\phi_v} = \int_{4\pi} A_N(\sigma)B(\sigma)e^{-j2\pi\mathbf{D}_\lambda\cdot\sigma}\,d\Omega. \tag{2.22}$$

Then by separating the real and imaginary parts we obtain

$$\int_{4\pi} A_N(\sigma)B(\sigma)\cos(2\pi\mathbf{D}_\lambda\cdot\sigma)\,d\Omega = |\mathscr{V}|\cos\phi_v, \tag{2.23}$$

$$\int_{4\pi} A_N(\sigma)B(\sigma)\sin(2\pi\mathbf{D}_\lambda\cdot\sigma)\,d\Omega = -|\mathscr{V}|\sin\phi_v, \tag{2.24}$$

and

$$r(\mathbf{D}_\lambda, s_0) = A_0\,\Delta v|\mathscr{V}|\cos(2\pi\mathbf{D}_\lambda\cdot s_0 - \phi_v). \tag{2.25}$$

Thus the output of the correlator can be expressed in terms of a fringe pattern of period corresponding to that for a hypothetical point source at the position s_0. The modulus and phase of $\mathscr{V}$ are equal to the amplitude and phase of the fringes, the phase being measured relative to the response for the hypothetical source. As defined above, $\mathscr{V}$ has the dimensions of flux density (W m^{-2} Hz^{-1}) which is consistent with its Fourier transform relationship with B. Some authors have defined $\mathscr{V}$ as a normalized, dimensionless quantity, in which case it is necessary to reintroduce the brightness scale in the resulting image. The coordinate-free form of Eq. (2.22) allows us to defer the choice of coordinate system until Chapter 4, where the conditions under which the Fourier transform relationship between $\mathscr{V}$ and B is obtained will be discussed. The bandwidth has been assumed to be small in deriving Eq. (2.25), and the effect of a finite bandwidth on the synthesized map is examined in Chapter 6.

Two assumptions of a fundamental nature have been used in this chapter and these should be emphasized before closing. First, the source is in the far field of the interferometer. That is, the incoming wavefront from any point on the source is sufficiently plane over the extent of the baseline so that the incoming rays are parallel at the two antennas. The requirement is most severe

[†] In formulating the fundamental Fourier transform relationship in synthesis mapping, which follows from Eq. (2.22), we use the negative exponent to derive the complex visibility function (or mutual coherence function) from the brightness distribution, and the positive exponent for the inverse operation. From a physical viewpoint the choice is purely arbitrary, and the literature contains examples of both this and the reverse convention. Our choice follows the precedents set by Born and Wolf (1959) in their derivation of the van Cittert–Zernike theorem and by Bracewell (1958) in his detailed formulation of complex visibility.

when the source direction is normal to the baseline. Then if R_s is the distance of the source, the angle between the rays is D/R_s and this we assume to be small compared with the fringe width λ/D. Thus we require

$$R_s \gg \frac{D^2}{\lambda}, \tag{2.26}$$

which for the longest terrestrial baselines and shortest wavelengths is invalid only for objects within the solar system. Interferometry can still be done in the near field, but a correction factor involving R_s must be introduced: see Appendix 3.1. The second assumption enters in the integration over the source in Eq. (2.21), which requires that the contributions to the correlator output from different parts of the source are independent and can be combined by summation. This requirement is met if the source is spatially incoherent, that is, the radiated signals from any two points on the source are uncorrelated. Spatial coherence is discussed further in the following chapter, but we note here that most cosmic sources are spatially incoherent (Hagfors and Moran 1970), with the possible exception of pulsars, which may have some degree of coherence.

REFERENCES

Born, M., and E. Wolf, *Principles of Optics*, Pergamon Press, London, 1959 (and later eds.).

Bracewell, R. N., Radio Interferometry of Discrete Sources, *Proc. IRE*, **46** 97–105, 1958.

Bracewell, R. N., *The Fourier Transform and Its Applications*, McGraw-Hill, New York, 1965 (2nd ed., 1978).

Bracewell, R. N. and J. A. Roberts, Aerial Smoothing in Radio Astronomy, *Aust. J. Phys.*, **7**, 615–640, 1954.

Champeney, D. C., *Fourier Transforms and Their Physical Applications*, Academic Press, London, 1973.

Hagfors, T. and J. M. Moran, Detection and Estimation Practices in Radio Astronomy, *Proc. IEEE*, **58**, 743–759, 1970.

3

FURTHER THEORY OF THE INTERFEROMETER RESPONSE

In Chapter 2 we examined the response of an interferometer in terms of a sinusoidal fringe pattern and showed how the relationship between the fringe visibility $\mathcal{V}$ and the brightness distribution B of a radio source can be obtained. It is also possible to derive the same results by examining the coherence functions of the radiation. This alternative approach involves a more general examination of the properties of the radiation field and extends concepts, such as the degree of coherence of the source, which are fundamental to the theory of interferometry but were only briefly mentioned in the earlier discussion. There are also a number of Fourier transform and convolution relationships that involve characteristics of the electric field and of the receiving equipment. These provide further insights. However, for the reader who is mainly interested in rapidly obtaining an overview of practical observing details, it is possible to proceed directly from Chapter 2 to Chapter 4 with no essential gap in the logical development. The present chapter is based on analyses by Swenson and Mathur (1968, 1969) and the treatment of partially coherent sources developed by Drane and Parrent (1962), Beran and Parrent (1964), and Mandel and Wolf (1965).

3.1 PROPERTIES OF THE INCIDENT RADIATION

Incident Field

As we have already mentioned in Chapter 1, the radiation processes can be considered stationary and ergodic. To simplify the discussion at this point we treat the electric field as a scalar quantity, that is, we ignore the polarization properties. In practice this means that only the component of the field is considered for which the polarization parameters are matched to those of the receiving antennas. No loss in generality is involved, and we shall discuss in Chapter 4 how the antenna polarization should be chosen to examine fully the incident field. Also, without loss of generality, we confine our discussion largely to one angular dimension. An extension to two dimensions is given in Appendix 3.1.

The incident radiation from the sky is a function of the direction of arrival and of time. The electric field of this radiation can be represented by $E(\xi, t)$, where t is the time variable and ξ is the direction cosine of the angle of arrival with respect to the direction of the interferometer baseline. The angular distribution of the incident radiation is, therefore, denoted by $E(\xi, t)$ within the range $-1 \leq \xi \leq 1$. If the direction of incidence makes an angle θ with the normal to the baseline, then $\xi = \sin\theta$. In practice we are usually interested in a small angular range limited by the source or the antenna beams. It simplifies the necessary equations if the reference direction is defined so that ξ is small and linear with angle within this range of interest. Thus we first consider a small range centered on $\theta = 0$ and extend the result to an arbitrary value of θ in Section 3.3.

The field $E(\xi, t)$ can be expressed in terms of the spatial-frequency spectrum $e(u, t)$ by

$$E(\xi, t) = \int_{-\infty}^{\infty} e(u, t) e^{j2\pi\xi u} \, du, \tag{3.1}$$

where u is the spatial frequency as introduced in Section 2.3. The inverse Fourier transform with respect to ξ and u gives

$$e(u, t) = \int_{-\infty}^{\infty} E(\xi, t) e^{-j2\pi\xi u} \, d\xi, \tag{3.2}$$

where

$$E(\xi, t) = 0, \qquad |\xi| \geq 1. \tag{3.3}$$

In addition to Eq. (3.3), we have invoked the assumption that $E(\xi, t)$ is non-negligible only for small ξ; if this condition is not met $E(\xi, t)$ should be multiplied by a factor $(1 - \xi^2)^{-1/2}$ in Eqs. (3.1) and (3.2). (The equivalent two-dimensional factor is discussed in Section 4.1)

The time variation of the radiation can be analyzed in terms of the temporal-frequency spectrum $\hat{E}(\xi, \nu)$, which is the Fourier transform of $E(\xi, t)$ with respect to ν and t. However, $E(\xi, t)$, a statistically stationary random process, is assumed to extend from infinite past into infinite future. Such a function does not satisfy the conditions for the existence of a Fourier transform. Nevertheless, the powerful techniques of Fourier analysis can still be used by considering the truncated function defined below:

$$E_T^r(\xi, t) = \begin{cases} E(\xi, t), & |t| \leq T \\ 0 & |t| > T. \end{cases} \tag{3.4}$$

In terms of this truncated function we define a complex analytic signal

$$E_T(\xi, t) = E_T^r(\xi, t) + jE_T^i(\xi, t). \tag{3.5}$$

The real part represents the radiation received. The real and imaginary parts form a Hilbert transform pair, and although the real part is truncated, the imaginary part is not. The function in Eq. (3.5) is known as the *analytic signal* and a detailed discussion of it can be found elsewhere (Beran and Parrent 1964, Born and Wolf 1965, Bracewell 1965). We have introduced it to follow the general practice in coherence theory, but a detailed knowledge of this concept is not required to proceed further. Now

$$\hat{E}_T(\xi, \nu) = \int_{-\infty}^{\infty} E_T(\xi, t)e^{-j2\pi\nu t}\, dt \tag{3.6}$$

and

$$E_T(\xi, t) = \int_{-\infty}^{\infty} \hat{E}_T(\xi, \nu)e^{j2\pi\nu t}\, d\nu. \tag{3.7}$$

The Fourier transform relationships are shown in Fig. 3.1. A word about the notation used is in order. It has been chosen to emphasize the Fourier transform relationships, and for some quantities the symbol used in other chapters is different. Upper and lowercase letters are used to denote Fourier pairs with respect to ξ and u. A circumflex ($\hat{}$) is used when the quantity is in the temporal-frequency (ν) domain; no circumflex is used for a quantity in the

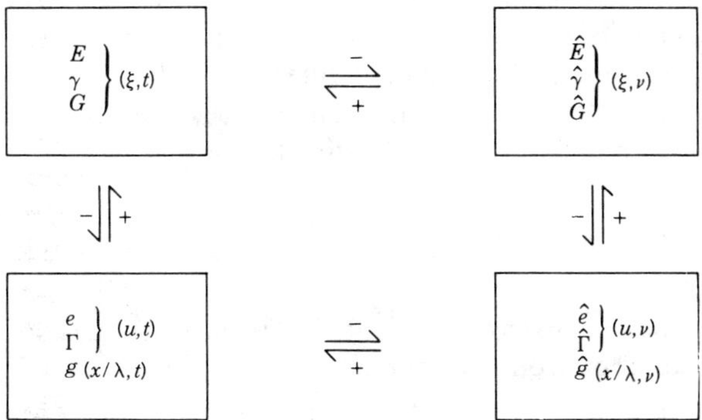

Figure 3.1   Notation for Fourier pairs used in Chapter 3. The + and − signs indicate the sign of the exponential term in the Fourier kernel.

time (t) domain. The positive and negative signs on the arrows indicate the sign of the exponential in the integral [this is made clear by comparing the above notation with Eqs. (3.1), (3.2) and (3.6), (3.7)].

Source Coherence Function

In deriving the response to a source of finite angular dimensions in Section 2.4 the source was assumed to be incoherent, that is, the signals received from different points on the source are uncorrelated. Most natural, cosmic, radio sources are believed to be of this type. However, to examine the implications of the incoherence assumption it is instructive to derive the response for the general case of a partially coherent source. As a measure of coherence we therefore introduce the *source coherence function* γ. This is defined in terms of the cross correlation of signals received from two different points on the source at two different times:

$$\lim_{T \to \infty} \frac{1}{2T} \int_{-T}^{T} E(\xi_1, t) E^*(\xi_2, t - \tau) \, dt. \tag{3.8}$$

In view of the existence conditions for the Fourier transform it is advantageous to use the truncated functions. The source coherence function is therefore defined as follows:

$$\gamma(\xi_1, \xi_2, \tau) = \lim_{T \to \infty} \frac{1}{2T} \int_{-\infty}^{\infty} E_T(\xi_1, t) E_T^*(\xi_2, t - \tau) \, dt$$

$$= \left\langle E_T(\xi_1, t) E_T^*(\xi_2, t - \tau) \right\rangle, \tag{3.9}$$

where the asterisk denotes the complex conjugate. Note that the limits of integration are $\pm \infty$. $\gamma(\xi_1, \xi_2, \tau)$ is very similar to the coherence function of a source or object discussed by Drane and Parrent (1962) and Beran and Parrent (1964), except that as used by those authors the position coordinates are measured linearly on a surface of the source, whereas we measure points on a very distant source in terms of direction cosines. Because of the assumption of stationarity, γ is a function of the time difference τ though not of time t.

If $\xi_1 = \xi_2 = \xi$, the source coherence function becomes the time autocorrelation function of the radiation from the direction ξ:

$$\gamma(\xi, \tau) = \lim_{T \to \infty} \frac{1}{2T} \int_{-\infty}^{\infty} E_T(\xi, t) E_T^*(\xi, t - \tau) \, dt. \tag{3.10}$$

When $\tau = 0$, the autocorrelation function is proportional to the average brightness of radiation from the source,

$$\gamma(\xi, 0) = \lim_{T \to \infty} \frac{1}{2T} \int_{-\infty}^{\infty} |E_T(\xi, t)|^2 \, dt. \tag{3.11}$$

The complex degree of coherence of an extended source is defined as the normalized source coherence function,

$$\gamma_N(\xi_1, \xi_2, \tau) = \frac{\gamma(\xi_1, \xi_2, \tau)}{\sqrt{\gamma(\xi_1, 0)\gamma(\xi_2, 0)}}. \tag{3.12}$$

It is easily shown by using the Schwarz inequality that $0 \le |\gamma_N(\xi_1, \xi_2, \tau)| \le 1$. The extreme values of 0 and 1 correspond to the cases of complete incoherence and complete coherence, respectively. When dealing with extended sources of arbitrary spectral width, it is possible that, for a given pair of points ξ_1 and ξ_2, $|\gamma_N(\xi_1, \xi_2, \tau)|$ is zero for one value of τ and unity for another value. Therefore, a more careful definition of complete coherence and incoherence is necessary. The following definitions will be used (Parrent 1959):

1. The radiations from the directions ξ_1 and ξ_2 are completely coherent (incoherent) if $|\gamma_N(\xi_1, \xi_2, \tau)| = 1$ (0) for all values of τ.
2. An extended source is coherent (incoherent) if the radiations from all pairs of directions ξ_1, ξ_2 within the source are coherent (incoherent).

In all other cases the extended source is said to be partially coherent.

Consider now the coherence function of the distant field $e(u, t)$ of a source measured, say, at the ground level, u being a linear coordinate measured in wavelengths in a direction normal to $\xi = 0$. This coherence function is

$$\Gamma(u_1, u_2, \tau) = \lim_{T \to \infty} \frac{1}{2T} \int_{-\infty}^{\infty} e_T(u_1, t) e_T^*(u_2, t - \tau) \, dt$$

$$= \langle e_T(u_1, t) e_T^*(u_2, t - \tau) \rangle. \tag{3.13}$$

This is the quantity most frequently referred to as the *mutual coherence function* and commonly denoted by $\Gamma_{12}(\tau)$. As in Eq. (3.1), u can also be interpreted as a spatial frequency, so $\Gamma(u_1, u_2, \tau)$ can be described as the spatial-frequency spectrum of $\gamma(\xi_1, \xi_2, \tau)$. By using the Fourier transform relationships between $E(\xi, t)$ and $e(u, t)$ one obtains

$$\Gamma(u_1, u_2, \tau) = \int_{-\infty}^{\infty} \int_{-\infty}^{\infty} \gamma(\xi_1, \xi_2, \tau) e^{-j2\pi(\xi_1 u_1 - \xi_2 u_2)} \, d\xi_1 \, d\xi_2 \tag{3.14}$$

and

$$\gamma(\xi_1, \xi_2, \tau) = \int_{-\infty}^{\infty} \int_{-\infty}^{\infty} \Gamma(u_1, u_2, \tau) e^{j2\pi(\xi_1 u_1 - \xi_2 u_2)} \, du_1 \, du_2. \tag{3.15}$$

Note that from Eq. (3.9) the restrictions on $E(\xi, t)$ in Eq. (3.3) and the small-ξ assumption apply also to $\gamma(\xi_1, \xi_2, \tau)$. The relationships in Eqs. (3.14) and

(3.15) take on a simple form in the important case of complete incoherence. In radio astronomy the sources are generally assumed to be incoherent. This means that the radiation from one point on a source is statistically independent of the radiation from another point and the degree of coherence becomes zero. The coherence function in this case can be expressed as (Parrent 1959)

$$\gamma(\xi_1, \xi_2, \tau) = \gamma(\xi_1, \tau)\delta(\xi_1 - \xi_2). \tag{3.16}$$

Using this relation in conjunction with Eqs. (3.14) and (3.15) we find that the self-coherence function of a completely incoherent source and its spatial-frequency spectrum are Fourier transforms of each other:

$$\Gamma(u, \tau) = \int_{-\infty}^{\infty} \gamma(\xi, \tau)e^{-j2\pi\xi u}\, d\xi \tag{3.17}$$

$$\gamma(\xi, \tau) = \int_{-\infty}^{\infty} \Gamma(u, \tau)e^{j2\pi\xi u}\, du, \tag{3.18}$$

where $u = u_1 - u_2$. It is clear that the spatial-frequency spectrum of a completely incoherent source is stationary in the spatial-frequency plane, that is to say, $\Gamma(u, \tau)$ is independent of u_1 and u_2 and depends only on their difference. u can be interpreted as the spacing component of two sample points between which the coherence of the field is measured. However, u_1 and u_2, which are the conjugate variables of ξ_1 and ξ_2 in the Fourier transformation, have no obvious physical significance of their own. If $\tau = 0$

$$\gamma(\xi, 0) = \langle |E_T(\xi)|^2 \rangle, \tag{3.19}$$

which is the one-dimensional brightness distribution of the source B_1 introduced in Chapter 2. Again for $\tau = 0$, from Eq. (3.13),

$$\Gamma(u, 0) = \langle e_T(u_1, t)e_T^*(u_1 - u, t) \rangle. \tag{3.20}$$

$\Gamma(u, 0)$ is measured between points along a line normal to the direction $\xi = 0$. As measured with an interferometer, it is a form of the complex visibility $\mathscr{V}$ introduced in Chapter 2, the origin of phase being the $\xi = 0$ direction. From Eqs. (3.17) and (3.19)

$$\Gamma(u, 0) = \int_{-\infty}^{\infty} \langle |E_T(\xi)|^2 \rangle e^{-j2\pi\xi u}\, d\xi, \tag{3.21}$$

which is the basic relationship between visibility and brightness. The nonzero value of $\Gamma(u, 0)$ in Eq. (3.21) also illustrates that an incoherent source always gives rise to a partially coherent radiation field in space.

The Fourier transform relationship between the source brightness distribution and $\Gamma(u, 0)$ is related to the van Cittert–Zernike theorem of optics

(Born and Wolf 1965): a derivation of this theorem is given in Appendix 3.1. This derivation is similar to that in Section 2.4 but relates directly to the mutual coherence function. Both the van Cittert–Zernike theorem and the Wiener–Khinchin relation, discussed in Chapter 2 and Appendix 3.2, involve frequency spectra, but one concerns spatial frequencies and the other temporal frequencies. The van Cittert–Zernike theorem is valid only for completely incoherent sources. For cases of partial coherence the more general equation (3.14) applies.

3.2   PROPERTIES OF THE RECEIVING SYSTEM

The basic features of the receiving system for an interferometer are shown in Fig. 3.2. A very general receiving system is considered here. The two antenna outputs pass through filters H_1 and H_2 and are then fed to the correlator, which gives the time average of their product. The characteristics of the three basic constituents, the antennas, the filters, and the correlator, are discussed below.

Components of the Receiving System

Antennas.   The receiving antennas are characterized by their reception patterns (the question of polarization can be neglected in this discussion as explained in Section 3.1). The voltage reception pattern $\hat{G}(\xi, \nu)$, which should not be confused with the power gain, has the dimension *length* and is more convenient than the power reception pattern (dimension *area*) when the

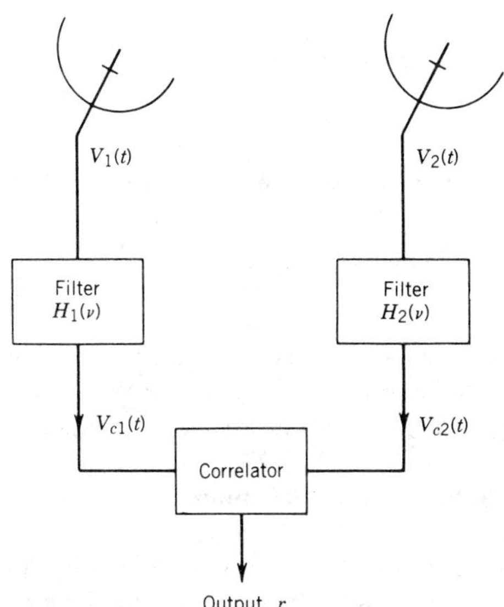

Figure 3.2   Basic features of a two-element interferometer.

antennas are not identical. It is a function of the geometry of the antenna, the frequency of operation, and the aperture field illumination $\hat{g}(x, \nu)$, x being a coordinate of position within the aperture. The following Fourier transform relationships hold:

$$\hat{G}(\xi, \nu) = \int_{-\infty}^{\infty} \hat{g}(x, \nu) \exp\left(j2\pi\xi\frac{x}{\lambda} \right) d(x/\lambda) \qquad (3.22)$$

and

$$\hat{g}(x,\nu) = \int_{-\infty}^{\infty} \hat{G}(\xi, \nu) \exp\left(-j2\pi\xi\frac{x}{\lambda} \right) d\xi, \qquad (3.23)$$

where λ is the wavelength (see, e.g., Silver 1949, Booker and Clemmow 1950). $\hat{G}(\xi, \nu)$, the voltage reception pattern of the antenna, is the same as the voltage radiation pattern by the reciprocity theorem.

Filters.   The output of the antenna is conventionally mixed with a local oscillator signal and is then fed to one or more IF amplifiers. The IF amplifiers incorporate filters that define the overall frequency response. Various delays arise in the system at various stages. However, for the purpose of this discussion a detailed consideration of these delays or of the mixing techniques (see Chapter 6) is unnecessary. The whole system can be represented by a filter having a frequency response $\hat{H}(\nu)$. The output of the filter $\hat{V}_c(\nu)$ is related to the input $\hat{V}(\nu)$ by

$$\hat{V}_c(\nu) = \hat{V}(\nu)\hat{H}(\nu). \qquad (3.24)$$

In the time domain this relationship is, by the convolution theorem,

$$V_c(t) = \int_{-\infty}^{\infty} H(t')V(t - t')dt' = H(t) * V(t). \qquad (3.25)$$

Correlator.   The correlator produces the cross correlation of the two voltages fed to it. If $V_1(t)$ and $V_2(t)$ are the input voltages, then the correlator output is

$$r(\tau) = \lim_{T \to \infty} \frac{1}{2T} \int_{-\infty}^{\infty} V_{1T}(t)V_{2T}^*(t - \tau) \, dt. \qquad (3.26)$$

The subscript T indicates the truncated function, as discussed earlier. τ is the time by which voltage V_2 is delayed with respect to voltage V_1. In the similar equation (2.5), V was considered to be real so the conjugate was unnecessary.

Response of the Receiving System to the Incident Radiation

The properties of the incident radiation and of the receiving system will now be combined to show the overall response of the system. It will then be seen how the properties of the radiation can be interpreted in terms of the system responses. We consider the general case of a partially coherent source.

Antenna Output.   For an antenna with angular response pattern $G(\theta, \nu)$ pointed in the direction θ, the response to an input angular distribution $\hat{E}(\theta, \nu)$ is

$$\hat{V}(\theta, \nu) = \int_{-\infty}^{\infty} \hat{E}(\theta'', \nu)\hat{G}(\theta'' - \theta, \nu)\,d\theta'', \tag{3.27}$$

which is the voltage response form of the cross correlation in Expression (2.14). Equation (3.27) is useful when the antennas are not pointed at the source under observation, as in a drift scan. Here, however, we write $\hat{G}(\xi, \nu)$ for the response to radiation from direction ξ, so the antenna output voltage becomes

$$\hat{V}(\nu) = \int_{-\infty}^{\infty} \hat{E}(\xi, \nu)\hat{G}(\xi, \nu)\,d\xi. \tag{3.28}$$

This form is appropriate when the antennas are pointed toward the direction of interest, as occurs when they track the source with zero error. Each antenna pattern is defined with reference to its own phase center, that is, in applying Eq. (3.22) to Eq. (3.28), x is measured from the center of the antenna.

Filter Output.   The outputs of the two filters are by Eq. (3.24)

$$\hat{V}_{c1}(\nu) = \hat{V}_1(\nu)\hat{H}_1(\nu) \tag{3.29}$$

and

$$\hat{V}_{c2}(\nu) = \hat{V}_2(\nu)\hat{H}_2(\nu). \tag{3.30}$$

Correlator Output.   The correlator output is obtained by using Eq. (3.26):

$$r(\tau) = \lim_{T\to\infty} \frac{1}{2T}\int_{-\infty}^{\infty} V_{c1T}(t)V_{c2T}^*(t - \tau)\,dt. \tag{3.31}$$

The delay τ of the signal $V_{c2}(t)$ with respect to $V_{c1}(t)$ is caused, in general, by the geometry of the source with respect to the two antennas and by the various instrumental delays. The system may contain an adjustable delay unit so that the overall delay can be controlled.

In order to relate $r(\tau)$ to the input signal $\hat{E}(\xi, \nu)$, Eqs. (3.7) and (3.28)–(3.30) are used in Eq. (3.31), and we obtain

$$r(\tau) = \lim_{T\to\infty} \frac{1}{2T}\int_{-\infty}^{\infty} dt \int_{-\infty}^{\infty} d\nu_1 \int_{-\infty}^{\infty} d\nu_2 \int_{-\infty}^{\infty} d\xi_1 \int_{-\infty}^{\infty} d\xi_2$$

$$\times \hat{E}_T(\xi_1, \nu_1)\hat{E}_T^*(\xi_2, \nu_2)\hat{H}_1(\nu_1)\hat{H}_2^*(\nu_2)\hat{G}_1(\xi_1, \nu_1)$$

$$\times \hat{G}_2^*(\xi_2, \nu_2)e^{j2\pi\nu_1 t}e^{-j2\pi\nu_2(t-\tau)}. \tag{3.32}$$

Note that the use of separate response functions $\hat{G}_1$ and $\hat{G}_2$ for the two antennas can accommodate different antenna designs, or different pointing-offset errors, or both. It is seen that the t integral in Eq. (3.32) results in $\delta(\nu_1 - \nu_2)$. This enables one to perform one of the ν integrations to yield

$$r(\tau) = \int_{-\infty}^{\infty} d\nu \int_{-\infty}^{\infty} d\xi_1 \int_{-\infty}^{\infty} d\xi_2 \left[\lim_{T \to \infty} \frac{1}{2T} \hat{E}_T(\xi_1, \nu) \hat{E}_T^*(\xi_2, \nu) \right]$$

$$\times \hat{H}_1(\nu) \hat{H}_2^*(\nu) \hat{G}_1(\xi_1, \nu) \hat{G}_2^*(\xi_2, \nu) e^{j2\pi\nu\tau}. \tag{3.33}$$

If we write

$$\hat{\gamma}(\xi_1, \xi_2, \nu) = \lim_{T \to \infty} \frac{1}{2T} \left[\hat{E}_T(\xi_1, \nu) \hat{E}_T^*(\xi_2, \nu) \right], \tag{3.34}$$

which follows from Eq. (3.9) by the cross-correlation relation given in Appendix 3.2 of this chapter, then

$$\hat{\gamma}(\xi_1, \xi_2, \nu) = \int_{-\infty}^{\infty} \gamma(\xi_1, \xi_2, \tau) e^{-j2\pi\nu\tau} d\tau \tag{3.35}$$

and

$$\gamma(\xi_1, \xi_2, \tau) = \int_{-\infty}^{\infty} \hat{\gamma}(\xi_1, \xi_2, \nu) e^{j2\pi\nu\tau} d\nu. \tag{3.36}$$

In terms of Eq. (3.34), Eq.(3.33) yields

$$r(\tau) = \int_{-\infty}^{\infty} d\nu \int_{-\infty}^{\infty} d\xi_1 \int_{-\infty}^{\infty} d\xi_2 \, \hat{\gamma}(\xi_1, \xi_2, \nu) \hat{H}_1(\nu) \hat{H}_2^*(\nu)$$

$$\times \hat{G}_1(\xi_1, \nu) \hat{G}_2^*(\xi_2, \nu) e^{j2\pi\nu\tau}. \tag{3.37}$$

Equation (3.37) is the relationship between the output of a correlator interferometer and the source coherence function of a polychromatic partially coherent source in terms of the antenna reception patterns and filter responses. Results obtained by other authors, including Drane and Parrent (1962) and MacPhie (1964), can be recovered from this equation by considering the appropriate special cases.

The output is influenced by a number of factors, namely, the distribution of radiation in the sky, the size, illumination, and direction of mechanical pointing of the antennas, the separation of the antennas, and the filter responses.

3.3 RESPONSE TO A COMPLETELY INCOHERENT SOURCE

The incoherent source is the case of main interest in radio astronomy. The radiation from different points of a completely incoherent source is uncorrelated and the source coherence function is given by Eq. (3.16). Using this in Eq. (3.37) we obtain

$$r(\tau) = \int_{-\infty}^{\infty} \int_{-\infty}^{\infty} \hat{\gamma}(\xi, \nu) \hat{H}_1(\nu) \hat{H}_2^*(\nu) \hat{G}_1(\xi, \nu) \hat{G}_2^*(\xi, \nu) e^{j2\pi\nu\tau} \, d\nu \, d\xi.$$

$$(3.38)$$

The above result can be further simplified for special cases of antennas and filters.

Let the antennas be identical and let the filters have identical responses $\hat{H}(\nu)$ and center frequencies ν_0. Then we have

$$r(\tau) = \int_{-\infty}^{\infty} \int_{-\infty}^{\infty} \hat{\gamma}(\xi, \nu) |\hat{G}(\xi, \nu)|^2 |\hat{H}(\nu)|^2 e^{j2\pi\nu\tau} \, d\nu \, d\xi. \qquad (3.39)$$

In any measurement of $r(\tau)$ the delay τ is made up of the geometrical delay τ_g and the instrumental delay τ_i. In spectral-line measurements it is appropriate to vary τ_i such that $r(\tau)$ is measured over a range of τ. Then Fourier transformation with respect to τ and ν yields $\hat{r}(\nu)$, the cross power spectrum.

For a point ξ, the geometrical delay τ_g is $D\xi/c$, where D is the separation between the phase centers of the two antennas and c is the velocity of light. We define a direction ξ_i for which the signal delays to the correlator inputs are equal. Then

$$\tau = \tau_g - \tau_i = \frac{D(\xi - \xi_i)}{c}. \qquad (3.40)$$

If the filter passband $\Delta\nu$ is sufficiently small, the functions $\hat{\gamma}(\xi, \nu)$ and $|\hat{G}(\xi, \nu)|^2$ do not vary significantly in the passband and Eq. (3.39) reduces to the form

$$r = \int_{-\infty}^{\infty} \hat{\gamma}(\xi, \nu_0) \hat{P}(\xi) \, d\xi. \qquad (3.41)$$

The function $\hat{P}(\xi)$ is the product of three functions: the antenna power reception pattern $|\hat{G}(\xi, \nu_0)|^2$, the bandwidth pattern (or delay pattern) defined by

$$\mathscr{B}(\xi_i - \xi, \Delta\nu, D) = \int_{-\Delta\nu/2}^{\Delta\nu/2} |\hat{H}(\nu_0 + \nu')|^2 \exp\left[\frac{j2\pi(\xi - \xi_i)D\nu'}{c}\right] d\nu',$$

$$(3.42)$$

where $v' = v - v_0$, and the interference (fringe) pattern $F(\xi_i, \xi, D) = \exp[\,j2\pi(\xi - \xi_i)\,Dv_0/c]$. Note that because the signals are represented by the analytic signal in this chapter we do not include a negative frequency band as in Eq. (2.8), and the fringes are represented by the real part of an exponential function. The bandwidth pattern has a peak in the direction ξ_i: its effect upon the field of view is determined by the bandwidth and the baseline. The antenna pattern limits the field of view in a manner determined by the size and illumination of the antennas. The interference pattern provides the fringes and determines the resolution of the instrument. Illustrations of these functions can be found in Figs. 2.3, 2.5, and 2.6.

We now consider the case where the source extent and the antenna beamwidth are very small, but centered on an arbitrary direction θ_0 instead of $\theta = 0$. Then, as in Chapter 2, we define

$$\theta' = \theta_0 - \theta, \tag{3.43}$$

where θ' is very small so we may write

$$\xi \simeq \xi_0 - \xi' \cos \theta_0. \tag{3.44}$$

ξ' is the sine of the angle measured from the direction θ_0. The instrumental delay τ_i is then adjusted to the value $D\xi_0/c$ so that the delay pattern also has a peak in the direction ξ_0. Again as in Chapter 2 we write

$$u = \frac{D \cos \theta_0}{\lambda_0} = \frac{Dv_0 \cos \theta_0}{c}, \tag{3.45}$$

and recall that the spatial frequency u is equal to the component of the baseline (measured in wavelengths at the center frequency v_0) normal to the reference direction θ_0. Equation (3.41) can now be written

$$r(u) = \int_{-\infty}^{\infty} \hat{\gamma}(\xi', v_0)\hat{P}(\xi')\,d\xi'. \tag{3.46}$$

The output is, of course, a function of ξ_0, v_0, and the antenna aperture illuminations in addition to u. But the u dependence is emphasized because the output is a measure of the spatial-frequency spectrum of the source brightness distribution.

If the bandwidth is narrow enough that the bandwidth pattern is much wider than the antenna pattern for all baselines used, then

$$r(u) = \text{const.} \int_{-\infty}^{\infty} \hat{\gamma}(\xi', v_0)|\hat{G}(\xi', v_0)|^2 e^{-j2\pi\xi' u}\,d\xi', \tag{3.47}$$

where the constant preserves dimensional homogeneity. In the exponent we have used Eqs. (3.40), (3.44), and $\xi_i = \xi_0$. If, further, one is concerned with isolated sources of small extent (small compared with the antenna pattern), then $|\hat{G}(\xi', \nu_0)|^2$ is essentially constant over the range of ξ' for which $\hat{\gamma}(\xi', \nu_0)$ is nonzero. The correlator output as a function of the spatial frequency u is

$$r(u) = \text{const.} \int_{-\infty}^{\infty} \hat{\gamma}(\xi', \nu_0) e^{-j2\pi\xi' u} \, d\xi' = \text{const.} \Gamma(u, \nu_0), \qquad (3.48)$$

which is a form of the self-coherence function: compare Eq. (3.48) with Eq. (3.17). $\Gamma(u, \nu_0)$ is a form of the complex visibility function $\mathscr{V}(u)$ which has been defined in Chapter 2. Here it is measured also as a function of frequency rather than time delay as in Eq. (3.17). The brightness distribution can be reconstructed from measurements of $\hat{\Gamma}(u, \nu_0)$ by

$$\hat{\gamma}(\xi', \nu_0) = \int_{-\infty}^{\infty} \hat{\Gamma}(u, \nu_0) e^{j2\pi\xi' u} \, du. \qquad (3.49)$$

In cases where the antennas track the source, but the source is not small compared with the beams, the brightness reconstruction expression becomes

$$\hat{\gamma}(\xi', \nu_0) |\hat{G}_N(\xi', \nu_0)|^2 = \int_{-\infty}^{\infty} \hat{\Gamma}(u, \nu_0) e^{j2\pi\xi' u} \, du, \qquad (3.50)$$

where $\hat{G}_N$ is the voltage reception pattern normalized to unity on the beam axis. The output of the correlator as described up to this point is represented by the real part of r. In Chapter 6 we introduce the complex correlator, the outputs of which correspond to both the real and imaginary parts of r.

In studies of spectral lines it is necessary to measure the brightness distribution as a function of frequency. One method of obtaining the required spectral data is from the measurement of r over a range of τ. The delay in Eq. (3.40) can be rewritten

$$\tau = \frac{D(\xi - \xi_0)}{c} + \tau' \qquad (3.51)$$

for the delay-tracking angle θ_0 and an additional delay offset τ' introduced at the correlator. Thus, Eq. (3.39) becomes

$$r(\tau') = \int_{-\infty}^{\infty} \int_{-\infty}^{\infty} \hat{\gamma}(\xi, \nu'') |\hat{G}(\xi, \nu'')|^2 |\hat{H}(\nu'')|^2 e^{j2\pi D\nu''(\xi - \xi_0)/c}$$

$$\times e^{j2\pi\nu''\tau'} \, d\nu'' \, d\xi, \qquad (3.52)$$

where we have changed the frequency variable to ν''. The Fourier transform of Eq. (3.52) is

$$r(\nu) = \int_{-\infty}^{\infty} r(\tau') e^{-j2\pi\nu\tau'} \, d\tau'. \qquad (3.53)$$

Substituting Eq. (3.52) into Eq. (3.53) and noting that from Eq. (3.44) $D\nu(\xi - \xi_0)/c = -\xi'u$, we obtain

$$r(u, v) = \int_{-\infty}^{\infty} \hat{\gamma}'(\xi', v) e^{-j2\pi\xi'u} \, d\xi', \tag{3.54}$$

where

$$\hat{\gamma}'(\xi', v) = \hat{\gamma}(\xi', v)|\hat{G}(\xi', v)|^2|\hat{H}(v)|^2. \tag{3.55}$$

Thus

$$\hat{\gamma}(\xi', v) = \frac{\displaystyle\int_{-\infty}^{\infty} r(u, v) e^{j2\pi\xi'u} \, du}{|\hat{G}(\xi', v)|^2|\hat{H}(v)|^2}. \tag{3.56}$$

This is the desired brightness as a function of angle and frequency.

3.4   RESPONSE TO A COMPLETELY COHERENT SOURCE

It has been shown by Parrent (1959) that an extended source can be completely coherent if, and only if, it is monochromatic. As examples of such a source the reader may visualize the aperture of a distant, large antenna, or an ensemble of smaller antennas all driven by the same monochromatic signal. If a completely coherent source is observed with an interferometer the output voltage of each interferometer antenna is a sinusoid in time. The relative phase of the outputs of the two interferometer antennas depends on the geometry of the source and its position relative to the antennas, and the correlator output voltage will be a fringe pattern similar to that for an incoherent source. Now if one antenna is moved to vary the baseline, the amplitude of the signal voltage at its output may change since, as suggested by the examples mentioned above, the radiation from the source may be highly directional. However, the received voltage will remain a sinusoid of the same frequency as that received by the other antenna. Thus the modulus of the normalized correlation of the signals as measured by the two antennas will remain unity, and very little information can be gained about the brightness distribution of an extended coherent source from interferometer measurements of this type. However, the radiation pattern of such a source can be explored by examining the outputs of the individual interferometer antennas, the range of spacings required being similar to that required to resolve an incoherent source of the same angular dimensions. This last statement follows from a consideration of the Fourier transform relationships between the voltage radiation pattern and the field distribution of a coherent source [Eq. (3.22)] and between the visibility and brightness distribution of an incoherent source [Eq. (3.50)].

3.5 SPACE–FREQUENCY EQUIVALENCE

In this section we compare the effects on the directional response of an interferometer to an incoherent source, of (1) the frequency response of the receiving system (the bandwidth pattern) and (2) the antenna aperture distribution (the reception pattern). This subject has been discussed by Kock and Stone (1958) and Swenson and Mathur (1969). Here we show that an exact equivalence exists between wideband operation of an interferometer with isotropic antennas, and narrowband operation with directional antennas. The equivalence applies only to the one-dimensional angular response measured in a plane containing the baseline and the source. It does not apply to the sensitivity of an interferometer (MacPhie 1979).

The response of an interferometer has been shown in Eq. (3.41) to be the product of three functions: the antenna reception pattern, the bandwidth pattern, and the fringe pattern. Consider first a system in which the antenna patterns are isotropic with voltage response G_i. The response of the interferometer is governed by the fringe and bandwidth functions and is of the form

$$\hat{P}_1(\xi') = G_i^2 \int_{-\infty}^{\infty} \hat{H}_1(\nu)\hat{H}_2^*(\nu)\exp\left(\frac{-j2\pi\xi'D\nu\cos\theta_0}{c}\right) d\nu, \qquad (3.57)$$

where ξ' is measured relative to the direction θ_0 for which the instrumental and geometrical delays are equal. Equation (3.57) can also be expressed as

$$\hat{P}_1(\xi') = \mathscr{P}_1(\xi')\exp\left[-j\left(\frac{2\pi\xi'D\nu_0\cos\theta_0}{c}\right)\right], \qquad (3.58)$$

where

$$\mathscr{P}_1(\xi') = G_i^2 \int_{-\infty}^{\infty} \hat{H}_1(\nu_0 + \nu')\hat{H}_2^*(\nu_0 + \nu')\exp\left(\frac{-j2\pi\xi'D\nu'\cos\theta_0}{c}\right) d\nu'.$$

$$(3.59)$$

Here $\nu' = \nu - \nu_0$ and ν_0 is the center frequency of the interferometer passband.

Now consider an interferometer with directive antennas pointed in the direction θ_0, and let the bandwidth $\Delta\nu$ be small enough that the bandwidth pattern is much wider than the antenna beams and can be neglected. The response is then governed by the antenna and fringe patterns and is of the form

$$\hat{P}_2(\xi') = \hat{G}_1(\xi', \nu_0)\hat{G}_2^*(\xi', \nu_0)\Delta\nu \exp\left(\frac{-j2\pi\xi'D\nu_0\cos\theta_0}{c}\right), \qquad (3.60)$$

which can be expressed as

$$\hat{P}_2(\xi') = \hat{\mathscr{P}}_2(\xi')\Delta\nu\,\exp\left[-j\left(\frac{2\pi\xi' D\nu_0\cos\theta_0}{c}\right)\right], \tag{3.61}$$

where

$$\hat{\mathscr{P}}_2(\xi') = \hat{G}_1(\xi', \nu_0)\hat{G}_2^*(\xi', \nu_0). \tag{3.62}$$

By appropriate choice of the frequency response functions in Eq. (3.57) and the reception patterns in Eq. (3.60), it is possible to obtain equivalence between the responses $\hat{P}_1(\xi')$ and $\hat{P}_2(\xi')$. This is most easily demonstrated by making use of the Fourier transform relationship between the voltage reception pattern and aperture illumination function of an antenna in Eq. (3.22). Thus, using the convolution relationship for Fourier transforms, we can write

$$\mathscr{P}_2(\xi') = \int_{-\infty}^{\infty} \hat{g}_1(x) * \hat{g}_2^*(x)\exp\left(\frac{j2\pi\xi' x\nu_0}{c}\right) d\left(\frac{x\nu_0}{c}\right), \tag{3.63}$$

where $\hat{g}_1(x)$ and $\hat{g}_2(x)$ are the aperture illumination functions. The required equivalence occurs when $\hat{\mathscr{P}}_1(\xi')$ and $\hat{\mathscr{P}}_2(\xi')$ differ by no more than a constant factor. From Eqs. (3.59) and (3.63) the condition for this is

$$\hat{H}_1\left[\nu_0\left(1 - \frac{x}{D\cos\theta_0}\right)\right]\hat{H}_2^*\left[\nu_0\left(1 - \frac{x}{D\cos\theta_0}\right)\right] = \text{const.}\,\hat{g}_1(x) * \hat{g}_2^*(x).$$

$$\tag{3.64}$$

As an example, consider a system employing filters with identical rectangular passbands. It is clear from Eq. (3.57) that with isotropic antennas the interferometer pattern will then have an envelope of the form $[\sin(K\xi')]/K\xi'$ where K is a constant. To achieve the same pattern with a monochromatic system, one can employ a pair of antennas, one isotropic and the other a uniformly illuminated linear aperture. More generally, since convolution of any function with the delta function produces no change in the function, Eq. (3.64) is satisfied if one of the antennas is isotropic and the other has an aperture illumination function of the same form as the product of the frequency response functions. In practice, one can control both the shape of the frequency response functions and the aperture illumination functions to achieve suppression of sidelobes.

Frequency-passband shaping to control the response pattern of an interferometer has useful applications in situations in which the sizes or numbers of antennas available are inadequate to provide the desired pattern (Goldstein 1959, Douglas et al. 1973). The application to very-long-baseline interferometry (bandwidth synthesis) is discussed in Chapter 9. In multielement arrays the effect of the frequency response is complicated by the different antenna spacings and is considered in Chapter 6.

APPENDIX 3.1 VAN CITTERT–ZERNIKE THEOREM

We now show by means of coherence arguments that interferometer measurements on the ground can be used to map the brightness distribution of a cosmic source. We follow the demonstration of Born and Wolf (1959, Chapter 10) but use somewhat different notation to conform to radio astronomy practice. This analysis illustrates an alternative approach to the derivation of the relationship between the interferometer response and the source brightness distribution. Consider a source S emitting radiation from each element ds of its surface, and consider two detectors (antennas) in a distant plane assumed in this discussion to be parallel with the plane tangent to the celestial sphere at O in Fig. A3.1. The source is assumed to be small in angular extent and the distances R_1, R_2, and R from the source to the antenna plane are assumed to be very much larger than any other dimension in the system. The antennas and associated amplifiers have bandwidth $\Delta\nu$ much smaller than their center frequency ν_0. The radiation mechanism is assumed such that radiation from any source element ds_m is statistically independent of that from any other element ds_n, that is, the source is incoherent. The time average of the electric

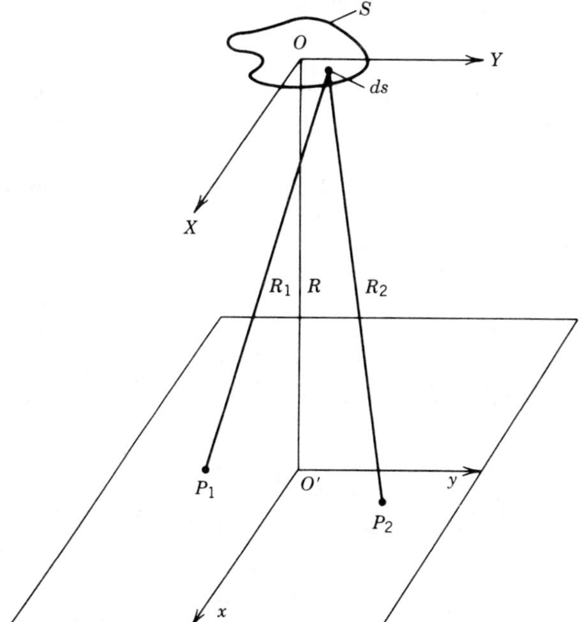

Figure A3.1 Geometry of source S and field measurement points P_1 and P_2.

field is zero. All elements of the system are assumed copolarized so that a scalar treatment is appropriate.

Each element ds of S produces an electric field $E_1(s,t)$ at P_1 and field $E_2(s,t)$ at P_2. The total electric field at each receiving antenna is the integral over all elements of the source. In terms of analytic signals we have:

$$e_{P1}(t) = \int_S E_1(s,t)\,ds, \qquad e_{P2}(t) = \int_S E_2(s,t)\,ds. \qquad (A3.1)$$

From the definition of the mutual coherence function in Eq. (3.13) we obtain

$$\Gamma_{12}(\tau) = \lim_{T \to \infty} \frac{1}{2T} \int_{-\infty}^{\infty} e_{P1}(t)\,e_{P2}^*(t - \tau)\,dt. \qquad (A3.2)$$

The integrand contains terms of the type $E_1(s_m, t)\, E_2^*(s_n, t - \tau)$. Upon integrating with respect to s, all terms for which $m \neq n$ become zero because of the condition of statistical independence mentioned above. The mth terms can be written

$$E_1(s_m, t) = \mathscr{E}_m\left(t - \frac{R_{m1}}{c}\right) \frac{\exp\left[-j2\pi\nu(t - R_{m1}/c)\right]}{R_{m1}}$$

$$(A3.3)$$

$$E_2(s_m, t) = \mathscr{E}_m\left(t - \frac{R_{m2}}{c}\right) \frac{\exp\left[-j2\pi\nu(t - R_{m2}/c)\right]}{R_{m2}},$$

where $\mathscr{E}_m$ is the complex amplitude at the source of the radiation from the mth element s_m of S and R_{m1} and R_{m2} are the distances from this element to P_1 and P_2, respectively.

For $\tau = 0$ the time-averaged product of the fields at P_1 and P_2 due to s_m can be written

$$\langle e_{P1}(t)\,e_{P2}^*(t)\rangle_m$$

$$= \left\langle \mathscr{E}_m\left(t - \frac{R_{m1}}{c}\right) \mathscr{E}_m^*\left(t - \frac{R_{m2}}{c}\right)\right\rangle \frac{\exp\left[-j2\pi\nu\left(t - \dfrac{R_{m1}}{c}\right)\right]\exp\left[j2\pi\nu\left(t - \dfrac{R_{m2}}{c}\right)\right]}{R_{m1}R_{m2}}$$

$$= \left\langle \mathscr{E}_m(t)\,\mathscr{E}_m^*\left(t - \frac{R_{m2} - R_{m1}}{c}\right)\right\rangle \frac{\exp\left[j2\pi\nu(R_{m1} - R_{m2})/c\right]}{R_{m1}R_{m2}}.$$

$$(A3.4)$$

If the quantity $(R_{m2} - R_{m1})/c$ is small compared with the reciprocal receiver bandwidth $(\Delta\nu)^{-1}$, we can neglect it within the sharp brackets of Eq. (A3.4). Then

$$\langle e_{P1}(t)e_{P2}^*(t)\rangle_m = \frac{\langle\mathscr{E}_m(t)\mathscr{E}_m^*(t)\rangle\exp\left[j2\pi\nu(R_{m1}-R_{m2})/c\right]}{R_{m1}R_{m2}}.$$

$$(A3.5)$$

The quantity $\langle\mathscr{E}_m(t)\mathscr{E}_m^*(t)\rangle$ is a measure of the time-averaged intensity (brightness) B_m of the surface of the source at point m. Integration of Eq. (A3.5) over the source gives the mutual coherence function of the fields at points P_1 and P_2

$$\Gamma_{12}(0) = \int_S \frac{B(X,Y)\exp\left[j2\pi\nu(R_1-R_2)/c\right]}{R_1R_2}\,ds, \qquad (A3.6)$$

where R_1 and R_2 are the distances from ds to P_1 and P_2, respectively, and X and Y are linear coordinates in the source plane. Note that $(R_1-R_2)/c$ is the differential travel time for distances R_1 and R_2. With reference to Fig. A3.1, the distance from s at (X,Y) to P_1 at (x_1,y_1) is given by

$$R_1^2 = (x_1-X)^2+(y_1-Y)^2+R^2, \qquad (A3.7)$$

or, when only the first terms in R_1 are retained in a binomial expansion,

$$R_1 \simeq R + \frac{1}{2R}\left[(x_1-X)^2+(y_1-Y)^2\right], \qquad (A3.8)$$

which in practice requires that $X/R \ll 1$ and $Y/R \ll 1$. A similar expression holds for R_2, so that

$$R_1-R_2 \simeq \frac{(x_1^2+y_1^2)-(x_2^2+y_2^2)}{2R}+\frac{(x_2-x_1)X+(y_2-y_1)Y}{R}.$$

$$(A3.9)$$

This is the usual approximation of Fraunhofer diffraction theory. The denominator of the integrand of Eq. (A3.6) can be approximated adequately by R^2. Substituting $u=(x_1-x_2)\nu/c$, $\upsilon=(y_1-y_2)\nu/c$, $\xi=X/R$, $\eta=Y/R$,

$$\delta = \frac{x_1^2+y_1^2-(x_2^2+y_2^2)}{2R}, \qquad (A3.10)$$

and noting that $B(\xi,\eta)$ is zero outside S, and that $d\xi\,d\eta = ds/R^2$, we obtain

$$\Gamma_{12}(0) = e^{j2\pi\nu\delta/c}\int_{-\infty}^{\infty}\int_{-\infty}^{\infty} B(\xi,\eta)e^{-j2\pi(u\xi+\upsilon\eta)}\,d\xi\,d\eta. \qquad (A3.11)$$

The quantity $2\pi\nu\delta/c$ is a phase shift and is caused by the path differences to the two antennas. This phase term is usually small and can be neglected when the source is in the far field of the interferometer, which is the condition in Eq. (2.26):

$$R \gg \frac{D^2}{\lambda} = R_0,$$

(A3.12)

where D is the baseline length, $|P_1 - P_2|$. If the source is at a known distance closer than R_0, then the phase term can be compensated. This may sometimes be necessary in solar system studies. The quantities ξ and η are small and can be regarded as angles on the celestial sphere measured from O', or direction cosines relative to the x and y axes. Thus from Eq. (A3.11) the mutual coherence function $\Gamma_{12}(0)$ in the antenna plane is the Fourier transform of the brightness distribution $B(\xi, \eta)$ of the source. This is a form of the van Cittert–Zernike theorem of optics. The converse is also true: if the mutual coherence function can be measured, the brightness distribution can be obtained from it by inverse Fourier transformation.

The restriction stated following Eq. (A3.4) that $(R_{m2} - R_{m1})/c$ be less than $(\Delta \nu)^{-1}$ can be written

$$\frac{\Delta \nu}{\nu} < \frac{1}{\xi_d u} \quad \text{and} \quad \frac{\Delta \nu}{\nu} < \frac{1}{\eta_d v},$$

(A3.13)

where ξ_d and η_d are the maximum angular dimensions of the source. This is the requirement that the source is within the bandwidth pattern of the interferometer. Thus the field of view defines the maximum bandwidth that can be used. Observations covering wider bands can be achieved by superposition of responses using different narrowband filter channels.

APPENDIX 3.2   CORRELATION, CONVOLUTION, AND THE WIENER–KHINCHIN RELATION

The response of a correlator is given in Eq. (3.26). In practice, the correlation is sampled for a finite time period, $2T$, which is usually a few minutes or seconds but is very large compared with both the period and the reciprocal bandwidth of the waveforms. The integral in Eq. (3.26) without the factor $(2T)^{-1}$ is commonly referred to as the cross correlation of the two waveforms (e.g., Champeney, 1973) and represented by the pentagram symbol:

$$V_{1T} \bigstar V_{2T} = \int_{-\infty}^{\infty} V_{1T}(t) V_{2T}^*(t - \tau)\, dt.$$

(A3.14)

This integral can be expressed as a convolution in the following way:

$$V_{1T} \bigstar V_{2T} = \int_{-\infty}^{\infty} V_{1T}(t) V_{2T-}^*(\tau - t)\, dt = V_{1T} * V_{2T-}^*,$$

(A3.15)

where $V_{2T_-}(t) = V_{2T}(-t)$. Now the ν, t Fourier transforms are denoted as follows: $\hat{V}_{1T}(t) \rightleftharpoons V_{1T}(\nu)$ and $\hat{V}_{2T}(t) \rightleftharpoons V_{2T}(\nu)$, whence $V_{2T_-}^*(t) \rightleftharpoons \hat{V}_{2T}^*(\nu)$. Thus from the convolution theorem

$$V_{1T} \star V_{2T} \rightleftharpoons \hat{V}_{1T}\hat{V}_{2T}^*. \tag{A3.16}$$

The right-hand side of Eq. (A3.16) is known as the cross power spectrum of $V_{1T}(t)$ and $V_{2T}(t)$. The cross power spectrum is a function of frequency, and we see that it is the Fourier transform of the cross correlation which is a function of τ. This is a useful result, and in the case where $V_{1T} = V_{2T}$ it becomes the Wiener–Khinchin relation.

REFERENCES

Beran, M. J. and G. B. Parrent, Jr., *Theory of Partial Coherence*, Prentice-Hall, Englewood Cliffs, New Jersey, 1964; reprinted by Society of Photo-Optical Instrtumentation Engineers, 1974.

Booker, H. G. and P. C. Clemmow, The Concept of an Angular Spectrum of Plane Waves, and Its Relation to that of Polar Diagram and Aperture Distribution, *Proc. IEE*, **97**, 11–17, 1950.

Born, M. and E. Wolf, *Principles of Optics*, Pergamon Press, London, 1959 (and later eds.).

Bracewell, R. N., *The Fourier Transform and its Applications*, McGraw-Hill, New York, 1965 (2nd ed. 1978).

Champeney, D. C., *Fourier Transforms and Their Physical Applications*, Academic Press, London, 1973.

Douglas, J. N., F. N. Bash, F. D. Ghigo, G. F. Moseley, and G. W. Torrence, First Results from the Texas Interferometer: Positions of 605 Discrete Sources, *Astron. J.*, **78**, 1–17, 1973.

Drane, C. J. and G. B. Parrent, Jr., On the Mapping of Extended Sources with Nonlinear Correlation Antennas. *IRE Trans. Antennas Propag.*, **AP-10**, 126–130, 1962.

Goldstein, S. J., Jr., The Angular Size of Short-Lived Solar Radio Disturbances, *Astrophys. J.*, **130**, 393–399, 1959.

Kock, W. E. and J. L. Stone, Space–Frequency Equivalence, *Proc. IRE*, **46**, 499–500, 1958.

MacPhie, R. H., On the Limitations of the Concept of Space–Frequency Equivalence, *Radio Sci.*, **14**, 1185–1187, 1979.

MacPhie, R. H., On the Mapping by a Cross Correlation Antenna System of Partially Coherent Radio Sources, *IEEE Trans. Antennas Propag.*, **AP-12**, 118–124, 1964.

Mandel, L., and E. Wolf., Coherence Properties of Optical Fields, *Rev. Mod. Phys.*, **37**, 231–287, 1965.

Parrent, G. B., Jr., Studies in the Theory of Partial Coherence, *Opt. Acta*, **6**, 285–296, 1959.

Silver, S., *Microwave Antenna Theory and Design*, Radiation Laboratory Series, Vol. 12, McGraw-Hill, New York, 1949, p. 174.

Swenson, G. W., Jr. and N. C. Mathur. The Interferometer in Radio Astronomy, *Proc. IEEE*, **56**, 2114–2130, 1968.

Swenson, G. W., Jr. and N. C. Mathur. On the Space–Frequency Equivalence of a Correlator Interferometer, *Radio Sci.*, **4**, 69–71, 1969.

4

GEOMETRICAL RELATIONSHIPS AND OTHER PRACTICAL CONSIDERATIONS

In this chapter we start to examine some of the practical aspects of interferometry. These include baselines, antenna mounts and beamshapes, and the response to polarized radiation, all of which involve geometrical considerations and coordinate systems. The discussion is concentrated upon earth-based arrays with tracking antennas, which illustrate well the principles involved, although other systems such as those that include one or more antennas in earth orbit are also important. We begin by considering the coordinate systems used to express the Fourier transform relationship between visibility and brightness and examining the approximations and limitations encountered in practical synthesis mapping. At the end of the chapter we touch upon the discrete Fourier transform, the fast algorithm of which is widely used in mapping for reasons of economy in computation. The uniform sampling required for these procedures imposes a characteristic aliasing on the overall response of the interferometer or array.

4.1 COORDINATE SYSTEMS FOR THE FOURIER TRANSFORM RELATIONSHIP

In the preceding chapters we have shown that a Fourier transform relationship exists between the radio brightness distribution of the sky and the complex visibility as a function of baseline vector. This relationship can be envisioned as resulting from the sinusoidal pattern on the sky as described in Chapter 2, or from the measurement of the mutual coherence of the radiation field by the spaced antennas as described in Chapter 3. The relationship has been expressed in Eq. (2.22) in the following form:

$$\mathcal{V}(\mathbf{D}_\lambda) = \int_{4\pi} A_N(\boldsymbol{\sigma}) B(\boldsymbol{\sigma}) e^{-j2\pi\mathbf{D}_\lambda \cdot \boldsymbol{\sigma}} \, d\Omega \tag{4.1}$$

where $\mathcal{V}$ is the complex fringe visibility, $\mathbf{D}_\lambda$ is the baseline vector of the two antennas, $A_N(\boldsymbol{\sigma})$ is the normalized antenna area, $B(\boldsymbol{\sigma})$ is the radio brightness, $d\Omega$ is an element of solid angle on the sky, and $\boldsymbol{\sigma} = \mathbf{s} - \mathbf{s}_0$ as in Fig. 2.9.

Recall from Chapters 2 and 3 that the assumptions required in the application of Eq. (4.1) are the incoherence and the far-field conditions for the source. The geometry that we now consider is illustrated in Fig. 4.1. The two antennas track the center of the field to be mapped. They are assumed to be identical,

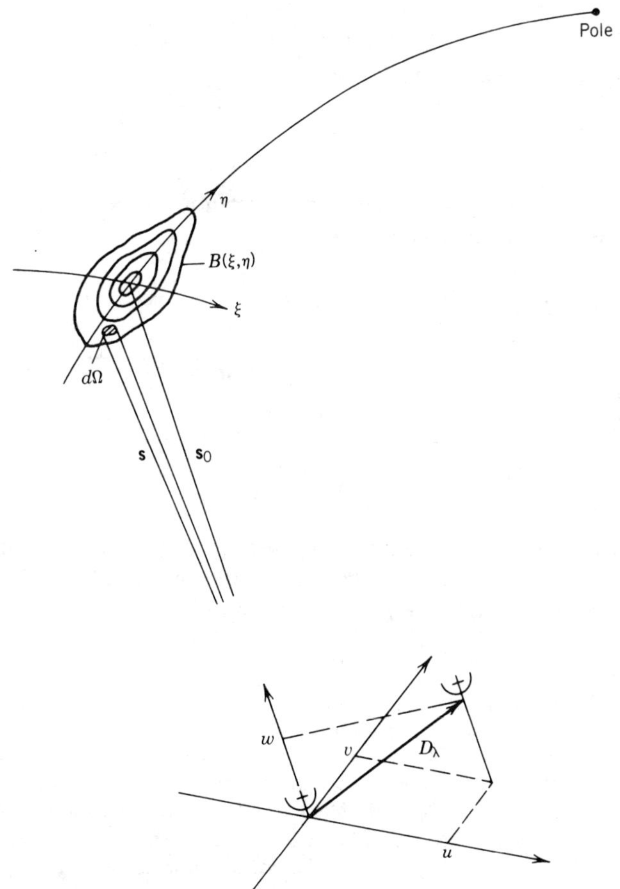

Figure 4.1   Geometrical relationship between a source under observation $B(\xi, \eta)$ and an interferometer or one antenna pair of an array. The spacing vector $\mathbf{D}_\lambda$ of the antennas has components (u, v, w).

but if they differ $A_N(\boldsymbol{\sigma})$ is the geometric mean of the beam patterns of the two antennas. The magnitude of the baseline vector is measured in wavelengths at the center frequency of the observing band, and the baseline has components (u, v, w) in a right-handed coordinate system where u and v are measured in a plane normal to the direction of interest, v toward the north as defined by the plane through the origin, the source, and the pole, and u toward the east. The component w is measured in the direction $\mathbf{s}_0$ which is the center of the field of view. This point defines a *phase reference position* for the observation, and the phase of the visibility function $\mathscr{V}$ is measured relative to the phase for a hypothetical point source at the field center. On Fourier transformation, the phase reference position becomes the origin of the derived brightness distribution $B(\xi, \eta)$, where ξ and η are direction cosines measured with respect to the axes u and v. In terms of these coordinates we obtain

$$\left. \begin{aligned}
\mathbf{D}_\lambda \cdot \mathbf{s}_0 &= w \\[2mm]
\mathbf{D}_\lambda \cdot \mathbf{s} &= \left(u\xi + v\eta + w\sqrt{1 - \xi^2 - \eta^2} \right) \\[2mm]
d\Omega &= \frac{d\xi\, d\eta}{\sqrt{1 - \xi^2 - \eta^2}} ,
\end{aligned} \right\} \tag{4.2}$$

where $\sqrt{1 - \xi^2 - \eta^2}$ is equal to the third direction cosine ζ measured with respect to the w axis.[†] Thus from Eq. (4.1):

$$\mathscr{V}(u, v, w) = \int_{-\infty}^{\infty} \int_{-\infty}^{\infty} A_N(\xi, \eta) B(\xi, \eta)$$

$$\times \exp\left\{ -j2\pi \left[u\xi + v\eta + w\left(\sqrt{1 - \xi^2 - \eta^2} - 1 \right) \right] \right\}$$

$$\times \frac{d\xi\, d\eta}{\sqrt{1 - \xi^2 - \eta^2}} . \tag{4.3}$$

Note that measurement of angular position with respect to the direction $\mathbf{s}_0$ introduces a factor $e^{j2\pi w}$ on the right-hand side of Eq. (4.3). The function $A_N B$ in Eq. (4.3) is zero for $\xi^2 + \eta^2 \geq 1$, and in practice it usually falls to very

---

[†] The expression for $d\Omega$ is obtained by considering the unit sphere centered on the (u, v, w) origin. A point P on the sphere with coordinates (u, v, w) is projected onto the (u, v) plane at $u = \xi$, $v = \eta$, and the increments $d\xi$, $d\eta$ define a column of square cross section running through $(u, v, 0)$ parallel to the w axis. The column makes an angle $\cos^{-1}\zeta$ with the normal to the spherical surface at P, and $d\Omega$ is equal to the surface area intersected by the column, which is $d\xi\, d\eta/\zeta$, or $d\xi\, d\eta/\sqrt{1 - \xi^2 - \eta^2}$. Alternatively, the solid angle can be expressed in polar coordinates as $d\Omega = \sin\theta\, d\theta\, d\phi$, where θ and ϕ are the polar and azimuthal angles in the (u, v, w) plane, that is, $\theta = \sin^{-1}\sqrt{\xi^2 + \eta^2}$ and $\phi = \tan^{-1}\eta/\xi$. Calculation of the Jacobian of the transformation from (θ, ϕ) coordinates to (ξ, η) coordinates gives the result $d\Omega = d\xi\, d\eta/\sqrt{1 - \xi^2 - \eta^2}$ (Apostol 1962).

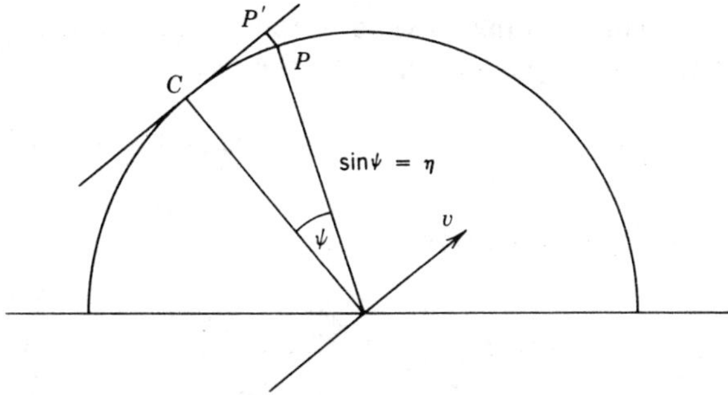

Figure 4.2   Mapping of the celestial sphere onto a plane, shown in one dimension. The position of the point P is measured in terms of the direction cosine η with respect to the v axis. When projected onto a plane surface with a scale linear in η, P appears at P' at a distance from the field center C proportional to $\sin \psi$.

low values for directions outside the field to be mapped as a result of the antenna beam pattern, the bandwidth pattern, or the finite size of the source. Thus we can extend the limits of integration to $\pm \infty$. Note, however, that Eq. (4.3) requires no small-angle assumptions. The reason why we use direction cosines rather than a linear measure of angle in interferometer theory is that they occur in the exponential term of this fundamental relationship.

The coordinate system (ξ, η) defined above is chosen because it is a convenient one in which to present a brightness distribution. It corresponds to the projection of the celestial sphere onto a plane that is a tangent at the field center, as shown in Fig. 4.2. The distance of any point in the map from the (ξ, η) origin is proportional to the sine of the corresponding angle on the sky, so for small fields distances on the map are closely proportional to the corresponding angles. The same relationship usually applies to the field of an optical telescope. For a detailed discussion of relationships on the celestial sphere and tangent planes see König (1962).

If all the measurements could be made with the antennas in a plane normal to the w direction so that $w = 0$, Eq. (4.3) would reduce to an exact two-dimensional Fourier transform. In general this is not possible, and we now consider ways in which the transform relationship can be applied. Recall first that the basis of the synthesis-mapping process is the measurement of $\mathscr{V}$ over a wide range of u and v. For a ground-based array this can be achieved by varying the length and direction of the antenna spacing and also by tracking the field-center position as the earth rotates. The rotation causes the projection of $\mathbf{D}_\lambda$ to move across the (u, v) plane. Thus an observation often lasts for 6–12 hr. As the earth's rotation carries the antennas through space, the baseline vector remains in a plane only if $\mathbf{D}_\lambda$ has no component parallel to the rotation axis, that is, the baseline is an east–west line on the earth's surface. In the general case there is a three-dimensional distribution of the measurements of $\mathscr{V}$. The simplest form of the transform relationship that can then be used is based on an approximation that is valid so long as the synthesized field is not too large. If ξ and η are small enough that the term

$$\left(\sqrt{(1 - \xi^2 - \eta^2)} - 1\right)w \simeq -\tfrac{1}{2}(\xi^2 + \eta^2)w \tag{4.4}$$

can be neglected, Eq. (4.3) becomes

$$\mathcal{V}(u,v,w) \simeq \mathcal{V}(u,v,0) = \int_{-\infty}^{\infty}\int_{-\infty}^{\infty} \frac{A_N(\xi,\eta)B(\xi,\eta)}{\sqrt{1 - \xi^2 - \eta^2}}\, e^{-j2\pi(u\xi + v\eta)}\, d\xi\, d\eta. \tag{4.5}$$

Thus for a restricted range of ξ and η, $\mathcal{V}(u,v,w)$ is approximately independent of w, and for the inverse transform we can write

$$\frac{A_N(\xi,\eta)B(\xi,\eta)}{\sqrt{1 - \xi^2 - \eta^2}} = \int_{-\infty}^{\infty}\int_{-\infty}^{\infty} \mathcal{V}(u,v)e^{j2\pi(u\xi + v\eta)}\, du\, dv. \tag{4.6}$$

With the above approximation it is usual to omit the w dependence and write the visibility as the two-dimensional function $\mathcal{V}(u,v)$. Note that the factor $\sqrt{1 - \xi^2 - \eta^2}$ in Eqs. (4.5) and (4.6) can be absorbed into the function $A_N(\xi,\eta)$, if desired. The approximation in Eq. (4.5) introduces a phase error equal to 2π times the neglected term, that is, approximately $\pi(\xi^2 + \eta^2)w$. Limitation of this error to some tolerable value places a restriction on the size of the synthesized field, which can be roughly estimated as follows. If the antennas track the source under observation down to low elevation angles, the values of w can approach the maximum spacings $(D_\lambda)_{\text{max}}$ in the array, as shown in Fig. 4.3. Also if the spatial frequencies measured are evenly distributed out to the maximum spacing, the synthesized beamwidth θ_b is approximately equal to $(D_\lambda)_{\text{max}}^{-1}$. Thus the maximum phase error is approximately $\pi(\theta_f/2)^2\theta_b^{-1}$, where θ_f is the width of the synthesized field. The condition that

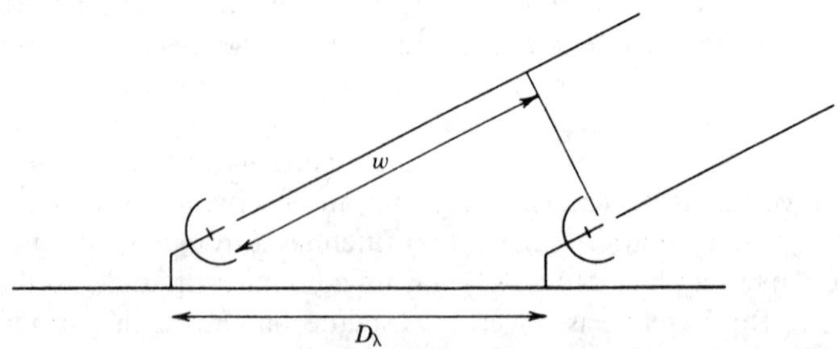

Figure 4.3   When observations are made at a low angle of elevation, and at an azimuth close to that of the baseline, the spacing component w becomes comparable to the baseline length D_λ.

no phase errors can exceed, say, 0.1 rad then requires that

$$\theta_f < \tfrac{1}{3}\sqrt{\theta_b}\,,\tag{4.7}$$

where the angles are measured in radians. For example, if $\theta_b = 1$ arcsec, $\theta_f < 2.5$ arcmin. Most synthesis mapping in astronomy has been performed within this restriction, but ways of obtaining larger maps will be discussed.

We now return to the case of arrays with east–west spacings only, and discuss further the conditions for which we can put $w = 0$, and the resulting effects. Let us first rotate the (u, v, w) coordinate system about the u axis until the w axis points toward the pole as shown in Fig. 4.4. We indicate by primes the quantities measured in this system. The (u', v') axes lie in a plane parallel to the earth's equator. The east–west antenna spacings contain components in this plane only (i.e. $w' = 0$), and as the earth rotates the spacing vectors sweep out circles concentric with the (u', v') origin. From Eq. (4.3) we can write

$$\mathscr{V}(u', v') = \int_{-\infty}^{\infty}\int_{-\infty}^{\infty} A_N(\xi', \eta')\, B(\xi', \eta')\, e^{-j2\pi(u'\xi' + v'\eta')}\, \frac{d\xi'\, d\eta'}{\sqrt{1 - \xi'^2 - \eta'^2}}\,,\tag{4.8}$$

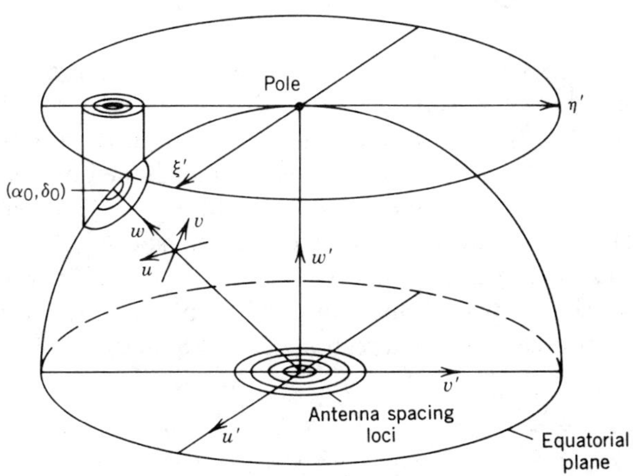

Figure 4.4   The (u', v', w') coordinate system for an east–west array. The (u', v') plane is the equatorial plane and the antenna-spacing vectors trace out arcs of concentric circles as the earth rotates. Note that the directions of the u' and v' axes are chosen so that the v' axis lies in the plane containing the pole, the observer, and the point under observation (α_0, δ_0). In Fourier transformation from the (u', v') to the (ξ', η') planes the celestial hemisphere is mapped as a projection onto the tangent plane at the pole. The (u, v, w) coordinates for observation in the direction (α_0, δ_0) are also shown.

where (ξ', η') are direction cosines measured with respect to (u', v'). Equation (4.8) holds for the whole hemisphere above the equatorial plane. The inverse transformation yields

$$\frac{A_N(\xi', \eta') B(\xi', \eta')}{\sqrt{1 - \xi'^2 - \eta'^2}} = \int_{-\infty}^{\infty} \int_{-\infty}^{\infty} \mathscr{V}(u', v') e^{j2\pi(u'\xi' + v'\eta')} \, du' \, dv'. \quad (4.9)$$

In this mapping the hemisphere is projected onto the tangent plane at the pole, as shown in Fig. 4.4. In practice, however, a map is usually confined to a small area within the antenna beams. In the vicinity of such an area, centered at right ascension and declination (α_0, δ_0), angular distances in the map are compressed by a factor $\sin \delta_0$ in the η' dimension. Also in mapping the (α_0, δ_0) vicinity it is convenient if the origin of the angular position variables is shifted to (α_0, δ_0) to facilitate use of the fast algorithm for Fourier transformation (see Section 4.10). Expansion of the scale and shift of the origin can be accomplished by the coordinate transformation

$$\xi = \xi', \qquad \eta'' = (\eta' - \cos \delta_0)\operatorname{cosec} \delta_0. \quad (4.10)$$

If we write $F(\xi', \eta')$ for the left-hand side of Eq. (4.9),

$$F(\xi', \eta') \rightleftharpoons \mathscr{V}(u', v'), \quad (4.11)$$

and

$$F\left[\xi', (\eta' - \cos \delta_0)\operatorname{cosec} \delta_0\right] \rightleftharpoons |\sin \delta_0| \mathscr{V}(u', v'\sin \delta_0) e^{-j2\pi v'\cos \delta_0}.$$

$$(4.12)$$

Equation (4.12) follows from the behavior of Fourier pairs with change of variable and involves the application of the similarity and shift theorems (e.g., Bracewell 1965). The coordinates $(u', v'\sin \delta_0)$ on the right-hand side of Eq. (4.12) represent the projection of the equatorial plane onto the (u, v) plane which is normal to the direction (α_0, δ_0). In the (u, v, w) system $u = u'$ and $v = v'\sin \delta_0$. The coordinate w for these observations is equal to $-v'\cos \delta_0$, so $e^{-j2\pi v'\cos \delta_0}$ in Eq. (4.12) is the same factor that we found in Eq. (4.3) as the phase term that is introduced when the visibility phase is measured relative to that for a point source in the w direction. Thus Eq. (4.9) becomes

$$\frac{A_N(\xi, \eta'') B(\xi, \eta'')}{\sqrt{1 - \xi^2 - \eta''^2}} = \int_{-\infty}^{\infty} \int_{-\infty}^{\infty} \mathscr{V}(u', v'\sin \delta_0) |\sin \delta_0| e^{-j2\pi v'\cos \delta_0}$$

$$\times e^{j2\pi(u'\xi' + v'\eta')} \, du' \, dv'$$

$$= \int_{-\infty}^{\infty} \int_{-\infty}^{\infty} \mathscr{V}(u, v) e^{j2\pi(u\xi + v\eta'')} \, du \, dv. \quad (4.13)$$

A similar analysis is given by Brouw (1971).

The derivation of Eq. (4.13) from Eq. (4.9) involves a redefinition of the η coordinate, but no approximations. Equation (4.13) is of the same form as Eq. (4.6) in which the term in Eq. (4.4) was neglected. Thus if we apply the mapping scheme of Eq. (4.6), which is based on omitting this term, to observations made with an east–west array, the phase errors introduced distort the map in a way that corresponds exactly to the change of definition of the η variable to η''. Since η'' is derived from a direction cosine measured from the v' axis in the equatorial plane, there is a progressive change in the north–south angular scale over the map. The factor $\csc\delta_0$ in Eq. (4.10) establishes the correct angular scale at the center of the map, but this simple correction is acceptable only for small fields. The crucial point to note here is that when visibility data measured in a plane are projected into (u, v, w) coordinates, w is a linear function of u and v, (and a linear function of v alone for east–west baselines). Hence the phase error $\pi(\xi^2 + \eta^2)w$ is linear in u and v. Phase errors of this kind have the effect of introducing position shifts in the resulting map, but there remains a one-to-one correspondence between points in the map and on the sky. The effect is simply to produce a predictable, and hence correctable, distortion of the coordinates.

It is clear from Fig 4.4 that if all the measurements lie in the (u', v') plane, then the values of v in the (u, v) plane become seriously foreshortened for directions close to the celestial equator. To obtain two-dimensional resolution in such directions requires components of antenna spacing parallel to the earth's axis. The design of such arrays is discussed in Chapter 5. If the earth's rotation is used to improve the (u, v) coverage the observations cannot lie in a plane, so we return to the three-dimensional equation (4.3) or to the problem of the restriction of the synthesized field that results from the approximation in Eq. (4.5). Several approaches have been suggested and are outlined as follows.

1.   Large maps can be constructed as mosaics of smaller ones, but care must be taken to avoid the effects of aliasing of false detail into the smaller maps. The small maps should all be derived as projections onto the same tangent plane.

2.   In Eq. (4.3) the term $\sqrt{1 - \xi^2 - \eta^2}$ can be replaced by the third direction cosine ζ defined with respect to the w axis. The measured quantity, that is, the visibility relative to that for a point source at the phase reference position, then becomes

$$\mathcal{V}(u, v, w) = \int_{-\infty}^{\infty} \int_{-\infty}^{\infty} \zeta^{-1} A_N(\xi, \eta) B(\xi, \eta) e^{-j2\pi[u\xi + v\eta + w(\zeta - 1)]} \, d\xi \, d\eta.$$

$$(4.14)$$

The small-field approximation would here appear as $(\zeta - 1) = 0$. The inverse relationship can be written as a three-dimensional Fourier transform with integration with respect to v, u, and w, and this leads to $A_N B$ as a function of

(ξ, η, ζ). A two-dimensional brightness distribution can then be taken on the unit sphere in (ξ, η, ζ).

3.   Most connected-element arrays are constructed on ground sufficiently flat that at any instant all the baselines are approximately in a plane, and they remain close to this plane for observations of short duration. As pointed out in the discussion of east–west baselines, visibility measurements made in a plane can be projected into the (u, v) plane with only a calculable distortion of the coordinates in the resulting map. It is therefore possible to divide a long observation into a series of shorter ones, to each of which a separate coordinate correction is applied before they are combined.

4.   If the brightness distribution within the area to be mapped is known in advance to some degree of accuracy, this information can be used to derive $\mathscr{V}(u, v, 0)$ from the measured $\mathscr{V}(u, v, w)$ with improved accuracy. Thus a map with some distortion can be obtained from Eq. (4.6) and used to improve the estimation of $\mathscr{V}(u, v, 0)$. A more accurate map is then obtained, and, in principle, this process can be iterated until the desired accuracy is obtained.

Methods 1 to 3 above have been discussed by Clark (1982) and methods 3 and 4 by Bracewell (1979, 1984a). Up to the mid-1980s these methods have been little more than points for discussion, since   most measurements have been performed within the restrictions for which Eq. (4.6) is adequate. Model calculations based on method 3 have been made by Hudson (1977).

4.2   ANTENNA SPACING COORDINATES AND (u, v) LOCI

Various coordinate systems are used to specify the relative positions of the antennas in an array, and of these one of the more convenient for terrestrial arrays is shown in Fig. 4.5. A right-handed Cartesian coordinate system is used where X and Y are measured in a plane parallel to the earth's equator, X in the meridian plane (defined as the plane through the poles of the earth and the reference point in the array), Y toward the east, and Z is measured toward the north pole. In terms of hour angle H and declination δ, (X, Y, Z) are measured toward $(H = 0, \delta = 0)$, $(H = -6^{\mathrm{h}}, \delta = 0)$ and $(\delta = 90°)$, respectively. If $(X_\lambda, Y_\lambda, Z_\lambda)$ are the components of $\mathbf{D}_\lambda$ in the (X, Y, Z) system, the components (u, v, w) are given by

$$
\begin{bmatrix} u \\ v \\ w \end{bmatrix} = \begin{bmatrix} \sin H & \cos H & 0 \\ -\sin \delta \cos H & \sin \delta \sin H & \cos \delta \\ \cos \delta \cos H & -\cos \delta \sin H & \sin \delta \end{bmatrix} \begin{bmatrix} X_\lambda \\ Y_\lambda \\ Z_\lambda \end{bmatrix}. \tag{4.15}
$$

Here (H, δ) are usually the hour angle and declination of the phase reference position. (In VLBI observations it is customary to set the X axis in the Greenwich meridian, in which case H is measured with respect to that

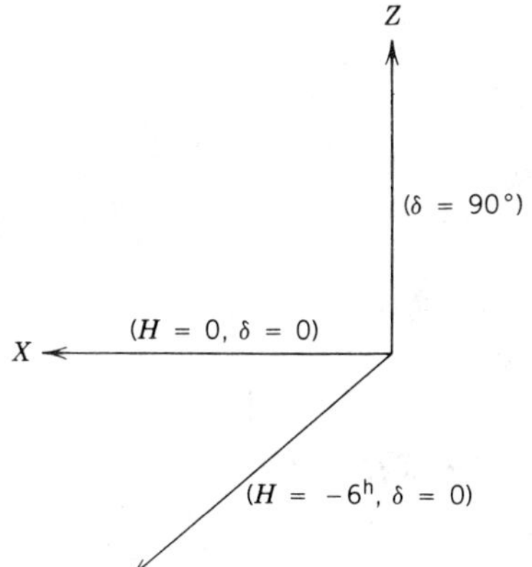

Figure 4.5 The (X, Y, Z) coordinate system for specification of relative positions of antennas. Directions of the axes are specified in terms of hour angle H and declination δ.

meridian rather than a local one.) The elements of the transformation matrix above are the direction cosines of the (u, v, w) axes with respect to the (X, Y, Z) axes and can easily be derived from the relationships in Fig. 4.6. Another method of specifying the baseline vector is in terms of its length, D, and the hour angle and declination, (h, d), of the intersection of the baseline direction with the northern celestial hemisphere. The coordinates in the

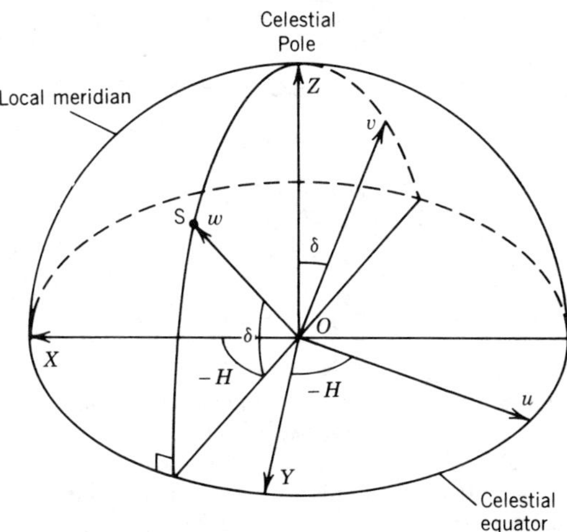

Figure 4.6 Relationships between the (X, Y, Z) and (u, v, w) coordinate systems. The (u, v, w) system is defined for observation in the direction of the point S which has hour angle and declination H and δ. As shown S is in the eastern half of the hemisphere and H is therefore negative. The direction cosines in the transformation matrix in Eq. (4.15) follow from the relationships in this diagram. The relationship in Eq. (4.16) can also be derived if we let S represent the direction of the baseline and put the baseline coordinates (h, d) for (H, δ).

(X, Y, Z) system are then given by

$$\begin{bmatrix} X \\ Y \\ Z \end{bmatrix} = D \begin{bmatrix} \cos d \cos h \\ -\cos d \sin h \\ \sin d \end{bmatrix}. \tag{4.16}$$

The coordinates in the (u, v, w) system are, from Eqs. (4.15) and (4.16),

$$\begin{bmatrix} u \\ v \\ w \end{bmatrix} = D_\lambda \begin{bmatrix} \cos d \sin(H - h) \\ \sin d \cos \delta - \cos d \sin \delta \cos(H - h) \\ \sin d \sin \delta + \cos d \cos\delta \cos(H - h) \end{bmatrix}. \tag{4.17}$$

The (D, h, d) system was used more widely in the earlier literature, particularly for instruments involving only two antennas: see, for example, Rowson (1963) for an early application of Eq. (4.17) to a tracking interferometer.

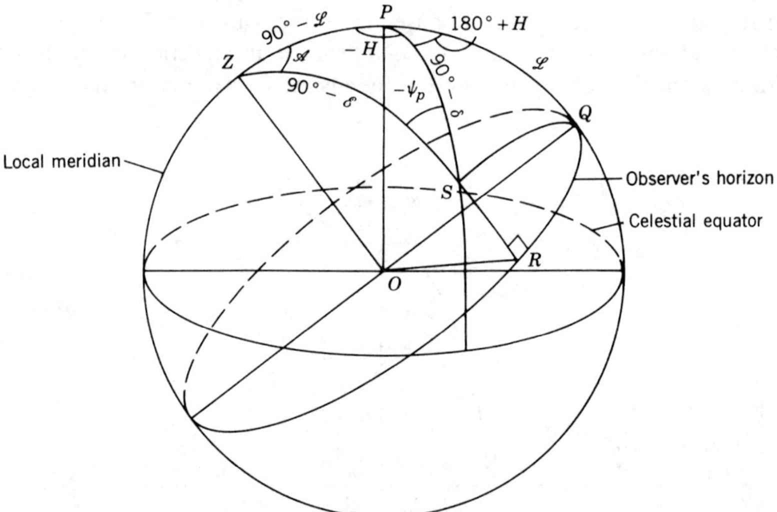

Figure 4.7   Relationship between the celestial coordinates (H, δ) and the elevation and azimuth $(\mathscr{E}, \mathscr{A})$ of a point S as seen by an observer at latitude $\mathscr{L}$. P is the celestial pole and Z the observer's zenith. The parallactic angle ψ_p is the position angle of the observer's vertical on the sky measured from north toward east. The lengths of the arcs measured in terms of angles subtended at the center of the sphere O are as follows:

$$ZP = 90° - \mathscr{L} \quad PQ = \mathscr{L} \qquad SR = \mathscr{E} \quad RQ = \mathscr{A}$$
$$SZ = 90° - \mathscr{E} \quad SP = 90° - \delta \quad SQ = \cos^{-1}(\cos \mathscr{E} \cos \mathscr{A})$$

The three equations in (A4.1) can be obtained by application of the sine and cosine rules for spherical triangles to ZPS and PQS. Note that with S in the eastern half of the observer's sky, as shown, H and ψ_p are negative.

When the (X, Y, Z) components of a new baseline are first established, the usual practice is to determine the elevation $\mathscr{E}$, azimuth $\mathscr{A}$, and length of the baseline by field surveying techniques. The relationship between $(\mathscr{E}, \mathscr{A})$ and other coordinate systems is shown in Fig. 4.7 and Appendix 4.1. For latitude $\mathscr{L}$, using Eqs. (4.16) and (A4.2), we obtain

$$\begin{bmatrix} X \\ Y \\ Z \end{bmatrix} = D \begin{bmatrix} \cos \mathscr{L} \sin \mathscr{E} - \sin \mathscr{L} \cos \mathscr{E} \cos \mathscr{A} \\ \cos \mathscr{E} \sin \mathscr{A} \\ \sin \mathscr{L} \sin \mathscr{E} + \cos \mathscr{L} \cos \mathscr{E} \cos \mathscr{A} \end{bmatrix}. \tag{4.18}$$

Examination of Eqs. (4.15) or (4.17) shows that the locus of the projected antenna-spacing components u and v defines an ellipse with hour angle as the variable. Thus from Eq. (4.15),

$$u^2 + \left(\frac{v - Z_\lambda \cos \delta_0}{\sin \delta_0} \right)^2 = X_\lambda^2 + Y_\lambda^2, \tag{4.19}$$

where (H_0, δ_0) is the phase reference position. In the (u, v) plane Eq. (4.19) defines an ellipse with the semimajor axis equal to $\sqrt{X_\lambda^2 + Y_\lambda^2}$, and the semiminor axis equal to $\sin \delta_0 \sqrt{X_\lambda^2 + Y_\lambda^2}$ as in Fig. 4.8a. The ellipse is centered on the v axis at $(u, v) = (0, Z_\lambda \cos \delta_0)$. The arc of the ellipse that is traced out during any observation depends upon the azimuth, elevation, and latitude of the baseline, the declination of the source, and the range of hour angle covered,

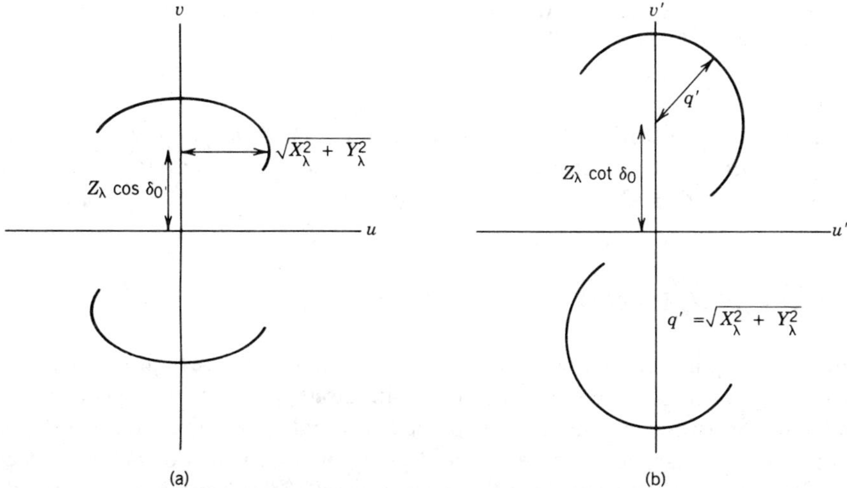

Figure 4.8   (a) Spacing-vector locus in the (u, v) plane from Eq. (4.19). (b) Spacing-vector locus in the (u', v') plane from Eq. (4.22). The lower arc in each diagram represents the locus of conjugate values of visibility. Unless the source is circumpolar the cutoff at the horizon limits the lengths of the arcs.

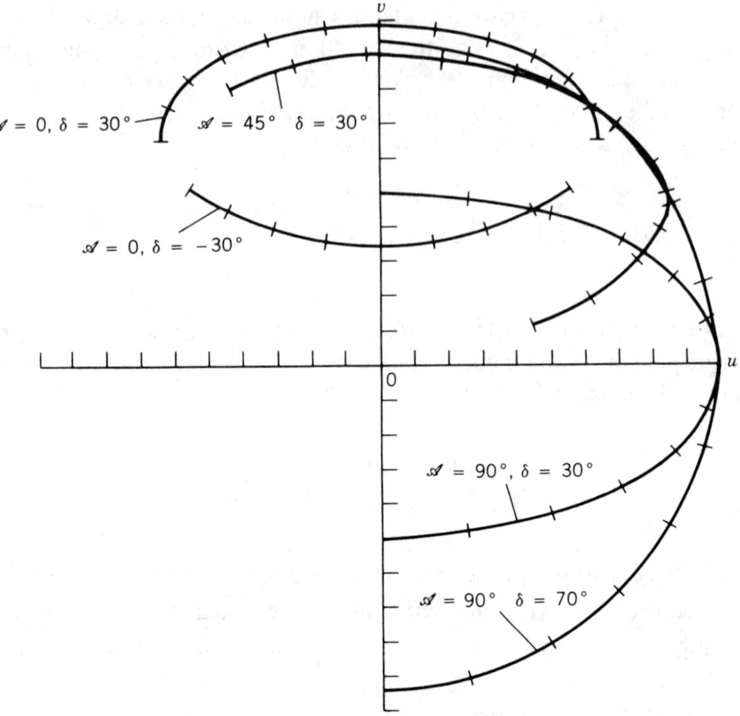

Figure 4.9 Examples of (u, v) loci to show the variation with baseline azimuth $\mathscr{A}$ and observing declination δ (the baseline elevation $\mathscr{E}$ is zero). The baseline length in all cases is equal to the length of the axes measured from the origin. The tracking range is -4 to $+4$ hr for $\delta = -30°$, and -6 to $+6$ hr in all other cases. Marks along the loci indicate 1-hr intervals in tracking. Note the change in ellipticity for east–west baselines ($\mathscr{A} = 90°$) with $\delta = 30°$ and with $\delta = 70°$. The loci are calculated for latitude $40°$.

as illustrated in Fig. 4.9. Since $\mathscr{V}(-u, -v) = \mathscr{V}^*(u, v)$, any observation supplies simultaneous measurements on two arcs, which are part of the same ellipse only if $Z_\lambda = 0$.

4.3 (u', v') PLANE

The (u', v') plane, which was introduced in discussing the response of an array with east–west baselines, is also useful in discussing certain aspects of the behavior of arrays in general. This plane is normal to the direction of the pole and can be envisaged as the equatorial plane of the earth. For non east–west baselines we can also consider the projection of the spacing vectors onto the (u', v') plane. All such vectors sweep out circular loci as the earth rotates. The spacing components in the (u', v') plane are derived from those in the (u, v) plane by the simple transformation $u' = u$, $v' = v \operatorname{cosec} \delta_0$. In terms of the components of the baseline $(X_\lambda, Y_\lambda, Z_\lambda)$ for two antennas, we obtain from Eq. (4.15)

$$u' = X_\lambda \sin H_0 + Y_\lambda \cos H_0 \tag{4.20}$$

$$v' = -X_\lambda \cos H_0 + Y_\lambda \sin H_0 + Z_\lambda \cot \delta_0. \tag{4.21}$$

The loci are circles centered on $(0, Z_\lambda \cot \delta_0)$, with radii q' given by

$$q'^2 = u'^2 + (v' - Z_\lambda \cot \delta_0)^2 = X_\lambda^2 + Y_\lambda^2, \tag{4.22}$$

as shown in Fig. 4.8b. The projected spacing vectors that generate the loci rotate with constant angular velocity ω_e, the rotation velocity of the earth, which is easier to visualize than the elliptic motion in the (u, v) plane. In particular, problems involving the effect of time such as the averaging of visibility data are conveniently dealt with in the (u', v') plane. Examples of its use will be found in Section 4.5 and in Chapters 6 and 14. In Fourier transformation the conjugate variables of (u', v') are (ξ', η'), where $\xi' = \xi$ and $\eta' = \eta \sin \delta_0$, that is, the map plane is compressed by a factor $\sin \delta_0$ in the η direction.

4.4 FRINGE FREQUENCIES

The component w of the baseline represents the path difference to the two antennas for a plane wave incident from the phase reference position. The corresponding time delay is w/ν_0, where ν_0 is the center frequency of the observing band. The relative phase of the signals at the two antennas changes by 2π radians when w changes by unity. Thus the frequency of the oscillations at the output of the correlator that combines the signals is

$$\frac{dw}{dt} = \frac{dw}{dH}\frac{dH}{dt} = -\omega_e[X_\lambda \cos \delta \sin H + Y_\lambda \cos \delta \cos H] = -\omega_e u \cos \delta,$$

$$\tag{4.23}$$

where $dH/dt = \omega_e = 2\pi/(24 \times 60^2) = 7.27 \times 10^{-5}$ radians per sidereal second. The sign of dw/dt indicates whether the phase is increasing or decreasing with time. The above result applies to the case where the signals suffer no time-varying instrumental phase changes between the antennas and the correlator inputs. In an array in which the antennas track a source, time delays to compensate for the space path differences w are usually applied under computer control to maintain correlation of the signals. If an exact compensating delay were introduced in the radiofrequency section of the receivers the relative phases of the signals at the correlator input would remain constant, and the correlator output would show no fringes. However, the compensating delays are usually introduced at an intermediate frequency of which the band center ν_d is usually much less than the radiofrequency ν_0. The adjustment of the compensating delay introduces a rate of phase change $2\pi\nu_d(dw/dt)/\nu_0 = -\omega_e u(\cos\delta)\nu_d/\nu_0$. The resulting fringe frequency at the correlator output is

$$\nu_f = \frac{dw}{dt}\left(1 \mp \frac{\nu_d}{\nu_0}\right) = -\omega_e u\cos\delta\left(1 \mp \frac{\nu_d}{\nu_0}\right), \qquad (4.24)$$

where the negative sign refers to upper-sideband reception and the positive sign to lower-sideband: these distinctions and the double-sideband case are explained in Chapter 6. Note that from Eq. (4.17) the right-hand side of Eq. (4.24) is equal to $-\omega_e D\cos d\cos\delta\sin(H - h)(\nu_0 \mp \nu_d)/c$.

4.5 VISIBILITY FREQUENCIES

As explained in Section 4.1, the phase of the complex visibility is measured with respect to that of a hypothetical point source at the phase reference position. The fringe-frequency variations do not appear in the visibility function, but slower variations occur that depend on the position of the radiating sources within the field. We now examine the maximum temporal frequency of the visibility variations. Consider a point source represented by the delta function $\delta(\xi_1, \eta_1)$. The visibility function is the Fourier transform of $\delta(\xi_1, \eta_1)$ which is

$$e^{-j2\pi(u\xi_1 + v\eta_1)} = \cos 2\pi(u\xi_1 + v\eta_1) - j\sin 2\pi(u\xi_1 + v\eta_1). \qquad (4.25)$$

This expression represents two sets of sinusoidal corrugations—one real and one imaginary. These corrugations are shown in (u', v') coordinates in Fig. 4.10, where the arguments of the trigonometric functions in Eq. (4.25) become $2\pi(u'\xi_1 + v'\eta_1\sin\delta_0)$. The frequency of the corrugations in terms of cycles per unit distance in the (u', v') plane is ξ_1 in the u' direction, $\eta_1\sin\delta_0$ in the v' direction, and

$$r_1' = \sqrt{\xi_1^2 + \eta_1^2\sin^2\delta_0} \qquad (4.26)$$

in the direction of most rapid variations. Expression (4.26) is maximized at the pole and then becomes equal to r_1 which is the radial distance of the source from the (ξ, η) origin. For any antenna pair the spatial-frequency locus in the (u', v') plane is a circle of radius q' generated by a vector rotating with angular velocity ω_e, q' being defined in Eq. (4.22). From Fig. 4.10 it is clear that the temporal variation of the measured visibility is greatest at the point P and is equal to $r_1'\omega_e q'$. This is a useful result, since if r_1 represents a position at the

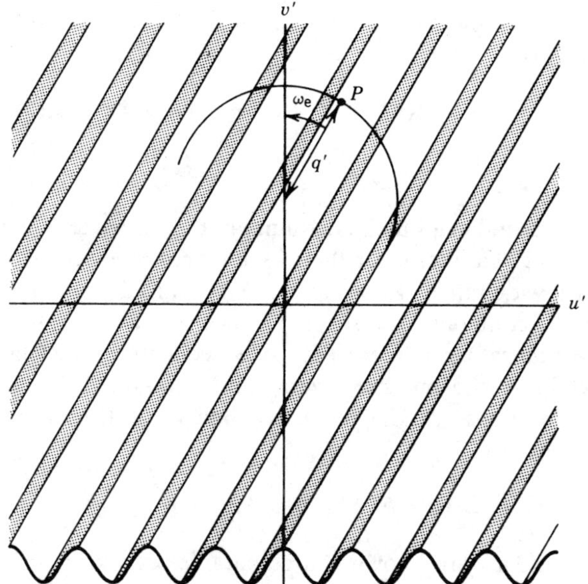

Figure 4.10 The (u', v') plane showing sinusoidal corrugations that represent the visibility of a point source. For simplicity only the real part of the visibility is included. The most rapid variation in the visibility is encountered at the point P where the direction of the spacing locus is normal to the ridges in the visibility. ω_e is the rotation velocity of the earth.

edge of the field to be mapped it indicates that to follow the most rapid variations the visibility must be sampled at time intervals sufficiently small compared with $(\omega_e r_1' q')^{-1}$. Also we may wish to alternate between two frequencies or polarizations during an observation, and these changes must be made on a similarly short time scale. Note that this requirement is also covered by the sampling theorem in Section 4.11.

4.6 CALIBRATION OF THE BASELINE

The position parameters (X, Y, Z) for each antenna relative to a common reference point can usually be established to a few centimeters or millimeters by a conventional engineering survey. Except at long wavelengths, the accuracy required is greater than this. We must be able to compute the phase at any hour angle for a point source at the phase reference position to an accuracy of, say, 1° and subtract it from the observed phase. This reference phase is represented by the factor $e^{j2\pi w}$ in Eq. (4.3) and it is therefore necessary to calculate w to 1/360 of the observing wavelength. The baseline parameters can be obtained to the required accuracy from observations of calibration sources for which the positions are accurately known. The phase of such a calibrator observed at the phase reference position (H_0, δ_0) should ideally be zero.

However, if practical uncertainties are taken into account the measured phase is, from Eq. (4.15),

$$2\pi \Delta w + \phi_{\text{in}} = 2\pi(\cos \delta_0 \cos H_0 \, \Delta X_\lambda - \cos \delta_0 \sin H_0 \, \Delta Y_\lambda + \sin \delta_0 \, \Delta Z_\lambda) + \phi_{\text{in}},$$

$$(4.27)$$

where the prefix Δ indicates the uncertainty in the associated quantity, and ϕ_{in} is an instrumental phase term for the two antennas involved. If a calibrator is observed over a wide range of hour angle, ΔX_λ and ΔY_λ can be obtained from the even and odd components, respectively, of the phase variation with H_0. To measure ΔZ_λ calibrators at more than one declination must be included. A possible procedure is to observe several calibrators at different declinations repeating a cycle of observations for several hours. For the k th observation we can write from Eq. (4.27)

$$a_k \, \Delta X_\lambda + b_k \, \Delta Y_\lambda + c_k \, \Delta Z_\lambda + \phi_{\text{in}} = \phi_k, \qquad (4.28)$$

where a_k, b_k, and c_k are known source parameters and ϕ_k is the measured phase. The calibrator source position need not be accurately known since the phase measurements can be used to estimate both the source positions and the baselines. Techniques for this analysis are discussed in Chapter 12. In practice, the instrumental phase ϕ_{in} will vary slowly with time: instrumental stability is discussed in Chapter 7. Also there will be atmospheric phase variations which are discussed in Chapter 13. These effects set the final limit on the attainable accuracy in observing both calibrators and sources under investigation.

Measurement of baseline parameters to an accuracy of order 1 part in 10^7 (e.g., 3 mm in 30 km) implies timing accuracy of order $10^{-7} \, \omega_e^{-1} \simeq 1$ msec. Timekeeping is discussed in Chapter 12.

4.7   ANTENNA MOUNTS

In discussing the dependance of the measured phase upon the baseline components we have ignored any effects introduced by the antennas, which is tantamount to assuming that the antennas are identical and their effects on the signals cancel out. This, however, is only approximately true. In most synthesis arrays the antennas must have collecting areas of tens or hundreds of square meters for reasons of sensitivity. These large structures must be capable of accurately tracking a radio source across the sky. Tracking antennas are almost always mounted on equatorial or altazimuth mounts, as illustrated in Fig. 4.11. In an equatorial mount the polar axis is parallel to the earth's axis of rotation, and to track a source requires only that the antenna be turned about the polar axis at the sidereal rate. Equatorial mounts are mechanically more difficult to construct than altazimuth ones and are found mainly on antennas built prior to the introduction of computers for control and coordinate conversion.

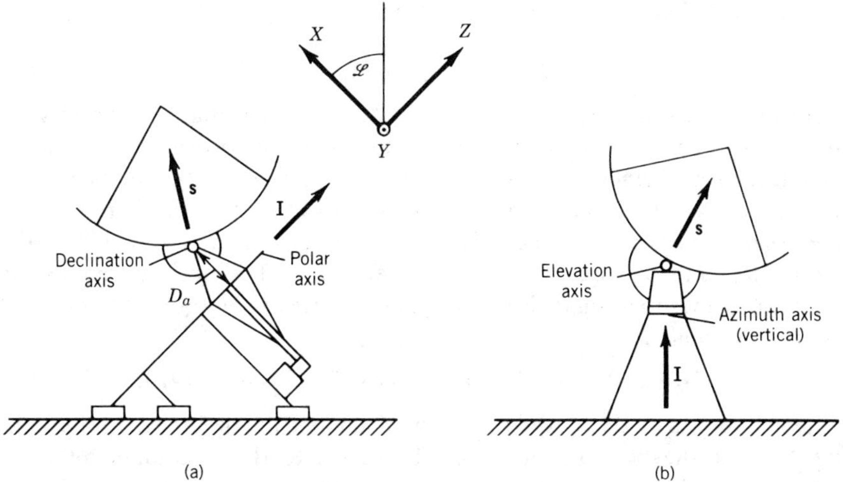

Figure 4.11 Schematic diagrams of antennas on (a) equatorial and (b) altazimuth mounts. In the positions shown the declination and elevation axes are normal to the plane of the page. In the equatorial mount there is a distance D_a between the two rotation axes, but in the altazimuth mount the axes often intersect, as shown.

In most tracking arrays used in radio astronomy the antennas are circularly symmetrical reflectors. A desirable feature is that the axis of symmetry of the reflecting surface should intersect both the rotation axes of the mount. If this is not the case, pointing motions will cause the antenna to have a component of motion along the direction of the beam. It is then necessary to take account of phase changes associated with small pointing corrections which differ from one antenna to another. In most antenna mounts, however, whether of equatorial or altazimuth type, the reflector axis intersects the rotation axes with sufficient precision that phase errors of this type are negligible.

It is convenient though not essential that the two rotation axes of the mount should intersect. The intersection point then provides an appropriate reference point for defining the baseline between antennas, since whatever direction the antenna points its aperture plane is always the same distance from that point as measured along the axis of the beam. In most large, equatorially mounted antennas the polar and declination axes do not intersect. In many cases there is an offset of several meters between the polar and declination axes. Wade (1970) has considered the implication of this offset for high-accuracy phase measurements and shown that it is necessary to take account of variations in the offset distance and in the accuracy of alignment of the polar axis. These results can be obtained as follows. Let $\mathbf{l}$ and $\mathbf{s}$ be unit vectors in the direction of the polar axis and the direction of the source under observation, respectively, and let $\mathbf{D}_a$ be the spacing vector between the two axes measured perpendicular to $\mathbf{l}$: see Fig. 4.11a. The quantity that we need to compute is the projection of $\mathbf{D}_a$ in the direction of observation, $\mathbf{D}_a \cdot \mathbf{s}$. Since $\mathbf{D}_a$ is perpendicular to $\mathbf{I}$, the cosine of the angle between $\mathbf{D}_a$ and $\mathbf{s}$ is $\sqrt{1-(\mathbf{I}\cdot\mathbf{s})^2}$. Thus,

$$\mathbf{D}_a\cdot\mathbf{s}=D_a\sqrt{1-(\mathbf{I}\cdot\mathbf{s})^2},\qquad(4.29)$$

where D_a is the magnitude of $\mathbf{D}_a$. In the (X,Y,Z) coordinate system in which the baseline components are measured, $\mathbf{I}$ has direction cosines (I_X,I_Y,I_Z) and $\mathbf{s}$ has direction cosines given by the transformation matrix on the right-hand side of Eq. (4.16), h and d being replaced by H and δ which refer to the direction of observation. If the polar axis is correctly aligned to within about 1 arcmin, I_X and I_Y are of order 10^{-3} and $I_Z\simeq1$. Thus we can use the direction cosines to evaluate Eq. (4.29), and ignoring second-order terms in H_X and I_Y we obtain

$$\mathbf{D}_a\cdot\mathbf{s}=D_a(\cos\delta-I_X\sin\delta\cos H+I_Y\sin\delta\sin H).\qquad(4.30)$$

If the magnitude of $\mathbf{D}_a$ is expressed in wavelengths, the difference in the values of $\mathbf{D}_a\cdot\mathbf{s}$ for the two antennas must be added to the w component of the baseline given by Eq. (4.15) when calculating the reference phase at the field center. To do this it is first necessary to determine the unknown constants in Eq. (4.30), which can be done by adding a term of the form $2\pi(\alpha\cos\delta_0+\beta\sin\delta_0\cos H_0+\gamma\sin\delta_0\sin H_0)$ to the right-hand side of Eq. (4.27) and extending the solution to include α, β, and γ. The result then represents the differences in the corresponding mechanical dimensions of the two antennas. Note that the terms in I_X and I_Y in Eq. (4.30) are important only when D_a is large. If D_a is no more than one wavelength it should be possible to ignore them.

The above result can be extended to the case of an altazimuth mount by letting $\mathbf{I}$ represent the direction of the azimuth axis as in Fig. 4.11b. Then $I_X=\cos(\mathscr{L}+\varepsilon)$, $I_y=\sin\varepsilon'$, and $I_Z=\sin(\mathscr{L}+\varepsilon)$, where $\mathscr{L}$ is the latitude and ε and ε' are, respectively, the tilt errors in the XZ plane and the plane containing the Y axis and the local vertical. The errors again should be quantities of order 10^{-3}. In many altazimuth mounts the axes are designed to intersect, and D_a represents only a structural tolerance. Thus we assume that D_a is small enough to allow terms in I_YD_a and εD_a to be ignored, and evaluation of Eq. (4.29) gives

$$\mathbf{D}_a\cdot\mathbf{s}=D_a\left[1-(\sin\mathscr{L}\sin\delta+\cos\mathscr{L}\cos\delta\cos H)^2\right]=D_a\cos\mathscr{E},$$

$$(4.31)$$

where $\mathscr{E}$ is the elevation of direction $\mathbf{s}$: see Eq. (A4.1) of Appendix 4.1. Correction terms of this form can be added to the expressions for the baseline calibration and for w.

4.8   BEAMWIDTH AND BEAMSHAPE EFFECTS

The interpretation of data taken with arrays containing antennas with non-identical beamwidths is not always a straightforward matter. Each antenna pair responds to an effective brightness distribution which is the product of the actual sky brightness and the geometric mean of the normalized beam profiles. If different pairs of antennas respond to different effective distributions then, in principle, the Fourier transform relationship between $B(\xi, \eta)$ and $\mathcal{V}(u, v)$ cannot be applied to the ensemble of observations. Mixed arrays are frequently used in VLBI when it is necessary to make use of antennas at widely separated locations which have been constructed for other purposes. However, in VLBI studies the source structure under investigation is very small compared with the widths of the antenna beams so the differences in the beams can usually be ignored. If cases arise where different beams are used and the source is not small compared with beamwidths, it is possible to restrict the measurements to the field defined by the narrowest beam by convolution of the visibility data with an appropriate function in the (u, v) plane.

A similar problem to that of unmatched beams occurs if the antennas have altazimuth mounts and the beam contours are not circularly symmetrical about the nominal beam axis. As a point in the sky is tracked using an altazimuth mount the beam rotates with respect to the sky about this nominal axis. This rotation does not occur for equatorial mounts. The angle between the vertical at the antenna and the direction of north at the point being observed (defined by the great circle through the point and the north pole) is the parallactic angle ψ_p in Fig. 4.7. Application of the sine rule to the spherical triangle ZPS gives

$$\frac{-\sin \psi_p}{\cos \mathcal{L}} = \frac{-\sin H}{\cos \mathcal{E}} = \frac{\sin \mathcal{A}}{\cos \delta}, \tag{4.32}$$

which can be combined with Eqs. (A4.1) or (A4.2) to express ψ_p as a function of $(\mathcal{A}, \mathcal{E})$ or (H, δ). If the beam has elongated contours, and width comparable to the source under observation, rotation of the beam causes the effective brightness distribution to vary with hour angle. However, in most tracking arrays the antenna beams are sufficiently circularly symmetrical that this problem rarely arises.

4.9 POLARIMETRY

Parameters Defining Polarization

Polarization measurements are very important in radio astronomy. For example, most synchrotron radiation shows a small degree of polarization which increases with angular resolution and helps to reveal the distribution of the magnetic fields within the source. As noted in Chapter 1, this polarization is generally linear (plane) and can vary in magnitude and position angle over the source. Polarization of radio emission also results from the Zeeman effect in atoms and molecules, cyclotron radiation and plasma oscillations in the solar atmosphere, and Brewster angle effects at planetary surfaces. The measure of polarization that is almost universally used in astronomy is the set of four parameters introduced by Sir George Stokes in 1852. We assume here that readers have some familiarity with the concept of Stokes parameters or can refer to one of numerous texts that describe them (e.g., Kraus and Carver 1973, Born and Wolf 1959).

Stokes parameters are related to the amplitudes of the components of the electric field, E_x and E_y, resolved in two perpendicular directions normal to the direction of propagation from a radiating point. Thus if E_x and E_y are represented by $\mathscr{E}_x(t)\cos[2\pi\nu t + \delta_x(t)]$ and $\mathscr{E}_y(t)\cos[2\pi\nu t + \delta_y(t)]$, respectively, Stokes parameters are defined as follows:

$$I = \left\langle \mathscr{E}_x^2(t) \right\rangle + \left\langle \mathscr{E}_y^2(t) \right\rangle$$

$$Q = \left\langle \mathscr{E}_x^2(t) \right\rangle - \left\langle \mathscr{E}_y^2(t) \right\rangle$$

$$U = 2\left\langle \mathscr{E}_x(t)\mathscr{E}_y(t)\cos\left[\delta_x(t) - \delta_y(t)\right] \right\rangle \qquad (4.33)$$

$$V = 2\left\langle \mathscr{E}_x(t)\mathscr{E}_y(t)\sin\left[\delta_x(t) - \delta_y(t)\right] \right\rangle,$$

where the angular brackets denote the expectation or time average. This averaging is necessary because in radio astronomy we are dealing with fields that vary with time in random manner. Of the four parameters, I is a measure of the total power in the wave, Q and U represent the linearly polarized component, and V represents the circularly polarized component. Stokes parameters can be converted to a measure of polarization with a more direct physical interpretation as follows:

$$m_\ell = \frac{\sqrt{Q^2 + U^2}}{I} \qquad (4.34)$$

$$m_c = \frac{V}{I} \qquad (4.35)$$

$$m_t = \frac{\sqrt{Q^2 + U^2 + V^2}}{I} \qquad (4.36)$$

$$\psi = \tfrac{1}{2}\tan^{-1}\left(\frac{U}{Q}\right), \qquad 0 \leq \psi \leq \pi, \qquad (4.37)$$

where m_ℓ, m_c, and m_t are the degrees of linear, circular, and total polarization, respectively, and ψ is the position angle of the plane of linear polarization. For monochromatic signals $m_t = 1$ and the polarization can be fully specified by just three parameters. For random signals such as those of cosmic origin $m_t \leq 1$ and all four parameters are required. The properties of the

Stokes parameters that make them particularly useful is that all four have the dimensions of flux density or brightness, and they propagate in the same manner as the electromagnetic field. Thus they can be determined by measurement or calculation at any point along a wave path and their relative magnitudes define the state of polarization at that point. Stokes parameters combine additively for independent waves.

In considering the response of interferometers and arrays up to this point we have ignored the question of polarization. This simplification has been justified by the assumption that we have been dealing with completely unpolarized radiation for which only the parameter I is nonzero. In that case the response of an interferometer with identically polarized antennas is proportional to the total flux density of the radiation. As will be shown below, in the more general case the response is proportional to a linear combination of two or more of Stokes parameters, the combination being determined by the antenna polarizations. By observing with different states of polarization of the antennas it is possible to separate the responses to the four parameters and determine the corresponding components of the visibility. The variation of each of the parameters over the source can thus be mapped individually, and the polarization of the radiation emitted at any point can be determined. Note that while I, which measures the total radio brightness, is always positive, Q, U, and V can take both positive and negative values depending on the position angle or sense of rotation of the polarization. There are, of course, alternative methods of describing the polarization state of a wave, of which the coherency matrix is perhaps the most important. Ko (1967a, b) has analyzed the response of an interferometer in terms of the coherency matrix and pointed out some advantages of this approach. These include a somewhat simpler derivation of the interferometer response if the antenna characteristics are expressed by a similar matrix. Nevertheless, the classical treatment in terms of Stokes parameters has remained essentially universal in its usage by astronomers, and we therefore follow it here.

Interferometer Response in Terms of Stokes Parameters

The polarization of an antenna in either transmission or reception can be described most generally by stating that the electric vector of a transmitted signal traces out an elliptical locus in the wavefront plane. Most antennas are designed so that the ellipse approximates a line or circle, corresponding to linear or circular polarization, in the central part of the main beam. However, exact linear or circular responses are never achieved in practice. As shown in Fig. 4.12a, the essential characteristics of the polarization ellipse are given by the position angle ψ of the major axis, and the axial ratio which it is convenient to express as the tangent of an angle χ, $-\pi/4 \leq \chi \leq \pi/4$. A practical antenna of arbitrary polarization can be modeled in terms of two idealized dipoles as shown in Fig. 4.12b. Consider *transmitting* with this antenna by applying a signal waveform to the terminal A. The signals to the

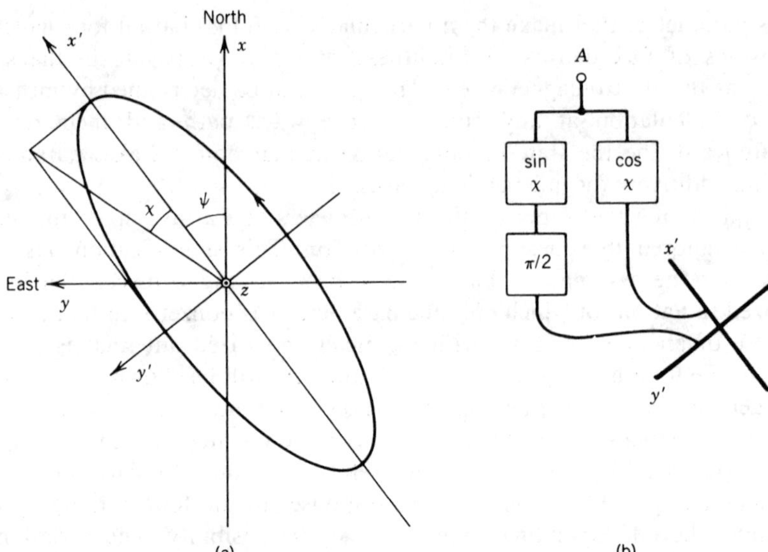

Figure 4.12 (a) Description of the general state of polarization of an antenna in terms of the characteristics of the ellipse generated by the electric vector in the transmission of a sinusoidal signal. The position angle ψ of the major axis is measured with respect to the x axis which points toward the direction of north on the sky. For an approaching wave the arrow on the ellipse indicates the direction of right handed polarization. (b) Model antenna that radiates the electric field in (a) when a signal is applied to the terminal A. Cos χ and sin χ indicate the voltage responses of the units shown, and $\pi/2$ indicates a phase lag.

dipoles pass through networks with voltage responses proportional to cos χ and sin χ, and the signal to the y' dipole also passes through a network that introduces a $\pi/2$ phase lag. Thus the antenna produces field components of amplitude $\mathscr{E}_{x'}$ and $\mathscr{E}_{y'}$ in phase quadrature along the directions of the major and minor axes of the ellipse. If the antenna input is a radiofrequency sine wave $V_0\cos 2\pi\nu t$, then the field components are:

$$\mathscr{E}_{x'}\cos 2\pi\nu t \propto V_0\cos\chi\,\cos(2\pi\nu t)$$

$$\mathscr{E}_{y'}\sin 2\pi\nu t \propto V_0\sin\chi\,\sin(2\pi\nu t). \tag{4.38}$$

In the above equations the y' component lags the x' component by $\pi/2$. If $\chi = \pi/4$ the radiated electric vector traces a circular locus with the sense of rotation from the x' axis to the y' axis (i.e., counterclockwise in Fig. 4.12a). Then a wave propagating in the z' direction as defined for a right-handed coordinate system (i.e., toward the reader in Fig. 4.12a) is right circularly polarized in the IEEE (1977) definition. (This definition is not universal, and in most optics texts such a wave would be defined as left circularly polarized.) The International Astronomical Union (IAU 1973) has adopted the IEEE definition and also stated that the position angle of the electric vector on the sky should be measured from north through east with reference to the system of right ascension and declination. Note that Stokes parameters in Eq. (4.33) specify only the field in the (x, y) plane, and to determine whether a circularly polarized wave is left- or right-handed the direction of propagation must be given. For Eq. (4.33) and the accompanying expressions for E_x and E_y, a wave travelling in the z direction in right-handed coordinates is right circularly polarized for positive V.

In reception an electric vector that rotates in a clockwise direction in Fig. 4.12 produces a voltage in the y' dipole that leads the voltage in the x' dipole by $\pi/2$ in phase, and the two signals therefore combine in phase at A. For a counterclockwise direction the signals interfere destructively. Thus the antenna transmits right-handed polarization in the positive z direction, and for signals incident from the same direction it receives right-handed waves. To receive a right-handed wave propagating in the positive z direction the polarity of one of the dipoles must be reversed, which changes χ to $-\pi/4$.

To determine the interferometer response we begin by considering the output of the antenna modeled in Fig. 4.12b. We redefine the field components in complex form: $E_x(t) = \mathcal{E}_x(t)e^{j[2\pi\nu t + \delta_x(t)]}$, $E_y(t) = \mathcal{E}_y(t)e^{j[2\pi\nu t + \delta_y(t)]}$. The received signal voltage expressed in complex form is proportional to

$$V' = E_{x'}\cos\chi - jE_{y'}\sin\chi, \tag{4.39}$$

where the factor $-j$ represents the $\pi/2$ phase lag applied to the y' signal. Now we need to specify the polarization of the incident wave in terms of Stokes parameters. In astronomy these parameters are usually defined using axes in the directions of north and east on the sky, which are represented by x and y in Fig. 4.12a. In terms of the field in the x and y directions the components of the field in the x' and y' directions are

$$E_{x'}(t) = \left[\mathcal{E}_x(t)e^{j\delta_x(t)}\cos\psi + \mathcal{E}_y(t)e^{j\delta_y(t)}\sin\psi\right]e^{j2\pi\nu t}$$

$$\tag{4.40}$$

$$E_{y'}(t) = \left[-\mathcal{E}_x(t)e^{j\delta_x(t)}\sin\psi + \mathcal{E}_y(t)e^{j\delta_y(t)}\cos\psi\right]e^{j2\pi\nu t}.$$

Derivation of the response at the output of the correlator for antennas m and n of an array involves straightforward manipulation of some rather lengthy expressions which will not be reproduced here. The steps are as follows:

1. Substitute $E_{x'}$ and $E_{y'}$ from Eqs. (4.40) into Eq. (4.39) to obtain the output of one antenna.

2. Indicate values of ψ, χ, and V' for the two antennas by subscripts m and n and calculate the correlator output

$$r_{mn} = G_{mn}\langle V'_m V'^*_n\rangle, \tag{4.41}$$

where G_{mn} is an instrumental gain factor.

3. Substitute Stokes parameters for $\mathcal{E}_x, \mathcal{E}_y, \delta_x, \delta_y$ using Eqs. (4.33) as follows:

$$\left\langle \left(\mathcal{E}_x e^{j\delta_x}\right)\left(\mathcal{E}_x e^{j\delta_x}\right)^* \right\rangle = \left\langle \mathcal{E}_x^2 \right\rangle = \tfrac{1}{2}(I + Q)$$

$$\left\langle \left(\mathcal{E}_y e^{j\delta_y}\right)\left(\mathcal{E}_y e^{j\delta_y}\right)^* \right\rangle = \left\langle \mathcal{E}_y^2 \right\rangle = \tfrac{1}{2}(1 - Q)$$

$$\left\langle \left(\mathcal{E}_x e^{j\delta_x}\right)\left(\mathcal{E}_y e^{j\delta_y}\right)^* \right\rangle = \left\langle \mathcal{E}_x \mathcal{E}_y e^{j(\delta_x - \delta_y)} \right\rangle = \tfrac{1}{2}(U + jV)$$

$$\left\langle \left(\mathcal{E}_x e^{j\delta_x}\right)^* \left(\mathcal{E}_y e^{j\delta_y}\right) \right\rangle = \left\langle \mathcal{E}_x \mathcal{E}_y e^{-j(\delta_x - \delta_y)} \right\rangle = \tfrac{1}{2}(U - jV).$$

$$(4.42)$$

The result is

$$
\begin{aligned}
r_{mn} = \tfrac{1}{2} G_{mn} \big\{ &I\left[\cos(\psi_m - \psi_n)\cos(\chi_m - \chi_n) + j\sin(\psi_m - \psi_n)\sin(\chi_m + \chi_n)\right] \\
&+ Q\left[\cos(\psi_m + \psi_n)\cos(\chi_m + \chi_n) + j\sin(\psi_m + \psi_n)\sin(\chi_m - \chi_n)\right] \\
&+ U\left[\sin(\psi_m + \psi_n)\cos(\chi_m + \chi_n) - j\cos(\psi_m + \psi_n)\sin(\chi_m - \chi_n)\right] \\
&- V\left[\cos(\psi_m - \psi_n)\sin(\chi_m + \chi_n) + j\sin(\psi_m - \psi_n)\cos(\chi_m - \chi_n)\right] \big\}.
\end{aligned}
$$

$$(4.43)$$

This is a general and very useful formula that applies to all cases. It was originally derived by Morris, Radhakrishnan, and Seielstad (1964). The derivation is also given by Weiler (1973) whose result differs from that given by Morris et al. in the sign of V, that is, in the matter of the convention for the sense of rotation. With identical antennas set to receive right circularly polarized radiation $\chi_m = \chi_n = -\pi/4$ and in Eq. (4.43) the term in V is positive. Thus in Eq. (4.43) positive V represents right circular polarization incident from the sky.

The symbols I, Q, U and V in Eq. (4.43) refer to the values of the parameters as they are measured by the spaced antennas, that is, they refer to the corresponding visibility values. These are the quantities required in mapping, and we derive them from the correlator output values by using Eq. (4.43). This equation is considerably simplified when the nominal polarization characteristics of practical antennas are inserted. First consider the case where both antennas are identically polarized. Then $\chi_m = \chi_n$, $\psi_m = \psi_n$, and Eq. (4.43) becomes

$$r_{mn} = \tfrac{1}{2} G_{mn}\left[I + Q\cos 2\psi_m \cos 2\chi_m + U\sin 2\psi_m \cos 2\chi_m - V\sin 2\chi_m\right]. \quad (4.44)$$

If the antennas are linearly polarized $\chi_m = 0$ and

$$r_{mn} = \tfrac{1}{2}G_{mn}\left[I + Q\cos 2\psi_m + U\sin 2\psi_m\right]. \qquad (4.45)$$

Let us assume that the antennas are paraboloids and that the feeds can be rotated to vary the position angle ψ of the E vector. For ψ equal to $0°$, $45°$, $90°$, and $135°$ the output is proportional to $(I + Q)$, $(I + U)$, $(I - Q)$, and $(I - U)$, respectively. By rotating the feeds to these positions I, Q, and U, but not V, can be measured. In many cases circular polarization is negligibly small, and the inability to measure V is then not a serious problem. However, Q and U are often only a few percent of I, and in attempting to measure them with identical feeds one faces the usual problems of measuring a small difference in two much larger quantities. The same is true if one attempts to measure V using identical circular feeds for which $\chi = \pm\pi/4$ and the response is proportional to $(I \mp V)$. A common practice is therefore to use oppositely polarized feeds to measure Q, U, or V.

With oppositely polarized feeds we write $\psi_n = \psi_m + \pi/2$ and $\chi_m = -\chi_n$. For linear polarization the χ terms are zero and the planes of polarization orthogonal. The antennas are then described as cross polarized, as typified by crossed dipoles. For such a case $\chi_m = \chi_n = 0$ and Eq. (4.43) becomes

$$r_{mn} = \tfrac{1}{2}G_{mn}\left[-Q\sin 2\psi_m + U\cos 2\psi_m \pm jV\right]. \qquad (4.46)$$

Then for ψ_m equal to $0°$ and $45°$ the response is proportional to $(U + jV)$ and $(-Q + jV)$. If V is assumed to be zero, this suffices to measure the polarized component. If both antennas contain a pair of cross-polarized feeds, the outputs of which go to two separate receiving channels at each antenna, four correlators can be used for each antenna pair to provide responses for both crossed and parallel pairs as follows:

Position angles	Stokes parameters measured	
$0°,0°$	$I+Q$	
$90°,90°$	$I-Q$	
$0°,90°$	$U+jV$	Position I
$90°,0°$	$U-jV$	
$45°,45°$	$I+U$	
$135°,135°$	$I-U$	
$45°,135°$	$-Q+jV$	Position II
$135°,45°$	$-Q-jV$	

$$(4.47)$$

Thus if the feeds can be periodically rotated through 45°, as indicated in positions I and II above, Q, U, and V can be measured without taking differences between responses involving I. This change of angle must be done sufficiently rapidly to follow variations in the visibility discussed in Section 4.5. The use of oppositely polarized antennas to suppress the response to I is particularly important for measurement of V since it is usually such a small percentage of I: see, for example, Weiler and Raimond (1976).

A scheme for polarization measurement in which linearly polarized feeds are not rotated during the observations has been used with the Westerbork synthesis radio telescope (Weiler 1973). In this array the outputs of a series of movable antennas are correlated with those from a series of fixed antennas. The position angles of the planes of polarization for the movable antennas are 45° and 135° and those of the fixed antennas 0° and 90°. These angles result in the following responses:

$$
\begin{array}{ll}
\text{Position angle} & \text{Stokes parameters measured} \\
0°,45° & (I + Q + U + jV)/\sqrt{2} \\
0°,135° & (-I - Q + U + jV)/\sqrt{2} \\
90°,45° & (I - Q + U - jV)/\sqrt{2} \\
90°,135° & (I - Q - U + jV)/\sqrt{2}
\end{array}
\tag{4.48}
$$

Although the responses are reduced by a factor of $\sqrt{2}$ relative to those in (4.47), there is no loss in sensitivity since each parameter appears at all four correlator outputs. The parameters Q, U, and V appear in combination with I in each case, but it is not necessary to rotate the feeds. Note, however, that this scheme is not applicable to an array in which signals from all possible pairs are correlated, unless the feeds are rotatable.

An additional complication occurs if linearly polarized antennas are used with altazimuth mounts since the plane of polarization rotates relative to the sky as the antennas track, unless the feeds are rotated to compensate for this effect. The position angle of the antenna relative to the sky varies as the parallactic angle given by Eq. (4.32), and the response at the correlator with an arbitrary value of ψ can be derived from Eq. (4.43). Conway and Kronberg (1969) have pointed out that the rotation of the position angle with altazimuth mounts can be used to advantage, since it helps to distinguish between instrumental effects and the true polarization of a source as discussed further in Chapter 10. Note also that in VLBI there may be large differences between the parallactic angles at the widely spaced antenna locations.

Opposite circularly polarized feeds offer certain advantages for measurements of linear polarization. In determining the responses we will retain an arbitrary position angle ψ_m for antenna m to show the effect of rotation caused, for example, by an altazimuth antenna mount. If the antennas provide simultaneous outputs for opposite senses of rotation (R and L) and four correlators are used for each antenna pair, the outputs are proportional to

Sense of rotation m, n	Stokes parameters measured
R, R	$I + V$
L, L	$I - V$
R, L	$(U - jQ)e^{-j2\psi_m}$
L, R	$-(U + jQ)e^{j2\psi_m}$

$$(4.49)$$

Here we have made $\psi_{\text{left}} = \psi_{\text{right}} + \pi/2$, and $\chi = -\pi/4$ for right circular polarization and $\chi = \pi/4$ for left circular. The feeds need not be rotated during an observation, and the responses to Q and U are separated from those to I. The expressions in (4.49) can be simplified by choosing values of ψ_{right} such as $-\pi/4$ or 0. For example, if $\psi_{\text{right}} = 0$ the sum of the RL and LR responses is a measure of Stokes parameter U. Again the effects of the rotation of the position angle with altazimuth mounts, and the differences in parallactic angle with very long baselines, must be taken into account. Conway and Kronberg (1969) appear to have been the first to use an interferometer with circularly polarized antennas to measure linear polarization in weakly polarized sources. Circularly polarized antennas have since become widely used in radio astronomy.

Instrumental Polarization

The responses with the various combinations of linearly and circularly polarized antennas discussed above are derived on the assumption that the polarization is exactly linear or circular and that the position angles of the linear feeds are exactly determined. This is not the case in practice, and the polarization ellipse can never be maintained as a perfect circle or straight line. The nonideal characteristics of the antennas cause an unpolarized source to appear polarized and are therefore referred to as instrumental polarization. The effect of these deviations from ideal behavior can be calculated from Eq. (4.43) if the deviations are known. The result is that in the expressions in (4.47), (4.48), and (4.49) the responses given are only the major terms, and all four Stokes parameters are, in general, involved. For example, consider the case of crossed linear feeds with nominal position angles 0° and 90°. Let the actual values of ψ and χ be such that $(\psi_m + \psi_n) = \pi/2 + \Delta\psi^+$, $(\psi_m - \psi_n) = -\pi/2 + \Delta\psi^-$, $\chi_m + \chi_n = \Delta\chi^+$, $\chi_m - \chi_n = \Delta\chi^-$. Then from Eq. (4.43)

$$r_{mn} \simeq \tfrac{1}{2}G_{mn}\left[I(\Delta\psi^- - j\Delta\chi^+) - Q(\Delta\psi^+ - j\Delta\chi^-) + U + jV\right]. \quad (4.50)$$

Generally, antennas can be adjusted so that the Δ terms are no more than about 3°, and here we have assumed that they are small enough that their cosines can be approximated by unity, their sines by the angles, and products of two sines by zero. Also, Q, U, and V are generally less then 10% of I, so products of Q, U, and V with Δ terms, such as the Q term in Eq. (4.50), can often be neglected. It is desirable that the instrumental polarization be small enough that such approximations can be made, otherwise the calibration is more complicated. In general, instrumental polarization is different for each antenna or antenna pair and must be corrected in the visibility data before they are combined into a map. A discussion of the case for circularly polarized antennas is given by Conway and Kronberg (1969).

With the above approximations the main effect of the nonideal antenna parameters is to add to each response in (4.47), (4.48), and (4.49) an instrumental term $P_{nmp}I$, where P_{nmp} is complex, of magnitude a few percent, and different for each polarization combination, p, of each antenna pair, (m, n). These instrumental terms are important only in those responses that do not otherwise involve I, since if I is already included they are absorbed in the normal response term involving I. They occur for both crossed-linear or opposite-circular polarizations of the antennas and can be calibrated by observing an unpolarized source, or one with accurately known polarization. In practice, the instrumental terms may vary over the main beam of the antennas, and this adds complications if the source under investigation is wide or the antenna pointing is inaccurate. Thus it is usual to make polarization measurements near the beam center only, and also to assume that the instrumental polarization terms remain constant throughout the duration of an observation.

Of all measurements in radio astronomy, those of polarization perhaps require the most careful calibration. Instrumental polarization usually limits the accuracy to a few tenths of 1% of the unpolarized component. Calibration of polarization measurements is discussed further in Chapter 10, and Faraday rotation of the plane of polarization in Chapter 13.

4.10 DISCRETE FOURIER TRANSFORMATION

Although a detailed description of methods for computing radio maps from visibility data is the subject of Chapter 10, it is useful for the reader to have a brief introduction at this point as a help in understanding some of the intervening material. The discrete transform is very widely used in synthesis mapping because of computational advantages, but certain complications are introduced which are the subject of this section. With the discrete transform the functions $\mathscr{V}(u, v)$ and $B(\xi, \eta)$ are expressed as rectangular matrices of sampled values at uniform increments in the two variables involved. The main advantage of this system is that the fast Fourier transform (FFT) algorithm can be used to reduce the computation time. The FFT algorithm is discussed by Cooley and Tukey (1965), Cooley, Lewis, and Welch (1967), Cochran et al. (1967) and Brigham (1974). The rectangular grid points at which the map is obtained provide a convenient form for further data processing.

The two-dimensional form of the discrete transform for a Fourier pair f and g is defined by

$$f(m,n) = \frac{1}{MN} \sum_{k=0}^{M-1} \sum_{l=0}^{N-1} g(k,l) e^{-j2\pi km/M} e^{-j2\pi ln/N}, \qquad (4.51)$$

and the inverse is

$$g(k,l) = \sum_{m=0}^{M-1} \sum_{n=0}^{N-1} f(m,n) e^{j2\pi km/M} e^{j2\pi ln/N}. \qquad (4.52)$$

See, for example, Oppenheim and Schafer (1975). Note that the functions are periodic with periods of M samples in the k and m dimensions and N samples in the l and n dimensions. Evaluation of Eq. (4.51) or (4.52) by direct computation requires approximately $(MN)^2$ complex multiplications. In contrast, if M and N are powers of 2 the FFT algorithm requires only $\frac{1}{2}MN \log_2(MN)$ complex multiplications, a very important practical advantage. A further increase in speed by a factor of approximately 2 is offered by the fast Hartley transform (Bracewell 1984b). Here, however, we are concerned with properties of the process that are related to the discreteness rather than the speed.

To apply the discrete transform to the synthesis-mapping problem we first use an interpolation technique to obtain values of $\mathscr{V}(u,v)$ at points separated by Δu in u and by Δv in v as shown in Fig. 4.13. The dimensions of the (u,v) plane that contain these data are $M\Delta u$ by $N\Delta v$. In the (ξ, η) plane the points are spaced $\Delta\xi$ in ξ and $\Delta\eta$ in η and the map dimensions are $M\Delta\xi$ by $N\Delta\eta$. The dimensions in the two domains are related by

$$\Delta u = (M\Delta\xi)^{-1}$$

$$\Delta v = (N\Delta\eta)^{-1}$$

$$\Delta\xi = (M\Delta u)^{-1} \qquad (4.53)$$

$$\Delta\eta = (N\Delta v)^{-1}.$$

The point spacing in one domain is the reciprocal of the total dimension in the other domain. It is a common, though not universal, practice to make maps of equal dimensions in both coordinates, that is, $M = N$, $\Delta u = \Delta v$, and $\Delta\xi = \Delta\eta$.

The transformation between $\mathscr{V}(u,v)$ and $F(\xi, \eta)$, where F represents the left-hand side of Eq. (4.6), is obtained by substituting in Eq. (4.51) or (4.52) $g(k,l) = F(k\Delta\xi, l\Delta\eta)$, and $f(m,n) = \mathscr{V}(m\Delta u, n\Delta v)$. The relationship between the integral and discrete forms of the Fourier transform is found in numerous texts: see, for example, Rabiner and Gold (1975) or Papoulis (1977).

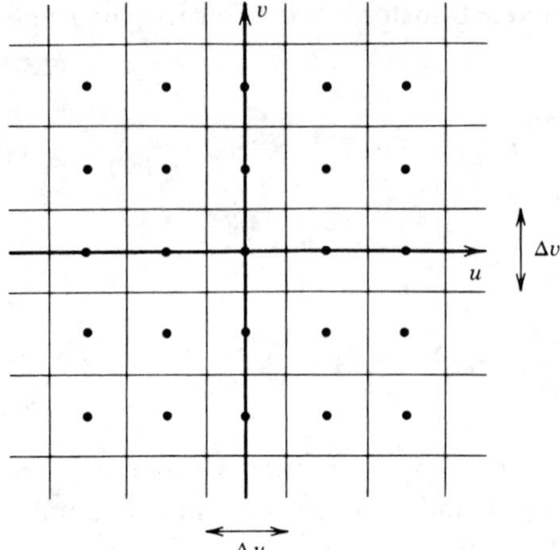

Figure 4.13 Points on a rectangular grid in the (u, v) plane at which the visibility is sampled for use with the discrete Fourier transform. As shown, the spacings Δu and Δv are equal. The division of the plane into cells for interpolation using the cell-averaging technique is also shown.

Note here that, by analogy with the integral relationships in Section 4.1, one would expect to write

$$\mathcal{V}(m \Delta u) = \sum_{k=-\infty}^{\infty} F(k \Delta \xi) e^{-j2\pi k \Delta u\, m\, \Delta \xi} \Delta \xi \qquad (4.54)$$

and

$$F(k \Delta \xi) = \sum_{m=-\infty}^{\infty} \mathcal{V}(m \Delta u) e^{j2\pi k \Delta u\, m\, \Delta \xi} \Delta u, \qquad (4.55)$$

where, for simplicity, only one dimension is included. From the relations in Eqs. (4.53) the exponentials will be found to be the same as those in Eqs. (4.51) and (4.52). Also the product $\Delta u\, \Delta \xi$ is equal to M^{-1}, so the symbols Δu and $\Delta \xi$ at the ends of Eqs. (4.54) and (4.55) can be replaced by a factor M^{-1} in either one of the two equations. Finally, the periodic nature of the functions allows the limits of the summations to be changed from $\pm \infty$ to 0 to $(M - 1)$. In this manner the discrete form of the Fourier transform is related to the integral form.

The Fourier transform relationship between the sampled functions requires that they be regarded as periodic, with periods $M\Delta u$ and $N\Delta v$ in u and v and $M\Delta \xi$ and $N\Delta \eta$ in ξ and η. This is illustrated in Fig. 4.14 where the area $ABCD$ represents the measured range of $\mathcal{V}(u, v)$ with respect to the origin O, and we must consider that the function is replicated at points such as P, Q, and R. Because the discrete variables m and n in Eq. (4.52) take positive values only, the function to be transformed is represented by the samples within the rectangle $OPQR$. In effect, the four quadrants of the original function in $ABCD$ have been rearranged. After transformation the brightness appears as a function of variables that again run in the positive direction only from zero, as in Eq. (4.51), so the output must be rearranged to obtain the

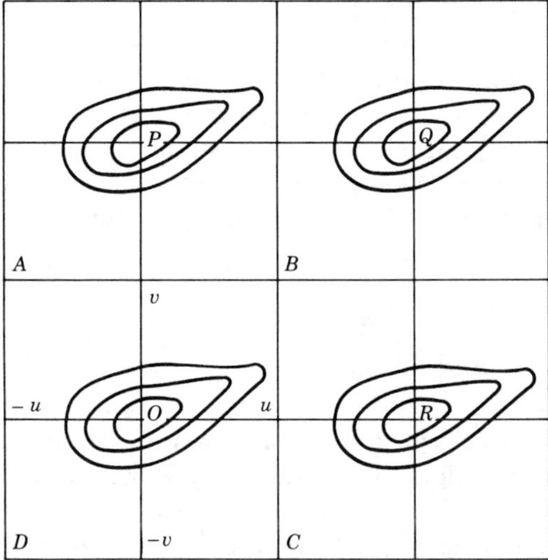

Figure 4.14　Rearrangement of a visibility function for input to the discrete Fourier transform. The (u, v) origin is at O. The contours represent the unsampled visibility which is measured in the range of u and v within the area $ABCD$. In the transformation of the sampled data the function is, in effect, replicated at points such as P, Q, and R. The input and output to the transformation process cover only positive values of u and v and are represented by the area $OPQR$.

origin at the center of the map. In practice, these rearrangements of the data can be made in the computer that performs the FFT.

Before leaving the subject of discrete Fourier transformation, we should look briefly at the problem of obtaining values of $\mathscr{V}(u, v)$ at the required points in the (u, v) plane. These points form a rectangular grid as shown in Fig. 4.13, while the measured visibility values lie on the elliptical spacing loci. A very simple means of obtaining values at the grid points is to divide the (u, v) plane into rectangular cells of dimensions Δu by Δv centered on the grid points as shown in Fig. 4.13. The mean of all measured points within each cell is then computed and assigned to the grid point at the cell center. Grid points for which there are no measurements within the cell are assigned a visibility value of zero. This procedure has been termed *cell averaging* (Thompson and Bracewell 1974), and in spite of obvious deficiencies it is an effective technique. More sophisticated procedures will be discussed in Chapter 10. The particular value of discussing cell averaging at this point is that it should help the reader to visualize what the distribution of measurements in the (u, v) plane should be. The spacing between the (u, v) loci should be comparable with the cell size to maximize the number of cells that are intersected by a locus. The averaging time for the data in the correlator should clearly be less than the time for the spacing vector to cross a cell. Minimization of the number of cells that do not contain visibility measurements is an important criterion in array design and is discussed in Chapter 5.

4.11   SAMPLING THEOREM

To illustrate further the requirement on the spacing of the points in discretely sampled data we consider the sampling theorem as applied to interferometry (e.g. Bracewell 1958). Figure 4.15a–c illustrates the sampling of the one-dimensional visibility function $\mathcal{V}(u)$. The sampling operation can be represented as multiplication of $\mathcal{V}(u)$ by the series of delta functions in Fig. 4.15b which can be written

$$\left[\frac{1}{\Delta u}\right]\mathrm{III}\left(\frac{u}{\Delta u}\right) = \sum_{k=-\infty}^{\infty} \delta(u - k\,\Delta u), \tag{4.56}$$

where the left-hand side is included to show how the series can be expressed in terms of the *shah* function III introduced by Bracewell and Roberts (1954). The series extends to infinity in both directions, and the delta functions are uniformly spaced with an interval Δu. The Fourier transform of Eq. (4.56) is the series of delta functions shown in Fig. 4.15e:

$$\mathrm{III}(\Delta u\,\xi) = \frac{1}{\Delta u} \sum_{m=-\infty}^{\infty} \delta\left(\xi - \frac{m}{\Delta u}\right). \tag{4.57}$$

In the ξ domain the Fourier transform of the sampled visibility is the convolution of the Fourier transform of $\mathcal{V}(u)$, which is the one-dimensional brightness $B_1(\xi)$, with Eq. (4.57). The result is the replication of $B_1(\xi)$ at intervals $(\Delta u)^{-1}$ shown in Fig. 4.15f. If $B_1(\xi)$ represents a source of finite dimensions, the replications of $B_1(\xi)$ will not overlap as long as $B_1(\xi)$ is nonzero only within a range of ξ that is no greater than $(\Delta u)^{-1}$. An example of overlapping replications is shown in Fig. 4.15g. The loss of information resulting from such overlapping is commonly referred to as *aliasing*, because the components of the function within the overlapping region lose their identity with respect to which end of the replicated function they properly belong. Avoidance of aliasing requires that the spacing of the delta functions in the ξ domain be large enough, and the requirement on the sampling interval Δu is that it should be no greater than the reciprocal of the interval in ξ within which $B_1(\xi)$ is nonzero. Note that this interval represents the width of the source as broadened by the finite resolution of the observations, rather than the true width of the source. This distinction is related to the phenomenon of leakage (Bracewell 1978).

The requirement for the restoration of a function from a set of samples, for example, deriving the function in Fig. 4.15a from the samples in Fig. 4.15c, is easily understood by considering the Fourier transforms in Fig. 4.15d and f. Interpolation in the u domain corresponds to removing the replications in the ξ domain, which can be achieved by multiplication of the function in Fig. 4.15f

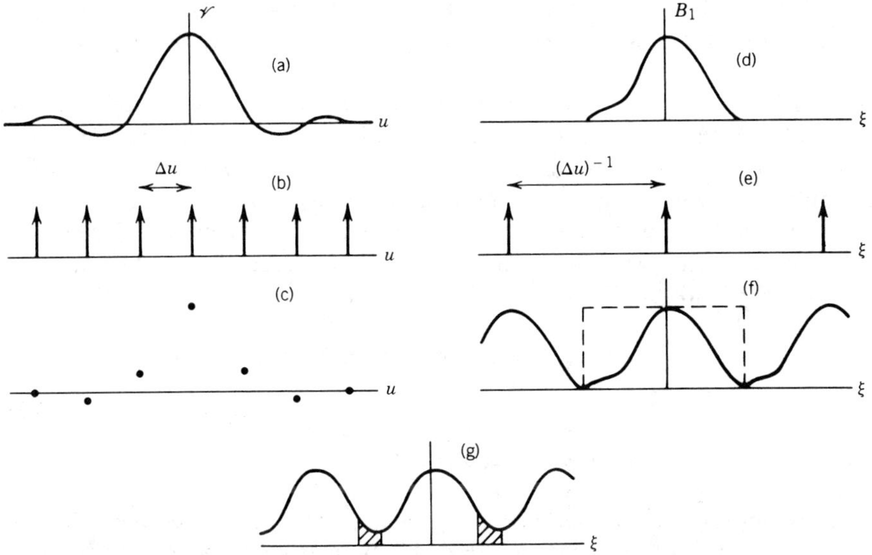

Figure 4.15 Illustration of the sampling theorem: (a) visibility function $\mathcal{V}(u)$, real part only; (b) sampling function in which the arrows represent delta functions; (c) sampled visibility function; (d) brightness function $B_I(\xi)$; (e) replication function; (f) replicated brightness function. Functions in (d), (e), and (f) are the Fourier transforms of those in (a), (b), and (c), respectively. (g) Replicated brightness function showing aliasing in shaded areas resulting from use of too large a sampling interval.

by the rectangular function indicated by the broken line. In the u domain this multiplication corresponds to convolution of the sampled values with the Fourier transform of the rectangular function, which is the unit-area sinc function

$$\frac{\sin(\pi u/\Delta u)}{\pi u}. \tag{4.58}$$

If aliasing is avoided, convolution with Expression (4.58) provides exact interpolation of the original function from the samples. Thus we can state, as a sampling theorem for the visibility, that if the brightness distribution is nonzero only within an interval of width ξ_w, $B_1(\xi)$ is fully specified by sampling the visibility function at points spaced $\Delta u = \xi_w^{-1}$ in u. The sampling theorem can be applied, of course, to any pair of quantities related by the Fourier transform. It is not confined to one-dimensional functions and can be applied to each coordinate for functions specified in a Cartesian system. In the case of the discrete Fourier transform, the spacing relationships in Eqs. (4.53) show that the visibility is adequately sampled by the grid points in u and v provided that the synthesized field is large enough to contain the entire brightness distribution to be observed.

APPENDIX 4.1 CONVERSION BETWEEN HOUR ANGLE–DECLINATION AND AZIMUTH–ELEVATION COORDINATES

Although the positions of cosmic sources are almost always specified in celestial coordinates (hour angle and declination) for purposes of observation, it is often convenient to convert to elevation and azimuth. The conversion formulas between hour angle and declination (H, δ) and elevation and azimuth $(\mathscr{E}, \mathscr{A})$ can be found in many reference books, for example, *Explanatory Supplement to the Astronomical Ephemeris*. For an observer at latitude $\mathscr{L}$ they are, for (H, δ) to $(\mathscr{A}, \mathscr{E})$:

$$\sin \mathscr{E} = \sin \mathscr{L} \sin \delta + \cos \mathscr{L} \cos \delta \cos H$$

$$\cos \mathscr{E} \cos \mathscr{A} = \cos \mathscr{L} \sin \delta - \sin \mathscr{L} \cos \delta \cos H \qquad (A4.1)$$

$$\cos \mathscr{E} \sin \mathscr{A} = -\cos \delta \sin H.$$

These equations can be derived by applying the sine and cosine rules for spherical triangles to the system in Fig. 4.7. Similarly for $(\mathscr{A}, \mathscr{E})$ to (H, δ) we have

$$\sin \delta = \sin \mathscr{L} \sin \mathscr{E} + \cos \mathscr{L} \cos \mathscr{E} \cos \mathscr{A}$$

$$\cos \delta \cos H = \cos \mathscr{L} \sin \mathscr{E} - \sin \mathscr{L} \cos \mathscr{E} \cos \mathscr{A} \qquad (A4.2)$$

$$\cos \delta \sin H = -\cos \mathscr{E} \sin \mathscr{A}.$$

Here azimuth is measured from north through east.

REFERENCES

Apostol, T. M., *Calculus, Vol. II*, Blaisdel Pub. Co., Waltham, Mass., 1962, p. 82.

Born, M. and E. Wolf, *Principles of Optics*, Pergamon Press, Oxford, 1959 (and later eds.).

Bracewell, R. N., Radio Interferometry of Discrete Sources, *Proc. IRE*, **46**, 97–105, 1958.

Bracewell, R. N., *The Fourier Transform and Its Applications*, McGraw-Hill, New York, 1965 (2nd ed. 1978).

Bracewell, R. N., Computer Image Processing, *Ann. Rev. Astron. Astrophys.*, **17**, 113–134, 1979.

Bracewell, R. N., Inversion of Nonplanar Visibilities, in *Indirect Imaging*, J. A. Roberts, Ed., Cambridge University Press, Cambridge, England, 1984a, pp. 177–183.

Bracewell, R. N., The Fast Hartley Transform, *Proc. IEEE*, **72**, 1010–1018, 1984b.

Bracewell, R. N. and J. A. Roberts, Aerial Smoothing in Radio Astronomy, *Aust. J. Phys.*, **7**, 615–640, 1954.

Brigham, E. O., *The Fast Fourier Transform*, Prentice Hall, Englewood Cliffs, New Jersey, 1974.

Brouw, W. N., Data Processing for the Westbrook Synthesis Radio Telescope, Doctoral Thesis, University of Leiden, 1971.

Clark, B. G., Wide Field Mapping, in *Synthesis Mapping, Proceedings of NRAO Workshop No. 5*, Socorro, New Mexico, June 21–25, 1982, A. R. Thompson and L. R. D'Addario, Eds.,

National Radio Astronomy Observatory, Green Bank, West Virginia, 1982.

Cochran, W. T., J. W. Cooley, D. L. Favin, H. D. Helms, R. A. Kaenel, W. W. Lang, G. C. Maling, D. E. Nelson, C. M. Rader, and P. D. Welch, What is the Fast Fourier Transform?, *IEEE Trans. Audio Electroacoust.*, **AU-15**, 45–55, 1967.

Conway, R. G. and P. P. Kronberg, Interferometer Measurement of Polarization Distribution in Radio Sources, *Mon. Not. R. Astron. Soc.*, **142**, 11–32, 1969.

Cooley, J. W., P. A. W. Lewis, and P. D. Welch, Historical Notes on The Fast Fourier Transform, *IEEE Trans. Audio Electroacoust.*, **AU-15**, 76–79, 1967.

Cooley, J. W. and J. W. Tukey, An Algorithm for the Machine Calculation of Complex Fourier Series, *Math. Comput.*, **19**, 297–301, 1965.

Explanatory Supplement to the Astronomical Ephemeris, H. M. Stationary Office, London, 1961, Ch. 2.

Hudson, J. A., An Analysis of Aberrations of the V.L.A. Radio Synthesis Telescope, Ph.D. Thesis, American University, Washington, DC, 1977.

IAU, *Trans. Int. Astron. Union*, **15B**, 166, 1973.

IEEE, Standard Definitions of Terms for Radio Wave Propagation, Std 211–1977, Institute of Electrical and Electronics Engineers, Inc., New York, 1977.

Ko, H. C., Coherence Theory of Radio-Astronomical Measurements, *IEEE Trans. Antennas Propag.*, **AP-15**, 10–20, 1967a.

Ko, H. C., Theory of Tensor Aperture Synthesis, *IEEE Trans. Antennas Propag.*, **AP-15**, 188–190, 1967b.

König, A., Astrometry with Astrographs, in *Astronomical Techniques, Stars and Stellar Systems*, Vol. 2, W. A. Hiltner, Ed., University of Chicago Press, Chicago, 1962, pp. 461–486.

Kraus, J. D. and K. R. Carver, *Electromagnetics*, McGraw-Hill, New York, 1973 (2nd ed.), p. 435.

Morris, D., V. Radhakrishnan, and G. A. Seielstad, On the Measurement of Polarization Distributions over Radio Sources, *Astrophys. J.*, **139**, 551–559, 1964.

Oppenheim, A. V. and R. W. Schafer, *Digital Signal Processing*, Prentice-Hall, Englewood Cliffs, New Jersey, 1975, Ch. 3.

Papoulis, A., *Signal Analysis*, McGraw-Hill, New York, 1977, p. 74.

Rabiner, L. R. and B. Gold, *Theory and Application of Digital Signal Processing*, Prentice-Hall, Englewood Cliffs, NJ, 1975, p. 50.

Rowson, B., High Resolution Observations with a Tracking Interferometer, *Mon. Not. R. Astron. Soc.*, **125**, 177–188, 1963.

Thompson, A. R. and R. N. Bracewell, Interpolation and Fourier Transformation of Fringe Visibilities, *Astron. J.*, **79**, 11–24, 1974.

Wade, C. M., Precise Positions of Radio Sources, I. Radio Measurements, *Astrophys. J.*, **162**, 381–390, 1970.

Weiler, K. W., The Synthesis Radio Telescope at Westerbork, Methods of Polarization Measurement, *Astron. Astrophys.*, **26**, 403–407, 1973.

Weiler, K. W. and E. Raimond, Aperture Synthesis Observations of Circular Polarization, *Astron. Astrophys.*, **52**, 397–402, 1976.

5

DESIGN OF ARRAYS

This chapter is concerned with the design of arrays, mainly from the viewpoint of how the configuration of the elements affects the performance in synthesis mapping. Of the antennas themselves, however, little need be said. The principal antenna characteristics relevant to synthesis mapping have been discussed already. Details of antenna design can be found in the books of Collin (1985), Johnson and Jasik (1984), Love (1978), Milligan (1985), and others.

In the following discussion it will be convenient to classify arrays into several broad categories. We distinguish between instruments in which the beams of the individual elements remain fixed during an observation, and those which operate by tracking a source over a range of hour angle. In nontracking arrays the antenna beams are usually directed toward the meridian at some desired declination and the rotation of the earth scans the beams in right ascension. Any source is observed for no more than a few minutes as it passes through the beams. During the 1950s and early 1960s most large radio telescopes were of this type. They operated at meter wavelengths where the flux densities of sources are high enough that large numbers can be detected and cataloged without long integration times. By the late 1970s much of the survey work had been completed, and many large transit instruments were superseded by centimeter- and millimeter-wavelength arrays designed for high-resolution studies of individual sources. These shorter-wavelength instruments were mainly of the tracking type. In considering their design it is useful to distinguish between one-dimensional configurations of antennas which are satisfactory for

114

observations at declinations greater than about 30° from the celestial equator, and two-dimensional configurations designed to be effective over the entire observable sky. Also considered are arrays designed for VLBI observations for which some of the practical constraints are different from those for connected-element arrays.

5.1 SPECTRAL SENSITIVITY FUNCTION OF A NONTRACKING ARRAY

Before discussing the design of arrays it is appropriate to introduce the *spectral sensitivity function* (Bracewell 1961, 1962), also called the spatial spectral sensitivity function, which is a measure of the sensitivity of an antenna or array to the spatial frequencies on the sky. This function can be understood through the diagram of antenna relationships in Fig. 5.1. The voltage reception pattern $G(\xi, \eta)$ of an antenna is the Fourier transform of the excitation (illumination) of the electric field across the aperture $\mathscr{E}(x_\lambda, y_\lambda)$, where x_λ and

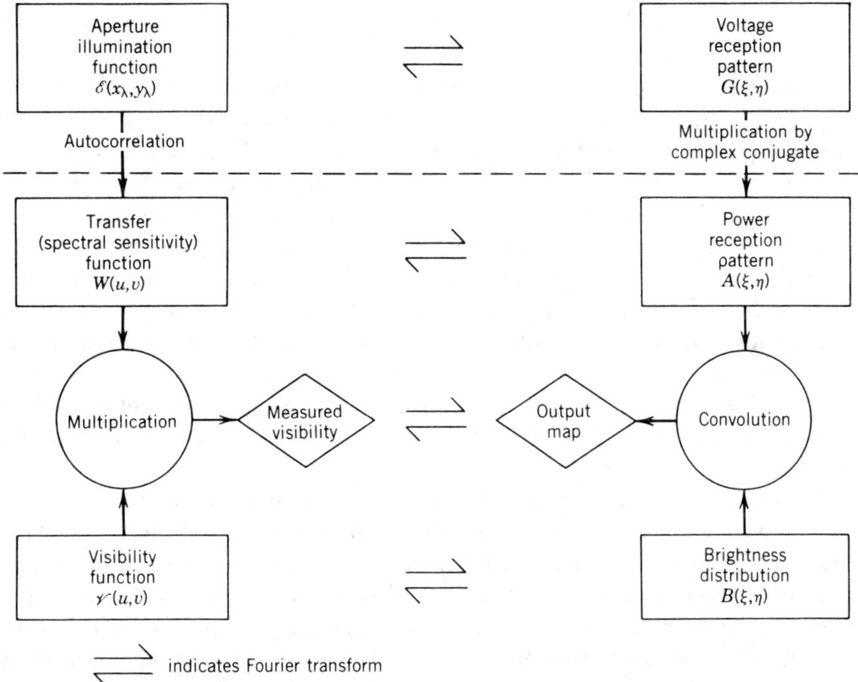

Figure 5.1 Relationships between antenna quantities for an incoherent field and a power-linear system. The functions shown also depend upon frequency ν. The whole diagram applies to single-aperture antennas but, in general, only the part below the broken line applies to correlator arrays.

y_λ are coordinates in the antenna aperture and are measured in wavelengths. Thus

$$G(\xi, \eta) \rightleftharpoons \mathscr{E}(x_\lambda, y_\lambda), \tag{5.1}$$

which follows from Fraunhofer diffraction theory and is expressed in one dimension in Eq. (3.22). The power reception pattern $A(\xi, \eta)$ is equal to $G(\xi, \eta)G^*(\xi, \eta)$ as indicated in Fig. 5.1. By the Wiener–Khinchin relation (see Appendix 3.2), $A(\xi, \eta)$ is the Fourier transform of the autocorrelation function of $\mathscr{E}(x_\lambda, y_\lambda)$. The two-dimensional autocorrelation function is indicated by the double pentagram symbol, and can be written as follows:

$$\mathscr{E} \star\!\star \mathscr{E}^* = \int_{-\infty}^{\infty} \int_{-\infty}^{\infty} \mathscr{E}(x_\lambda, y_\lambda) \mathscr{E}^*(x_\lambda - u, y_\lambda - v)\, dx_\lambda\, dy_\lambda. \tag{5.2}$$

u and v are measured in wavelengths, and, as explained in Chapter 2, they also represent spatial frequencies.

The autocorrelation function in Eq. (5.2) is the spectral sensitivity function $W(u, v)$:

$$W(u, v) = \mathscr{E}(x_\lambda, y_\lambda) \star\!\star \mathscr{E}^*(x_\lambda, y_\lambda). \tag{5.3}$$

The integral in Eq. (5.2) is proportional to the number of ways (suitably weighted) in which a specific spacing vector (u, v) can be found within the antenna aperture. Therefore, $W(u, v)$ is a measure of the sensitivity of the antenna to different spatial frequencies. The meaning of the spectral sensitivity function can also be explained in terms of the antenna response to a point source. Such a source can be represented by a two-dimensional delta function, for which the Fourier transform is a constant over the (u, v) plane. Thus, the Fourier transform of the response to a point source indicates those spatial frequencies that are accepted by the antenna, and their respective weights. In effect, the antenna acts as a spatial frequency filter. The power reception pattern $A(\xi, \eta)$ is an expression of the response to a point source, and, as indicated in Fig 5.1, the Fourier transform of $A(\xi, \eta)$ is equal to $W(u, v)$, the spectral sensitivity function.

Equation (5.3) holds for any antenna, or for any array of antennas in which the received voltages are summed linearly. It does not apply to a correlator interferometer or array, and we may illustrate the difference by considering the interferometer in Fig. 5.2a. First suppose that the output voltages from the two apertures are summed and fed to a power-measuring receiver, as in very early instruments. The three rectangular areas in Fig. 5.2b represent the autocorrelation function of the aperture distributions, that is, the spectral sensitivity. Now suppose that the two antennas are combined instead using a correlator. The output is represented only by the shaded areas in Fig. 5.2b, since the correlator forms only the cross products of signal components from the two apertures.

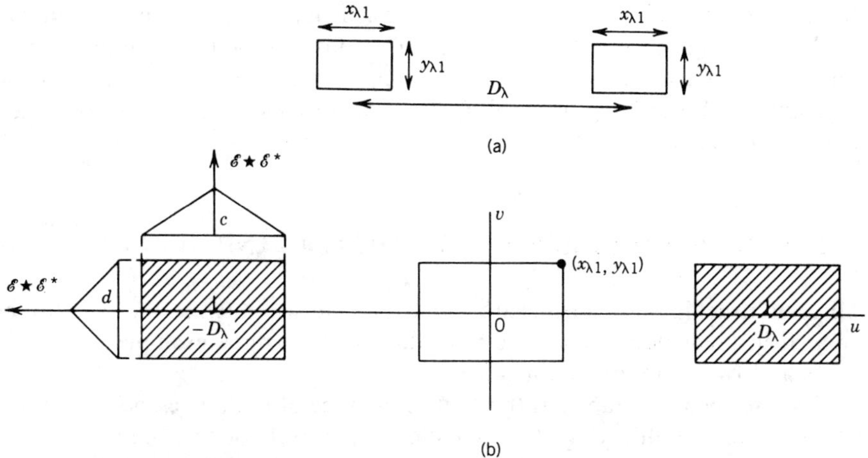

Figure 5.2 The two apertures in (a) represent a two-element interferometer, the spectral sensitivity function of which is shown in (b). The shaded areas contain the spectral sensitivity components that result from the cross correlation of the signals from the two antennas. If the field distribution is uniform over the apertures, the magnitude of the spectral sensitivity is linearly tapered. This is indicated by c and d which represent cross sections of the spectral sensitivity function.

The central rectangular area represents the correlation of each aperture with itself. For a multielement correlator array we take the sum of the cross correlations of the field distributions of the correlator-linked apertures.

The interpretation of the spectral sensitivity function as the Fourier transform of the point-source response can be applied to both the adding and correlator cases. For example, for the correlator implementation of the interferometer in Fig. 5.2a, the response to a point source is the Fourier transform of the function represented by the shaded areas. This Fourier transform is

$$\left[\frac{\sin(\pi x_{\lambda 1}\xi)}{\pi x_{\lambda 1}\xi}\right]^2\left[\frac{\sin(\pi y_{\lambda 1}\eta)}{\pi y_{\lambda 1}\eta}\right]^2\cos(2\pi D_\lambda\xi), \qquad (5.4)$$

where $x_{\lambda 1}$ and $y_{\lambda 1}$ are the aperture dimensions and D_λ is the aperture separation, all measured in wavelengths. The sinc-squared functions represent the power pattern of the uniformly illuminated rectangular apertures, and the cosine term represents the fringe pattern. The usefulness of the spectral sensitivity function is that it enables one to make a rapid estimate of the form of the response to a point source and to compare the responses of different arrangements of antennas. Bracewell (1961) has described a graphical method to determine the support[†] of the spectral sensitivity function. In early instruments the relative magnitude of the spectral sensitivity was controlled only by

---

[†] The support of a function is the domain over which the function is nonzero.

the field distribution over the antennas, but computer processing techniques enable the magnitude to be adjusted after an observation has been made. This may entail Fourier transformation of a map to the spatial-frequency domain, adjustment of weighting of the visibility function, and inverse transformation back to the sky domain.

5.2 NONTRACKING INTERFEROMETERS AND ARRAYS

Since nontracking instruments are largely of historical interest, they will not be treated in great detail. They are well illustrated by three examples, the Mills cross and two Cambridge instruments.

As described in Chapter 1, the Mills-cross type of radio telescope makes use of the voltage-multiplying characteristics of a correlator to combine the outputs of two antennas that produce narrow fan beams in orthogonal planes. The product of the fan beam responses is a narrow pencil beam, the axis of which is the line of intersection of the planes of the two fan beams. In this manner a narrow beam can be obtained very much more cheaply than by constructing a fully filled circular or rectangular aperture. Figure 5.3a shows the antenna arrangement of a cross. As in a conventional interferometer the correlator output integrated over the sky is zero, and in a cross the positive response of the main beam is offset by a negative response in low but wide-angle sidelobes. The development of this type of instrument can be followed in papers by Mills and Little (1953), Mills (1963), and Mills et al. (1958, 1963). The spectral sensitivity function of a cross is shown in Fig. 5.3b. The shaded area represents the cross correlation of the two arms. The area of spectral sensitivity can easily

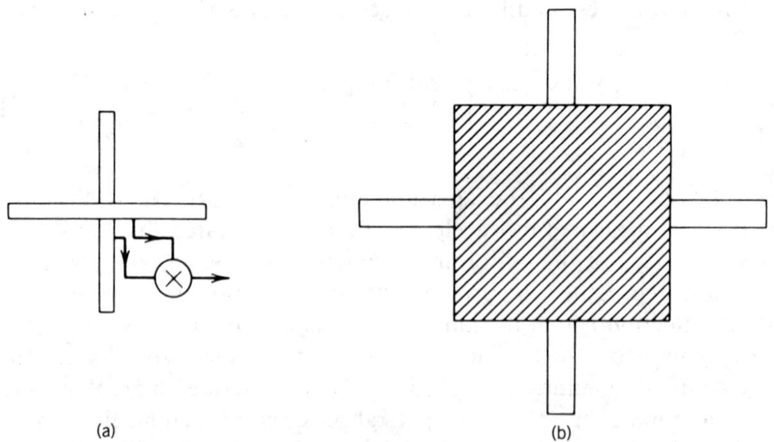

Figure 5.3 The aperture (a) and the spectral sensitivity function (b) of a Mills-cross radio telescope. The shaded area in (b) results from the cross correlation of the voltages from the two arrays in (a). Part (a) is in the aperture (x, y) plane and part (b) in the (u, v) plane.

be seen to be equivalent to that of a rectangular aperture of dimensions equal to half the lengths of the two arms in the corresponding directions. The weighting of the spectral sensitivity is uniform over the rectangular area if the excitation is uniform along the arms. Therefore, it is usual to apply a Gaussian or similar taper to each arm to reduce the excitation to about 10% at the ends, since the strong sidelobes that would otherwise result from the sharp discontinuity at the boundaries of the spectral sensitivity are highly undesirable. The beam of a Mills cross can be directed to different declinations on the meridian by adjusting the phasing of the dipoles on the north–south arm. The sensitivity is determined by the areas of the arms, which can be designed so that the limits imposed by sensitivity and by confusion of sources occur at nearly the same flux density level.

A variation on the cross in which half of one of the arms is removed is shown in Fig. 5.4a. The spectral sensitivity function of this T-shaped arrangement is shown in Fig. 5.4b. The shaded area representing the cross correlation of the two elements is of the same dimensions as that of the cross from which the T was derived. This equivalence between a cross and a T can be explained by noting that in the T the two arrays no longer have a common electrical center, and an interferometer fringe term is introduced into the response. Consider, for simplicity, a cross in which the excitation of the elements along the two arms is uniform. Then an arm of length $2L_\lambda$ wavelengths has a voltage reception pattern proportional to $\sin(2\pi L_\lambda \xi)/2\pi L_\lambda \xi$, where ξ is the direction cosine measured relative to the long axis of the arm. If half of the arm is removed, the remaining half has a reception pattern proportional to $\sin(\pi L_\lambda \xi)/\pi L_\lambda \xi$ and its electrical center is displaced by $L_\lambda/2$ from that of the other arm. Thus the combined reception pattern is proportional to

$$\frac{\sin(\pi L_\lambda \xi)\cos(\pi L_\lambda \xi)}{\pi L_\lambda \xi} = \frac{\sin(2\pi L_\lambda \xi)}{2\pi L_\lambda \xi}. \tag{5.5}$$

An early disadvantage of the T was that it required more careful phase

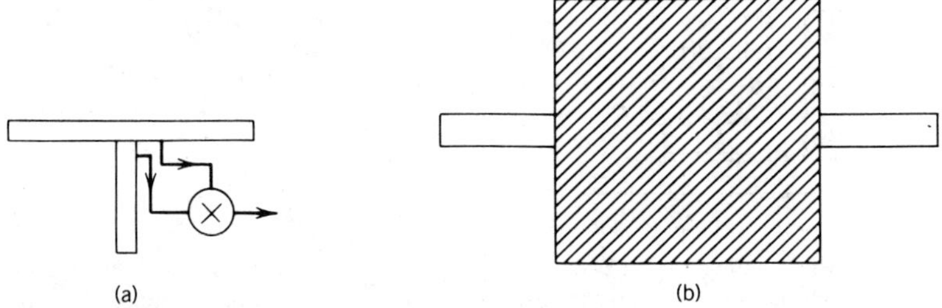

(a)   (b)

Figure 5.4   The aperture (a) and the spectral sensitivity function (b) of a T-antenna that results from omitting half of the north–south arm of the cross in Fig. 5.3. The shaded area in (b) results from the cross correlation of the voltages from the two arrays and contains all the frequency components in the corresponding function for the cross.

adjustment when used with the relatively simple phase-switching receivers used in the 1960s. Also, the width of the half-length arm must be doubled to obtain the same collecting area, that is, the same sensitivity. Although the characteristics of the T array were realized soon after the cross was developed, the cross was more commonly used in early instruments.

One of the first T-shaped apertures was developed at Cambridge, England. Two Cambridge instruments used to synthesize high-resolution observations are shown in Fig. 5.5. In each one a small antenna was moved to a new position between successive scans, and the records from the various positions were combined by computer to synthesize the response with the complete north–south aperture. Consider the response in terms of the (u, v, w) coordinate system as defined in Section 4.1, with the w axis in the zenith direction and the v axis in the meridian. The resolution in the ξ direction is provided in real time by the long east–west antenna in each case, and the sky brightness is synthesized in the η direction only. The n positions of the movable antenna are indicated by the coordinate v, and for angles close to the meridian ($\xi = 0$) the synthesized brightness can be derived from Eq. (4.6):

$$A(\eta)B'(t, \eta) = \sqrt{1 - \eta^2} \sum_{i=1}^{n} w_i \left[\mathcal{V}_R'(t, v_i)\cos(2\pi v_i\eta) - \mathcal{V}_I'(t, v_i)\sin(2\pi v_i\eta) \right].$$

$$(5.6)$$

Here $A(\eta)$ is the geometric mean of the power reception patterns of the two antennas in the η direction, B' is the measured brightness, and $\mathcal{V}_R'$ and $\mathcal{V}_I'$ are the real and imaginary parts of the visibility. In these functions the variable ξ is replaced by the time of meridian transit. w_i is a weighting factor that can be chosen to simulate the desired excitation taper in the north–south aperture. A similar expression is given by Blythe (1957). In plotting the derived bright-

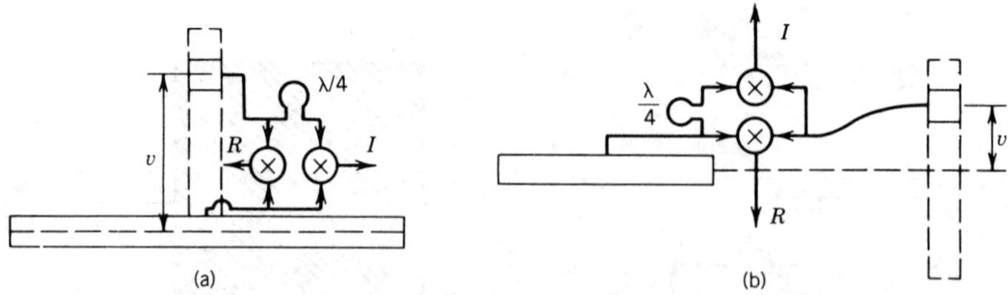

Figure 5.5 Schematic diagrams of two instruments in each of which a small antenna is moved to different positions between successive observations to synthesize the response that would be obtained with a full aperture corresponding to the rectangle shown by the broken line. The outputs of the two correlators R and I represent the real and imaginary parts of the visibility. Instruments of both types, the T-array (a), and the two-element interferometer (b), were constructed at the Mullard Radio Astronomy Observatory, Cambridge, England.

ness profiles, t is interpreted, of course, as the corresponding right ascension. The real part of $\mathcal{V}'$ is obtained by connecting the antennas in phase to the inputs of a correlation receiver. The imaginary part is obtained by introducing a $\pi/2$ phase shift into one signal path. The two parts can be measured by using two correlators connected to the antennas through a hybrid junction, or by time sharing with a single correlator and switching the $\pi/2$ phase shifter in and out of one signal path. These schemes perform the function of a complex correlator, which is discussed in Chapter 6.

In the operation of the two synthesis instruments in Fig. 5.5, n scans of duration 24 hr each are required to complete an observation. Since the small antenna has a beamwidth n times wider than the long aperture that it synthesizes, the time taken to survey a large area of the sky is the same as would be required if the full north–south aperture were used. With the full aperture the sensitivity for a given observing time would be $\sqrt{n}$ times greater than with the movable antenna, but with the latter each point in the sky area surveyed is scanned n times, so the same signal-to-noise ratio is obtained in either case. This equality of observing times and sensitivity does not always hold, for example, when a small antenna is moved in two dimensions to cover a large aperture. Observations of the above type were first discussed by Blythe (1957), Ryle, Hewish, and Shakeshaft (1959), and Ryle and Hewish (1960). In the last of these references the characteristics of several combinations of antennas are discussed in some detail.

The use of full apertures in the east–west direction in which the sky is automatically scanned by earth rotation, and synthesized apertures in the north–south direction, is an effective scheme for covering a large area of sky in a single series of moves of the small antenna. Synthesis mapping in one dimension is less demanding in computer capacity than synthesis in two dimensions and was better suited to the relatively primitive computers that were available when these instruments were in use.

5.3 PHASED ARRAYS AND CORRELATOR ARRAYS

In the development of mapping arrays in radio astronomy, some of the instruments were operated wholly or in part as phased arrays. It is useful, therefore, to examine the basic distinction between phased and correlator arrays. A schematic diagram of a four-element phased array feeding a square-law detector is shown in Fig. 5.6a. The signals from the antennas are combined in a branching network of transmission lines. If the voltages at the antenna outputs are V_1, V_2, V_3, and V_4, the output of the square-law detector is

$$\langle (V_1 + V_2 + V_3 + V_4)^2 \rangle. \tag{5.7}$$

Then if the signal path from each antenna to the detector is of the same

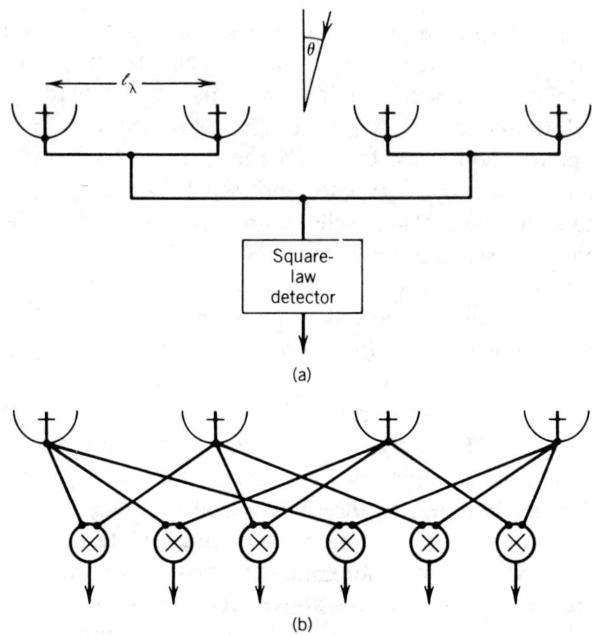

(a)

(b)

Figure 5.6 Simple four-element linear array, connected in (a) as a phased array feeding a square-law detector, and in (b) as a correlator array. ℓ_λ is the unit antenna spacing measured in wavelengths, and θ indicates the angle of incidence of a signal from a distant source.

electrical length, the signals combine in phase when the direction of the incoming radiation is given by

$$\theta = \sin^{-1}\left(\frac{n}{\ell_\lambda}\right), \tag{5.8}$$

where n is an integer including zero and ℓ_λ is the spacing interval measured in wavelengths. The position angles of the maxima, which represent the beam pattern of the array, can be varied by inserting phase shifters at the antenna outputs, and the beam pattern can thereby be scanned across the sky.

Now consider the correlator array in Fig. 5.6b. The correlator outputs generate cross products of the signal voltages of the form $V_m V_n$, and together their outputs contain all the terms obtained by expanding Expression (5.7), except for the self-products of the form V_m^2. The omission of such terms, which represent spatial frequencies near the (u, v) origin, has already been encountered in the example in Fig. 5.2 and does not have a major effect on the response of an array. The outputs of the correlators can be combined to produce maxima similar to those of the phased array. If one were to introduce a phase shift at the output of one of the correlator-array antennas, the result would appear as a corresponding change in the phase of the visibility measured with the correlators connected to that antenna. Thus, the effect of an antenna phase shift can be simulated by changing the measured visibility phase. In effect, the beam-scanning action is accomplished by combining measured visibility data in a computer with appropriate variations in the phase. This, of course, is what happens in computing the Fourier transform of the visibility function. The correlator array gathers data at a much greater rate than the phased array, unless the latter is equipped with a more complex signal-combining network that allows many beams to be formed simultaneously. It is interesting to note, however, that for the restricted problem of measuring the flux density of a point source of known position, the phased array would have greater sensitivity than the corresponding correlator array by a factor of $\sqrt{n_a/(n_a - 1)}$, n_a being the number of antennas, because of the inclusion of the self-product terms.

In the design of a synthesis array the principal concern is the sensitivity to different spatial frequencies. This is the same for the two arrays in Fig. 5.6, except for the missing self-product terms of the correlator array. In some cases, however, the method of interconnection of the antennas limits the pairs between which the cross correlation of the signals can be measured. For example, this occurs when two phased arrays are interconnected by a single correlator.

5.4　SPECTRAL SENSITIVITY AND TRANSFER FUNCTIONS OF A TRACKING ARRAY

The spectral sensitivity function of a tracking interferometer is illustrated in Fig. 5.7, in which the two shaded areas in (b) represent the support of the cross-correlation function of the two apertures of an east–west interferometer for a source on the meridian. As the source moves in hour angle, the changing (u, v) coverage is represented by a band centered on the spacing (i.e. spatial-frequency) locus of the two antennas. Recall from Section 4.2 that the locus for an earth-based interferometer is an arc of an ellipse, and that since $\mathcal{V}(-u, -v) = \mathcal{V}^*(u, v)$, any pair of antennas measures visibility along two arcs symmetric about the (u, v) origin, both of which are included in the spectral sensitivity function.

The tracking motion of the antennas introduces an effect which is illustrated in Fig. 5.8. Consider the radiation received by small areas of the apertures of two antennas centered at points A_1 and B. If the antennas remain fixed the corresponding signal components combine to form a component of the fringe pattern with frequency $\omega_e(u + \Delta u)\cos\delta$, as explained in Section 4.4. Now if the antennas track the source, the point B has a component of motion toward the source equal to $\omega_e \Delta u \cos\delta$ wavelengths per second, and this must be subtracted from the expression for the fringe frequency in the nontracking case. The fringe frequency when tracking thus becomes $\omega_e u \cos\delta$, the same as

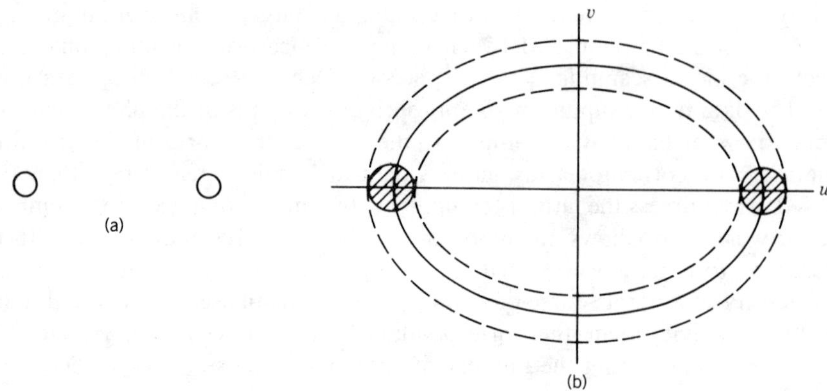

Figure 5.7 (a) The aperture of an east–west, two-element interferometer. The corresponding spectral sensitivity function for cross-correlated signals is shown by the shaded areas in (b). If the antennas track the source, the spacing vector traces out an elliptical locus (the solid line) in the (u, v) plane. The area between the broken lines in (b) indicates the spatial frequencies for which the visibility function contributes to the measured values. The spacing between the broken lines is determined by the cross correlation of the antenna apertures.

for the central points A_1 and A_2 of the apertures. So if the antennas track, the contributions from all pairs of points within the apertures appear at the same fringe frequency in the correlator output. Such contributions cannot be separated by Fourier analysis of the correlator output voltage, although they can in the nontracking case. Correspondingly in the (ξ, η) domain the antenna beams remain fixed with respect to the source, and thus the antenna pattern cannot be measured by examining the correlator output. One cannot, for example, distinguish between a source of brightness $B(\xi, \eta)$ observed with normalized antenna reception patterns $A_N(\xi, \eta)$ and a source of brightness $A_N(\xi, \eta)B(\xi, \eta)$ observed with isotropic antennas.

To accommodate the above effect it is usual to include the antenna patterns as a modification to the brightness distribution, which then becomes

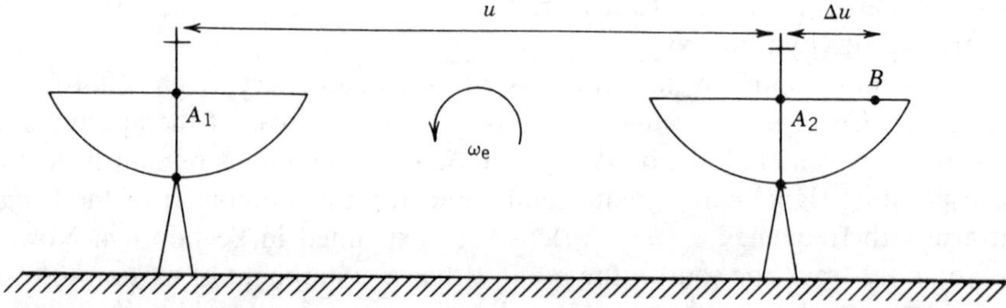

Figure 5.8   Illustration of the effect of tracking on the fringe frequency of the correlator output components. For simplicity the diagram shows an east–west baseline and antennas pointing at a source at meridian transit. The curved arrow indicates the tracking motion of the antennas.

$A_N(\xi, \eta) B(\xi, \eta)$, and to use a form of the spectral sensitivity function in which the aperture distributions of the antennas are represented by two-dimensional delta functions of x and y. For any given instant of time the resulting cross correlations are then represented by a series of delta functions of u and v, weighted in proportion to the magnitude of the response. As the earth rotates these delta functions generate a function of which the support is the ensemble of spacing loci, and the magnitude at any point is proportional to the instrumental response. This form of the spectral sensitivity function is usually called the *transfer function* of the array. It is, of course, a function of the declination and hour-angle range of the observation. Frequently, no distinction is made between the function and its support. This can be justified in practice because the magnitude of the response to any spatial frequency can be adjusted by weighting the visibility data. The term transfer function arises from an approximate analogy with network theory and refers to the spatial-frequency filtering action of an array.

Consider observation of a source $B(\xi, \eta)$, for which the visibility function is $\mathscr{V}(u, v)$, with antenna patterns $A_N(\xi, \eta)$. Then if $W(u, v)$ is the transfer function the measured visibility is

$$\left[\mathscr{V}(u, v) ** \overline{A}_N(u, v) \right] W(u, v), \tag{5.9}$$

where the double asterisk indicates two-dimensional convolution and the bar denotes the Fourier transform. The Fourier transform of Expression (5.9) gives the measured brightness (we assume that the factor $\sqrt{1 - \xi^2 - \eta^2}$ in Eq. (4.6) can be neglected):

$$\left[B(\xi, \eta) A_N(\xi, \eta) \right] ** \overline{W}(\xi, \eta). \tag{5.10}$$

If we observe a point source at the (ξ, η) origin, where $A_N = 1$, Expression (5.10) becomes the point-source response $b_0(\xi, \eta)$:

$$b_0(\xi, \eta) = \left[{}^2\delta(\xi, \eta) A_N(\xi, \eta) \right] ** \overline{W}(\xi, \eta) = \overline{W}(\xi, \eta). \tag{5.11}$$

Thus the point-source response is the Fourier transform of the transfer function. In the tracking case the spatial frequencies that contribute to the measurement are represented by $W(u, v) ** \overline{A}_N(u, v)$, as should be clear from Fig. 5.7. It will be seen that the support of $\overline{A}_N(u, v)$ is twice as wide as the corresponding aperture in the (x, y) domain.

5.5   ONE-DIMENSIONAL TRACKING ARRAYS

High-resolution mapping of individual sources usually involves tracking in hour angle to obtain sufficient sensitivity and adequate coverage of the (u, v) plane. In designing an array for such observations a principal concern is to obtain measurements of visibility that are distributed as uniformly as possible in u and v. We have seen in Section 4.2 that for pairs of antennas with east–west spacings, the loci in the (u, v) plane are a series of concentric ellipses. If the antenna spacings are chosen to increase in uniform increments, the density of the loci within the (u, v) plane varies only by the ratio of axes of the ellipses, which is equal to the sine of the declination. East–west linear arrays containing spacings at multiples of a basic interval have therefore found wide use in radio astronomy, particularly for observations at $|\delta| \gtrsim 30°$.

In the simplest type of linear array the antennas are spaced at uniform intervals ℓ_λ as in Fig. 5.6. This type of array is sometimes known as a grating array by analogy with an optical diffraction grating. If there are n_a antennas, such an array output contains $(n_a - 1)$ combinations with the unit spacing, $(n_a - 2)$ with twice the unit spacing, and so on. Thus short spacings are highly redundant, and one is led to seek other ways to configure the antennas to provide larger numbers of different spacings for a given n_a.

A type of array of some historical importance is shown in Fig 5.9. Here the antennas are divided into two phased subarrays, the signals from which are combined in a single correlator. An examination of Fig. 5.9 will show that combinations of antennas containing one from each subarray can be found for all multiples of ℓ_λ up to $16\ell_\lambda$ compared with $7\,\ell_\lambda$ for a simple eight-element grating. Arrays of the type shown in Fig. 5.9 are referred to as compound interferometers, following the designation of Covington and Broten (1957). They were commonly used during the 1960s with a phase-switching receiver to combine the outputs of the two arrays. The response consists of fan-beam maxima at spacings given by Eq. (5.8), each beam having a profile approximating a sinc function. If the separation of the centers of the two subarrays is an odd multiple of $\ell_\lambda/2$, the responses at the correlator are of alternate polarity for adjacent beams; if the spacing is an even multiple of $\ell_\lambda/2$ the responses are all the same polarity. Both types of design can be found among instruments described by Labrum et al. (1963), Picken and Swarup (1964), and Thompson and Krishnan (1965). These instruments were real-time, beam-forming devices, and the output from the single correlator was well adapted to the limited computing facilities of the period.

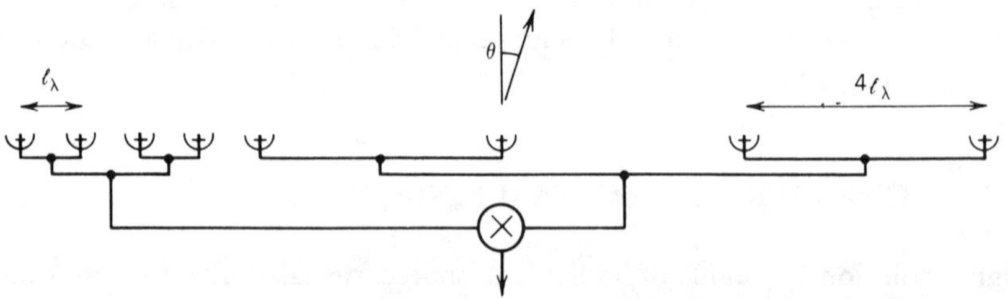

Figure 5.9 An eight-element array with a single correlator in which the antennas are divided into two phased subarrays. The correlator output contains the responses of antenna pairs with spacings at all multiples of ℓ_λ from 1 to 16.

The use of a separate correlator for each antenna pair, or for each non-redundant spacing, has now become the general practice. This not only increases the sensitivity but allows phase and amplitude calibration to be applied in the computer. Phase adjustment in the computer is more convenient, and usually more accurate, than adjusting the signal phases of the individual elements as must be done in the two four-element subarrays in Fig. 5.9. Finally, the use of multiple correlators allows much greater flexibility in array design. For up to four antennas linear arrays with no redundant spacings exist. The four-element configuration was first used by Arsac (1955) and is shown in Fig. 5.10a. The six possible pair combinations all have different spacings. With more than four antennas there is always either some redundancy or missing spacings. A five-element, *minimum-redundancy* configuration devised by Bracewell (1966) is shown in Fig. 5.10b. Moffet (1968) lists examples of minimum-redundancy arrays of up to 11 elements and points out that two classes may be defined. There are restricted arrays in which all spacings up to the maximum spacing, $n_{max}\ell_\lambda$ (i.e., the total length of the array), are present, and general arrays in which all spacings up to some particular value are present, and also some longer ones. Examples for eight elements are shown in Fig. 5.11. Arrays of the general type can provide slightly better performance but may be more than twice as long as the restricted configurations. A study in number theory by Leech (1956) indicates that for large numbers of elements the redundancy factor, defined as $\frac{1}{2}n_a(n_a - 1)/n_{max}$, approaches $\frac{4}{3}$. Minimum-redundancy solutions for up to 11 antennas are quoted by Moffet, and solutions for larger arrays are discussed by Ishiguro (1980). Only a few minimum-redundancy arrays have been constructed, a notable example being the one in Fig. 5.10b (Bracewell et al. 1973). This type of array is not the most convenient if future extension is envisaged, since if the total number of antennas is increased the whole layout may have to be changed to maintain minimum redundancy.

The ability to move a small number of elements between observations, adds greatly to the range of performance of an array. Figure 5.12 shows the arrangement of the three antennas of the Cambridge One-Mile telescope (Ryle 1962). Antennas 1 and 2 are fixed and their outputs are correlated with that

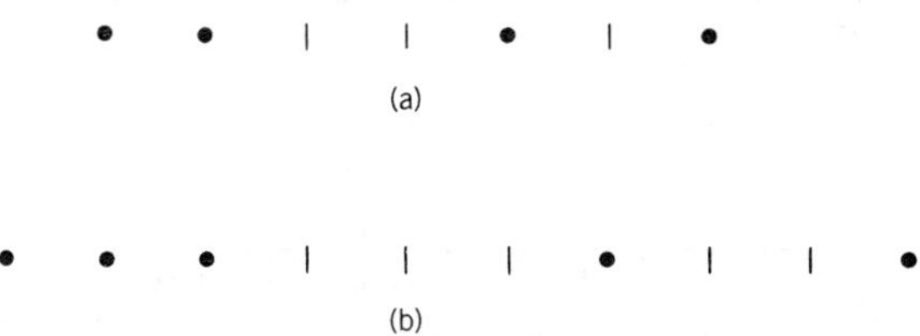

(a)

(b)

Figure 5.10 (a) Arsac's (1955) configuration containing all spacings up to six times the unit spacing, and (b) Bracewell's (1966) configuration containing all spacings up to nine times the unit spacing. Filled circles represent antennas. Part (a) contains no redundant spacings and (b) contains only one, the unit spacing.

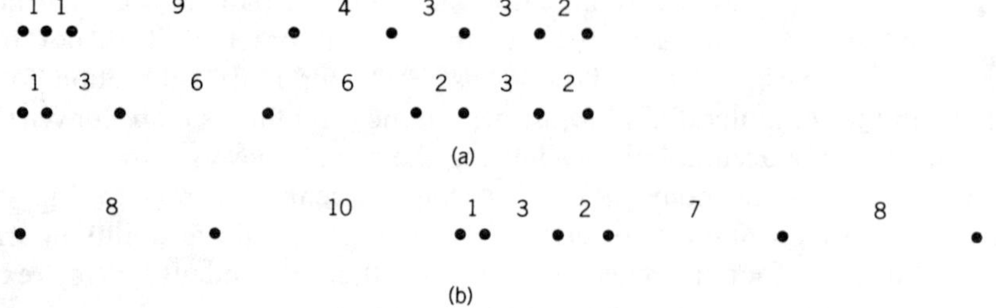

(a)

(b)

Figure 5.11   Eight-element, minimum-redundancy linear arrays: the numbers indicate spacings in multiples of the unit spacing. (a) Two restricted arrays that uniformly cover the range of 1 to 23 times the unit spacing. (b) A general array that uniformly covers 1 to 24 times the unit spacing. This array has a length of 39 times the unit spacing. The extra spacings are 8, 31 (twice), and 39 times the unit spacing. From Moffet (1968), © 1968 IEEE.

from antenna 3 which can be moved on a rail track. In each position of antenna 3 the source under observation is tracked for 12 hr, and visibility data are obtained over two spacing loci in the (u, v) plane. The observation is repeated as antenna 3 is moved progressively along the track, and the increments in the position of this antenna determine the spacing of the elliptical loci in the (u, v) plane. From the sampling theorem (Section 4.11), the required spacing is the reciprocal of the angular width of the source under investigation. Thus the ability to vary the incremental spacing adds versatility to the array and reduces the number of antennas required. The configuration of a larger instrument of this type, the Westerbork synthesis radio telescope (Baars and Hooghoudt 1974, Högbom and Brouw 1974), is shown in Fig. 5.13. Here ten fixed antennas are combined with four movable ones, and the time required to complete a sequence of observations is much less than with the three-element array.

The sampling of the visibility function at points on concentric, equispaced ellipses in the (u, v) plane results in the introduction of ringlobe responses. These may be understood by noting that for a linear array the instantaneous spacings are represented in one dimension by a series of δ functions, as shown in Fig. 5.14. If the array contains all multiples of the unit spacing up to $N\ell_\lambda$, and if the corresponding visibility measurements are combined with equal

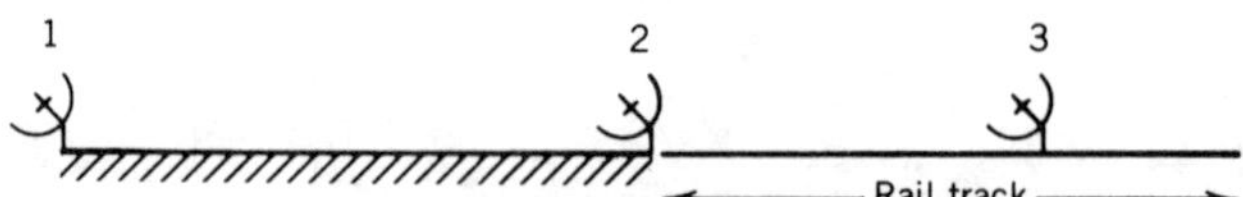

Figure 5.12   The Cambridge One-Mile radio telescope. Antennas 1 and 2 are at fixed locations and the signals they receive are each correlated with the signal from antenna 3 which can be located at various positions along a rail track. The fixed antennas are 762 m apart and the rail track is a further 762 m long. The unit spacing is equal to the increment of the position of antenna 3, and all multiples up to 1524 m can be obtained.

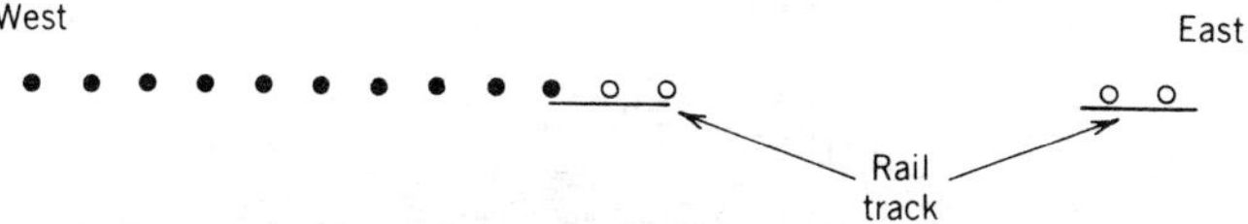

Figure 5.13   Antenna configuration of the Westerbork synthesis radio telescope. The ten filled circles represent antennas at fixed locations and the four open circles represent antennas that are movable on rail tracks. Forty correlators are used to combine the signals from each of the fixed antennas with the signals from each of the movable ones. The diameter of the antennas is 25 m and the spacing of the fixed antennas is 144 m. As originally constructed the array contained only the 12 western antennas, and the two at the east end were added later to double the range of spacings.

weights, the instantaneous response is a series of fan beams each with a profile of sinc-function form, as in Fig. 5.14b. This follows from the Fourier transform relationship for a truncated series of delta functions:

$$\sum_{n=-N}^{N} \delta(u - n\ell_\lambda) \rightleftharpoons \frac{\sin\left[(2N + 1)\pi\ell_\lambda\xi\right]}{\pi\ell_\lambda\xi} * \sum_{m=-\infty}^{\infty} \delta\left(\xi - \frac{m}{\ell_\lambda}\right),$$

$$(5.12)$$

where the left-hand side represents the spacings in the u domain and the right-hand side the beam pattern in the ξ domain. As the earth's rotation causes the spacing vectors to sweep out ellipses in the (u, v) plane, the corresponding rotation of the array relative to the sky can be visualized as causing a central fan beam to rotate into a narrow pencil beam, while its neighbors give rise to lower-level, ring-shaped responses concentric with the central beam, as in Fig. 5.15. This general argument gives the correct spacing of the ringlobes, the profile of which is modified from the sinc-function form (Bracewell and Thompson 1973). If the transfer function in the (u', v') plane is a series of circular delta functions of radius $q', 2q', \ldots, Nq'$, the profile of the

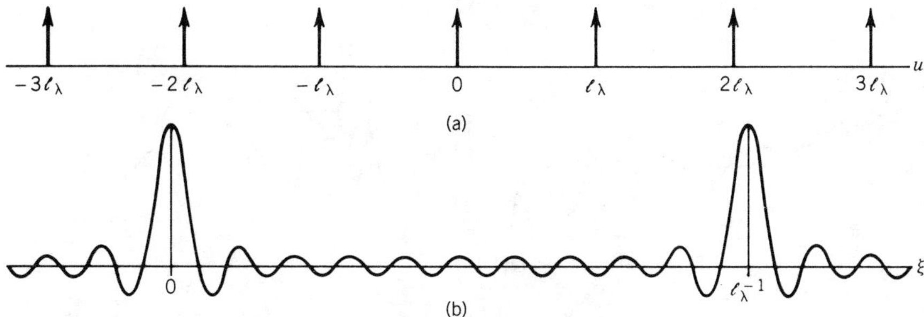

Figure 5.14   Part of a series of δ functions representing the instantaneous distribution of spacings for a uniformly spaced, linear array. (b) Part of the corresponding series of fan beams that constitute the instantaneous response. Parts (a) and (b) represent the left and right sides of Eq. (5.12), respectively.

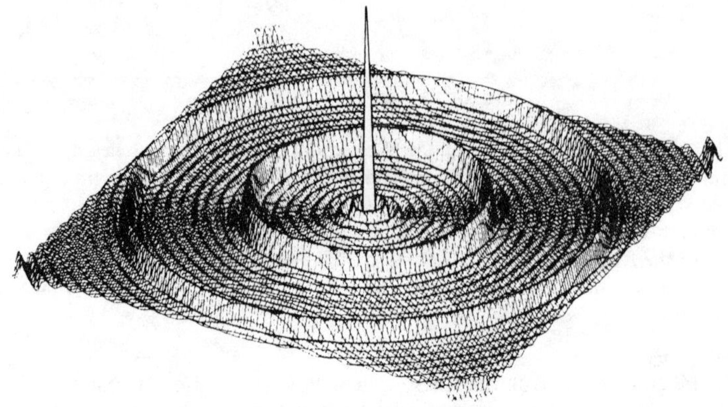

Figure 5.15   Example of ringlobes. The response of an array for which the support of the transfer function is a series of nine circles concentric with the (u, v) origin. The radii of these circles are consecutive integral multiples of the unit antenna spacing. The weighting corresponds to the principal response discussed in Chapter 10. From Bracewell and Thompson (1973).

k^{th} ringlobe is of the form

$$\text{sinc}^{1/2}\left[2\left(N + \tfrac{1}{2}\right)\left(q'r' - k\right)\right], \tag{5.13}$$

where $r' = \sqrt{\xi'^2 + \eta'^2}$. The function $\text{sinc}^{1/2}(\chi)$ is plotted in Fig. 5.16 and is the half-order derivative of $\sin(\pi\chi)/\pi\chi$. It can be computed using Fresnel integrals. The form and relative amplitudes of ringlobes in a map are modified by the receiving bandwidth (see Chapter 6) and by the responses of the antennas.

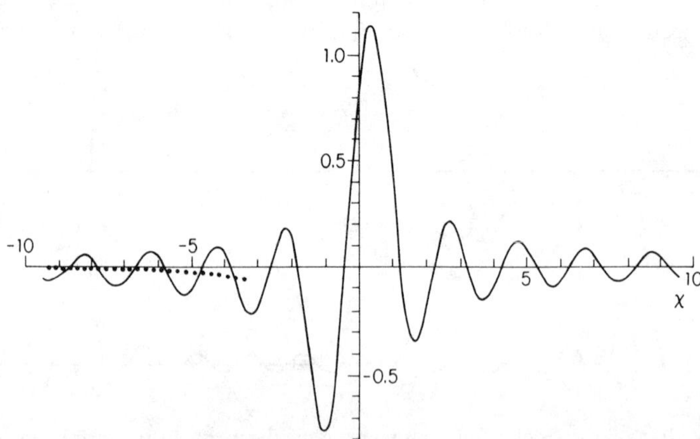

Figure 5.16   Cross section of a ringlobe in the principal response to a point source of an east–west array with uniform increments in antenna spacing. The left-hand side is the inside of the ring and the right the outside. The dotted line indicates a negative mean level of the oscillations on the inner side. From Bracewell and Thompson (1973).

As mentioned above, the application of the sampling theorem to the choice of incremental spacing requires that the latter be no greater than the reciprocal of the source width. In terms of ringlobes, this condition ensures that the minimum ringlobe spacing is no less than the source width. Thus the main beam response to a source is not overlapped by a ringlobe response to the same source. In arrays like those in Figs. 5.12 and 5.13, ringlobes can be effectively suppressed if the movable antennas are positioned in steps slightly less than the antenna diameter, in which case the ringlobe lies outside the primary antenna beam. Note, however, that the first spacing cannot be less than the antenna diameter, and the missing low-spacing measurements may have to be obtained by other means (see Chapter 10). Ringlobes can also be removed by image-processing techniques such as the CLEAN algorithm which is described in Chapter 11. Thus although ringlobes are prominent unwanted features in some radio maps, the danger of their erroneous interpretation as astronomical structure can usually be made very small.

5.6 TWO-DIMENSIONAL TRACKING ARRAYS

As noted in Section 4.1, the spacing loci for an east–west array lie in the equatorial plane and become severely foreshortened for observations near the celestial equator. For two-dimensional mapping of sources near the celestial equator, a configuration of antennas is required in which the Z component of the antenna spacing, as defined in Section 4.2, is comparable to the X and Y components. This is achieved by including spacings with azimuths other than east–west. The optimum configuration is then two dimensional. An array located at an intermediate latitude and designed to operate at low declinations can cover the sky from the pole to declinations of about $30°$ into the opposite celestial hemisphere. This range includes about 70% of the total sky, that is, almost three times as much as that of an east–west array. Furthermore, the low declination range is essential for solar system studies.

In two dimensions there exist no known minimum-redundancy solutions like those for linear arrays. The choice of an effective configuration of antennas is largely empirical. Since large arrays are expensive, economic factors are important. For connected-element arrays practical considerations of land acquisition and power and signal distribution point toward simple configurations, such as a series of linear subarrays extending outward from some central point where the signal-processing hardware can be located. A cross fulfills these requirements, and two early instruments, both designed for solar observations, were crossed arrays of small tracking paraboloids (Christiansen and Mathewson 1958, Bracewell and Swarup 1961). In both cases the antennas were equally spaced along the arms in order to provide the necessary spacings in two dimensions; a combination of two minimum-redundancy linear arrays would not give the desired result.

The considerations involved in the optimization of a two-dimensional array are well illustrated by the design of the Very Large Array (VLA) of the

National Radio Astronomy Observatory (Thompson et al. 1980, Napier, Thompson, and Ekers 1983). This instrument is located at latitude 34°N in New Mexico and is therefore able to track objects as far south as $-30°$ for almost 7 hr without going below 10° in elevation. Performance specifications called for mapping with full resolution down to at least $-20°$ declination and for obtaining a map in no more than 8 hr of observation without moving antennas to new locations. Comparison of the performance of various antenna configurations was accomplished by computing the transfer function for tracking over the hour angle range ± 4 hr at various declinations. The (u, v) plane was divided into cells of dimensions appropriate for mapping, and the percentage of holes, that is, cells not intersected by a spacing locus, was computed. To take account of tapering which may be applied in computing a map, a weighted sum of the number of holes was used in certain cases, the weighting being a Gaussian tapering function. The basic concern was to minimize sidelobes in the synthesized beam, but computing the beam for each configuration and declination took much more computer time than computing the percentage of holes in the (u, v) plane. Sample calculations of the beam showed a sufficiently consistent relationship between the sidelobe levels and the percentage of holes that it was possible to use the latter quantity for performance comparisons (NRAO 1967, 1969).

Various possibilities were explored, and a three-armed, equiangular, Y-shaped array was the best simple arrangement that was found. It has all the advantages of a cross, and can be considered as derived from one by omitting the largely redundant half of one arm to obtain a T, and then optimizing the angular spacing of the three remaining arms. Inverting the Y has no effect on the beam, but if the antennas have the same radial disposition on each arm, the performance near zero declination is improved by rotating the array so that the nominal north or south arm makes an angle of about 5° with the north–south direction. Without this rotation the baselines between corresponding antennas on the other two arms are exactly east–west, and for $\delta = 0°$ the spacing loci degenerate to straight lines that are coincident with the u axis in a highly redundant manner. The total number of antennas, 27, was chosen from a consideration of (u, v) coverage and sidelobe levels, and resulted in peak sidelobes at least 16 dB below the main beam response, except at $\delta = 0°$ where earth rotation is least effective. The 27 antennas provide 351 pair combinations.

The positions of the antennas along the arms provided another set of variables that could be adjusted to optimize the spacing loci. One approach to this problem was found in a pseudodynamic computation technique by Mathur (1969), in which arbitrarily chosen initial conditions are adjusted by a computer until an optimum (u, v) coverage is reached. The configuration obtained by Mathur is shown in Fig. 5.17a. Subsequently, a more analytical approach was attempted by Chow (1972), who represented the distribution of baselines by a flat surface with an elliptical boundary and examined the volume swept out in space by this surface as the antennas track a source. This analysis led to

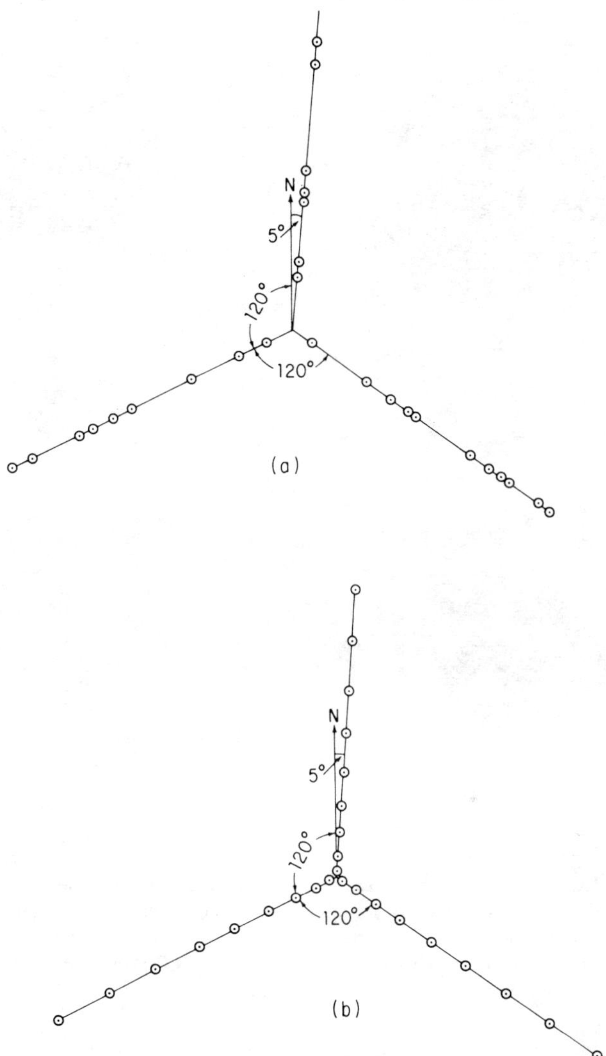

Figure 5.17   (a) Proposed antenna configuration for the VLA that resulted from Mathur's (1969) computer-optimized design. (b) Power-law design adopted for the VLA. From Napier, Thompson, and Ekers (1983), © 1983 IEEE.

the conclusion that a power-law spacing in which the distance of the nth antenna on an arm is proportional to n^{α} would provide good (u, v) coverage. Comparison of the power-law spacing with $\alpha \simeq 1.7$ and Mathur's empirically optimized result showed the two to be essentially equal in performance. The power-law scheme was chosen for the construction of the VLA for important practical reasons. A requirement of the design was that four sets of antenna

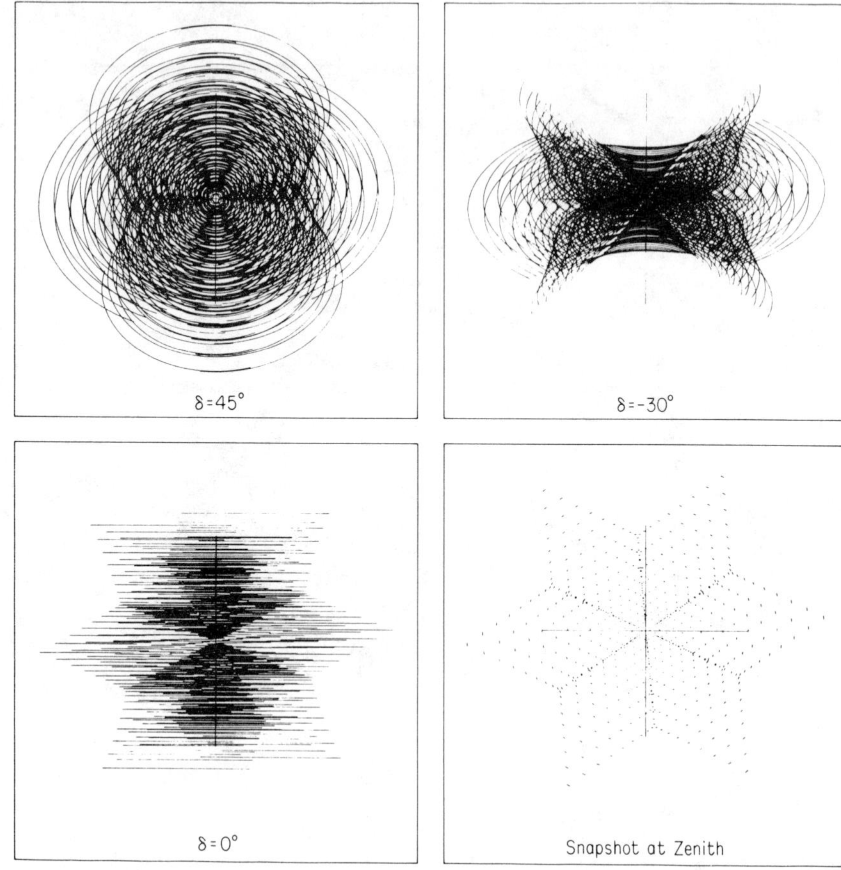

Figure 5.18   Transfer functions in the (u, v) plane obtained with the power-law configuration of Fig. 5.17b. The range of hour angle is ± 4 hr or as limited by a minimum pointing elevation of $9°$, and ± 5 min for the snapshot. The lengths of the (u, v) axes from the origin represent the maximum distance of an antenna from the array center, that is, 21 km for the largest configuration. From Napier, Thompson, and Ekers (1983), © 1983 IEEE.

stations be provided to vary the scale of the spacings in four steps, and thereby provide a choice of resolution and field of view for different astronomical objects. By making α equal to the logarithm to the base 2 of the scale factor between configurations, the location of the nth station for one configuration coincides with that of the $2n$th station for the next smaller configuration. The total number of antenna stations required was thereby reduced from 108 to 72. Another advantage of the power-law spacing is that the array can be enlarged by adding more antennas to the arms without changing the existing stations. The power-law configuration for the VLA is shown in Fig. 5.17b and plots of the transfer function in Fig. 5.18. The snapshot in Fig. 5.18 shows the instantaneous (u, v) coverage, which is adequate for mapping simple structure in strong sources without extensive tracking. Note that the effects of sidelobes can be greatly reduced by processing techniques described in Chapter 11, and images of high contrast thereby obtained. Nevertheless, the percentage of unsampled cells in the (u, v) plane remains a good measure of the efficiency with which the array gathers visibility data.

One other two-dimensional configuration that should be mentioned is a circle in which the antennas are uniformly spaced around the circumference, or an ellipse if it is necessary to compensate for forshortening at some declination range. Such an array provides a good distribution of spacings in both angle and distance out to a limit imposed by the diameter of the circle. Within this limit the (u, v) loci of the transfer function are more uniformly distributed than, for example, those of the VLA. The ring is less convenient for use as a general purpose instrument than an array with open-ended arms because it is more difficult to include provision to change the scale of the configuration. For observations of a specific object a fixed scale can be used, and an array of this type was constructed at Culgoora, Australia, for observations of the sun (Wild 1967). This was a multibeam, scanning, phased array rather than a correlator array, consisting of 96 antennas arranged around a circle of diameter 3 km and operating at 80 and 160 MHz. To suppress unwanted sidelobes of the beam, Wild (1965) devised an ingenious phase-switching scheme termed J^2 synthesis. The Culgoora array has also been analyzed as a correlator array by Swenson and Mathur (1967), and this study illustrates some possibilities of the ring configuration.

Other large instruments with interesting configurations include the Australia telescope (Frater 1984) and the array at Ootacamund (Swarup 1984).

5.7 VLBI ARRAYS

For arrays of antennas that operate in the very-long-baseline mode the criteria for adequate coverage of the (u, v) plane are essentially identical to those discussed above. The main differences in the layout of antennas result from the practical operating details. Since there is no attempt to interconnect the antennas directly, except by computer links for control purposes, simple geometrical configurations offer no particular advantages. More important factors involve proximity to existing observatories for technical support services, and access to transportation centers for return of tapes to the correlator facility. Ranges of hour angle and declination that are simultaneously observable from such widely spaced locations must also be considered. VLBI arrays thus tend to be flexible in configuration. Any suitable radio telescope with an atomic frequency standard, phase-locked oscillators, and appropriate receiving and recording systems can be used in a VLBI experiment. Several organized networks have been set up to coordinate joint experiments between different observations. Instruments that have frequently been involved in the U.S. network are listed in Table 5.1.

TABLE 5.1   Principal Antennas Used During the 1970s and 1980s in the U.S. VLBI Network

Antenna Diameter	Location of Antenna	Institution
37 m	Haystack Observatory, Westford, Massachusetts	North-East Radio Observatory Corporation
43 m	Green Bank, West Virginia	National Radio Astronomy Observatory
37 m	Vermilion River Observatory, Illinois	University of Illinois
18 m	North Liberty, Iowa	University of Iowa
26 m	George R. Agassiz Station, Fort Davis, Texas	Harvard College Observatory
26 m	Hat Creek Observatory, California	University of California
40 m	Owens Valley Radio Observatory, California	California Institute of Technology
26 m	Maryland Point, Maryland	U.S. Naval Observatory
64 m	Goldstone, California	Jet Propulsion Laboratory

For the first two decades after the inception of the VLBI technique, observations were mainly joint ventures between different institutions. Consideration of arrays dedicated solely to VLBI occurred as early as 1975 (e.g., see Swenson and Kellermann 1975), but construction of such instruments did not begin for another decade. A study of antenna locations for a VLBI array has been described by Seielstad, Swenson, and Webber (1979). To obtain a single number as a measure of the performance of any configuration, the transfer function was computed for a number of declinations. The fraction of appropriately sized (u, v) cells intersected by the spacing loci was then weighted in proportion to the area of sky at each declination and averaged. Other studies have involved computing the response to a model source, synthesizing a map, and comparing the result with the model. In such studies the closure phase and related techniques that are described in Chapter 11 can be used, and an estimate of the dynamic range, that is, the brightness ratio of the minimum reliable detail to the maximum in the map, provides a measure of the quality of the configuration. The use of closure phase introduces another consideration into array design: the efficiency of this method increases with the number of antennas in the array, as discussed in Chapter 11. The minimum desirable number of antennas is not sharply defined but is in the range five to ten (see Fig. 11.4).

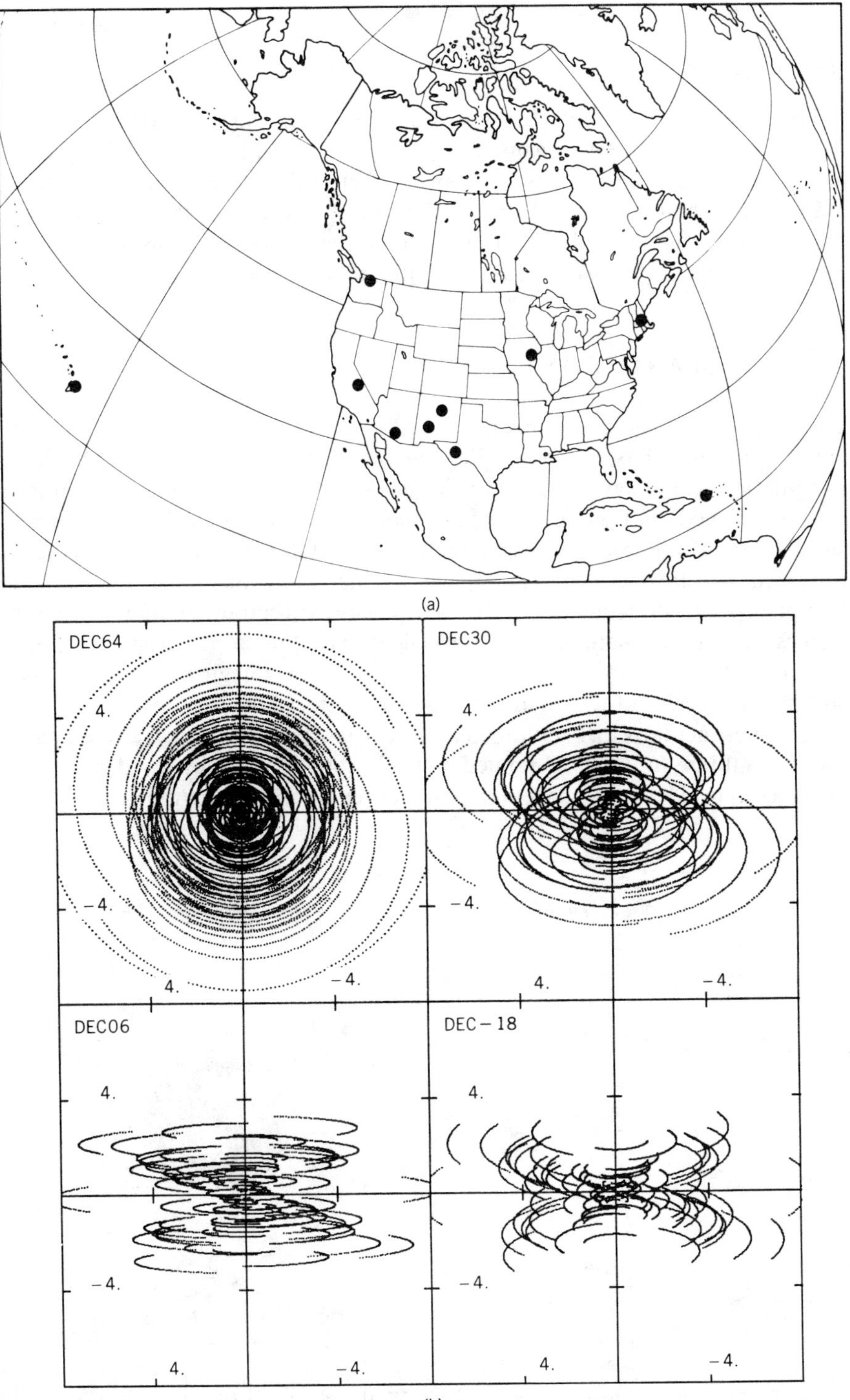

Figure 5.19 Very-long-Baseline Array in the United States: (a) locations of the 10 antennas, and (b) transfer functions (spacings in thousands of kilometers) for declinations of 64°, 30°, 6°, and −18°, the observing time at each antenna being determined by an elevation limit of 10°. From Walker (1984).

The design of a Canadian array and the Very-Long-Baseline Array (VLBA) of the United States (Kellermann and Thompson 1985), based on the considerations outlined above, are described by Walker (1984). The antenna locations and (u, v) coverage for the VLBA are shown in Fig. 5.19. Studies for this array also indicate that an additional site within the southern hemisphere could greatly enhance the (u, v) coverage at southern declinations.

5.8   VLBI ANTENNAS IN SPACE

A logical step in the development of VLBI from ground-based arrays is the addition of antennas in space (Preston et al. 1983, Burke 1984). Such a combination would have several obvious advantages. Higher angular resolution can be achieved, and the ultimate limit may be set by interstellar scintillation (see Chapter 13). Because of a space station's orbital motion the (u, v) plane is well filled, and this permits higher dynamic range in the maps. The coverage in the north–south direction, which is usually poor with ground-based arrays for sources at low declination, can be excellent. The projected baselines change more rapidly than those of ground-based arrays, so imaging of sources that vary on time scales of less than a day would be possible.

An experimental VLBI station on a low earth orbit vehicle such as the Space Shuttle could be fully self-contained with an atomic frequency standard and a tape recorder. The (u, v) coverage on a source at 0° declination is shown in

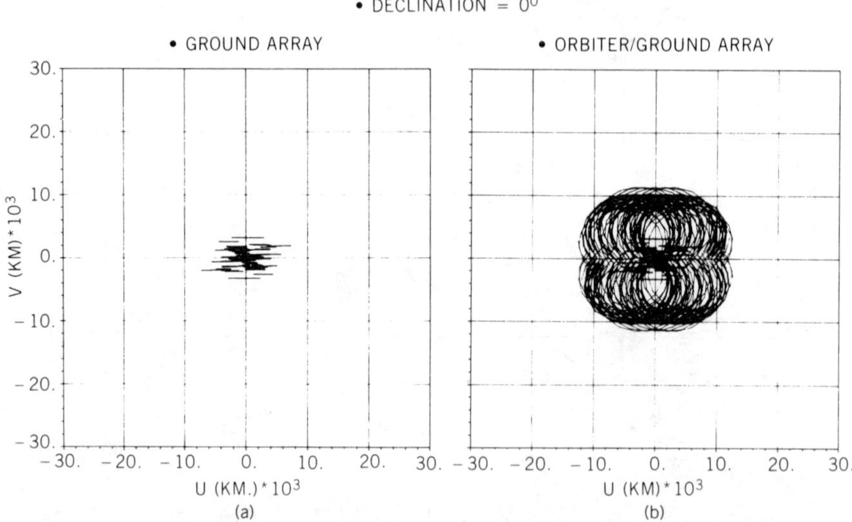

Figure 5.20   Transfer functions for a ground-based VLBI array and a source at declination 0°. (b) Transfer functions for the same source and the same array combined with an antenna in low earth orbit, and a 16-hr observing period. From Preston et al., in *Very Long Baseline Interferometry Techniques*, F. Biraud, Ed., Cepadues, France, 1983.

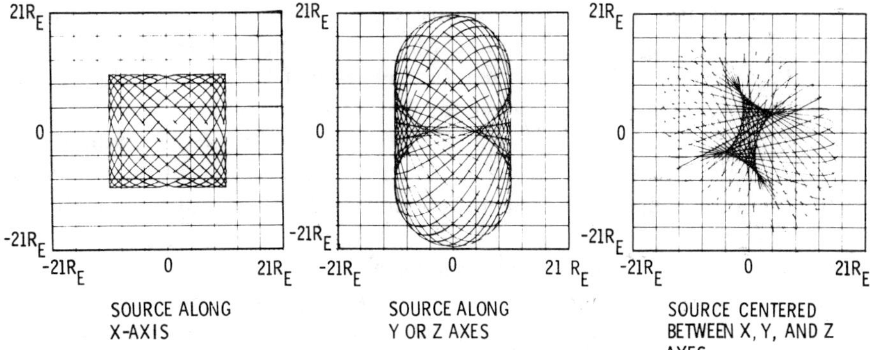

Figure 5.21 Transfer functions for two antennas on satellites with circular orbits of radius approximately ten times the earth's radius R_E. The orbits lie in the XY and XZ planes of a rectangular coordinate system. The satellite periods differ by 10% and the observing period is approximately 20 days. From Preston et al., in *Very Long Baseline Interferometry Techniques*, F. Biraud, Ed., Cepadues, France, 1983.

Fig. 5.20. The different locations of the ground-based antennas is an important factor in filling in the (u, v) coverage. For higher orbits, free-flying spacecraft may be used. The (u, v) coverage is similar to that shown in Fig. 5.20 but on a larger scale. With a high orbit, it would be feasible to transfer the time and frequency from the ground and send the data back on a telemetry link. A spacecraft in an orbit with a perigee altitude of 4000 km and apogee altitude of 15,000 km would complement nicely a ground-based array. For higher orbits there would be holes in the (u, v) plane, which could be filled by using multiple spacecraft. An example of the (u, v) coverage obtained between two spacecraft in circular orbits, with orthogonal planes that have periods differing by 10%, is shown in Fig. 5.21. A high degree of phase stability could be achieved on these baselines because they are outside the earth's atmosphere.

5.9 THEORETICAL RELATIONSHIPS AND APERTURE SYNTHESIS

The basic relationships between certain characteristics of antenna systems differ in important details for correlator arrays and for filled-aperture antennas. To review these effects we return to Fig. 5.1, which applies to radiation from an incoherent source. The entire diagram can be applied to mapping with a single antenna in which the output is measured by a receiver that is linear in power. The right-hand column shows the more conventional engineering viewpoint in which the output map is the convolution of the brightness distribution and the power reception pattern, the latter being the square of the modulus of the voltage reception pattern. In the left-hand column the equivalent processes in the (u, v) and antenna–aperture planes are shown.

For a correlator interferometer or array the relationships below the broken line in Fig. 5.1 apply, but those above it, in general, do not. In a correlator array the absence of the autocorrelation component for the individual antennas, as explained in Fig. 5.2, precludes the autocorrelation relationship at the top left in Fig. 5.1. However, this deficiency can be corrected by adding data from a single antenna. In a tracking array the Fourier transform of the point-source response is the transfer function, and the form of this function in which the support is a series of discrete elliptical arcs does not fulfill the mathematical requirements of an autocorrelation function. Correspondingly, the power reception pattern may have negative sidelobes and thus cannot be the squared modulus of any function. In general, therefore, the concepts of an equivalent aperture and a voltage reception pattern do not apply to a tracking correlator array.

Ryle, Hewish, and Shakeshaft (1959) and Ryle and Hewish (1960) have discussed observations with a nontracking correlator interferometer in which all spatial frequencies within a central area of the (u, v) plane are measured. In these cases an equivalent aperture can be defined, and the term *aperture synthesis* was introduced to describe them. In the subsequent development of synthesis mapping, the term aperture synthesis has often been extended to describe instruments for which no equivalent aperture is readily definable.

REFERENCES

Arsac, J., Nouveau Réseau Pour l'Observation Radioastronomique de la Brillance de la Soleil à 9530 Mc/s, C. R. *Acad. Sci.*, **240**, 942–945, 1955.

Baars, J. W. M. and B. G. Hooghoudt, The Synthesis Radio Telescope at Westerbork. General Layout and Mechanical Aspects. *Astron. Astrophys.*, **31**, 323–331, 1974.

Blythe, J. H., A New Type of Pencil Beam Aerial for Radio Astronomy. *Mon. Not. R. Astron. Soc.*, **117**, 644–651, 1957.

Bracewell, R. N., Interferometry and the Spectral Sensitivity Island Diagram, *IRE Trans. Antennas Propag.*, **AP-9**, 59–67, 1961.

Bracewell, R. N., Radio Astronomy Techniques, in *Handbuch der Physik*, Vol. 14, S. Flugge, Ed., Springer-Verlag, Berlin, 1962, pp. 42–129.

Bracewell, R. N., Optimum Spacings for Radio Telescopes with Unfilled Apertures, in *Progress in Scientific Radio*, Report on the 15th General Assembly of U.R.S.I., Publication 1468 of the National Academy of Sciences, Washington, DC, 1966, pp. 243–244.

Bracewell, R. N., R. S. Colvin, L. R. D'Addario, C. J. Grebenkemper, K. M. Price, and A. R. Thompson, The Stanford Five-Element Radio Telescope, *Proc. IEEE*, **61**, 1249–1257, 1973.

Bracewell, R. N. and G. Swarup, The Stanford Microwave Spectroheliograph Antenna, a Microsteradian Pencil Beam Interferometer, *IRE Trans. Antennas Propag.*, **AP-9**, 22–30, 1961.

Bracewell, R. N. and A. R. Thompson, The Main Beam and Ringlobes of an East–West Rotation-Synthesis Array, *Astrophys. J.*, **182**, 77–94, 1973.

Burke, B. F., Orbiting VLBI: A Survey, in *VLBI and Compact Radio Sources*, R. Fanti, K. Kellermann, and G. Setti, Eds., D. Reidel, Dordrecht, Holland, 1984.

Chow, Y. L., On Designing a Supersynthesis Antenna Array, *IEEE Trans. Antennas Propag.*, **AP-20**, 30–35, 1972.

Christiansen, W. N. and D. S. Mathewson, Scanning the Sun With a Highly Directional Array, *Proc. IRE*, **46**, 127–131, 1958.

Collin, R. E., *Antennas and Radiowave Propagation*, McGraw-Hill, New York, 1985.

Covington, A. E. and N. W. Broten, An Interferometer for Radio Astronomy with a Single-Lobed Radiation Pattern, *IRE Trans. Antennas Propag.*, **AP-5**, 247–255, 1957.

Frater, R. H., The Australia Telescope – The Radio Astronomer's Highway to the Future, *Proc. Astron. Soc. Aust.*, **5**, 440–445, 1984.

Högbom, J. A. and W. N. Brouw, The Synthesis Radio Telescope at Westerbork. Principles of Operation, Performance and Data Reduction. *Astron. Astrophy.*, **33**, 289–301, 1974.

Ishiguro, M., Minimum Redundancy Linear Arrays for a Large Number of Antennas, *Radio Sci.*, **15**, 1163–1170, 1980.

Johnson, R. C. and H. Jasik, Eds., *Antenna Engineering Handbook*, McGraw-Hill, New York, 1984.

Kellermann, K. I. and A. R. Thompson, The Very Long Baseline Array, *Science*, **229**, 123–130, 1985.

Labrum, N. R., E. Harting, T. Krishnan, and W. J. Payten, A Compound Interferometer with a 1.5 Minute of Arc Fan Beam, *Proc. IRE Aust.*, **24**, 148–155, 1963.

Leech, J., On Representation of 1,2..., *n* by Differences, *J. London Math. Soc.*, **31**, 160–169, 1956.

Love, A. W., Ed., *Reflector Antennas*, IEEE Press, The Institute of Electrical and Electronics Engineers, Inc., New York, 1978.

Mathur, N. C., A Pseudodynamic Programming Technique for the Design of Correlator Super-synthesis Arrays, *Radio Sci.*, **4**, 235–244, 1969.

Milligan, T. A., *Modern Antenna Design*, McGraw-Hill, New York, 1985.

Mills, B. Y., Cross-Type Radio Telescopes, *Proc. IRE Aust.*, **24**, 132–140, 1963.

Mills, B. Y., R. E. Aitchison, A. G. Little, and W. B. McAdam, The Sydney University Cross-Type Radio Telescope, *Proc. IRE Aust.*, **24**, 156–165, 1963.

Mills, B. Y. and A. G. Little, A High Resolution Aerial System of a New Type, *Aust. J. Phys.*, **6**, 272–278, 1953.

Mills, B. Y., A. G. Little, K. V. Sheridan, and O. B. Slee, A High Resolution Radio Telescope for Use at 3.5 m, *Proc. IRE*, **46**, 67–84, 1958.

Moffet, A. T., Minimum-Redundancy Linear Arrays, *IEEE Trans. Antennas Propag.*, **AP-16**, 172–175, 1968.

Napier, P. J., A. R. Thompson, and R. D. Ekers, The Very Large Array: Design and Performance of a Modern Synthesis Radio Telescope, *Proc. IEEE*, **71**, 1295–1320, 1983.

National Radio Astronomy Observatory, *A Proposal for a Very Large Array Radio Telescope*, National Radio Astronomy Observatory, Green Bank, West Virginia, Vol. 1, Jan 1967, and Vol. 3, Jan. 1969.

Picken, J. S. and G. Swarup, The Stanford Compound-Grating Interferometer, *Astron. J.*, **69**, 353–356, 1964.

Preston, R. A., B. F. Burke, R. Doxsey, J. F. Jordan, S. H. Morgan, D. H. Roberts, and I. I. Shapiro, The Future of VLBI Observatories in Space, in *Very Long Baseline Interferometry Techniques*, F. Biraud, Ed., Cepadues, Toulouse, France, 1983.

Ryle, M., The New Cambridge Radio Telescope, *Nature*, **194**, 517–518, 1962.

Ryle, M. and A. Hewish, The Synthesis of Large Radio Telescopes, *Mon. Not. R. Astron. Soc.*, **120**, 220–230, 1960.

Ryle, M., A. Hewish, and J. R. Shakeshaft, The Synthesis of Large Radio Telescopes by the Use of Radio Interferometers, *IRE Trans. Antennas Propag.*, **7**, S120–S124, 1959.

Seielstad, G. A., G. W. Swenson Jr. and J. C. Webber, A New Method of Array Evaluation

Applied to Very Long Baseline Interferometry, *Radio Sci.*, **14**, 509–517, 1979.

Swarup, G., The Ooty Synthesis Radio Telescope: First Results, *J. Astrophys. Astron.*, **5**, 139–148, 1984.

Swenson, G. W., Jr. and K. I. Kellermann, An Intercontinental Array – A Next-Generation Radio Telescope, *Science*, **188**, 1263–1268, 1975.

Swenson, G. W., Jr. and N. C. Mathur, The Circular Array in the Correlator Mode, *Proc. IREE Aust.*, **28**, 370–374, 1967.

Thompson, A. R., B. G. Clark, C. M. Wade, and P. J. Napier, The Very Large Array, *Astrophys. J. Suppl.*, **44**, 151–167, 1980.

Thompson, A. R. and T. Krishnan, Observations of the Six Most Intense Radio Sources with a 1.0′ Fan Beam, *Astrophys. J.*, **141**, 19–33, 1965.

Walker, R. C., VLBI Array Design, in *Indirect Imaging*, J. A. Roberts, Ed., Cambridge University Press, Cambridge, England, 1984, pp. 53–65.

Wild, J. P., A New Method of Image Formation with Annular Apertures and an Application in Radio Astronomy, *Proc. R. Soc. A*, **286**, 499–509, 1965.

Wild, J. P., Ed., *Proc. IREE Aust.*, Special Issue on The Culgoora Radioheliograph, Vol. 28, No. 9, 1967.

6

RESPONSE OF THE RECEIVING SYSTEM

This chapter is concerned with the response of the basic receiving system that accepts the signals from the antennas, amplifies and filters them, and measures the cross correlations for the various antenna pairs. We show how the basic parameters of the system affect the output. Some of the effects were introduced in earlier chapters, and here we present a more detailed development that leads to consideration of system design in Chapters 7 and 8.

6.1 FREQUENCY CONVERSION, FRINGE ROTATION, AND COMPLEX CORRELATORS

Frequency Conversion

In practically all receiving systems in radio astronomy the frequencies of the signals received at the antennas are changed by mixing with a local oscillator signal. This feature, referred to as frequency conversion (or *heterodyne* frequency conversion), enables the major part of the signal processing to be performed at intermediate frequencies that are most appropriate for amplification, transmission, filtering, delaying, recording, and similar processes.

Frequency conversion takes place in a mixer, in which the signal to be converted plus a local oscillator waveform are applied to a circuit element with a nonlinear voltage–current response. This element may be a diode as shown in Fig. 6.1a. The current i through the diode can be expressed as a power series in the applied voltage V:

$$i = a_0 + a_1 V + a_2 V^2 + a_3 V^3 + \cdots . \tag{6.1}$$

Now let V consist of the sum of a local oscillator voltage $b_1 \cos(2\pi \nu_{LO} t + \theta_{LO})$ and a signal, of which one Fourier component is $b_2 \cos(2\pi \nu_s t + \phi_s)$. The second-order term in V then gives rise to a product in the mixer output of the form

$$b_1 b_2 \cos(2\pi \nu_{LO} t + \theta_{LO})$$

$$\times \cos(2\pi \nu_s t + \phi_s) = \tfrac{1}{2} b_1 b_2 \cos\left[2\pi(\nu_s + \nu_{LO})t + \phi_s + \theta_{LO}\right]$$

$$+ \tfrac{1}{2} b_1 b_2 \cos\left[2\pi(\nu_s - \nu_{LO})t + \phi_s - \theta_{LO}\right].$$

$$\tag{6.2}$$

Thus the current through the diode contains components at the sum and difference of ν_s and ν_{LO}. Other terms in (6.1) lead to other combinations, such as $3\nu_{LO} \pm \nu_s$, but the filter H shown in Fig. 6.1 passes only the wanted output spectrum, and with proper design unwanted combinations can be prevented from falling within the filter passband. Usually the signal voltage is much smaller than the local oscillator voltage, so harmonics and intermodulation products of the signal are small compared with the wanted terms containing ν_{LO}.

In most cases of frequency conversion the signal frequency is being reduced, and the second term on the right-hand side in Eq. (6.2) is the important one. The filter H then defines an intermediate frequency (IF) band centered on ν_0 as shown in Fig. 6.1b. Signals from within the bands centered on $\nu_{LO} - \nu_0$ and

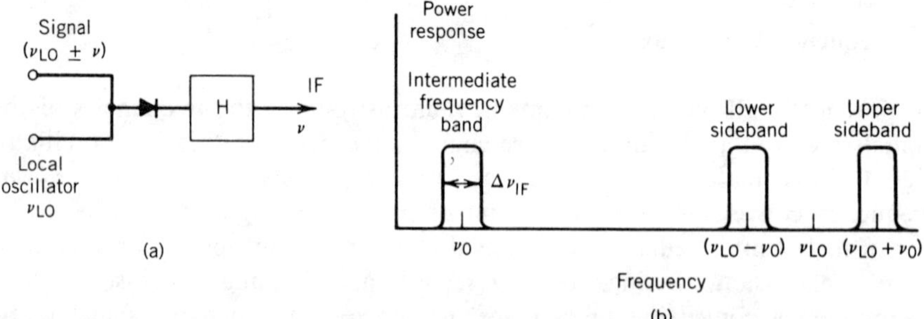

Figure 6.1   Frequency conversion in a radio receiving system. (a) Simplified diagram of a mixer and a filter H that defines the intermediate frequency (IF) band. The nonlinear element shown is a diode. (b) Signal spectrum showing upper and lower sidebands that are converted to the intermediate frequency. ν_0 is the center of the IF band.

$\nu_{LO} + \nu_0$ are converted and admitted by the filter. These bands are known as the lower and upper sidebands, as shown, and if only a single sideband is wanted the other can often be removed by a suitable filter or narrowband amplifier inserted before the mixer. In other cases both sidebands are accepted, resulting in a double-sideband response.

Response of a Single-Sideband System

Figure 6.2 shows a basic receiving system for two antennas, m and n, of a synthesis array. Here we are interested in the effects of frequency conversion which were omitted from the discussion in Chapters 2 and 3. The time difference τ_g between the arrival at the antennas of the signals from a radio source varies continuously as the earth rotates and the antennas track the

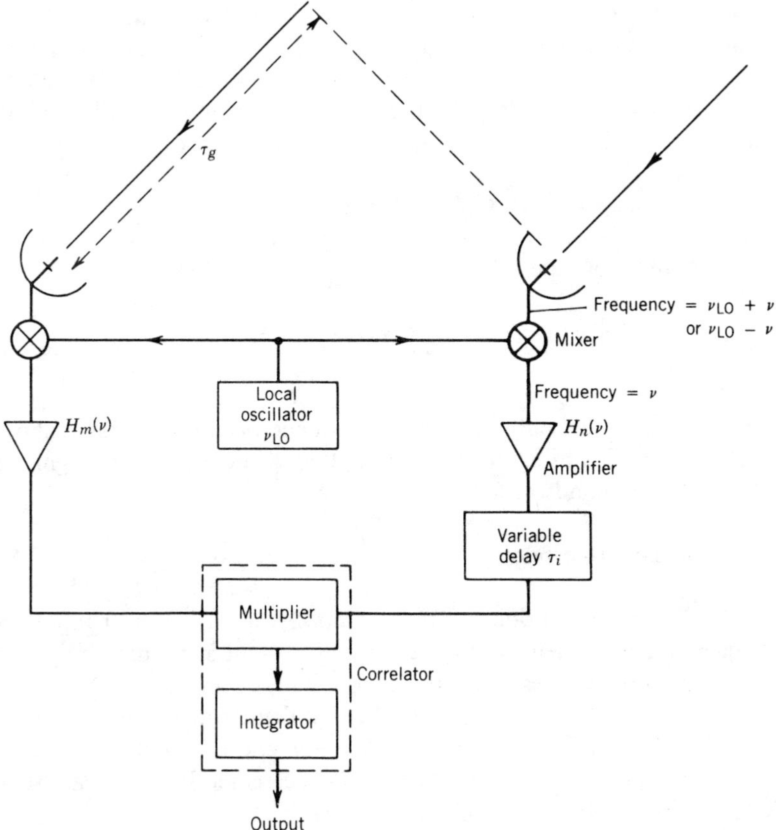

Figure 6.2   Basic receiving system for two antennas of a synthesis array. The variable delay τ_i is continuously adjusted under computer control to compensate for the geometrical path delay τ_g. The frequency response functions $H_m(\nu)$ and $H_n(\nu)$ represent the overall bandpass characteristics of the amplifiers and filters in the signal channels.

source across the sky. An instrumental delay τ_i which compensates for the geometrical delay τ_g can be inserted under computer control so that the signals arrive simultaneously at the correlator, as discussed in Section 2.2. The receiving channels through which the signals pass contain amplifiers and filters, the overall frequency responses of which are $H_m(\nu)$ and $H_n(\nu)$ for antennas m and n. Here ν represents a frequency at the correlator input, the corresponding frequency at the antenna being $\nu_{LO} \pm \nu$. The signals in the receiving system result from cosmic noise and system noise; we consider the common case in which these processes are independent of frequency within the receiver passband. The spectra at the correlator inputs are thus determined mainly by the response of the receiving system. Let ϕ_m be the phase change in the signal path through antenna m resulting from τ_g and the local oscillator phase, and let ϕ_n be the corresponding phase change in the signal for the path through antenna n, including τ_i. ϕ_m and ϕ_n represent the phases of the signal at the correlator inputs, and negative values of these parameters indicate phase lag (signal delay). The response to a source for which the visibility is $\mathcal{V}(u, v) = |\mathcal{V}|e^{j\phi_v}$ is most easily obtained by returning to Eq. (2.25) and replacing the phase difference $2\pi \mathbf{D}_\lambda \cdot \mathbf{s}_0$ by the more general term $\phi_n - \phi_m$. Then the response at the correlator output resulting from a frequency band of width $d\nu$ can be written as

$$dr = \mathrm{Re}\left\{ A_0|\mathcal{V}|H_m(\nu)H_n^*(\nu)e^{j(\phi_n - \phi_m - \phi_v)}d\nu \right\}, \tag{6.3}$$

and the response from the full system passband is

$$r = \mathrm{Re}\left\{ A_0|\mathcal{V}|\int_0^\infty H_m(\nu)H_n^*(\nu)e^{j(\phi_n - \phi_m - \phi_v)}d\nu \right\}, \tag{6.4}$$

where we have included only the positive frequencies and taken the real part of the resulting complex integral. We assume that $\mathcal{V}$ does not vary significantly over the observing bandwidth.

Upper-Sideband Reception

For upper-sideband reception a filter or amplifier at the receiver input selects frequencies in a band defined by the correlator input spectrum (frequency ν) plus ν_{LO}. We now express the phases ϕ_m and ϕ_n in terms of the phases encountered by the signals in Fig. 6.2. The signal entering antenna m traverses the geometrical delay τ_g at a frequency $\nu_{LO} + \nu$, and thus suffers a phase shift $2\pi(\nu_{LO} + \nu)\tau_g$. At the mixer its phase is also decreased by the local oscillator phase θ_m. Thus we obtain

$$\phi_m(\nu) = -2\pi(\nu_{LO} + \nu)\tau_g - \theta_m. \tag{6.5}$$

The phase of the signal entering antenna n is decreased by the local oscillator phase θ_n, and the signal then traverses the instrumental delay τ_i at a frequency ν, thus suffering a shift $2\pi\nu\tau_i$. The total phase shift for antenna n is

$$\phi_n(\nu) = -2\pi\nu\tau_i - \theta_n. \qquad (6.6)$$

From Eqs. (6.4), (6.5), and (6.6) the correlator output is

$$r_u = \mathrm{Re}\left\{ A_0|\mathscr{V}|e^{j[2\pi\nu_{\mathrm{LO}}\tau_g + (\theta_m - \theta_n) - \phi_v]}\int_0^\infty H_m(\nu)H_n^*(\nu)e^{j2\pi\nu\Delta\tau}d\nu \right\}.$$

$$(6.7)$$

The real part of the integral in Eq. (6.7) is one-half the Fourier transform of the (Hermitian) cross power spectrum $H_m(\nu)H_n^*(\nu)$ with respect to the delay compensation error $\Delta\tau = \tau_g - \tau_i$, which introduces a linear phase slope across the band.[†] For example, if the passbands are rectangular with identical phase responses, that is,

$$|H_m(\nu)| = |H_n(\nu)| = \begin{cases} H_0, & |\nu - \nu_0| < \Delta\nu_{\mathrm{IF}}/2 \\ 0, & |\nu - \nu_0| > \Delta\nu_{\mathrm{IF}}/2, \end{cases} \qquad (6.8)$$

then,

$$\int_0^\infty H_m(\nu)H_n^*(\nu)e^{j2\pi\nu\Delta\tau}d\nu = H_0^2\Delta\nu_{\mathrm{IF}}\left[\frac{\sin(\pi\Delta\nu_{\mathrm{IF}}\Delta\tau)}{\pi\Delta\nu_{\mathrm{IF}}\Delta\tau}\right]e^{j2\pi\nu_0\Delta\tau}, \qquad (6.9)$$

where $\Delta\nu_{\mathrm{IF}}$ is the width and ν_0 is the center frequency of the IF band at the correlator input. In the general case we define an instrumental gain factor $G_{mn} = |G_{mn}|e^{j\phi_G}$ as follows:

$$A_0\int_0^\infty H_m(\nu)H_n^*(\nu)e^{j2\pi\nu\Delta\tau}d\nu = G_{mn}(\Delta\tau)e^{j2\pi\nu_0\Delta\tau}$$

$$= |G_{mn}(\Delta\tau)|e^{j(2\pi\nu_0\Delta\tau + \phi_G)}. \qquad (6.10)$$

The variation of G_{mn} with $\Delta\tau$ causes the delay pattern effect discussed in earlier chapters. The phase ϕ_G results from the difference in the phase responses of the amplifiers and filters. The local oscillator phases θ_m and θ_n are not included within the general instrumental phase term ϕ_G because they enter into the upper and lower sidebands with different signs.

[†] Here we assume that the source is sufficiently close to the center of the field being mapped that the condition $\Delta\tau = 0$ maintains zero delay error. The effect of the variation of the delay error across a wider field of view is considered in Section 6.3.

Substituting Eq. (6.10) into Eq. (6.7) we obtain, for upper-sideband reception,

$$r_u = |\mathcal{V}||G_{mn}(\Delta\tau)|\cos\left[2\pi(\nu_{\mathrm{LO}}\tau_g + \nu_0\Delta\tau) + (\theta_m - \theta_n) - \phi_v + \phi_G\right].$$

$$(6.11)$$

The term $2\pi\nu_{\mathrm{LO}}\tau_g$ in the cosine function results in a quasisinusoidal oscillation as the source moves through the fringe pattern. The phase of this oscillation depends on the delay error $\Delta\tau$, the relative phases of the local oscillator signals, the phase responses of the signal channels, and the phase of the visibility function. The frequency of the output oscillation $\nu_{\mathrm{LO}}\, d\tau_g/dt$ is often referred to as the natural fringe frequency. As noted in Chapter 4 the oscillations result because the signals traverse the delays τ_g and τ_i at different frequencies, that is, at the input radiofrequency for τ_g and at the intermediate frequency for τ_i. Thus, even if these two delays are identical they introduce different phase shifts, and they increase or decrease progressively as the earth rotates.

Lower-Sideband Reception

Consider now the situation where the frequencies accepted from the antenna are those in the lower sideband, at ν_{LO} minus the correlator input frequencies. The phases are

$$\phi_m = 2\pi(\nu_{\mathrm{LO}} - \nu)\tau_g + \theta_m \tag{6.12}$$

and

$$\phi_n = -2\pi\nu\tau_i + \theta_n. \tag{6.13}$$

The signs of these terms and of ϕ_v differ from those in the upper-sideband case because increasing the phase of the signal at the antenna here decreases the phase at the correlator. The expression for the correlator output is

$$r_\ell = \mathrm{Re}\left\{ A_0|\mathcal{V}|e^{-j[2\pi\nu_{\mathrm{LO}}\tau_g + (\theta_m - \theta_n) - \phi_v]}\int_0^\infty H_m(\nu)H_n^*(\nu)e^{j2\pi\nu\Delta\tau}\,d\nu\right\}.$$

$$(6.14)$$

Proceeding as in the upper-sideband case we obtain

$$r_\ell = |\mathcal{V}||G_{mn}(\Delta\tau)|\cos\left[2\pi(\nu_{\mathrm{LO}}\tau_g - \nu_0\Delta\tau) + (\theta_m - \theta_n) - \phi_v - \phi_G\right].$$

$$(6.15)$$

Multiple Frequency Conversion

In an operational system the signals may undergo several frequency conversions between the antennas and the correlators. Operation with multiple frequency conversions is essentially the same as with the systems considered above. A frequency conversion in which the output is at the lower sideband (i.e., the local oscillator frequency minus the input frequency) results in a reversal of the signal spectrum, and frequencies at the high end at the input appear at the low end at the output, and vice versa. If there is no net reversal (i.e., an even number of lower-sideband conversions), Eq. (6.11) applies, except that ν_{LO} must be replaced by a combination of local oscillator frequencies sometimes known as the signed-sum of the local oscillator frequencies because some frequencies enter it with positive signs and some with negative ones. Similarly, the oscillator phase terms θ_m and θ_n are replaced by corresponding combinations of oscillator phases. If there is à net reversal of the frequency band Eq. (6.15) applies with similar modifications.

Delay Tracking and Fringe Rotation

Adjustment of the compensating delay τ_i of Fig. 6.2 is usually accomplished under computer control, the required delay being a function of the antenna positions and the position of the phase center of the field under observation. This can be acheived by designating the antenna farthest from the source (i.e., with the greatest geometrical delay) as the delay reference, setting its compensating delay to zero, and adjusting the delays of all other antennas as required. However, the delay reference can be at any convenient point.

To control the frequency of the sinusoidal fringe variations in the correlator output, a continuous phase change can be inserted into one of the local oscillator signals. Equations (6.11) and (6.15) show that the fringe frequency can be reduced to zero by causing $\theta_m - \theta_n$ to vary at a rate that maintains constant, modulo 2π, the term $[2\pi\nu_{LO}\tau_g + (\theta_m - \theta_n)]$. Note that the phase changes required to slow the fringe oscillations to zero frequency, plus the phase changes introduced by the compensating delay, are exactly those required by the factor $e^{j2\pi w}$ in Eq. (4.3) which represents the fringe phase at the field center. Reduction of the output frequency reduces the quantity of data to be processed, since each correlator output must be sampled at least twice per cycle of the output frequency (the Nyquist rate) to preserve the information, as discussed in Section 8.2. With antenna spacings required for angular resolution of millarcsecond order, the natural fringe frequency, $\nu_{LO}\, d\tau_g/dt$, can exceed 100 kHz. For an array with more than one antenna pair it is possible to reduce each output frequency to the same fraction of its natural frequency, or to zero. Reduction to zero frequency is generally the preferred practice and is often referred to as *fringe stopping*. Some special technique such as the use of a complex correlator, described in the following subsection, is then required to extract the amplitude and phase of the output.

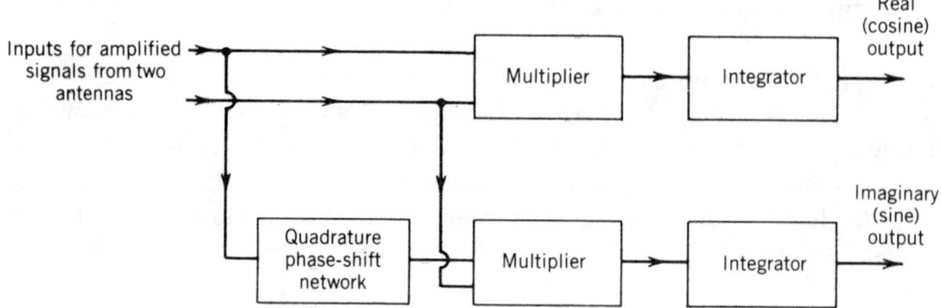

Figure 6.3   Use of two correlators to measure the real and imaginary parts of the visibility. This device is called a complex correlator.

Simple and Complex Correlators

A method of measuring the amplitude and phase of the correlator output signal when the fringe frequency at the correlator output is reduced to zero is shown in Fig. 6.3. Two correlators are used, one that multiplies the signals in the manner considered above, and another that has a quadrature phase-shift network in one input. This network shifts the phase of each frequency component in the input band by $\pi/2$, and the output is thus the Hilbert transform of the input. For signals of finite bandwidth the phase shift is not, of course, equivalent to a delay. The phase shift can also be effected by feeding the signal into two separate mixers and converting it with two local oscillators in phase quadrature. The output of the second correlator can be obtained by replacing $H_m(\nu)$ by $H_m(\nu)e^{-j\pi/2}$. From Eq. (6.10) the result is to add $-\pi/2$ to ϕ_G, and thus in Eqs. (6.11) and (6.15) the cosine function is replaced by $\pm$ sine. The combination of two correlators and the quadrature network is usually referred to as a *complex correlator*, and the two outputs as the cosine and sine, or real and imaginary, outputs. (When necessary to emphasize the distinction we can refer to a single multiplier and integrator as a *simple* or *single-multiplier* correlator.) In normal operation the compensating delay is adjusted so that $\Delta\tau = 0$ and the fringe rotation maintains the condition $2\pi\nu_{LO}\tau_i + (\theta_m - \theta_n) = 0$. Thus the cosine and sine outputs represent the real and imaginary parts of $G_{mn}\mathscr{V}(u, v)$. With the use of the complex correlator, the rotation of the earth which sweeps the fringe pattern across the source is no longer a necessary feature in the measurement of visibility. Another very important feature of the complex correlator is that the noise fluctuations in the cosine and sine outputs are independent, as we discuss in Section 6.2.

Response of a Double-Sideband System

A double-sideband receiving system is one in which both the upper- and lower-sideband responses are accepted. From Eqs. (6.11) and (6.15) the output is

$$r_d = r_u + r_\ell = 2|\mathcal{V}||G_{mn}(\Delta\tau)|\cos(2\pi\nu_0\Delta\tau + \phi_G)$$

$$\times \cos\left[2\pi\nu_{\mathrm{LO}}\tau_g + (\theta_m - \theta_n) - \phi_v\right]. \tag{6.16}$$

There is a significant difference from the single-sideband cases. The phase of the fringe-frequency term is no longer dependent on $\Delta\tau$ or ϕ_G, but instead these quantities appear in the term that controls the fringe amplitude:

$$|G_{mn}(\Delta\tau)|\cos(2\pi\nu_0\Delta\tau + \phi_G). \tag{6.17}$$

Thus, as shown in Fig. 6.4, the cross correlation (fringe amplitude) falls off more rapidly because of the cosine term than it does in the single-sideband case in which it depends only upon $G_{mn}(\Delta\tau)$. The required precision in matching the geometrical and instrumental delays is correspondingly increased. The lack of dependance of the fringe phase on the phase response of the signal channel occurs because the latter has equal and opposite effects on the signals from the two sidebands.

The response of a double-sideband system with a complex correlator is Eq. (6.16) for the cosine output, and for the sine output it is obtained by replacing ϕ_G by $\phi_G - \pi/2$:

$$(r_d)_{\mathrm{sine}} = 2|\mathcal{V}||G_{mn}(\Delta\tau)|\sin\left[2\pi\nu_0\Delta\tau + \phi_G\right]$$

$$\times \cos\left[2\pi\nu_{\mathrm{LO}}\tau_g + (\theta_m - \theta_n) - \phi_v\right]. \tag{6.18}$$

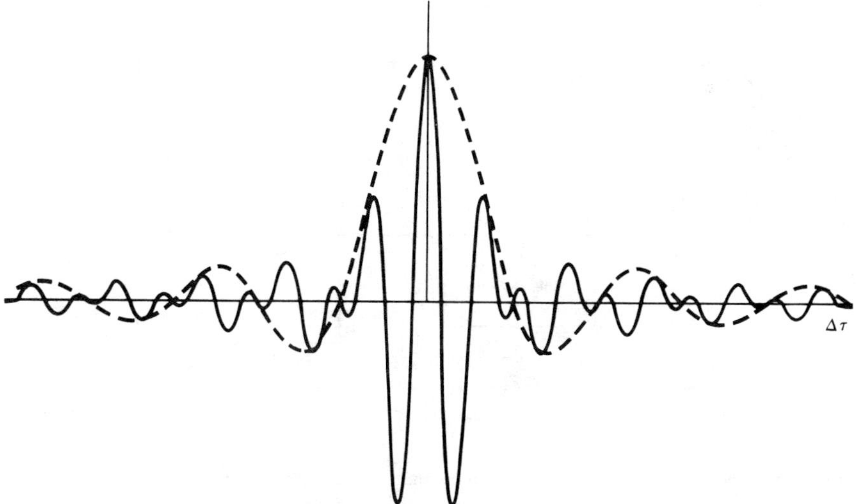

Figure 6.4 Example of the variation of the fringe amplitude as a function of $\Delta\tau$ for a double-sideband system (full line). The centers of the two sidebands are separated by three times the IF bandwidth, that is, $\nu_0 = 1.5\,\Delta\nu_{\mathrm{IF}}$, and the IF response is rectangular. The broken line shows the equivalent function for a single-sideband system with the same IF response.

If the term $2\pi\nu_0\Delta\tau + \phi_G$ is adjusted to maximize one of the two outputs, the other will be zero. Thus for continuum observations in which the signal is of equal strength in both sidebands, the complex correlator offers no increase in sensitivity. However, it can be useful for observations in the sideband-separation mode described later.

The response with multiple frequency conversions is more complicated for a double-sideband interferometer than for a single-sideband one and is illustrated by considering the system in Fig. 6.5. The signal phase terms are determined by considerations similar to those described in the derivation of Eqs. (6.5) and (6.6). Thus we obtain

$$\phi_m = \mp 2\pi(\nu_1 \pm \nu_2 \pm \nu)\tau_g \mp \theta_{m1} - \theta_{m2} \tag{6.19}$$

and

$$\phi_n = -2\pi(\nu_2 + \nu)\tau_{i1} - 2\pi\nu\tau_{i2} \mp \theta_{n1} - \theta_{n2}, \tag{6.20}$$

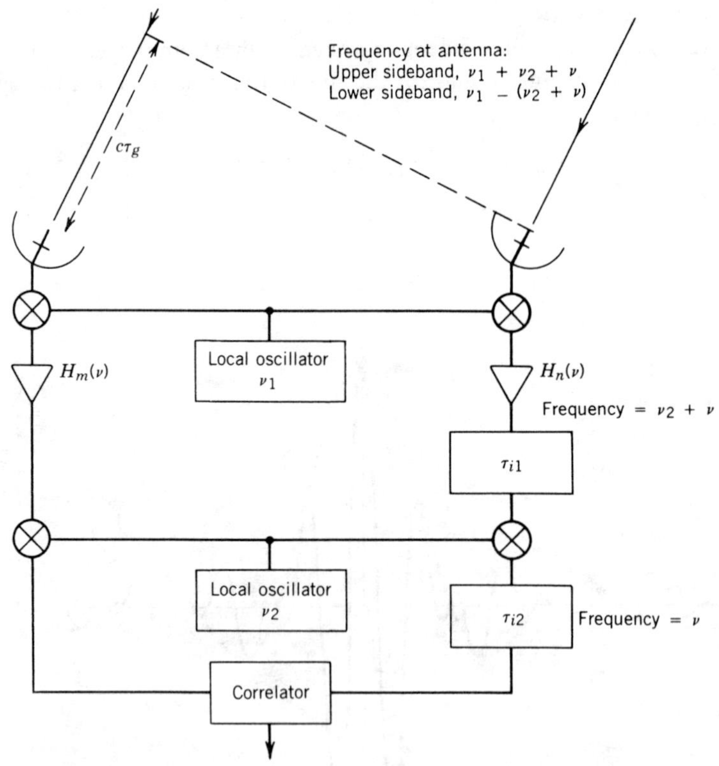

Figure 6.5 Receiving system for two antennas that incorporates two frequency conversions, the first being double sideband and the second upper sideband. Two compensating delays τ_{i1} and τ_{i2} are included so that in deriving the response for a double-sideband system the effect of the position of the delay relative to the first mixer can be investigated. In practice only one compensating delay is required. The overall frequency responses H_m and H_n are specified as functions of ν which is the corresponding frequency at the correlator input.

where the upper signs correspond to upper-sideband conversion at all mixers and the lower signs to lower-sideband conversion at the first mixer in each channel and upper sideband at the other mixers. Then proceeding as in previous examples we obtain for the double-sideband case

$$r_d = 2|\mathscr{V}||G_{mn}(\Delta\tau)|\cos\left\{2\pi\left[\nu_2(\tau_{i1} - \tau_g) - \nu_0\Delta\tau\right] - (\theta_{m2} - \theta_{n2}) - \phi_G\right\}$$

$$\times \cos\left[2\pi\nu_1\tau_g + (\theta_{m1} - \theta_{n1}) - \phi_v\right], \tag{6.21}$$

where $\Delta\tau = \tau_g - \tau_{i1} - \tau_{i2}$. Note that the phase of the output fringe pattern, given by the second cosine term, depends only upon the phase of the first local oscillator, and fringe rotation must be applied at this oscillator. The first cosine term in Eq. (6.21) affects the fringe amplitude and two cases must be considered, as follows:

1. If the compensating delay is τ_{i1}, at the intermediate frequency immediately following the double-sideband mixer, and $\tau_{i2} = 0$, we see that in the first cosine function $\tau_{i1} - \tau_g \simeq 0$. ϕ_G should be small if the frequency responses of the two channels are similar, and it is necessary only to equalize θ_{m2} and θ_{n2} to maximize the amplitude of the fringe-frequency term. This is similar to the single conversion case in Eq. (6.16).

2. If the compensating delay is τ_{i2}, which is located after the second mixer, and $\tau_{i1} = 0$, a continuously varying phase shift is required in θ_{m2} or θ_{n2} to keep the value of the first cosine function close to unity as τ_g varies. This phase shift does not affect the phase of the output fringe oscillations, only the amplitude (see, e.g., Wright et al. 1973).

Relative Advantages of Double- and Single-Sideband Systems

The principal reason for using double-sideband reception in interferometry is that in certain cases the lowest receiver noise temperatures are obtained by using input stages that are inherently double-sideband devices. At millimeter and shorter wavelengths it is difficult to make low-noise amplifiers, and receiving systems often use a mixer as the input stage followed by a low-noise IF amplifier, both components usually being cryogenically cooled to minimize the system noise. If a filter is placed between the antenna and the mixer to cut out one sideband, the received signal power is halved, and there will also be insertion loss in the filter. There is no similar reduction in the receiver noise, so for best sensitivity both sidebands are retained. Also, as a historical note, double-sideband systems were used at centimeter wavelengths during the 1960s and early 1970s with the degenerate type of parametric amplifier as the low-noise input stage. These amplifiers were inherently double-sideband devices (see, e.g., de Jager and Robinson 1961), and their use in interferometry is discussed by Vander Vorst and Colvin (1966). The lack of dependence of the output phase on the phase response of the amplifiers has been considered an advantage of double-sideband systems, particularly before the development of solid-state amplifiers (see, e.g., Read 1961).

Double-sideband systems have a number of disadvantages. Increased accuracy of delay setting is required; instrumental phase adjustments including an additional tracking phase shifter may be required; interpretation of spectral-line data is complicated if there are lines in both sidebands; and the width of the interference-free band of the radio spectrum required is doubled. Also the smearing effect of a finite bandwidth, to be discussed in Section 6.3, is increased. These problems have stimulated the development of schemes by which the correlations for upper and lower sidebands are separated.

Sideband Separation

To illustrate the method by which the responses for the two sidebands can be separated in the correlator output of a double-sideband receiving system, we examine the sum of the upper- and lower-sideband responses from which Eq. (6.16) was obtained. This may be written

$$r_d = r_u + r_\ell = |\mathscr{V}||G_{mn}(\Delta\tau)|\Big\{\cos\big[2\pi(\nu_{LO}\tau_g + \nu_0\Delta\tau) + \theta_{mn} - \phi_v + \phi_G\big]$$

$$+ \cos\big[2\pi(\nu_{LO}\tau_g - \nu_0\Delta\tau) + \theta_{mn} - \phi_v - \phi_G\big]\Big\},$$

$$(6.22)$$

where $\theta_{mn} = \theta_m - \theta_n$. Equation (6.22) represents the real output of a complex correlator. We rewrite Eq. (6.22) as

$$r_d = |\mathscr{V}||G_{mn}|(\cos\Psi_u + \cos\Psi_\ell), \qquad (6.23)$$

where Ψ_u and Ψ_ℓ represent the corresponding expressions in Eq. (6.22). The expression for the imaginary output of the correlator is obtained by replacing ϕ_G by $\phi_G - \pi/2$. The responses considered above represent the normal output of the interferometer, which we call condition 1. Consider a second condition in which a $\pi/2$ phase shift is introduced into the first local oscillator signal of antenna m, so that θ_{mn} becomes $\theta_{mn} - \pi/2$. The correlator outputs for the two conditions are obtained from Eq. (6.22):

$$\left.\begin{array}{l} r_1 = |\mathscr{V}||G_{mn}|(\cos\Psi_u + \cos\Psi_\ell) \\ r_2 = |\mathscr{V}||G_{mn}|(\sin\Psi_u - \sin\Psi_\ell) \end{array}\right\} \text{condition 1} \qquad (6.24)$$

$$\left.\begin{array}{l} r_3 = |\mathscr{V}||G_{mn}|(\sin\Psi_u + \sin\Psi_\ell) \\ r_4 = |\mathscr{V}||G_{mn}|(-\cos\Psi_u + \cos\Psi_\ell) \end{array}\right\} \begin{array}{l}\text{condition 2} \\ (\theta_{mn} \to \theta_{mn} - \pi/2),\end{array} \qquad (6.25)$$

where r_1 and r_3 represent the real outputs of the correlator, and r_2 and r_4 the imaginary outputs. Thus the upper-sideband response, expressed in complex form, is

$$|\mathscr{V}||G_{mn}|(\cos\Psi_u + j\sin\Psi_u) = \tfrac{1}{2}\big[(r_1 - r_4) + j(r_2 + r_3)\big]. \qquad (6.26)$$

Similarly, for the lower sideband we have

$$|\mathscr{V}||G_{mn}|(\cos\Psi_\ell + j\sin\Psi_\ell) = \tfrac{1}{2}\big[(r_1 + r_4) - j(r_2 - r_3)\big]. \qquad (6.27)$$

If the $\pi/2$ phase shift is periodically switched into and out of the local oscillator signal, the upper- and lower-sideband responses can be obtained as indicated by Eqs. (6.26) and (6.27).

In VLBI observations a different technique for sideband separation may be used. In a double sideband VLBI system in which the sidebands are widely spaced in frequency, the data cannot be processed with a simple correlator, as discussed in this section under *Response of a Double-Sideband System*, because the correlation function would be too narrow in delay when the data sampling is done at the Nyquist rate (see Fig. 6.4). If the fringe rotation is set to stop the fringes in one sideband, the fringes resulting from the other sideband will generally have a sufficiently high frequency to be reduced to a negligible level by the time averaging within the correlator. In VLBI systems the fringe rotation is often applied during the playback operation, that is, after the first local oscillator. Fringe rotation then has the effect of reducing the fringe frequency for one sideband and increasing it for the other: this can be understood from Eq. (6.19) in which it is seen that the second local oscillator phase term has the same sign as the geometrical delay term for one sideband but the opposite sign for the other. The data are played back to the correlator twice, once for each sideband, with appropriate fringe rotation.

6.2   RESPONSE TO THE NOISE

The ultimate sensitivity of a receiving system is determined principally by the system noise. We now consider the response to the noise and the resulting threshold of sensitivity, beginning with the effect at the correlator output and the resulting uncertainty in the real and imaginary parts of $\mathscr{V}$. This leads to calculation of the rms noise level in a synthesized map in terms of the peak response to a source of given flux density. Finally, we consider the effect of noise in terms of the rms fluctuations in the amplitude and phase of $\mathscr{V}$.

Signal and Noise Processing in the Correlator

Consider an observation in which the field to be mapped contains only a point source of flux density S W m^{-2} Hz^{-1} located at the phase reference position. Let $V_m(t)$ and $V_n(t)$ be the waveforms at the correlator input from the signal channels of antennas m and n. The output is

$$r = \langle V_m(t)V_n(t)\rangle, \tag{6.28}$$

where all three functions are real, and the expectation denoted by the angular brackets is approximated in practice by a finite time average. To determine the relative power levels of the signal and noise components of r we determine their power spectra by first calculating the autocorrelation functions. The autocorrelation of the signal product in Eq. (6.28) is

$$\rho_r(t) = \langle V_m(t)V_n(t)V_m(t-\tau)V_n(t-\tau)\rangle. \tag{6.29}$$

This expression can be evaluated using the following fourth-order moment relation:[†]

$$\langle z_1 z_2 z_3 z_4\rangle = \langle z_1 z_2\rangle\langle z_3 z_4\rangle + \langle z_1 z_3\rangle\langle z_2 z_4\rangle + \langle z_1 z_4\rangle\langle z_2 z_3\rangle,$$

$$\tag{6.30}$$

where z_1, z_2, z_3, and z_4 are joint Gaussian random variables with zero mean. Thus

$$
\begin{aligned}
\rho_r(\tau) = {}& \langle V_m(t)V_n(t)\rangle\langle V_m(t-\tau)V_n(t-\tau)\rangle \\
& + \langle V_m(t)V_m(t-\tau)\rangle\langle V_n(t)V_n(t-\tau)\rangle \\
& + \langle V_m(t)V_n(t-\tau)\rangle\langle V_m(t-\tau)V_n(t)\rangle \\
= {}& \rho_{mn}^2(0) + \rho_m(\tau)\rho_n(\tau) + \rho_{mn}(\tau)\rho_{mn}(-\tau),
\end{aligned}
\tag{6.31}
$$

where ρ_m and ρ_n are the unnormalized autocorrelation functions of the two signals V_m and V_n respectively, and ρ_{mn} is their cross-correlation function. Each of the V terms is the sum of a signal component s and a noise component n, and to examine how these components contribute to the correlator output we substitute them in Eq. (6.31). Products of uncorrelated terms, that is, products of signal and noise voltages, or noise voltages from different antennas, have an expectation of zero, and omitting them we obtain

$$
\begin{aligned}
\rho_r(\tau) = {}& \langle s_m(t)s_n(t)\rangle\langle s_m(t-\tau)s_n(t-\tau)\rangle \\
& + \langle s_m(t)s_m(t-\tau) + n_m(t)n_m(t-\tau)\rangle\langle s_n(t)s_n(t-\tau) \\
& + n_n(t)n_n(t-\tau)\rangle + \langle s_m(t)s_n(t-\tau)\rangle\langle s_m(t-\tau)s_n(t)\rangle.
\end{aligned}
$$

$$\tag{6.32}$$

<hr>

[†] This relation is a special case of a more general expression for the expectation of the product of N such variables, which is zero if N is odd and a sum of pair products if N is even. A form of Eq. (6.30) can be found in the following texts: Lawson and Uhlenbeck (1950), p. 68; Middleton (1960), p. 343; Wozencraft and Jacobs (1965), p. 205.

To determine the effect of the frequency response of the receiving system on the various terms of $\rho(\tau)$ we need to convert them to power spectra. By the Wiener–Khinchin relation we should therefore examine the Fourier transforms of each of the terms on the right-hand sides of Eqs. (6.31) and (6.32).

The first term $\rho_{mn}^2(0)$ is a constant, and its Fourier transform is a delta function at the origin in the frequency domain, multiplied by $\rho_{mn}^2(0)$. From Eq. (6.32) we see that $\rho_{mn}^2(0)$ involves only the signal terms, which it is convenient to express as antenna temperatures. By the integral theorem of Fourier transforms $\rho_{mn}(0)$ is the infinite integral of the Fourier transform of $\rho_{mn}(\tau)$, and thus the Fourier transform of $\rho_{mn}^2(0)$ is

$$k^2 T_{Am} T_{An} \left[\int_{-\infty}^{\infty} H_m(\nu) H_n^*(\nu) \, d\nu \right]^2 \delta(\nu), \qquad (6.33)$$

where k is Boltzmann's constant, T_{Am} and T_{An} are the components of antenna temperature resulting from the source [see Eq. (1.5)], and $H_m(\nu)$ and $H_n(\nu)$ are the frequency responses of the signal channels.

The Fourier transform of the second term of Eq. (6.31) is the convolution of the transforms of ρ_m and ρ_n, that is,

$$k^2 (T_{Sm} + T_{Am})(T_{Sn} + T_{An}) \int_{-\infty}^{\infty} H_m(\nu) H_m^*(\nu) H_n(\nu' - \nu) H_n^*(\nu' - \nu) \, d\nu,$$

$$(6.34)$$

where T_{Sm} and T_{Sn} are the system temperatures. Note that the magnitude of this term is proportional to the product of the total noise temperatures.

The Fourier transform of the third term of Eq. (6.31) is the convolution of the transforms of $\rho_{mn}(\tau)$ and $\rho_{mn}(-\tau)$, and the latter is the complex conjugate of the former, since ρ_{mn} is real. Thus the Fourier transform of $\rho_{mn}(\tau)\rho_{mn}(-\tau)$ is

$$k^2 T_{Am} T_{An} \int_{-\infty}^{\infty} H_m(\nu) H_n^*(\nu) H_m^*(\nu' - \nu) H_n(\nu' - \nu) \, d\nu. \qquad (6.35)$$

In Expression (6.35), as in (6.33), only the antenna temperatures appear, since the receiver noise for different antennas makes no contribution to the cross correlation.

Expression (6.33) represents the signal power in the correlator output and Expressions (6.34) and (6.35) represent the noise. The effect of the time averaging at the correlator output can be modeled in terms of a filter that passes frequencies from 0 to $\Delta\nu_{\mathrm{LF}}$. The output bandwidth $\Delta\nu_{\mathrm{LF}}$ is less than the correlator input bandwidth by several or many orders of magnitude. Therefore, the spectral density of the output noise can be assumed to be equal to its value at zero frequency, that is, for $\nu' = 0$ in Expressions (6.34) and (6.35). From these considerations, and $H_m(\nu)$ and $H_n(\nu)$ being Hermitian, the ratio of the signal voltage to the rms noise voltage at the averaged correlator output is

$$\mathcal{R}_{sn} = \frac{\sqrt{T_{Am}T_{An}} \int_{-\infty}^{\infty} H_m(\nu)H_n^*(\nu)\,d\nu}{\sqrt{(T_{Am} + T_{Sm})(T_{An} + T_{Sn}) + T_{Am}T_{An}} \sqrt{2\,\Delta\nu_{LF}\int_{-\infty}^{\infty} |H_m(\nu)|^2|H_n(\nu)|^2\,d\nu}}$$

$$(6.36)$$

where $2\,\Delta\nu_{LF}$ is the equivalent bandwidth after averaging, negative frequencies being included. It is unusual for $\mathcal{R}_{sn}$, the estimate of the signal-to-noise ratio at the output of a simple correlator, to be required to an accuracy better than a few percent. Indeed, it is usually difficult to specify T_S to any greater accuracy since the effects of ground radiation and atmospheric absorption in T_S vary as the antennas track. Thus, it is usually satisfactory to approximate $H_m(\nu)$ and $H_n(\nu)$ by identical rectangular functions of width $\Delta\nu_{IF}$. Also, in sensitivity calculations one is concerned most often with sources near the threshold of detectability for which $T_A \ll T_S$. With these simplifications Eq. (6.36) becomes

$$\mathcal{R}_{sn} = \sqrt{\frac{T_{Am}T_{An}}{T_{Sm}T_{Sn}}} \sqrt{\frac{\Delta\nu_{IF}}{\Delta\nu_{LF}}} . \qquad (6.37)$$

Figure 6.6 shows the signal and noise spectra for the rectangular bandpass approximation. Note that the input spectra $|H_m(\nu)|^2$ and $|H_n(\nu)|^2$ contain both positive and negative frequencies and are symmetrical about the origin in ν. Thus, the output noise spectrum can be described as proportional to either the convolution or the cross-correlation function of $|H_m(\nu)|^2$ and $|H_n(\nu)|^2$.

The output bandwidth is related to the data averaging time τ_a since the averaging can be described as convolution with a rectangular function of unit area and width τ_a. The power response of the averaging circuit as a function of frequency is the square of the Fourier transform of the rectangular function, that is, $\sin^2(\pi\tau_a\nu)/(\pi\tau_a\nu)^2$. The equivalent bandwidth, including both positive and negative frequencies, is

$$2\,\Delta\nu_{LF} = \int_{-\infty}^{\infty} \frac{\sin^2(\pi\tau_a\nu)}{(\pi\tau_a\nu)^2}\,d\nu = \frac{1}{\tau_a} . \qquad (6.38)$$

Then from Eq. (6.37) we obtain

$$\mathcal{R}_{sn} = \sqrt{\left(\frac{T_{Am}T_{An}}{T_{Sm}T_{Sn}}\right)2\,\Delta\nu_{IF}\tau_a} . \qquad (6.39)$$

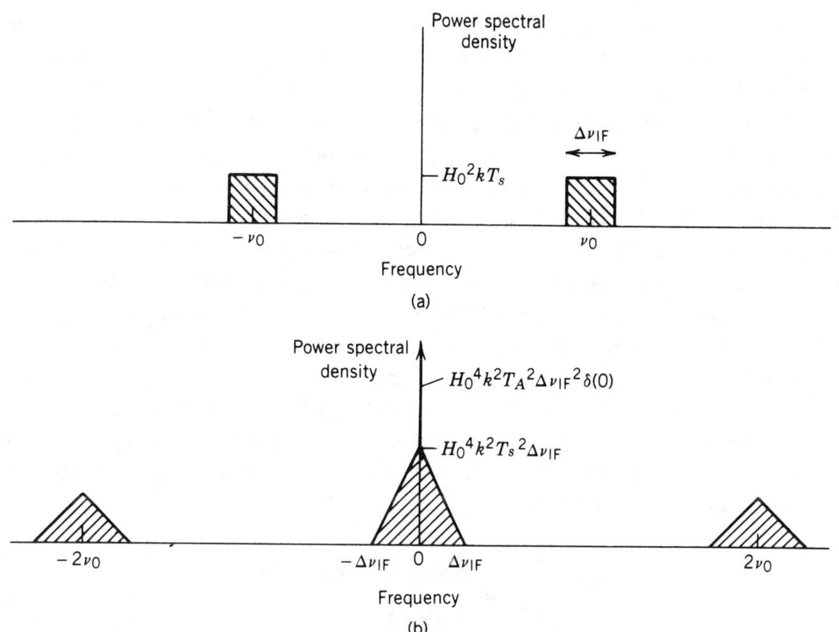

Figure 6.6   Spectra of (a) the input and (b) the output waveforms of a correlator. The input passbands are rectangular of width $\Delta\nu_{IF}$. Shown in (b) is the complete spectrum of signals generated in the multiplication process, including noise bands at twice the input frequency. Only frequencies very close to zero are passed by the averaging circuit at the correlator output. These include the wanted signal, the spectrum of which has the form of a delta function and is represented by the arrow. It is assumed that $T_A \ll T_S$.

Note that $2\,\Delta\nu_{IF}\tau_a$ is the number of independent samples of the signal waveform in time τ_a as discussed in Section 1.2.

If the source is unpolarized each antenna responds to half the total flux density S and the received power density is

$$kT_A = \tfrac{1}{2}AS, \tag{6.40}$$

where A is the effective collecting area of the antenna. For identical antennas and system temperatures we obtain from Eqs. (6.39) and (6.40)

$$\mathcal{R}_{sn} = \frac{AS}{kT_S}\sqrt{\frac{\Delta\nu_{IF}\tau_a}{2}}. \tag{6.41}$$

Similar derivations of this result can be found in the work of Blum (1959), Colvin (1961), and Tiuri (1964). Usually the result in Eq. (6.41), in which we have assumed $T_S \gg T_A$, is the one needed. At the other extreme, $T_A \gg T_S$, which may be encountered in observation of very strong sources such as the sun, $\mathscr{R}_{\rm sn} = \sqrt{\Delta\nu_{\rm IF}\tau_a}$ and is independent of the areas of the antennas because the signal-to-noise-ratio is determined by the fluctuations in signal level.

From Fig. 6.6 we can see how the factor $\sqrt{\Delta\nu_{\rm IF}\tau_a}$ in Eq. (6.41), which enables very high sensitivity to be achieved in radio astronomy, arises. The noise within the correlator results from beats between components in the two input bands and thus extends in frequency up to $\Delta\nu_{\rm IF}$. The triangular noise spectrum in Fig. 6.6 is simply proportional to the number of beats per unit frequency interval. However, only the very small fraction of this noise that falls within the output bandwidth is retained after the averaging. Note that the signal bandwidth $\Delta\nu_{\rm IF}$ that is important here is the bandwidth at the correlator input. In a double-sideband system this is only one-half of the total input bandwidth at the antenna.

One other factor that affects the signal-to-noise ratio should be introduced at this point. If the signals are quantized and digitized before entering the correlators, an efficiency factor η_Q related to the quantization must be included, and (6.41) becomes

$$\mathscr{R}_{\rm sn} = \frac{AS\eta_Q}{kT_S}\sqrt{\frac{\Delta\nu_{\rm IF}\tau_a}{2}}, \tag{6.42a}$$

or in terms of antenna temperature,

$$\mathscr{R}_{\rm sn} = \frac{T_A\eta_Q}{T_S}\sqrt{2\Delta\nu_{\rm IF}\tau_a}. \tag{6.42b}$$

Values of η_Q vary between 0.64 and 1 and are discussed in Chapter 8; see Table 8.1. In VLBI observing other losses affect the signal-to-noise ratio, and η_Q is replaced by a general loss factor η discussed in Section 9.7.

Noise in the Measurement of Complex Visibility

To understand precisely what $\mathscr{R}_{\rm sn}$ represents, note that in deriving Eqs. (6.42) no delay was introduced between the signal components at the correlator, and the phase responses of the signal channels were assumed to be identical. Thus the source is in the central fringe of the interferometer pattern, and in this particular case the response is the peak fringe amplitude which represents the modulus of the visibility. To express the rms noise level at the correlator output in terms of the flux density σ of an unresolved source for which the peak fringe amplitude produces an equal output, we put $\mathscr{R}_{\rm sn} = 1$ in Eq. (6.42a) and replace S by σ:

$$\sigma = \frac{\sqrt{2}\,kT_S}{A\eta_Q\sqrt{\Delta\nu_{\rm IF}\tau_a}}. \tag{6.43}$$

Consider the case of an instrument with a complex correlator in which the output oscillations are slowed to zero frequency as described earlier. The noise fluctuations in the real and imaginary outputs are uncorrelated as we now show. Suppose that the antennas are pointed at blank sky so that the only inputs to the correlators in Fig. 6.3 are the noise waveforms n_m, n_n, and n_m^H, where the last is the Hilbert transform of n_m produced by the quadrature phase shift. The expectation of the product of the real and imaginary outputs is $\langle n_m n_n n_m^H n_n \rangle$, which can easily be shown to be zero by using Eq. (6.30) and noting that the expectations $\langle n_m n_n \rangle$, $\langle n_m n_m^H \rangle$, and $\langle n_m^H n_n \rangle$ must all be zero. Thus the noise from the real and imaginary outputs is uncorrelated.

The signal and noise components in the measurement of the complex visibility are shown in Fig. 6.7 as vectors in the complex plane. Here $\mathscr{V}$ represents the visibility as it would be measured in the absence of noise, and $\mathbf{Z}$ represents the sum of the visibility and noise, $\mathscr{V} + \boldsymbol{\varepsilon}$. We consider $\mathbf{Z}$ and $\boldsymbol{\varepsilon}$ to be vectors whose components correspond to the real and imaginary parts of the corresponding quantities. The noise in both components of $\mathbf{Z}$ has an rms amplitude σ. In practice, we must combine the real and imaginary outputs of the correlator to measure the visibility, and the resulting rms uncertainty in the measurement is

$$\varepsilon_{\text{rms}} = \sqrt{\langle \mathbf{Z} \cdot \mathbf{Z} \rangle - \langle \mathbf{Z} \rangle^2} = \sqrt{\langle \boldsymbol{\varepsilon} \cdot \boldsymbol{\varepsilon} \rangle} = \sqrt{2}\,\sigma, \tag{6.44}$$

since $\langle \boldsymbol{\varepsilon} \cdot \boldsymbol{\varepsilon} \rangle = \langle \varepsilon_x^2 \rangle + \langle \varepsilon_y^2 \rangle = 2\sigma^2$, where ε_x and ε_y are the components of $\boldsymbol{\varepsilon}$. If the measurement is made using only a single-multiplier correlator one can periodically introduce a quadrature phase shift at one input thus obtaining real and imaginary outputs, each for half the observing time. Then the data are half that which would be obtained with a complex correlator, and the noise in the visibility measurement is greater by $\sqrt{2}$. The same result is obtained by recording the single-multiplier output with a nonzero fringe frequency and fitting a sine curve. If the position of an unresolved source is known, it is

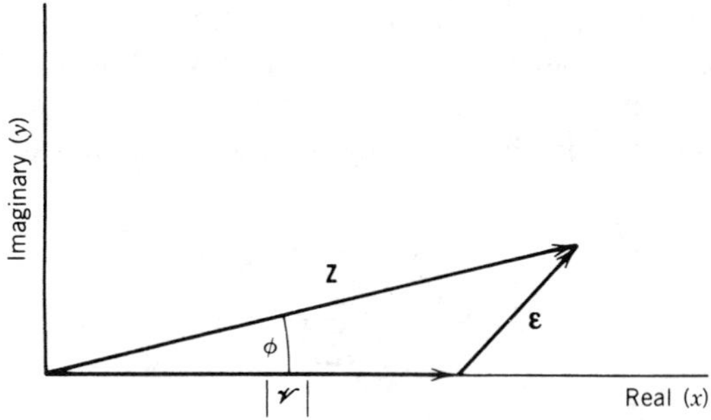

Figure 6.7   Complex quantity $\mathbf{Z}$, which is the sum of the modulus of the true complex visibility $|\mathscr{V}|$ and the noise $\boldsymbol{\varepsilon}$. The noise has real and imaginary components of rms amplitude σ. ϕ is the phase deviation resulting from the noise.

possible to stop the fringes to give the maximum output, and thus measure the visibility magnitude with the same sensitivity as when using a complex correlator. However, this does not measure the *complex* visibility and is not generally useful.

Signal-to-Noise Ratio in a Synthesized Map

Having determined the noise-induced error in the visibility, the next step is to consider the signal-to-noise ratio in a map. Consider an array with n_p antenna pairs and suppose that the visibility data are averaged for time τ_a and that the whole observation covers a time interval τ_0. The total number of independent data points in the (u, v) plane is therefore

$$n_d = n_p \frac{\tau_0}{\tau_a}. \tag{6.45}$$

In mapping an unresolved source at the field center for which the visibility data combine in phase, we should thus expect the signal-to-noise ratio in the map to be greater than that in Eq. (6.42) by a factor $\sqrt{n_p \tau_0 / \tau_a}$. This simple consideration gives the correct result for the case in which the data are combined with equal weights. We now derive the result for the more general case of arbitrarily weighted data.

The ensemble of measured data can be represented by

$$\sum_{i=1}^{n_d} {}^2\delta(u - u_i, v - v_i)(\mathcal{V}_i + \varepsilon_i) + {}^2\delta(u + u_i, v + v_i)(\mathcal{V}_i^* + \varepsilon_i^*),$$

$$\tag{6.46}$$

where ${}^2\delta$ is the two-dimensional delta function and ε_i is the complex noise contribution to the ith datum. Each datum appears at two points, reflected through the origin of the (u, v) plane. Before taking the Fourier transform of the data in Eq. (6.46), each data point is assigned a weight w_i. To simplify the calculation we assume that the source is unresolved and located at the phase reference point of the map, and therefore produces a constant real visibility $\mathcal{V}$ equal to its flux density S. The brightness at the center of the map is then

$$B_0 = \frac{\displaystyle\sum_{i=1}^{n_d} w_i(\mathcal{V} + \varepsilon_{Ri})}{\displaystyle\sum w_i}, \tag{6.47}$$

where ε_{Ri} is the real part of ε_i. Note that the imaginary part of ε_i vanishes at the map origin when the conjugate components are summed. For neighboring points in the map the same rms level of noise is distributed between the real and imaginary parts of ε. The expectation of B_0 is

$$\langle B_c \rangle = \mathcal{V} = S, \tag{6.48}$$

since $\langle \varepsilon_{Ri} \rangle = 0$. The variance of the estimate of the brightness, σ_m^2, is

$$\sigma_m^2 = \langle B_0^2 \rangle - \langle B_0 \rangle^2 = \frac{\sum w_i^2 \langle \varepsilon_{Ri}^2 \rangle}{\left(\sum w_i \right)^2}. \tag{6.49}$$

Equation (6.49) is derived directly from Eq. (6.47) using the fact that the noise terms from different (u, v) locations are uncorrelated, that is, $\langle \varepsilon_{Ri} \varepsilon_{Rj} \rangle = 0$, for $i \neq j$. We define the mean weighting factor w_{mean} and rms weighting factor w_{rms} by the equations

$$w_{\text{mean}} = \frac{1}{n_d} \sum w_i \tag{6.50}$$

and

$$w_{\text{rms}}^2 = \frac{1}{n_d} \sum w_i^2. \tag{6.51}$$

The noise contribution [see Eq. (6.44)] is the same for each (u, v) point and is equal to $\langle \varepsilon_{Ri}^2 \rangle = \sigma^2$, where σ is given by Eq. (6.43). Thus, the signal-to-noise ratio can be calculated from Eqs. (6.48), (6.49), (6.50), and (6.51) as

$$\frac{\langle B_0 \rangle}{\sigma_m} = \frac{S \sqrt{n_d}}{\sigma} \frac{w_{\text{mean}}}{w_{\text{rms}}}. \tag{6.52}$$

For an array with complex correlators we have from Eq. (6.43)

$$\frac{\langle B_0 \rangle}{\sigma_m} = \frac{A S \eta_Q \sqrt{n_d \Delta \nu_{\text{IF}} \tau_a}}{\sqrt{2}\, k T_S} \frac{w_{\text{mean}}}{w_{\text{rms}}}. \tag{6.53}$$

If combinations of all pairs of antennas are used, $n_p = \frac{1}{2} n_a (n_a - 1)$, where n_a is the number of antennas. Since $n_d = n_p \tau_0 / \tau_a$, we obtain

$$\frac{\langle B_0 \rangle}{\sigma_m} = \frac{A S \eta_Q \sqrt{n_a (n_a - 1) \Delta \nu_{\text{IF}} \tau_0}}{2 k T_S} \frac{w_{\text{mean}}}{w_{\text{rms}}}. \tag{6.54}$$

To express the rms noise level in terms of flux density we put $B_0/\sigma_m = 1$ in Eq. (6.54). S then represents the flux density of a point source for which the peak response is equal to the rms noise level, and we can write

$$S_{rms} = \frac{2kT_S}{A\eta_Q\sqrt{n_a(n_a - 1)}\,\Delta\nu_{IF}\tau_0}\,\frac{w_{rms}}{w_{mean}}. \tag{6.55}$$

The peak response to an extended source can be determined in similar units by convolving the source brightness with the synthesized beam, as discussed in Chapter 10, and the signal-to-noise ratio for such an observation thereby obtained. If all the weighting factors w_i are equal, $w_{mean}/w_{rms} = 1$, and this situation is commonly referred to as the use of *natural weighting*. In such a case the signal-to-noise ratio given by Eq. (6.54) is equal to the corresponding sensitivity for a total power receiver combined with an antenna of aperture $\sqrt{n_a(n_a - 1)}\,A$ which approaches $n_a A$ as n_a becomes large (see, e.g., Tiuri 1964).

A common practice is to determine the visibility values in a rectangular array spaced at equal intervals in the u and v directions as described in Section 4.10. Suppose we use the cell averaging technique, in which each data point contributes only to the visibility at the nearest grid point. For uniform weighting unit weight is assigned to the data in each grid cell. Thus, if there are n_ℓ data points contributing to the ℓth cell, each is given a weight of n_ℓ^{-1}. The mean weight from Eq. (6.50) is n_r/n_d, where n_r is the number of grid points, and the square of the rms weight from Eq. (6.51) is $(\sum_{\ell=1}^{n_r} n_\ell^{-1})/n_d$. In this case w_{mean}/w_{rms} is always less than unity unless every grid point has the same number of data points, and the signal-to-noise ratio is lower than for the case of natural weighting.

Although the signal-to-noise ratio depends on the choice of weighting factors, in practice this dependence is not critical. The use of natural weighting maximizes the sensitivity for detection of a point source in a largely blank field but can also substantially broaden the synthesized beam. The advantage in sensitivity is usually small. For example, if the density of data points is inversely proportional to the distance from the (u, v) origin, as is the case for an east–west array with uniform increments in antenna spacing, the weighting factors required to obtain effective uniform density of data result in $w_{mean}/w_{rms} = 2\sqrt{2}/3 = 0.94$. In this case the natural weighting results in an undesirable beam profile in which the response remains positive for large angular distances from the beam axis and dies away only slowly. With uniform weighting the half-power beamwidth is decreased by a factor of 0.78, and the sidelobes die away relatively rapidly.

Various methods of Fourier transformation of visibility data are reviewed in Chapter 10, and the results derived in Eqs. (6.54) and (6.55) can be applied to these by using the appropriate values of w_{mean} and w_{rms}. Convolution of the visibility data in the (u, v) plane to obtain values at points on a rectangular grid is a widely used process. In general the data at adjacent grid points are then not independent, and a tapering of the signal and noise is introduced into the map. Aliasing can also cause the signal-to-noise ratio to vary across the map. (These effects are explained in Fig. 10.6 and the associated discussion.) In such cases the results derived here apply near the origin of the map where the effects of tapering and aliasing are unimportant. The rms noise level over the map can be obtained by the application of Parseval's theorem to the noise in the visibility data.

In practice a number of factors that affect the signal-to-noise ratio are difficult to determine precisely. For example, T_S varies somewhat with antenna elevation. There are also a number of effects that can reduce the response to a source without reducing the noise, but these are important only for sources not near the (ξ, η) origin of a map. These include the smearing resulting from the receiving bandwidth and from visibility averaging discussed later in this chapter, and the effect of non-coplanar baselines discussed in Section 4.1.

Noise in Visibility Amplitude and Phase

In synthesis mapping we are usually concerned with data in the form of the real and imaginary parts of $\mathcal{V}$, but sometimes it is necessary to work with amplitudes and phases. Given that the real and imaginary parts of $\mathcal{V}$ are accompanied by Gaussian noise of standard deviation σ, what are the probability distributions of the amplitudes and phases? The answers are well known and we do not derive them here. The sum of the visibility and noise is represented by $\mathbf{Z} = Ze^{j\phi}$, where we choose the real axis so that the phase ϕ is measured with respect to the phase of $\mathcal{V}$, as in Fig. 6.7. Then for $T_A \ll T_S$ the probability distributions of the resulting amplitude and phase are

$$p(Z) = \frac{Z}{\sigma^2} \exp\left(-\frac{Z^2 + |\mathcal{V}|^2}{2\sigma^2}\right) I_0\left(\frac{Z|\mathcal{V}|}{\sigma^2}\right), \qquad Z > 0 \tag{6.56a}$$

$$p(\phi) = \frac{1}{2\pi} \exp\left(-\frac{|\mathcal{V}|^2}{2\sigma^2}\right)\left\{1 + \sqrt{\frac{\pi}{2}} \frac{|\mathcal{V}|\cos\phi}{\sigma} \exp\left(\frac{|\mathcal{V}|^2\cos^2\phi}{2\sigma^2}\right)\right.$$

$$\left. \times \left[1 + \operatorname{erf}\left(\frac{|\mathcal{V}|\cos\phi}{\sqrt{2}\,\sigma}\right)\right]\right\},$$

$$\tag{6.56b}$$

where I_0 is the modified Bessel function of zero order, erf is the error function, and σ is given by Eq. (6.43). The amplitude distribution is identical to that for a sine wave in noise and the derivation is given by Rice (1944, 1945), Vinokur (1965), and Papoulis (1965), of which the last two also derive the result for the phase. $p(Z)$ is sometimes referred to as the Rice distribution, and for $\mathcal{V} = 0$ it

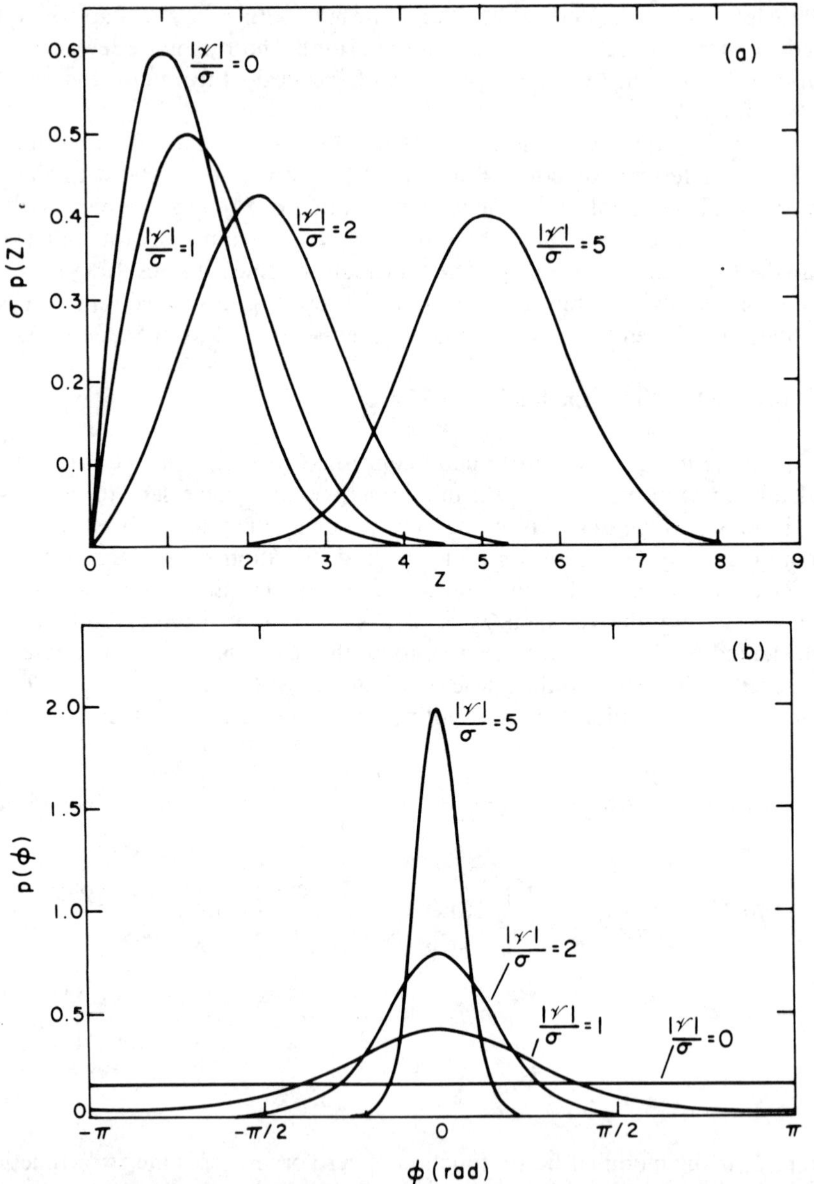

Figure 6.8 Probability distributions of (a) the amplitude, and (b) the phase of the measured complex visibility as functions of the signal-to-noise ratio. $|\gamma|$ is the modulus of the signal component. After Moran (1976).

166

reduces to the Rayleigh distribution. Curves of $p(Z)$ and $p(\phi)$ are given in Fig. 6.8. Comparison of the curves for $|\mathcal{V}|/\sigma = 0$ and 1 indicates that the presence of a weak signal is more easily detected by examining the visibility phase than by examining the amplitude.

Approximation for $p(Z)$ and $p(\phi)$ for the cases where $|\mathcal{V}|/\sigma \ll 1$ and $|\mathcal{V}|/\sigma \gg 1$ are given in Section 9.3. Expressions for the moments of Z and ϕ and their rms deviations are also given in that section. The rms phase deviation σ_ϕ is a particularly useful quantity, especially for astrometric and diagnostic work. The expression for σ_ϕ, valid for the case where $|\mathcal{V}|/\sigma \gg 1$, is $\sigma_\phi \simeq \sigma/|\mathcal{V}|$ [Eq. (9.53)]. This result is intuitively obvious from an examination of Fig. 6.7. By substituting Eq. (6.43) into this expression for σ_ϕ, setting $|\mathcal{V}|$ equal to the flux density of the source S, which is appropriate if the source is unresolved, and using Eq. (6.40) to relate the flux density and antenna temperature, we obtain

$$\sigma_\phi = \frac{T_S}{\eta_Q T_A \sqrt{2\,\Delta\nu_{\mathrm{IF}}\tau_a}}. \tag{6.57}$$

Equation (6.57) is valid for the conditions $T_S/\sqrt{2\,\Delta\nu_{\mathrm{IF}}\tau_a} \ll T_A \ll T_S$, which are the conditions most frequently encountered, and is useful for determining whether or not the noise in the phase measurements of an interferometer is due exclusively to receiver noise. Excess phase noise is often contributed by the atmosphere, by system instabilities, and in the case of VLBI by the frequency standards.

Discussion

The analysis of noise in interferometer systems is necessarily lengthy, so we end by drawing attention to the principal results. Relative sensitivities of various systems are given in Table 6.1.

The signal-to-noise ratio at the correlator output $\mathcal{R}_{sn}$ is given in full in Eq. (6.36), and for digital systems the quantization efficiency factor η_Q should be included. The more generally used weak-signal approximation is given by Eqs. (6.42). The interpretation of this result in terms of the rms error in the visibility is given in Eqs. (6.43) and (6.44) for the complex correlator case. The same sensitivity is obtained either by stopping the fringe oscillations or letting them run and fitting sinusoids to both the real and imaginary outputs using an appropriate least-squares technique. For a system with simple correlators the rms error is greater by $\sqrt{2}$. The signal-to-noise ratio in a synthesized map is given in Eq. (6.54) and the corresponding rms noise level in units of flux density in Eq. (6.55): both these results apply to complex correlator instruments.

In the double-sideband case it should be remembered that the precorrelator signal bandwidth $\Delta\nu_{\mathrm{IF}}$ is that of the IF response and is equal to the width of

TABLE 6.1   Relative Signal-to-Noise Ratios[a] for Several Types of Systems

Single sideband with complex correlator	1
Single sideband with simple correlator	$1/\sqrt{2}$
Double sideband with simple correlator	$1/\sqrt{2}$
One sideband of double-sideband system using complex correlator and sideband separation by $\pi/2$ switching of first LO phase [i.e., the right-hand side of Eq. (6.26) or (6.27)]	$1/2$
One sideband of a VLBI double-sideband system with complex correlator and sideband separation by fringe rotation	$1/2$

[a]Here the signal-to-noise ratio is the modulus of the complex visibility divided by the modulus of the rms noise ε (see Fig. 6.7). For double-sideband systems the double-sideband value of T_S is taken to be equal to T_S for single sideband systems.

one of the RF input sidebands, not the combined width of both. The definitions of T_A and T_S for the double-sideband case should not cause confusion if it is remembered that they are defined in Section 1.2 as the physical temperatures of a matched resistive load at the receiver input that would produce equivalent power levels in the IF stages, and that the noise power enters both sidebands. If the sideband responses are separated by using a complex correlator with quadrature phase switching of the local oscillator, the sensitivity is most easily determined for the case of a continuum source for which the flux density is the same in both sidebands. The noise components associated with each of the four correlator outputs r_1, r_2, r_3, and r_4 in Eqs. (6.24) and (6.25) are uncorrelated and have equal variance. Let us assume that the instrumental phases are adjusted so that only r_1 and r_3 contain signal components: see the discussion following Eq. (6.18). The signal components are represented by Eq. (6.16) and are of equal magnitude, but differ in phase by $\pi/2$. The output r_1 for a single multiplier correlator contains one signal component and one noise component. The output for one sideband, after sideband separation, as given by the right-hand side of Eq. (6.26) or (6.27), contains two signal components and four noise components. Each of these components is formed for only half the total observing time, because of the $\pi/2$ phase switching of the local oscillator. These considerations lead to the conclusion that the signal-to-noise ratio for the expressions in Eqs. (6.26) and (6.27) is less than that for the single multiplier correlator (r_1) by a factor of $\sqrt{2}$.

Finally, note that in many instruments two oppositely polarized signals (with crossed linear or opposite circular polarizations) are received and processed using separate IF amplifiers and correlators. For unpolarized sources, the overall signal-to-noise ratio is then $\sqrt{2}$ greater than the values derived above, which include only one signal from each antenna.

6.3 EFFECT OF BANDWIDTH

As seen in the preceding section, the sensitivity of a receiving system to a broadband cosmic signal increases with the system bandwidth. Here we are concerned with another effect of bandwidth, the restriction of the angular range over which fringes are detected, which we have already encountered in Chapters 2 and 3. This effect results from the variation of fringe frequency, in cycles per radian on the sky, with the received radiofrequency. If the monochromatic response is integrated over the bandwidth, the fringes are reinforced for directions close to that for which the time delays from the source to the correlator inputs are equal, but for other directions the fringes tend to cancel. As mentioned in Section 3.5 this limiting of the fringe envelope is sometimes exploited in meter-wavelength instruments in which it is difficult to make the primary beamwidth of the antenna small enough. At centimeter wavelengths the restriction of the field of view is often undesirable, although it is useful in VLBI.

Mapping in the Continuum Mode

With a two-antenna system the bandwidth restriction on the fringe envelope acts like an antenna beam that is narrow when measured in a plane containing the baseline. However, the width is inversely proportional to the antenna spacing, and in an array with baselines of different lengths and directions the beam analogy is inappropriate. The effect on the response of an array can best be described as follows. If we consider an array with a very narrow receiver bandwidth, we can write for the response

$$\mathscr{V}(u, v)W(u, v) \rightleftharpoons B(\xi, \eta) * * b_0(\xi, \eta). \tag{6.58}$$

The fringe visibility is multiplied by $W(u, v)$, the transfer function of the array for a particular observation, which incorporates any weighting applied in the data processing. The Fourier transform of the left-hand side of Eq. (6.58) gives the brightness distribution $B(\xi, \eta)$ convolved with the synthesized beam function $b_0(\xi, \eta)$. For simplicity we have omitted the primary antenna beam and minor effects related to use of the discrete Fourier transform. Note that the synthesized beam is defined here as the Fourier transform of $W(u, v)$. It is also commonly regarded as the response to a point source: since $W(u, v)$ is real and even the beam is symmetrical and we can ignore the distinction between the beam and its mirror image. Because the synthesized beam as defined above does not include the effects of bandwidth, or of the averaging time discussed later in this chapter, it should be regarded as the response to a point source at the (ξ, η) origin where these effects vanish.

Before considering the effect of the bandwidth on the synthesized map, we note the effect on the visibility. Consider an array in which the longest antenna spacing is D. For this longest spacing, and an RF bandwidth $\Delta\nu$, the visibility is reduced relative to that for a monochromatic signal by a factor

$$R'_b = \frac{\sin(\pi D \xi_1 \, \Delta v / c)}{\pi D \xi_1 \, \Delta v / c}, \tag{6.59}$$

where ξ_1 indicates the position of the source relative to the field center. This result follows from Eq. (2.3). Now the synthesized beamwidth θ_b is approximately equal to $\lambda_0 / D = c / v_0 D$, where v_0 is the observing frequency. (Note that in this section v_0 is the center frequency of the RF input band, not an IF band.) Thus we can write

$$R'_b \simeq \frac{\sin(\pi \, \Delta v \, \xi_1 / v_0 \theta_b)}{(\pi \, \Delta v \, \xi_1 / v_0 \theta_b)}. \tag{6.60}$$

If the parameter $\Delta v \, \xi_1 / v_0 \theta_b$ is equal to unity, $R'_b = 0$ and the measured visibility is reduced to zero. To keep R'_b close to unity we require $\Delta v \, \xi_1 / v_0 \theta_b \ll 1$.

We now examine the same effect in more detail by considering the distortion in the synthesized map. In operation in the continuum mode, the visibility data measured with bandwidth Δv are treated as though they were measured with a monochromatic receiving system tuned to the center frequency v_0. Thus for all frequencies within the bandwidth the assigned values of u and v are those appropriate to frequency v_0. At another frequency v within the passband the true spatial-frequency coordinates u_v and v_v are related to the assigned values u and v by

$$(u, v) = \left(\frac{v_0 u_v}{v}, \frac{v_0 v_v}{v} \right). \tag{6.61}$$

The contribution to the measured visibility from a narrow band of frequencies centered on v is

$$\mathcal{V}_v(u, v) = \mathcal{V}_v\left(\frac{u_v v_0}{v}, \frac{v_v v_0}{v} \right) \rightleftharpoons \left(\frac{v}{v_0} \right)^2 B\left(\frac{\xi v}{v_0}, \frac{\eta v}{v_0} \right), \tag{6.62}$$

where we have used the similarity theorem of Fourier transforms (e.g., Bracewell 1965). Thus the contribution to the measured brightness is the true brightness distribution scaled in (ξ, η) by a factor v / v_0 and in brightness by $(v / v_0)^2$. The derived brightness distribution is convolved with $b_0(\xi, \eta)$, the synthesized beam corresponding to frequency v_0. The beam does not vary with frequency since the same transfer function $W(u, v)$ is used to represent the whole frequency passband. The overall response is obtained by integrating over the passband with appropriate weighting and is

$$B_b(\xi, \eta) = \left[\frac{\int_0^\infty \left(\dfrac{\nu}{\nu_0}\right)^2 |H_{RF}(\nu)|^2 B\left(\dfrac{\xi\nu}{\nu_0}, \dfrac{\eta\nu}{\nu_0}\right) d\nu}{\int_0^\infty |H_{RF}(\nu)|^2 d\nu} \right] ** b_0(\xi, \eta).$$

$$(6.63)$$

Note that the integrals must be taken over the whole radiofrequency passband, denoted by the subscript RF, which includes both sidebands in the case of a double-sideband system. We assume that the passband function $H_{RF}(\nu)$ is identical for all antennas. The values of ξ and η in the brightness function in Eq. (6.63) are multiplied by the factor ν/ν_0 which varies as we integrate over the passband, being equal to unity at the band center. Thus one can envisage the integrals in the square brackets in Eq. (6.63) as a process of averaging a large number of maps, each with a different scale factor. The scale factors are equal to ν/ν_0, and the range of values of ν is determined by the observing passband. The maps are aligned at the origin and thus the effect of the integration over frequency is to produce a radial smearing of the brightness distribution before it is convolved with the beam. The response to a point source at position (ξ, η) is radially elongated by a factor equal to $\sqrt{\xi^2 + \eta^2}\, \Delta\nu/\nu_0$. For distances from the origin at which the elongation is large compared to the synthesized beamwidth, features on the sky become suppressed by the smearing, so there is an effective limitation of the field of view. The measured brightness is the smeared distribution convolved with the synthesized beam.

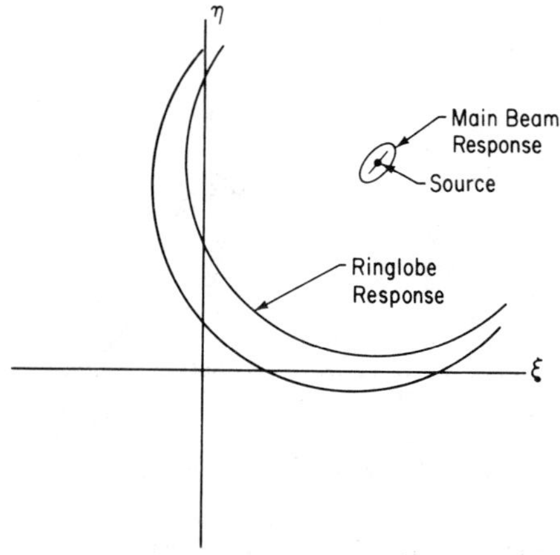

Figure 6.9   Radial smearing resulting from the bandwidth effect for a point source at (ξ_1, η_1). The effects on the responses of the main beam and a ringlobe (i.e. a sidelobe of the form in Fig. 5.15) are shown.

Details of the behavior of the derived brightness distribution can be deduced from Eq. (6.63). For example, suppose that the beam contains a circularly symmetrical sidelobe at a large distance from the beam axis, and that in a map the response to a distant source causes the sidelobe to fall near the origin. Is the sidelobe broadened near the origin? Since the distant source is elongated, the sidelobe will be smeared in a direction parallel to that of a line joining the source and the origin, as shown in Fig. 6.9. It will be broadened near the origin, but not at a point 90° around the sidelobe as measured from the source.

To estimate the magnitude of the suppression of distant sources, it is useful to calculate R_b, the peak response to a point source at a distance r_1 from the origin of the (ξ, η) plane, as a fraction of the response to the same source at the origin. Because the effect we are considering is a radial smearing, we need only consider the brightness along a radial line through the (ξ, η) origin as shown in Fig. 6.10a. We use idealized parameters; the bandpass is represented

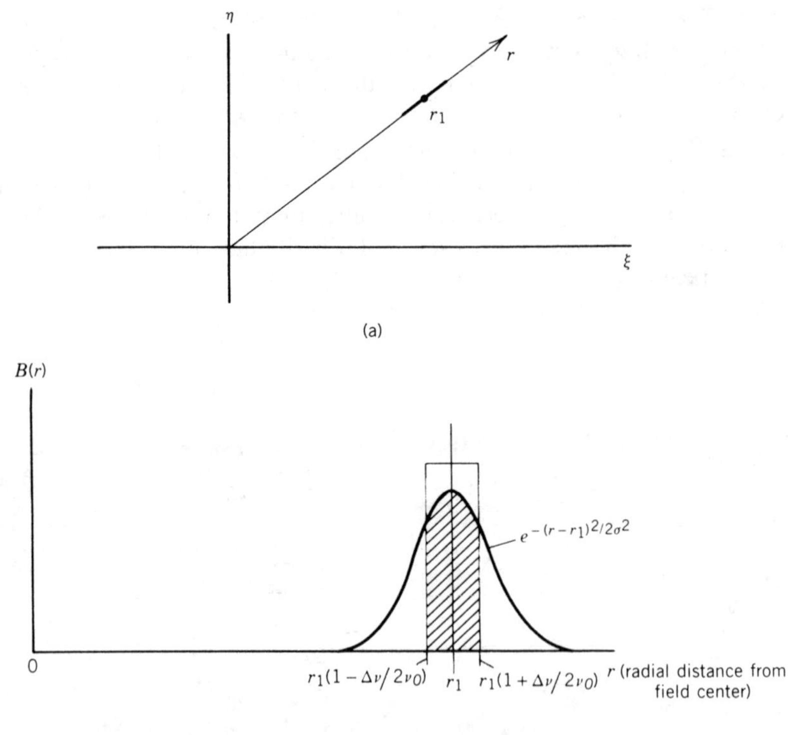

Figure 6.10   Response of an array with a broadband receiving system to a point source at distance r_1 from the origin of the (ξ, η) plane. (a) The point source (delta function) at r_1 becomes radially broadened into a rectangular function of unit area indicated by the heavy line. (b) Cross section of the brightness distribution in the r direction. The synthesized beam is represented by the Gaussian function. The peak brightness of the response to the source is proportional to the shaded area.

by a rectangular function of width $\Delta\nu$ and the synthesized beam by a circularly symmetrical Gaussian function of standard deviation $\sigma_b = \theta_b / \sqrt{8 \ln 2}$, where θ_b is the half-power beamwidth. For simplicity the factor $(\nu/\nu_0)^2$ in the integral in the numerator of Eq. (6.63) is omitted since fractional bandwidths rarely exceed 5%. The convolution becomes a one-dimensional (radial) process as shown in Fig. 6.10b. The radially elongated source is represented by a rectangular function from $r_1(1 - \Delta\nu/2\nu_0)$ to $r_1(1 + \Delta\nu/2\nu_0)$, normalized to unit area. The beam is represented by the function $e^{-r^2/2\sigma_b^2}$ which is normalized to unit amplitude on the beam axis. When the beam is centered on the source as shown in Fig. 6.10, R_b is given by

$$R_b = \frac{\nu_0}{r_1 \Delta\nu} \int_{r_1(1 - \Delta\nu/2\nu_0)}^{r_1(1 + \Delta\nu/2\nu_0)} e^{-(r-r_1)^2/2\sigma_b^2} \, dr = \sqrt{2\pi} \, \frac{\sigma_b \nu_0}{r_1 \Delta\nu} \mathrm{erf}\left(\frac{r_1 \Delta\nu}{2\sqrt{2}\,\sigma_b \nu_0} \right)$$

$$= 1.064 \frac{\theta_b \nu_0}{r_1 \Delta\nu} \mathrm{erf}\left(0.833 \frac{r_1 \Delta\nu}{\theta_b \nu_0} \right). \tag{6.64}$$

A curve of R_b as a function of the parameter $r_1 \Delta\nu/\theta_b \nu_0$, which is the distance of the source from the origin measured in beamwidths, multiplied by the fractional bandwidth, is shown in Fig. 6.11. When this parameter is unity, the

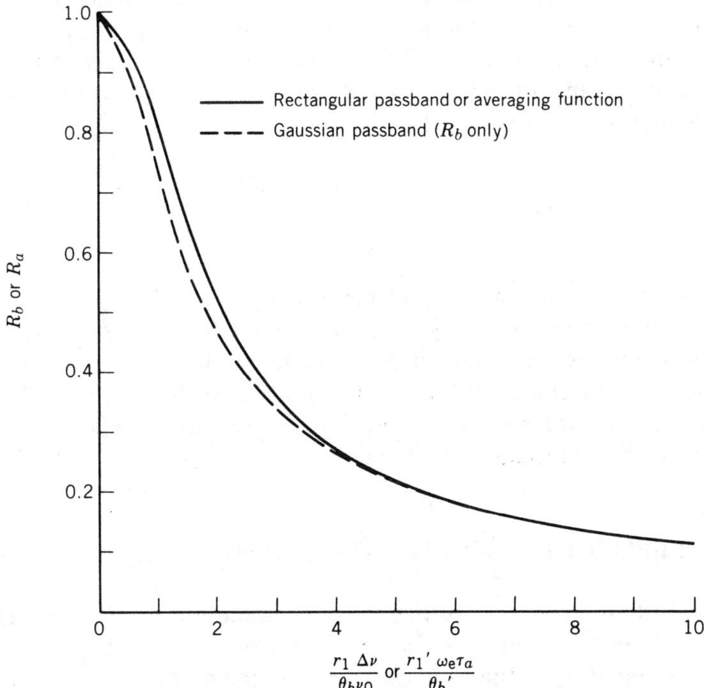

Figure 6.11   Relative amplitude of the peak response to a point source as a function of the distance from the field center and either the fractional bandwidth or the averaging time.

peak response to the point source is reduced by 19%. Note that from Eq. (6.60) the visibility measured on the longest baseline can then go to zero, if the baseline becomes parallel to r_1 and normal to the source direction.

If the receiving bandpass is represented by a Gaussian function of equivalent width $\Delta\nu$ (i.e., standard deviation $= \Delta\nu/2.5066$), the reduction factor becomes

$$R_b = \frac{1}{\sqrt{1 + (0.939 r_1 \Delta\nu/\theta_b \nu_0)^2}} . \tag{6.65}$$

A curve of this function is also included in Fig. 6.11. Note that R_b is not critically dependent on the passband shape.

Wide Field Mapping with a Multichannel System

Broadband maps can also be obtained by observing with a multichannel (spectral-line) system. In this case the passband is divided into a number of channels by using either a bank of narrowband filters or a digital spectral correlator. The visibility is measured independently for each channel, so the values of u and v can be scaled correctly and an independent map obtained for each channel. This scaling causes the transfer function to vary over the band, and at frequency ν the synthesized beam is $(\nu/\nu_0)^2 b_0(\xi\nu/\nu_0, \eta\nu/\nu_0)$, $b_0(\xi, \eta)$ being the monochromatic beam at frequency ν_0. The maps can be combined by summation, and if given equal weights the result for N channels is represented by

$$\cdot B(\xi, \eta) * * \left[\frac{1}{N} \sum_{i=1}^{N} \left(\frac{\nu_i}{\nu_0} \right)^2 b_0 \left(\frac{\xi\nu_i}{\nu_0}, \frac{\eta\nu_i}{\nu_0} \right) \right]. \tag{6.66}$$

In this case there is no smearing of the brightness distribution, but the beam suffers a radial smearing that has the desirable effect of suppressing distant sidelobes. Therefore, this mode of observation is well suited for mapping wide fields. The improvement in the beam is an example of space–frequency equivalence, since the receiving bandwidth is used to effect a corresponding spread in the spatial-frequency coverage.

6.4 EFFECT OF VISIBILITY AVERAGING

In most synthesis arrays the output of each correlator is averaged for consecutive time periods τ_a, and thus consists of a series of real or complex values spaced at intervals τ_a in time. Generally τ_a is in the range of a few seconds to a few minutes and the same value is used for all the correlators in the array. It is advantageous to make τ_a as long as possible to minimize the quantity of data to be processed. However, the effect of the averaging introduces a smearing in the map which increases with the distance from the (ξ, η) origin in a similar manner to the bandwidth effect.

We consider the case in which the instantaneous visibility data are combined with equal weights over the interval τ_a. As shown in Section 4.5, the visibility for a point source at a distance r_1 from the origin of the (ξ, η) plane varies sinusoidally with a maximum frequency $\omega_e r_1 D_E/\lambda$, where ω_e is the angular rotation velocity of the earth and D_E is the length of the equatorial component of the baseline vector. For an east–west baseline the synthesized beamwidth is $\theta_b \simeq \lambda/D_E$. Then for the maximum rate of variation of the visibility, the averaging results in a reduction of the measured visibility by a factor

$$R'_a \simeq \frac{\sin(\pi \omega_e \tau_a r_1/\theta_b)}{\pi \omega_e \tau_a r_1/\theta_b}. \tag{6.67}$$

Thus, the parameter $\omega_e \tau_a r_1/\theta_b$ should be small enough that R'_a is close to unity. Note that this parameter is equal to the rotation angle of the earth in time τ_a multiplied by the distance of the source from the origin, measured in beamwidths.

We now examine the effect of the averaging on the synthesized brightness distribution. In reducing the data, all visibility values within each interval τ_a are treated as though they applied to the time at the center of the averaging period. Thus, the measurements at the beginning of each averaging period, for example, enter into the visibility data with assigned values of u and v that apply to times $\tau_a/2$ later than the true values. The consequences can be visualized by considering the resulting map as consisting of the average of a large number of maps each with a different timing offset distributed progressively throughout the range $-\tau_a/2$ to $\tau_a/2$. Note that these timing offsets apply only to the assignment of (u, v) values and do not resemble a clock error that affects the whole receiving system.

To examine the result of a timing offset in the (u, v) assignment we use the concept of the (u', v') plane introduced in Section 4.3. The transformation $u' = u$, $v' = v \operatorname{cosec} \delta_0$, δ_0 being the phase reference declination, provides the projection of the (u, v) plane onto the earth's equatorial plane. The spacing loci become circular arcs generated by vectors rotating at angular velocity ω_e as shown in Fig. 6.12a. Consider first the case of an east–west linear array: then of the antenna spacing components (X, Y, Z) defined in Section 4.2 only Y is nonzero. The circular arcs of the spacing loci are centered on the $(u,' v')$ origin as in Fig. 6.12b, and a timing offset δt is equivalent to a rotation of the (u', v') axes through an angle $\omega_e \delta t$. In the (ξ', η') plane, in which $\xi' = \xi$ and $\eta' = \eta \sin \delta_0$, an exactly equal rotation occurs as can be seen by considering the effect for a point source at (ξ'_1, η'_1) in Fig. 6.13. We have encountered the visibility of a point source in Section 4.5: the source is represented by a delta function, and its visibility is the sum of two sets of sinusoidal corrugations, one

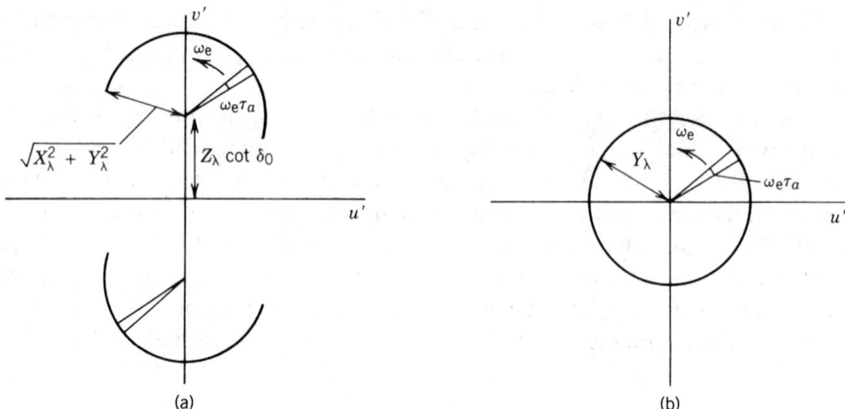

Figure 6.12   Spacing loci in the (u', v') plane, (a) for the general case and (b) for an east–west baseline. The angle $\omega_e \tau_a$ over which the averaging takes place is enlarged for clarity: for example, with a typical averaging time of 30 sec the angle would be 7.5 arcmin.

real and one imaginary:

$$\delta(\xi_1', \eta_1') \rightleftharpoons \cos 2\pi(u'\xi_1' + v'\eta_1') - j\sin 2\pi(u'\xi_1' + v'\eta_1'). \qquad (6.68)$$

The angle of the corrugations is related to the position angle $\psi' = \tan^{-1}(\eta_1'/\xi_1')$ of the point source as shown in Fig. 6.13. A change in ψ' causes an equivalent rotation of the corrugations, and vice versa. For an east–west array the time offsets therefore correspond to proportional rotations of the brightness in the

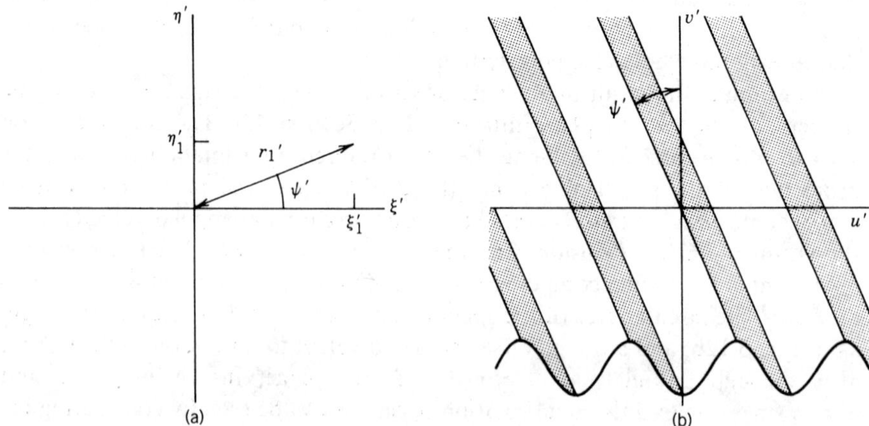

Figure 6.13   (a) Point source at (ξ_1', η_1') and (b) the real part of the corresponding visibility function. The ridges of the sinusoidal corrugations that represent the visibility in the (u', v') plane are orthogonal to the radius vector r_1' at the position of the source in the (ξ', η') plane.

(ξ', η') plane, and it should be clear that the effect of the time averaging is to produce a circumferential smearing similar to that resulting from the receiving bandwidth but orthogonal to it. If we express positions in the (ξ', η') plane in terms of the radial coordinates (r', ψ') shown in Fig. 6.13a, the map obtained from the averaged data can be expressed in terms of the sky brightness $B(r', \psi')$ by

$$B_a(r', \psi') = \left[\frac{1}{\omega_e \tau_a} \int_{-\omega_e \tau_a/2}^{\omega_e \tau_a/2} B(r', \psi')\,d\psi' \right] * * \, b_0(r', \psi'), \qquad (6.69)$$

where b_0 is the synthesized beam.

The fractional decrease in the peak response to the point source is most easily considered in the (ξ', η') plane. With an east–west baseline the contours of the synthesized beam are approximately circular in the (ξ', η') plane, as long as the observing time is approximately 12 hr, which results in spacing loci in the form of complete circles in the (u', v') plane. If we assume that the synthesized beam can be represented by a Gaussian function, as in the calculations for the bandwidth effect, the curve for the rectangular bandwidth in Fig. 6.11 can also be used for the averaging effect. In one case the spreading function is radial and width $r_1 \Delta\nu/\nu_0$, and in the other it is circumferential and of width $r_1'\omega_e\tau_a$. Thus for the averaging effect we can replace $r_1 \Delta\nu/\theta_b\nu_0$ in Eq. (6.64) and Fig. 6.11 (solid curve) by $r_1'\omega_e\tau_a/\theta_b'$, noting that $r_1' = \sqrt{\xi_1^2 + \eta_1^2\sin^2\delta_0}$ and θ_b', the synthesized beamwidth in the (ξ', η') plane, is equal to the east–west beamwidth in the (ξ, η) plane. Hence, for the decrease in the response to a point source resulting from averaging we can write

$$R_a = 1.064 \frac{\theta_b'}{r_1'\omega_e\tau_a} \mathrm{erf}\left(0.833 \frac{r_1'\omega_e\tau_a}{\theta_b'} \right). \qquad (6.70)$$

Generally, one chooses τ_a so that R_a is only slightly less than unity at any point in the map, in which case we can approximate the error function by the integral of the first two terms in the power series for a Gaussian function:

$$R_a \simeq 1 - \frac{1}{3}\left(\frac{0.833\omega_e\tau_a}{\theta_b'} \right)^2 \left(\xi_1^2 + \eta_1^2\sin^2\delta_0 \right). \qquad (6.71)$$

This is a useful formula for checking that τ_a is not too large.

Two aspects of the behavior predicted by Eq. (6.71) should be mentioned. First, if the source is on the η' axis the averaging has no effect for $\delta_0 = 0°$. This is because the ridges of the sinusoidal corrugations of the visibility function run parallel to the u' axis, and in the transformation $u' = u \, \mathrm{cosec}\, \delta_0$ the period of the variations in the v direction is expanded by a very large factor. In comparison the arc through which any spacing vector moves in time

τ_a is very small, and hence the averaging has negligible effect on the visibility amplitude. Second, for a source on the ξ' axis R_a is independent of δ_0. In this case the ridges of the corrugations run parallel to the v axis, and the expansion of the scale in the v direction has no effect on the sinusoidal period.

For arrays that contain baselines other than east–west, the centers of the corresponding loci in the (u', v') plane are offset from the origin, as in Fig. 6.12a, and a time offset is no longer equivalent to a simple rotation of axes. However, the lengths and orientations of the arcs over which the visibility is averaged remain unchanged, they are simply shifted parallel to the v' axis. Clearly, such an offset has only a small overall effect on the averaging of the visibility function of a point source, especially if the averaging arcs are distributed fairly uniformly in orientation. In the case of a source on the ξ' axis the shift of the origin of the spacing loci has precisely zero effect, since the shift is parallel to the ridges of the corrugations. Similarly, for a source near the pole the offset $Z_\lambda \cot \delta_0$ of the loci is very small and the situation is like that for east–west baselines. Thus, for most arrays the results in Eqs. (6.70) and (6.71) should provide a good estimate of the magnitude of the averaging effect.

REFERENCES

Blum, E. J., Sensibilité des Radiotélescopes et Récepteurs à Corrélation, *Ann. d'Astrophys.*, **22**, 140–163, 1959.

Bracewell, R. N., *The Fourier Transform and Its Applications*, McGraw-Hill, New York, 1965 (2nd ed. 1978).

Colvin, R. S., A Study of Radio Astronomy Receivers, Ph.D. Thesis, Stanford University, Stanford, California, 1961.

de Jager, J. T. and B. J. Robinson, Sensitivity of the Degenerate Parametric Amplifier, *Proc. IRE*, **49**, 1205–1206, 1961.

Lawson, J. L. and G. E. Uhlenbeck, *Threshold Signals*, Radiation Laboratory Series, Vol. 24, McGraw-Hill, New York, 1950.

Middleton, D., *An Introduction to Statistical Communication Theory*, McGraw-Hill, New York, 1960.

Moran, J. M., Very Long Baseline Interferometric Observations and Data Reduction, in *Methods of Experimental Physics*, Vol. 12C, M. L. Meeks, Ed., Academic Press, New York, 1976, pp. 228–260.

Papoulis, A., *Probability, Random Variables and Stochastic Processes*, McGraw-Hill, New York, 1965.

Read, R. B., Two-Element Interferometer for Accurate Position Determinations at 960 Mc, *IRE Trans. Antennas Propag.*, **AP-9**, 31–35, 1961.

Rice, S. O., Mathematical Analysis of Random Noise, *Bell Syst. Tech. J.*, **23**, 282–332, 1944, and **24**, 46–156, 1945. Reprinted in *Noise and Stochastic Processes*, N. Wax, Ed., Dover, New York, 1954.

Tiuri, M. E., Radio Astronomy Receivers, *IEEE Trans. Antennas Propag.*, **AP-12**, 930–938, 1964.

Vander Vorst, A. S. and R. S. Colvin, The Use of Degenerate Parametric Amplifiers in Interferometry, *IEEE Trans. Antennas Propag.*, **AP-14**, 667–668, 1966.

Vinokur, M., Optimisation dans la Recherche d'une Sinusoide de Période Connue en Présence de Bruit, *Ann. d'Astrophys.*, **28**, 412–445, 1965.

Wright, M. C. H., B. G. Clark, C. H. Moore, and J. Coe, Hydrogen-Line Aperture Synthesis at the National Radio Astronomy Observatory: Techniques and Data Reduction, *Radio Sci.*, **8**, 763–773, 1973.

Wozencraft, J. M. and I. M. Jacobs, *Principles of Communication Engineering*, Wiley, New York, 1965.

7

DESIGN OF THE ANALOG RECEIVING SYSTEM

The basic functions of the receiving system have been outlined in earlier chapters. Here we consider certain aspects of the system design in more detail. These mainly concern the equipment between the antennas and the correlators and, in particular, those characteristics of it that are critical to the accuracy of the visibility measurements. They include phase stability, frequency responses, spurious signals, and automatic level control. The analysis leads to specification of tolerances on system parameters that are consistent with the goals of sensitivity and accuracy. Analog systems only are included, and digital sampling, delaying, and correlating of signals are the subject of Chapter 8. In several cases the discussion is based on experience related to the development of the VLA (Thompson et al. 1980; Napier, Thompson, and Ekers 1983).

7.1 PRINCIPAL SUBSYSTEMS OF THE RECEIVING ELECTRONICS

We give only a brief description of the main features of a receiving system. Optimum techniques and components for implementation of the electronic hardware vary continuously as the state of the art advances, and descriptions in the literature provide examples of the techniques current at various times: see, for example, Read (1961), Elsmore, Kenderdine, and Ryle (1966), Baars et al. (1973), Bracewell et al. (1973), Wright et al. (1973), Welch et al. (1977),

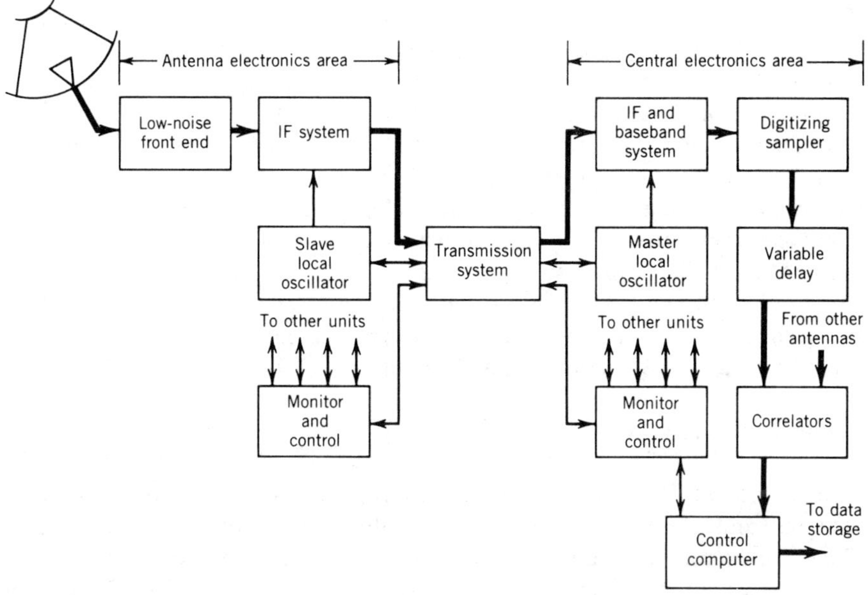

Figure 7.1   Simplified schematic diagram of the receiving system of a typical synthesis array. All the blocks indicate subsystems described in the text, except for the monitor and control blocks which comprise the digital communication system through which the computer monitors critical voltages, sets local oscillator frequencies, and so on. Monitor and control interconnections to the other blocks have been omitted for simplicity. Except for the master local oscillator and the computer, one of each block is required per antenna. The heavy line shows the path of the received signal.

Batty et al. (1982), Erickson, Mahoney, and Erb (1982), and Napier, Thompson, and Ekers (1983).

Figure 7.1 shows an example of a simplified schematic diagram of the receiving system of a large array of the linked-element type. For engineering convenience it is useful to divide the overall system into various subsystems which are likely to require the attention of specialists in different electronic areas. The principal subsystems are outlined below.

Low-Noise Input Stages

High sensitivity in a receiving system requires that the noise temperatures of the antenna and the input stages of the receiver be as low as possible. The overall system temperature T_S is related to the characteristics of the series of cascaded components through which the signal flows as follows (see, e.g., Kraus 1966):

$$T_S = T_A' + (L_1 - 1)T + T_{R1}L_1 + T_{R2}L_1G_1^{-1} + T_{R3}L_1(G_1G_2)^{-1} + \cdots$$

$$(7.1)$$

Here T_A' is the antenna temperature as defined in Section 1.2; L_1 is the power loss factor ($L_1 \geq 1$) of the transmission line connecting the antenna to the first stage of the receiver, and T is the temperature of the line; T_{Ri} is the noise temperature of the ith receiver stage and G_i is its power gain. If the first stage is a mixer, G_1 may be less than unity, and the second stage noise temperature then becomes very important. In modern receivers, minimizing the noise temperature usually involves cryogenic cooling of the amplifier or mixer stages from the input up to a point at which noise from succeeding stages is unimportant. The low-noise input stages are sometimes referred to as the front-end subsystem. For descriptions see, for example, Reid et al. (1973), Weinreb et al. (1977a), Weinreb, Fenstermacher, and Harris (1982), Casse, Woestenburg, and Visser (1982), and Phillips and Woody (1982).

Local Oscillator

As explained in the previous chapter, local oscillator signals are required at the antennas and often at other points along the signal paths to the correlators. The corresponding oscillator frequencies for different antennas must be maintained in phase synchronism to preserve the coherence of the signals. The phases of the oscillators at corresponding points on different antennas need not be identical, but the differences should be stable enough to permit calibration. Maintaining sychronism at different antennas requires transmitting one or more reference frequencies from a central master oscillator to the required points, where they may be used to phase lock other oscillators, sometimes referred to as slave oscillators. The frequencies required at the mixers can then be synthesized.

Special phase shifts to implement fringe rotation, as described in Section 6.1, or phase switching, to be described in Section 7.5, are required at certain mixers. Often these can best be implemented by digital synthesis techniques. For example, one method of producing a signal with accurately controlled phase and frequency offsets consists in dividing, say, a frequency of 50 MHz down to 100 kHz. The frequency divider is a cycle-counting circuit in which the output is the most significant bit. By loading an initial binary number into the counting register the phase of the output at 100 kHz can be controlled in steps of $1/500$ of a rotation. Furthermore, the input stage of the counter can be designed to count one input pulse as zero or as two on certain commands. These commands are then initiated by a variable rate-generating circuit and thereby result in accurately controlled frequency offsets. The output at 100 kHz can be added to an oscillator signal for frequency conversion by using the 100 kHz as a reference frequency in a phase-locked loop.

IF and Signal Transmission Subsystems

After amplification in the low-noise front-end stages, the signals pass through various IF amplifiers and a transmission system before reaching the correlators. Transmission between the antennas and a central location is effected by means of coaxial or parallel-wire lines, waveguide, optical fibers, or direct radiation by microwave radio link. Cables are generally used for small instruments, but for long distances the cable attenuation may require the use of too many line amplifiers, and lower-loss TE_{01}-mode waveguide (Weinreb et al. 1977b; Archer, Caloccia, and Serna 1980) or radio links have been used. Coaxial cables or waveguides can be buried at depths of 1–2 m to reduce temperature variations. Transmission frequencies vary from some tens or hundreds of megahertz in coaxial cable, tens of gigahertz in waveguide or radio links, to optical frequencies in fibers.

After arriving at the correlator location, the received signals are usually converted to a final intermediate frequency where bandwidth selection filters and compensating time delays are inserted. Phase errors, resulting from temperature effects in filters, and delay-setting errors can be minimized by using the lowest possible intermediate frequency at this point. Accordingly, the final IF amplifiers often have a *baseband* response defined by a lowpass filter. The response at the low-frequency end falls off at a frequency that is a few percent of the upper cutoff frequency. In the overall response of the receiver a sharp cutoff at the zero frequency of the baseband can be achieved by the use of a single-sideband mixer (see Appendix 7.1) for the conversion to baseband.

The transmission system may also be used to distribute the reference frequencies for the local oscillator subsystem. In that case signals must be transmitted in both directions between the antennas and the central location, and either time or frequency multiplexing can be used to separate the outgoing and incoming signals at any point.

Delay and Correlator Subsystems

The compensating delays and correlators can be implemented by either analog or digital techniques. An analog delay system may consist of a series of switchable delay units with a binary sequence of values in which the delay of the nth unit is $2^{n-1}\tau_0$, where τ_0 is the delay of the smallest unit. Such an arrangement, with N units, provides a range of delay from zero to $(2^N - 1)\tau_0$ in steps of τ_0. For delays up to about 1 μsec, lengths of coaxial cable can be used. For longer delays, cables become unwieldy and acoustic-wave devices can provide larger increments. Systems with analog delays usually have analog correlators. The design of analog multiplying circuits has been studied by Frater (1964) and Allen and Frater (1970). The output is averaged, digitized, and sent to a computer for further reduction. In spectral-line correlator systems of the analog type the final IF amplifier contains a bank of filters that are centered in frequency at equal increments across the signal band. The bandwidth of the filters is equal to the frequency increment. Each filter defines a signal channel, and a separate correlator is required for each channel of every antenna pair.

The development of digital circuitry capable of operating at clock frequencies of 100 MHz and higher has led to the practice of digitizing the final IF signal, so that the delay can be implemented digitally. Digital delays are discussed in Chapter 8. Their advantage is that greater precision can be achieved in the visibility measurements, since with analog delays it is very difficult to keep the bandpass response from varying as different units are switched into and out of the signal channels. It is also difficult to maintain accurate calibration of the delay of the longer analog units unless they are kept at a constant temperature. Correlators for both continuum and spectral-line observations are readily implemented in digital logic.

VLBI Systems

VLBI systems differ from those outlined above chiefly in the use of independent frequency standards for the local oscillators and in the recording of signals at the antennas rather than direct transmission to the correlators. These features are discussed in Chapter 9. Topics in the present chapter that are relevant to interferometers and arrays that use independent oscillators and recorders include the phase stability of filters, frequency responses of signal channels, and automatic level control.

7.2   LOCAL OSCILLATOR AND GENERAL CONSIDERATIONS OF PHASE STABILITY

Round-Trip Phase Measuring Schemes

Synchronizing of the oscillators at the antennas can be accomplished by phase locking them to a reference frequency that is transmitted out from a central master oscillator. Buried cables or waveguides offer the advantage of the greatest stability of the transmission path. At a depth of 1–2 m the diurnal temperature variation is almost entirely eliminated, but the annual variation is typically attenuated by a factor of 2–10 only. For a discussion of temperature variation in soil as a function of depth see the *Handbook of Geophysics and Space Environments* (USAF 1965). As an example, a 10-km-long buried cable with a temperature coefficient of length of 10^{-5} K^{-1} might suffer a diurnal temperature variation of 0.1 K, resulting in a change of 1 cm in electrical length. This change would cause a variation of 12° in the phase of a 1-GHz signal traversing the cable. An equal variation would occur in a 50-m length of cable running from the ground to the receiver enclosure on an antenna and subjected to a diurnal temperature variation of 20 K. Rotating joints and flexible cables can also contribute to significant phase variations.

Path length variations can be determined by monitoring the phase of a signal of known frequency that traverses the path. It is necessary for the signal to travel in two directions, that is, out from the master oscillator and back again, since the master provides the reference against which the phase must be measured. This technique is described as round-trip phase measurement. The phase changes at the antennas can be corrected in real time, they can be compensated in the data analysis, or two signals in the line may be sampled and combined to produce a signal with phase that is independent of variations in the line. As an illustration of the last procedure consider a signal applied to the near end of a loss-free transmission line that results in a voltage $V_0 \cos(2\pi\nu t)$ at the far end. At a distance ℓ from the far end the outgoing signal is $V_1 = V_0 \cos 2\pi\nu(t + \ell/v)$, where v is the phase velocity along the line. Suppose that the signal is reflected from the far end without change in phase. At the point ℓ the returned signal is $V_2 = V_0 \cos 2\pi\nu(t - \ell/v)$ and the total signal is

$$V_1 + V_2 = 2V_0 \cos(2\pi\nu t)\cos\left(\frac{2\pi\nu\ell}{v}\right). \tag{7.2}$$

The first cosine function in Eq. (7.2) represents the radiofrequency signal, the phase of which (modulo π) is independent of ℓ and also of line length variations. The second cosine function is a standing-wave amplitude term. Such a system cannot easily be implemented in practice because of attenuation and unwanted reflections, and thus more complicated schemes have evolved. In what follows we consider cable transmission, although the basic principles are applicable to other systems. Some general considerations, including the use of microwave links, are given by Thompson et al. (1968).

Swarup and Yang System

Several different round-trip schemes have been devised as instruments have developed, and one of the earliest of these was by Swarup and Yang (1961). A system based on this scheme is shown in Fig. 7.2. Part of the outgoing signal is reflected from a known reflection point at an antenna, and variation in the path length to the reflector is monitored by measuring the relative phase of the reflected component at the detector. The phase of the reflected signal is compared with that of a reference signal. The phase of the latter is variable by means of a movable probe which samples the outgoing signal. Since many other reflections may occur in the transmission line, it is necessary to identify the desired component. To do this one uses a modulated reflector, for example, a diode loosely coupled to the line and switched between conducting and nonconducting states by a square-wave voltage. A synchronous detector is used to separate the modulated component of the reflected signal.

An increase $\Delta\ell$ in the length of the transmission line is detected as a corresponding movement of $2\,\Delta\ell$ in the probe position. It results in an increase of $2\pi\Delta\ell\nu_1/v$ in the phase of the frequency ν_1 at the antenna. The corresponding changes in local oscillator phases and IF phases transmitted over the same path can be calculated and applied as a correction to the visibility phases. Alternatively, the correction can be applied directly to the signals through a phase shifter or a mechanical line extender. In the original application by Swarup and Yang the transmission line was part of a branching

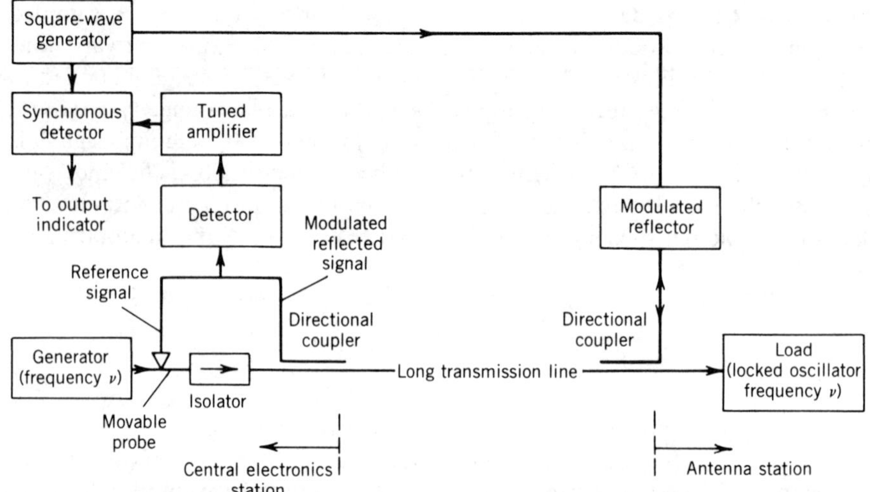

Figure 7.2   System for measuring variations in the electrical path length in a transmission line, based on the technique of Swarup and Yang (1961). The output of the synchronous detector is a sinusoidal function of the difference between the phases of the reference (outgoing) and reflected components at the detector. A null output is obtained when these signal phases are in quadrature, and the position of the probe for a null is thus a measure of the phase of the reflected signal. Because of the isolator in the line, the probe samples only the outgoing component of the signal.

feeder network to an array of antennas. Various developments of the modulated-reflector technique have been devised; see, for example, Legg (1965).

Frequency-Offset, Round-Trip System

A second scheme, shown in Fig. 7.3, is one in which the round-trip phase is measured directly. The signals traveling in opposite directions are at slightly different frequencies ν_1 and ν_2, which allows them to be separated easily. This type of system is widely used, and we examine its performance in some detail. Note that although directional couplers or circulators allow the signals going in the two directions in the line to be separated, the signal from the unwanted direction is suppressed by 20 to 30 dB only. An unwanted component at a level of -30 dB can cause a phase error of 1.8°. However, the frequency offset enables the signals to be separated with very high accuracy.

An oscillator at frequency ν_2 at an antenna is phase locked to the difference frequency of signals at ν_1 and $\nu_1 - \nu_2$ which travel to the antenna via a transmission line. The frequency ν_2 is returned to the master oscillator location for the round-trip phase comparison.

At the antenna the phases of the signals at frequencies ν_1 and $\nu_1 - \nu_2$ relative to their phases at the central location are $2\pi\nu_1 L/v$ and $2\pi(\nu_1 - \nu_2)L/v$, where L is the length of the cable. The phase of the ν_2 oscillator at the antenna is constrained by a phase-locked loop to equal the difference of these phases, that is, $2\pi\nu_2 L/v$. The phase change in the ν_2 signal in traveling back to the central location is $2\pi\nu_2 L/v$, and thus the measured round-trip phase is $4\pi\nu_2 L/v$ (modulo 2π, of course). Now suppose that the length of the line changes by a small fraction, β. The phase of the oscillator ν_2 at the antenna relative to the master oscillator changes to $2\pi\nu_2 L(1 + \beta)/v$. The required correction to the ν_2 oscillator is just half the change in the measured round-trip phase. The problem that arises is that several effects, including reflections and velocity dispersion in the transmission line, can cause an error in the round-trip phase correction. Such an error results in a phase offset of the oscillator at the antenna, which is not serious if it remains constant. However, in practice it is likely to vary with ambient temperature. The largest error usually results from reflections, and control of this error places an upper limit on the frequency difference $\nu_1 - \nu_2$. We now examine this limit.

Consider what happens if reflections occur at points A and B spaced apart a distance ℓ along the line as in Fig. 7.3. The complex voltage reflection coefficients at these points are ρ_A and ρ_B and their values will be assumed to be the same at frequencies ν_1 and ν_2. Signals ν_1 and ν_2, after traversing the cable, include components that have been reflected once at A and once again at B. The coefficients ρ_A and ρ_B are sufficiently small that components suffering more than one reflection at each point can be neglected. For the frequency ν_1

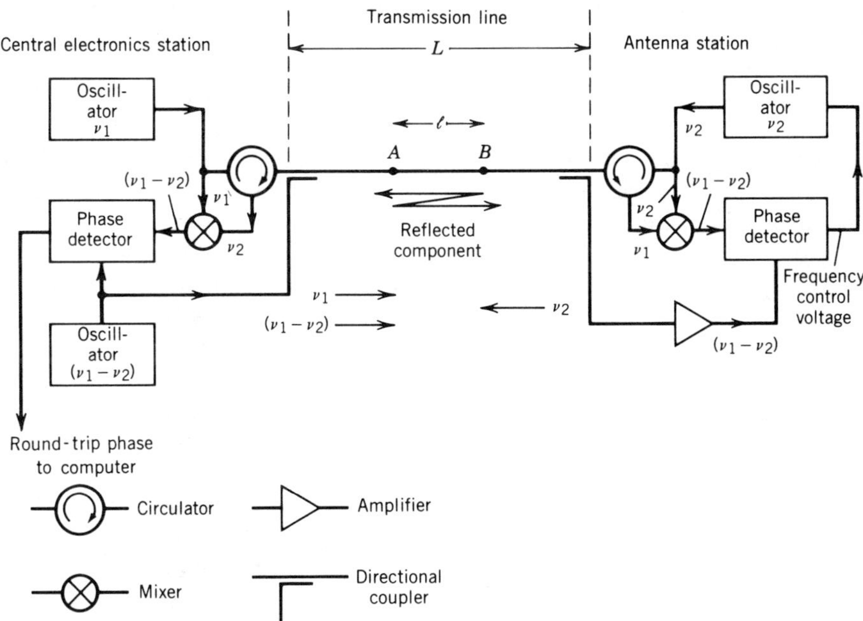

Figure 7.3   Phase-lock scheme for the oscillator ν_2 at the antenna. Frequencies ν_1 and $\nu_1 - \nu_2$ are transmitted to the antenna station where they provide the phase reference to lock the oscillator. ν_1 and ν_2 are almost equal, so $\nu_1 - \nu_2$ is small. A signal at frequency ν_2 is returned to the central station for the round-trip phase measurement.

arriving at the antenna, the amplitude of the reflected component relative to the unreflected one is

$$\Lambda = |\rho_A||\rho_B|10^{-\ell\alpha/10}, \qquad (7.3)$$

where α is the attenuation coefficient of the cable in decibels per unit length. Similarly, the phase of the reflected component relative to the unreflected one is, modulo 2π,

$$\theta_1 = 4\pi\ell v_1 v^{-1} + \phi_A + \phi_B, \qquad (7.4)$$

where ϕ_A and ϕ_B are the phase angles of ρ_A and ρ_B (i.e., $\rho_A = |\rho_A|e^{j\phi_A}$, etc.). Figure 7.4 shows a phasor representation of the reflected and unreflected components and their relative phase θ_1. The reflected component causes the resultant phase to be deflected through an angle ϕ_1 given by

$$\phi_1 \simeq \tan\phi_1 = \frac{\Lambda \sin\theta_1}{1 + \Lambda \cos\theta_1}. \qquad (7.5)$$

Similarly, the phase of the frequency v_2 is deflected through an angle ϕ_2, given by equations equivalent to Eqs. (7.4) and (7.5) with subscript 1 replaced by 2.

With the reflections at A and B, the round-trip phase for a line of length L is

$$4\pi v_2 L v^{-1} + \phi_1 + \phi_2. \qquad (7.6)$$

If the line length increases uniformly to $L(1 + \beta)$ the angles ϕ_1 and ϕ_2 vary in a nonlinear manner with ℓ and become $\phi_1 + \delta\phi_1$ and $\phi + \delta\phi_2$, respectively. The round-trip phase then becomes

$$4\pi v_2 L v^{-1}(1 + \beta) + \phi_1 + \delta\phi_1 + \phi_2 + \delta\phi_2. \qquad (7.7)$$

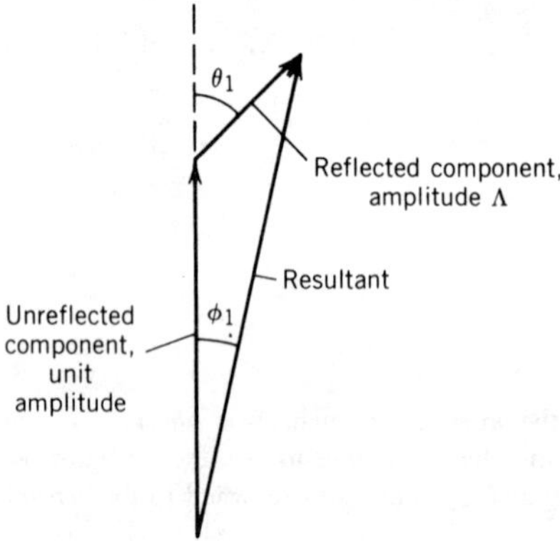

Figure 7.4 Phasor diagram of components at frequency v_1 transmitted by the cable.

(The effect of the reflection on the phase of the signal at frequency $\nu_1 - \nu_2$ has been neglected since the rate of change of phase with line length is very small.) The applied correction for the increase in line length is half the measured change in round-trip phase:

$$2\pi\nu_2\beta L\upsilon^{-1} + \tfrac{1}{2}(\delta\phi_1 + \delta\phi_2). \tag{7.8}$$

However, the exact correction would be equal to the change in the phase of ν_2 at the antenna which is

$$2\pi\nu_2\beta L\upsilon^{-1} + \delta\phi_2. \tag{7.9}$$

Consequently, the phase correction is in error by

$$\tfrac{1}{2}(\delta\phi_1 + \delta\phi_2) - \delta\phi_2 = \tfrac{1}{2}(\delta\phi_1 - \delta\phi_2). \tag{7.10}$$

If ν_1 and ν_2 were equal, the phase error would be zero. It is possible therefore to specify a maximum allowable frequency difference in terms of the maximum tolerable error.

The difference between the phase angles ϕ_1 and ϕ_2 is obtained from Eq. (7.5) as follows:

$$\phi_1 - \phi_2 = \frac{\partial\phi_1}{\partial\nu_1}(\nu_1 - \nu_2)$$

$$= \frac{4\pi\ell\upsilon^{-1}\Lambda\cos\theta_1(1 + \Lambda\cos\theta_1) + 4\pi\ell\upsilon^{-1}\Lambda^2\sin^2\theta_1}{(1 + \Lambda\cos\theta_1)^2}(\nu_1 - \nu_2).$$

$$\tag{7.11}$$

The reflected amplitude Λ must be much less than unity if phase errors are to be tolerable, so terms in Λ^2 can be omitted from the numerator in Eq. (7.11), and the denominator is approximately unity. Thus,

$$\phi_1 - \phi_2 \simeq 4\pi\ell\upsilon^{-1}\Lambda(\nu_1 - \nu_2)\cos\theta_1. \tag{7.12}$$

The variation of $\phi_1 - \phi_2$ with line length is given by

$$\delta\phi_1 - \delta\phi_2 = \beta\ell\frac{\partial}{\partial\ell}(\phi_1 - \phi_2)$$

$$= 4\pi\upsilon^{-1}\Lambda\big[\cos\theta_1 - 0.1\ell\alpha(\ln 10)\cos\theta_1 - 4\pi\upsilon^{-1}\ell\nu_1\sin\theta_1\big]$$

$$\times(\nu_1 - \nu_2)\beta\ell. \tag{7.13}$$

The maximum values of the terms in square brackets in Eq. (7.13) are dominated by the third term, which is of the order of the number of wavelengths in the line. If the two smaller terms are neglected, we obtain the magnitude of the phase error as follows:

$$\tfrac{1}{2}(\delta\phi_1 - \delta\phi_2) \simeq 8\pi^2 v^{-2}|\rho_A||\rho_B|\beta\ell^2 10^{-\alpha\ell/10} v_1(v_1 - v_2)\sin\theta_1. \quad (7.14)$$

The factor $\ell^2 10^{-\alpha\ell/10}$ has a maximum value at

$$\ell = 20(\alpha \ln 10)^{-1}. \quad (7.15)$$

This maximum occurs because for small values of ℓ the change in the angle θ with frequency or cable expansion is small, and for large values of ℓ the reflected component is greatly attenuated. The maximum value is equal to

$$[\ell^2 10^{-\alpha\ell/10}]_{\max} = 10.21\alpha^{-2}. \quad (7.16)$$

Curves of $\ell^2 10^{-\alpha\ell/10}$ are plotted in Fig. 7.5 for various values of α that correspond to good quality cables. It is evident that reducing the attenuation in a cable increases the error in the round-trip phase correction in Eq. (7.14), and it can be disadvantageous to use a cable with much less attenuation than can be tolerated from signal level considerations.

The type of reflections that may be encountered depends on the type of transmission line and how it is used. For example, consider a buried coaxial cable that runs along a set of stations used for a movable antenna. The principal cause of reflections in such a cable is the connectors that are inserted at the antenna stations. Unless the antenna is at the closest station, there are one or more interconnecting loops, where unused stations are bypassed, between the antenna and the master oscillator. If there are n connectors in the cable there are $N = n(n - 1)/2$ pairs between which reflections can occur. Also, if the phasors of the corresponding reflected components combine randomly, the overall rms error in the phase correction is, from Eq. (7.14),

$$\delta\phi_{\mathrm{rms}} = \sqrt{32}\,\pi^2 v^{-2}|\rho|^2\beta v_1(v_1 - v_2)F(\alpha, \ell), \quad (7.17)$$

where

$$F(\alpha, \ell) = \sqrt{\sum_{i=1}^{n} \sum_{k<i} \ell_{ik}^4 10^{-2\alpha\ell_{ik}/10}}, \quad (7.18)$$

the rms value has been used for $\sin\theta_1$, and the reflection coefficients are all approximated by an average magnitude $|\rho|$.

As an example, suppose that an interferometer is designed for observations near 100 GHz, and that it incorporates 10 antenna stations in a linear

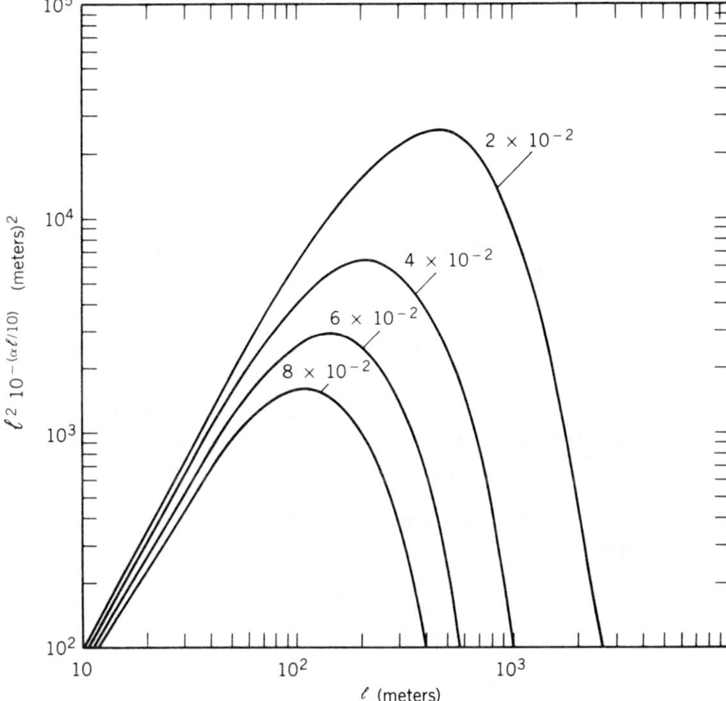

Figure 7.5 The function $\ell^2 10^{-\alpha\ell/10}$ plotted against ℓ for four values of the transmission-line attenuation, α dB m^{-1}. This function is a factor in the round-trip phase error given by Eq. (7.14).

configuration at approximately equal increments in distance up to 1 km from the master oscillator. The interconnecting oscillator cable carries a reference signal at $\nu_1 = 2$ GHz, and for this cable $|\rho| = 0.1$, $\alpha = 0.06$ dB m^{-1}, $v = 2.4 \times 10^8$ m sec^{-1}, and the temperature coefficient of electrical length is 10^{-5} K^{-1}. From Eq. (7.18) we find that $F(\alpha, \ell) = 1.1 \times 10^4$. For a temperature variation of 0.1 K in the cable, $\beta = 10^{-6}$. If phase errors at 100 GHz are required to be less than 1°, $\delta\phi_{\mathrm{rms}}$ must not exceed 0.02°, and from Eq. (7.17) ν_1 and ν_2 must not differ by more than 1.6 MHz.

Automatically Correcting System

An interesting variation on the round-trip scheme, shown in Fig. 7.6, was suggested by J. Granlund (NRAO 1967). It is particularly suitable for providing a stable reference frequency at a number of points along a linear array of antennas. Frequencies ν_1 and ν_2 are generated by stable oscillators and are injected at opposite ends of the transmission line. The frequency difference $\nu_1 - \nu_2$ is again very small. At an intermediate station the two signals are extracted by directional couplers and multiplied to form the sum frequency.

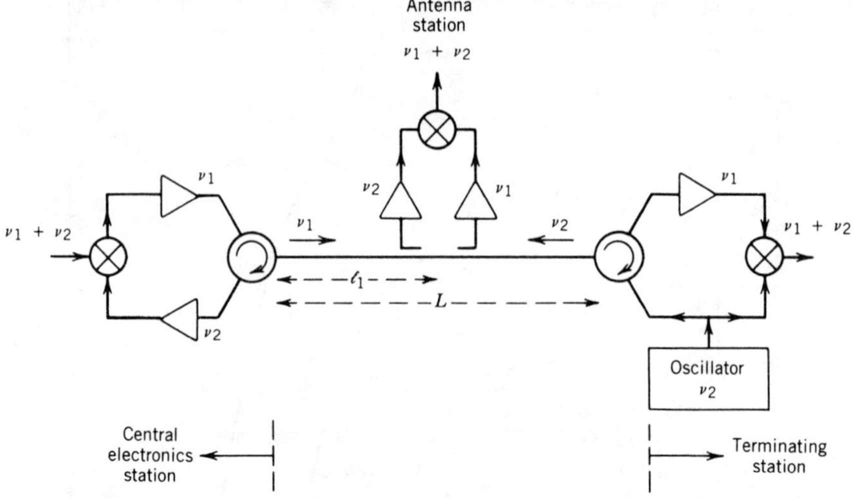

Figure 7.6   Scheme proposed by J. Granlund (NRAO 1967) for establishing a reference signal at frequency $\nu_1 + \nu_2$ at various stations along a transmission line. One such antenna station is shown.

The phase of this sum at the antenna station in Fig 7.6 is

$$2\pi\nu_1\ell_1 v^{-1} + 2\pi\nu_2(L - \ell_1)v^{-1} = 2\pi\nu_1 Lv^{-1} - 2\pi(\nu_1 - \nu_2)(L - \ell_1)v^{-1}.$$

$$(7.19)$$

For two points at positions ℓ_1 and ℓ_2 on the line the difference in the sum-frequency phases is

$$\Delta\phi = 2\pi(\nu_1 - \nu_2)(\ell_1 - \ell_2)v^{-1}. \qquad (7.20)$$

This difference, of course, would be zero if ν_1 and ν_2 were equal, but it is necessary to maintain a finite frequency difference because the directivity of the couplers alone is not usually sufficient to separate the two signals. The effect of the line-length variation is not measured explicitly in this case, but the correction occurs automatically except for the small term in Eq. (7.20). Reflections in the cable, of course, produce errors as described for the previous scheme and may be the limiting consideration for the frequency offset. A practical implementation of the scheme of Fig. 7.6 is described by Little (1969), and a calibration system for a circular array using a similar principle was developed by Morimoto (1965).

Phase-Locked Loops and Reference Frequencies

Some practical points in the implementation of local oscillator systems should be briefly mentioned. In two of the schemes described above, an oscillator at the antenna is controlled by a phase-locked loop. Details of the design of phase-locked loops are given, for example, by Gardner (1966), and here we mention only the choice of the natural frequency of the loop. Unless the natural frequency is about an order of magnitude less than the frequency at the inputs of the phase detector, the loop response may be fast enough to introduce undesirable phase modulation at the phase detector frequency. In the system in Fig 7.3, the frequency of the input signals to the phase detector is the offset frequency $\nu_1 - \nu_2$, an upper limit on which has been placed by consideration of the reflections in the line. Also, the bandwidth of the noise to which the loop responds is proportional to the natural frequency. These considerations place an upper limit on the natural frequency of the loop, which in turn limits the choice of the oscillator to be locked. An oscillator with an inherently poor phase stability (when unlocked) requires a loop with a higher natural frequency than a more stable oscillator. Crystal-controlled oscillators are highly stable and require loop frequencies of only a few hertz. They are especially suitable for long transmission lines because the noise bandwidth of the loop is correspondingly small. With crystal-controlled oscillators at the antennas it is possible to send out the reference frequency in bursts, rather than continuously. Signals traveling in opposite directions can then be separated by time multiplexing and no frequency offset is required. However, the change in impedance of the circuits at the ends of the cable when the direction of the signal is reversed becomes a limiting factor in the accuracy of the round-trip phase measurement. Systems of this type have been designed for several large arrays (Thompson et al. 1980; Davies, Anderson, and Morison 1980).

In addition to the establishment of a phase-locked oscillator at each antenna at a reference frequency (equal to ν in Fig. 7.2, ν_2 in Fig. 7.3, and $\nu_1 + \nu_2$ in Fig. 7.6), it is necessary to generate the multiples or submultiples of this frequency that are required for frequency conversions of the received signal. In frequency multiplication, phase errors increase in proportion to the frequency: see Eq. (9.125) and the associated discussion. Thus, to first order, the choice of reference frequency is unimportant. However, as the multiplication factor is increased, other effects such as noise or temperature variation in the frequency-muliplier components may become more important.

Minimization of phase variations in the frequency-multiplication circuit is largely a matter of reducing temperature-related effects, and in this regard the scheme depicted in Fig. 7.7 is worthy of mention. It may be useful to generate a "comb" spectrum consisting of many harmonics that can be used, for example, for tuning in discrete frequency intervals. This can be done by applying the fundamental frequency to a varactor diode, but the voltage at which the varactor goes into conduction varies with temperature, and hence so do the phases of the harmonics. In the circuit in Fig. 7.7, the input fundamental waveform at frequency ν is not applied directly to the harmonic generator but is used to lock an oscillator at frequency ν. The output at the oscillator frequency that is compared with the input frequency is taken after the varactor by selecting two adjacent harmonics and combining them in a mixer diode.

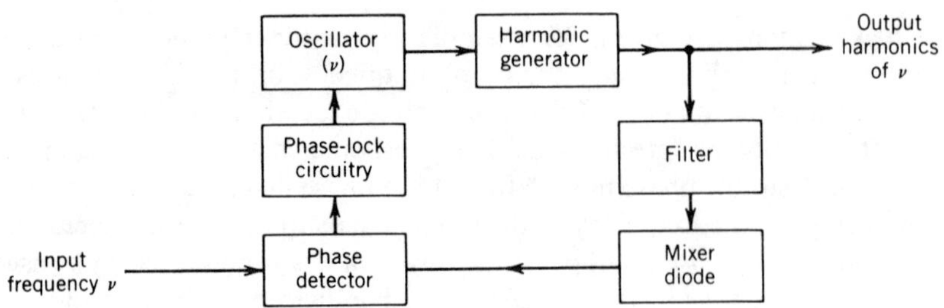

Figure 7.7 Scheme for generating a comb spectrum of harmonics of a frequency ν, in which phase changes in the harmonic generator are eliminated by enclosing it within a phase-locked loop. The filter passes two harmonics which combine in the mixer diode to generate a signal at frequency ν.

The phase-locked loop holds constant the phase of this output waveform relative to the input frequency ν, and adjusts the phase of the oscillator to compensate for the change in time of switch-on of the varactor.

Phase Stability of Filters

Tuned filters used for selecting local oscillator frequencies are also a source of temperature-related phase variations. The phase response ϕ of a filter changes by approximately $n\pi/2$ across the 3-db bandwidth, $\Delta\nu$, where n is the number of sections (poles). Thus, the rate of change of phase with frequency, measured at the center frequency ν_0, is

$$\left.\frac{\partial\phi}{\partial\nu}\right|_{\nu_0} = \frac{n\pi k_1}{2\,\Delta\nu}, \tag{7.21}$$

where k_1 is a constant of order unity that depends on the precise design of the filter. The center frequency varies with physical temperature T by

$$\frac{\partial\nu_0}{\partial T} = k_2\nu_0, \tag{7.22}$$

where k_2 is a constant related mainly to the expansion coefficients of the filter material. Thus the rate of variation of phase with temperature is given by

$$\frac{\partial\phi}{\partial T} = \left.\frac{\partial\phi}{\partial\nu}\right|_{\nu_0}\frac{\partial\nu_0}{\partial T} = nk_1k_2\left(\frac{\pi}{2}\right)\left(\frac{\nu_0}{\Delta\nu}\right). \tag{7.23}$$

The factor $\nu_0/\Delta\nu$ is the Q-factor of the filter. The combined constant k_1k_2 can be determined empirically and is typically of order 10^{-5} K^{-1} for tubular bandpass filters with center frequencies in the range 1 MHz to 1 GHz. Thus, for example, if one allows a 1-K temperature variation for such a filter and places an upper limit of 0.1° on its contribution to the phase variation, the fractional bandwidth must not be less than $n/110$, or 5.4% for a six-pole filter.

Filters of narrow fractional bandwidth should be avoided. Thus, to pick out a particular frequency from a series of closely spaced harmonics it may be preferable to use a phase-locked oscillator rather than a filter. The same considerations apply to filters that carry the received signal at the observing frequency or an intermediate frequency. Techniques such as the use of single-sideband mixers (see Appendix 7.1) can help to reduce the filtering required in a signal channel.

7.3 FREQUENCY RESPONSES OF THE SIGNAL CHANNELS

Optimum Response

The signals in a synthesis array usually pass through a number of amplifiers, filters, mixers, and transmission lines from the outputs of the antennas to the inputs of the correlators. The characteristics of these components are impressed on the signals, and therefore we should consider their effect on the sensitivity and accuracy of the visibility measurements. These characteristics can be specified largely in terms of the overall frequency response of the receiving channel. The important considerations are the optimum frequency response and the tolerances on the deviations of the channels of different antennas from this optimum response. The following discussion is based on an analysis by Thompson and D'Addario (1982).

First, assume that the astronomical signal and the receiver noise both have flat spectra over the width of an IF band or spectral channel. Then the spectrum of the signal delivered to the correlators from a given antenna is determined by the frequency response of the associated receiving equipment. If $H(\nu) = |H(\nu)|e^{j\phi(\nu)}$ is the complex voltage frequency response function, the output from the correlator for antennas m and n, resulting from cosmic signals, is proportional to

$$\int_0^\infty H_m(\nu) H_n^*(\nu)\, d\nu = \int_0^\infty |H_m(\nu)|\,|H_n(\nu)|\,e^{j(\phi_m - \phi_n)}\, d\nu, \qquad (7.24)$$

where the subscripts denote the antennas. We are concerned here only with the dependence of the signal-to-noise ratio of an observation on the frequency responses of the signal channels. In practice, the frequency responses are nonzero only within a limited frequency band of width $\Delta\nu$. From Eq. (6.36) we can define a factor $\mathscr{D}$ equal to the signal-to-noise ratio relative to that with identical rectangular responses of width $\Delta\nu$:

$$\mathscr{D} = \frac{\left| \int_0^\infty H_m(\nu) H_n^*(\nu)\, d\nu \right|}{\sqrt{\Delta\nu \int_0^\infty |H_m(\nu)|^2 |H_n(\nu)|^2\, d\nu}}. \qquad (7.25)$$

Here we have taken the modulus, rather than the real part, of the integral in the numerator, in order to represent the amplitude of the output of a complex correlator. The above expression has a maximum value if $|H_m(\nu)|$ and $|H_n(\nu)|$ are constant across the band $\Delta\nu$, that is, the amplitude response is a rectangular function, and if $\phi(\nu)$ is identical for both antennas. $\mathscr{D}$ is then equal to unity. Thus, in cases where the maximum usable bandwidth is limited, for example, by the frequency spectrum free of man-made signals, a rectangular passband that matches the available band yields the greatest obtainable sensitivity. Special circumstances, of course, may call for other passband shapes.

Of the other ways in which the receiving passband modifies the response of a synthesis array, the most important is the smearing of detail in the synthesized response which limits the field of sky that can be usefully mapped. This effect has been described in Section 6.3, and the way in which the frequency response enters the result is shown in Eq. (6.63). For a given sensitivity a rectangular passband results in the least smearing, since it is the most compact in the frequency dimension. In most cases the smearing is an undesirable feature, and the rectangular passband is again the best design.

An exact rectangular passband, of course, is only an ideal. In practice, the steepness of the sides of the passband must be determined by considerations such as the temperature coefficient of phase of the filters used. The response of a filter can be made to approximate a rectangular shape more closely as the number of poles increases, with a proportionate increase in $\partial\phi/\partial T$ as shown by Eq. (7.23). To examine the tolerable deviations of the actual passband responses, two effects must be considered. These are the decrease in the signal-to-noise ratio and the introduction of errors in determining gain factors for individual antennas in a manner that will be described.

Tolerances on Variation of the Frequency Response: Degradation of Sensitivity

We first consider the effects on the sensitivity. Equation (7.25) provides a degradation factor $\mathscr{D}$, which is the signal-to-noise ratio with frequency responses $H_m(\nu)$ and $H_n(\nu)$, expressed as a fraction of that which would be obtained with rectangular passbands of width $\Delta\nu$. In constructing a receiving system the usual goal is to keep the passband flat with steep edges, but, in practice, effects such as differential attenuation and reflections in cables introduce slopes and ripples in the frequency response which are not identical from one antenna to another. To examine these effects, $\mathscr{D}$ can be calculated for an initially rectangular passband with various distortions imposed. The distortions considered are the following.

1. Amplitude slope across the passband, with the logarithm of the amplitude varying linearly with frequency.

TABLE 7.1 Deviation of the Frequency Characteristic from an Ideal Rectangular Response, and Corresponding Expressions for $\mathscr{D}$ and G_{mn}

Frequency Response	Signal-to-Noise Degradation, $\mathscr{D}$	Antenna-Pair Gain, G_{mn}
Amplitude slope[a] $\|H(\nu)\|^2 = H_0^2 e^{\sigma(\nu - \nu_0)} \Pi\left(\dfrac{\nu - \nu_0}{\Delta\nu}\right)$	$\sqrt{\dfrac{4[e^{(\sigma_m + \sigma_n)\Delta\nu/2} - 1]}{\Delta\nu(\sigma_m + \sigma_n)[e^{(\sigma_m + \sigma_n)\Delta\nu/2} + 1]}}$	$\dfrac{2G_0}{\Delta\nu(\sigma_m + \sigma_n)}\left[e^{(\sigma_m + \sigma_n)\Delta\nu/4} - e^{-(\sigma_m + \sigma_n)\Delta\nu/4}\right]$
Sinusoidal ripple $H(\nu) = H_0\left[1 + \gamma e^{j2\pi(\nu - \nu_0)\tau}\right]\Pi\left(\dfrac{\nu - \nu_0}{\Delta\nu}\right)$	$\left[\dfrac{1 + 2\operatorname{Re}\{\gamma_m \gamma_n^*\} + \|\gamma_m \gamma_n\|^2}{1 + \|\gamma_m\|^2 + \|\gamma_n\|^2 + 2\operatorname{Re}\{\gamma_m \gamma_n^*\} + \|\gamma_m \gamma_n\|^2}\right]^{1/2}$ (see footnote b)	$G_0\left[1 + \dfrac{2}{\pi}(\gamma_m + \gamma_n^*) + \gamma_m \gamma_n^*\right]$ (see footnote c)
Center-frequency displacement[d] $H(\nu) = H_0\Pi\left(\dfrac{\nu - \delta\nu - \nu_0}{\Delta\nu}\right) \times$ $e^{jN\pi(\nu - \delta\nu - \nu_0)/\Delta\nu}$	$\sqrt{1 - \dfrac{\delta\nu_m - \delta\nu_n}{\Delta\nu}}$	$G_0\left(1 - \dfrac{\delta\nu_m - \delta\nu_n}{\Delta\nu}\right)e^{jN\pi(\delta\nu_m - \delta\nu_n)/\Delta\nu}$
Phase variation $H(\nu) = H_0\Pi\left(\dfrac{\nu - \nu_0}{\Delta\nu}\right)e^{j\phi(\nu)}$	$1 - \tfrac{1}{2}\langle\phi_{mn}^2\rangle$ $\phi_{mn}(\nu) = \phi_m(\nu) - \phi_n(\nu) - \langle\phi_m(\nu) - \phi_n(\nu)\rangle$ (see footnote e)	$G_0\left(1 - \tfrac{1}{2}\langle\phi_{mn}^2\rangle\right)$
Delay-setting error $H(\nu) = H_0 e^{j2\pi\nu\tau}\Pi\left(\dfrac{\nu - \nu_0}{\Delta\nu}\right)$	$\dfrac{\sin[\pi\Delta\nu(\tau_m - \tau_n)]}{\pi\Delta\nu(\tau_m - \tau_n)}$	$G_0\left(\dfrac{\sin\pi\Delta\nu(\tau_m - \tau_n)}{\pi\Delta\nu(\tau_m - \tau_n)}\right)e^{j\pi\Delta\nu(\tau_m - \tau_n)}$ (see footnote f)

[a] The unit rectangle function $\Pi(x)$ is equal to 1 for $\|x\| \leq \tfrac{1}{2}$ and zero for $\|x\| > \tfrac{1}{2}$. Parameters are as follows: H_0 and G_0, gain constants; σ, slope parameter; γ, relative amplitude of sinusoidal component; $\delta\nu$, frequency offset; τ, delay error.

[b] For integral values of $\Delta\nu\,\tau$.

[c] For $\Delta\nu\,\tau = \tfrac{1}{2}$.

[d] Linear phase response with difference $N\pi$ between passband edges.

[e] The brackets $\langle\ \rangle$ indicate a mean over the passband.

[f] Phase term corresponds to baseband response.

TABLE 7.2   Tolerances on Frequency Response Variations

Type of Variation	Criterion	
	2.5% Degradation in Signal-to-Noise Ratio	1% Maximum Gain Error
Amplitude slope	3.5 dB edge-to-edge	2.7 dB edge-to-edge
Sinusoidal ripple	2.9 dB peak-to-peak	2.0 dB peak-to-peak
Center-frequency displacement	$0.05\,\Delta\nu$	$0.007\,\Delta\nu$
Phase variation	$\phi_{mn} = 12.8°$ rms	$\phi_{mn} = 9.1°$ rms
Delay-setting error	$0.12/\Delta\nu$	$0.05/\Delta\nu$

2. Sinusoidal amplitude ripple: this could result from a reflection in a transmission line.

3. Displacement of the center frequency of the passband.

4. Variation in phase response as a function of frequency.

5. Delay-setting error, which introduces a component of phase linear with frequency.

Expressions for the frequency response involving the above effects are given in the first column of Table 7.1. The second column of the table gives the signal-to-noise degradation factor $\mathscr{D}$, and subscripts m and n indicate parameter values for particular antennas. The expressions in Table 7.1 have been used to derive the maximum tolerable passband distortion for each of the effects, allowing a loss in sensitivity of no more than $2.5\%(\mathscr{D} = 0.975)$. The resulting limits on the passband distortion are shown in Table 7.2.

Tolerances on Variation of the Frequency Response: Gain Errors

A second effect that sets limits on the deviations of the frequency responses results from errors that can be introduced in the calibration procedure. The correlator output for an antenna pair is

$$r_{mn} = G_{mn}\mathscr{V}_{mn},\qquad(7.26)$$

where $\mathscr{V}_{mn}$ is the source-dependent complex visibility from which the brightness map may be computed, and G_{mn} is related to the frequency responses of the signal channels and is proportional to Eq. (7.24). We suppose that these responses incorporate the characteristics of the antennas and electronics in such a way that G_{mn} is proportional to the correlator output for antenna pair

(m, n) when a point source of unit flux density at the field center is observed. In practice, the G_{mn} values may be determined from observations of calibration sources for which the visibilities are known. The measured antenna-pair gains can be used to correct the correlator output data directly, but there are advantages if, instead, they are used to determine (voltage) gain factors $g = |g|e^{i\phi}$ for the individual antennas such that

$$G_{mn} = g_m g_n^*. \tag{7.27}$$

Since, in a large array, there are many more correlatable antenna pairs than antennas [up to $n_a(n_a - 1)/2$ pairs for n_a antennas] not all the calibration data need be used. This adds important flexibility to the calibration procedure; for example, a source resolved at the longest spacings of an array can be used to determine the antenna gains from measurements at the shorter spacings only. The same principle leads to the important technique of adaptive calibration described in Chapter 11.

In general, the factoring in Eq. (7.27) requires that the frequency responses be identical for all antennas, or differ only by constant multiplicative factors. If this requirement is fulfilled, we can assign gain factors

$$g = \sqrt{\int_0^{\infty} |H(\nu)|^2 \, d\nu}\ . \tag{7.28}$$

In practice, the frequency responses differ, and an approximate solution to Eq. (7.27) can be obtained by choosing the g values to minimize

$$\sum |G_{mn} - g_m g_n^*|^2, \tag{7.29}$$

where the summation is taken over all antenna pairs (m, n) for which G_{mn} can be measured by observation of a calibration source. In calibrating subsequent observations of unknown sources $g_m g_n^*$ is then used in place of G_{mn} in Eq. (7.26) for all antenna pairs, whether or not they are directly calibrated. To avoid introducing errors with this scheme, the residuals

$$\varepsilon_{mn} = G_{mn} - g_m g_n^* \tag{7.30}$$

must be small, that is, the frequency responses must be sufficiently similar. Thus, we are concerned mainly with the deviations of the frequency responses from one another rather than from an ideal response.

By using model responses for groups of antennas, calculating the pair gains, the best-fit antenna gains, and the residuals, tolerances on the bandpass distortion can be assigned. Pair gains for the various distortions discussed earlier are given in the third column of Table 7.1. Tolerances corresponding to 1% maximum residuals are given in Table 7.2. The results depend to some extent on the distribution of distortions in the model responses, which for the results shown were chosen with the intention of maximizing the residuals

(Thompson and D'Addario 1982). The criteria of 2.5% loss in sensitivity and 1% maximum gain error used in Table 7.2 are arbitrary but represent tolerances that are generally realistic. Gain calibration errors are the more critical of the two effects. The results provide guidelines for the allowable deviation of individual responses from the design standard.

Delay-Setting Tolerances

Inaccuracies in adjustment of the compensating time delays in an array can result from two effects. There are errors in calculating the correct setting, which result from errors in calibration of antenna positions or of the delay devices. These errors can be reduced by using a calibration source that is close in position to the source being mapped. Tolerances on such errors are determined by the effects summarized in Table 7.2. There are also errors that result from the discretely adjustable nature of the delays. In analog systems, delay elements providing a binary sequence of values are switched in and out of the signal path. In digital systems the delay can be adjusted in steps governed by a train of timing pulses, as described in Chapter 8. In either case there is a minimum delay increment τ_0. If the delay for each antenna is readjusted whenever the magnitude of the error is equal to $\tau_0/2$, the probability distribution of the error is uniform from $-\tau_0/2$ to $\tau_0/2$. The errors for two antennas can generally be assumed to vary independently, so the differential delay error for any pair has a probability distribution equal to the convolution of the distributions for the individual antennas. This distribution is triangular with maximum errors $\pm\tau_0$ and an rms value of $\tau_0/\sqrt{6}$. The errors are independent, of course, for the observations of a source being mapped and for a nearby calibrator, and therefore they are not included in the gain calibration factors. Thus, they introduce a phase-error component in the visibility data equal to $2\pi\nu_d$ times the delay error, ν_d being the intermediate frequency at which the delay is inserted.

One method of eliminating the delay-step phase error is to make τ_0 equal to $1/\nu_d$. Adjustments of the delay then involve phase changes that are integral numbers of complete rotations of the phase. This technique requires that the IF bandwidth be small compared with the center frequency, so that the effective value of the latter does not vary significantly with bandpass distortions, and so that τ_0 is not a large fraction of the reciprocal bandwidth. The technique has been most useful therefore in some of the earlier arrays which had narrow receiving bandwidths. For instruments with wider bandwidths the phase errors can be made tolerable by making both τ_0 and ν_d small. If a baseband IF response is used, that is, one in which the passband is defined by a lowpass filter, ν_d is equal to $\Delta\nu/2$. This scheme is well suited for use with a digital delay system. It is used, for example, in the VLA, for which $\tau_0 = 1/32\Delta\nu$, and this results in an rms phase error of 2.3°.

The distributions of the delay errors discussed above refer, of course, to the instantaneous values, whereas the values that are critical to the mapping accuracy are the components in the averaged visibility data. The averaging time depends on the reduction technique, and in the simple case of cell averaging discussed in Section 4.10 it is the crossing time for a (u, v) cell. In a synthesis array the compensating delay for each antenna is adjusted to equalize the delay relative to some reference point on the array as the source moves across the sky. If the antenna spacings are large, the delay may change by several increments during most cell crossings, and the resulting phase errors are greatly decreased. However, at certain hour angles the rate of change of delay goes through zero. This can occur for two antennas simultaneously only if the two antennas and the delay reference position lie on a straight line. Thus the fraction of the observed data for which the phase errors are not reduced by the data averaging may be very small.

Implementation of Bandpass Tolerances

The tolerances summarized in Table 7.2 apply to the overall system from the antennas to the correlator inputs. In practice, the frequency response is determined mainly by filters in the late stages, immediately preceding the correlators or digital samplers. Specifications on such filters should provide for the required matching of responses and should include consideration of the temperature effects discussed in Section 7.2. The frequency selectivity of elements in the earlier stages can then be held to the minimum required for rejection of interfering signals, thus minimizing the effect on the overall response.

7.4 POLARIZATION-MISMATCH TOLERANCES

The response of two antennas to an unpolarized source is greatest when the antennas are identically polarized. Small variations in the polarization characteristics of one antenna relative to another occur as a result of mechanical tolerances. These variations lead to errors in the assignment of antenna gains in a manner similar to the variations in frequency responses. To examine this effect we calculate the response of two arbitrarily polarized antennas to a randomly polarized source, which is given by the term for the Stokes parameter I in Eq. (4.43). As an example we consider antennas with nominally identical circular polarization for which we can write $\chi_m = \pi/4 + \Delta\chi_m$ and $\chi_n = \pi/4 + \Delta\chi_n$. The parameters χ_m and χ_n are defined in Section 4.9, as are ψ_m and ψ_n which are also needed. The required response is

$$G_{mn} = G_0 \left[\cos(\psi_m - \psi_n)\cos(\Delta\chi_m - \Delta\chi_n) \right.$$

$$\left. + j\sin(\psi_m - \psi_n)\cos(\Delta\chi_m + \Delta\chi_n) \right]. \qquad (7.31)$$

Now $\psi_m - \psi_n$ and the Δ terms represent construction tolerances and are all small. Thus when we expand the trigonometric functions and retain only the first- and second-order terms, Eq. (7.31) becomes

$$G_{mn} = G_0\left\{1 - \tfrac{1}{2}\left[(\psi_m - \psi_n)^2 + (\Delta\chi_m - \Delta\chi_n)^2\right] + j(\psi_m - \psi_n)\right\}.$$

$$(7.32)$$

An analysis following the procedure for frequency responses in Section 7.3 can be made by assigning polarization characteristics to a model group of antennas and determining pair gains, best-fit antenna gains, and gain residuals. For simplicity, we assume that the spread of values is of similar magnitude for the parameters χ and ψ. A 1% maximum gain residual then results from a spread of $\pm 3.6°$ in χ and ψ. A value of $\Delta\chi = 3.6°$ corresponds to an axial ratio of 1.12 for the polarization ellipse, and it is not difficult to obtain feeds for which the deviation from circularity is within this value near the beam center. A similar analysis for linearly polarized antennas gives tolerances of the same order.

7.5 PHASE SWITCHING

Reduction of Responses to Spurious Signals

The technique of phase switching for a two-element interferometer has been described in Chapter 1, where it was explained as an early method of obtaining analog multiplication of signals. The principle is as indicated in Fig. 1.9. However, in later instruments the power-law detector is replaced by a correlator. Although more direct methods of signal multiplication are now used, phase switching is still useful to eliminate small offsets in correlator outputs that can result from imperfections in circuit operation or from spurious signals. The latter are difficult to eliminate entirely in any complicated receiving system, since combinations of harmonics of oscillator frequencies that fall within the observing frequency band or any intermediate frequency band may infiltrate the electronics. Such signals, at levels too low to detect by common test procedures, can be strong enough to produce unwanted components in the output. For an array of n_a antennas, a receiving bandwidth $\Delta\nu$, and an observing duration τ, signals at the limit of detectability are at a power level of order $(n_a\sqrt{\Delta\nu\tau})^{-1}$ relative to the noise; for example, 75 dB below the noise for $n_a = 27$, $\Delta\nu = 50$ MHz, and $\tau = 8$ hr. Similar effects can be produced by cross coupling of small amounts of noise from one channel or another, if this occurs after the compensating delays which can otherwise decorrelate such signals.

Since spurious signals produce components of the visibility that change only slowly with time, they show up as spurious detail near the origin of the map. If they enter the signal channel at a point that comes after the phase switch, so that they produce a component with no switch-frequency variation at the synchronous detector, they can be reduced by two or more orders of magnitude by phase switching.

Implementation of Phase Switching Using Walsh Functions

Consider the problem of phase switching a multielement array in which the products of the signals from all possible pairs of antennas are formed. Phase switching can be represented by multiplication of the received signals by periodic functions that alternate in time between values of $+1$ and -1. For the mth and nth antennas let these functions be $f_m(t)$ and $f_n(t)$. Synchronous detection of the correlator output for these two antennas requires a reference waveform $f_m(t)f_n(t)$, and any nonvarying components from the multiplier are reduced by a factor

$$\frac{1}{\tau} \int_0^\tau f_m(t)f_n(t)\,dt \tag{7.33}$$

after averaging for a time τ. This factor will be zero if $f_m(t)$ and $f_n(t)$ are orthogonal over the interval τ, or a submultiple of τ. In fact, the unwanted output components will not be exactly constant because the tracking of the compensating delays introduces slow changes in the phases with which the spurious signals are combined. However, the unwanted outputs will be strongly reduced by the synchronous detection as long as their variation is small over the minimum period of orthogonality of $f_m(t)$ and $f_n(t)$.

Implementation of phase switching on an n_a-element array calls for n_a mutually orthogonal, two-state waveforms. Square waves (Radamacher functions) can be used if only a small number of antennas are involved. Quadrature shifts in phase and factors of two in frequency can provide the required orthogonality. One antenna can remain unswitched, and the ratio of the highest to the lowest square-wave frequency is then 2^n, n being the smallest integer that is greater than or equal to $(n_a - 3)/2$. For large arrays this can be inconveniently large, for example, 4096 for $n_a = 27$.

An alternative that may be preferable is to use Walsh functions, which are rectangular waveforms in which the time interval between transitions from one state to the other is a varying but integral submultiple of a basic time interval, as in Fig. 7.8. For a description of Walsh functions see, for example, Harmuth (1969) or Beauchamp (1975). With Walsh functions the minimum time interval between transitions varies by a factor equal to the lowest power-of-two integer that is greater than or equal to $(n_a - 1)/2$, for example, 16 for $n_a = 27$.

Various systems of designating and ordering Walsh functions have been devised. We follow Harmuth (1972) and designate those with even symmetry as Cal (k, t) and those with odd symmetry as Sal (k, t). Here t is time expressed as a fraction of the time base, which is the interval at which the waveform repeats, and k is the sequency, which is equal to half the number of zero

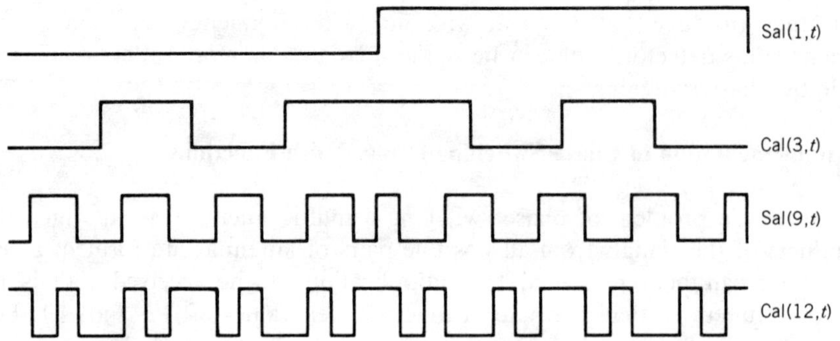

Figure 7.8   Four examples of Walsh functions, each of which repeats after the one cycle of the time base interval plotted above. Within this interval the Sal functions are odd, and the Cal functions are even. The first number in parentheses in the name of each function is the sequency, which is equal to half the number of zero crossings in the time base interval. t is time measured as a fraction of the same interval.

crossings within the time base. Walsh functions with nonidentical sequencies are orthogonal, and Cal and Sal functions of the same sequency are orthogonal but differ only by a time offset. Walsh functions can be derived in several ways. In one they are the products of a number of Radamacher functions. This number can be as low as 1, in which case the Walsh function is a square wave. In the examples in Fig. 7.8 Sal $(1, t)$ is a square wave, and Cal $(3, t)$ and Sal $(9, t)$ are each products of Sal $(1, t)$ and one other square wave. Another method of generating Walsh functions makes use of the Hadamard matrices, of which the one of lowest order is

$$H_2 = \begin{bmatrix} 1 & 1 \\ 1 & -1 \end{bmatrix}. \tag{7.34}$$

Higher-order matrices can be obtained by replacing each element of H_2 by the matrix H_2 multiplied by the element replaced. If this is performed twice, for example, we obtain

$$H_8 = \begin{bmatrix} 1 & 1 & 1 & 1 & 1 & 1 & 1 & 1 \\ 1 & -1 & 1 & -1 & 1 & -1 & 1 & -1 \\ 1 & 1 & -1 & -1 & 1 & 1 & -1 & -1 \\ 1 & -1 & -1 & 1 & 1 & -1 & -1 & 1 \\ 1 & 1 & 1 & 1 & -1 & -1 & -1 & -1 \\ 1 & -1 & 1 & -1 & -1 & 1 & -1 & 1 \\ 1 & 1 & -1 & -1 & -1 & -1 & 1 & 1 \\ 1 & -1 & -1 & 1 & -1 & 1 & 1 & -1 \end{bmatrix} \begin{matrix} \mathrm{Cal}(0, t) \\ \mathrm{Sal}(4, t) \\ \mathrm{Sal}(2, t) \\ \mathrm{Cal}(2, t) \\ \mathrm{Sal}(1, t) \\ \mathrm{Cal}(3, t) \\ \mathrm{Cal}(1, t) \\ \mathrm{Sal}(3, t). \end{matrix}$$

$$\tag{7.35}$$

The rows of the matrices correspond to the Walsh functions indicated, the signs being reversed for odd sequences in this particular generation process.

The waveform required at the phase detector is the product of the phase-switching functions at the two antennas involved. The product of two such Walsh functions is a Walsh function, the sequency of which is greater than, or equal to, the difference between the sequencies of the two original functions.

Interaction of Phase Switching with Fringe Rotation and Delay Adjustment

The effectiveness of phase switching in reducing the response to spurious signals depends on the point in the signal channel at which these unwanted signals are introduced. The three following cases illustrate the most important possibilities.

1.   The unwanted signal enters the antennas or some point in the signal channels which is ahead of the phase switching, the fringe rotation, and the compensating delays. The signal then suffers phase switching like the wanted signals and is not suppressed in the synchronous detection. Externally generated interference behaves in this manner, and its effect is discussed in Chapter 14.

2.   The unwanted signal enters after the phase switching but before the fringe rotation and delay compensation. The fringe-rotation phase shifts, designed to reduce to zero the fringe frequencies of the desired signals at the correlator output, act on the spurious signal and cause it to appear at the correlator output as a component at the natural fringe frequency for a point source at the phase reference position. This component then undergoes synchronous detection with a Walsh function. If the natural fringe frequency transiently matches the frequency of a Fourier component of this Walsh function, a spurious response can occur.

3.   The spurious signal enters after the phase switching and the fringe rotation but before the delay compensation. The signal then suffers phase shifts resulting from the changing of the compensating delay. The resulting component at the correlator output has a frequency equal to the natural fringe frequency that would occur if the observing frequency were equal to the intermediate frequency at which the compensating delays are introduced. Thus the oscillations are one to three orders of magnitude lower in frequency than the natural fringe frequency, and it is consequently easier to avoid coincidence with the frequency of a component of the Walsh function.

From the above considerations it is usually advantageous to perform both the phase switching and the fringe rotation as early in the signal channel as possible. Figure 7.9 shows, as an example, the phase-switching scheme used in the VLA from a description by Granlund, Thompson, and Clark (1978). The phase switching at the antenna is performed on a local oscillator, rather than on the full signal band, to avoid the possibility of introducing distortion in the frequency response. The signals are digitized at the output of the final IF amplifier and thereafter are delayed and multiplied digitally. Since digital

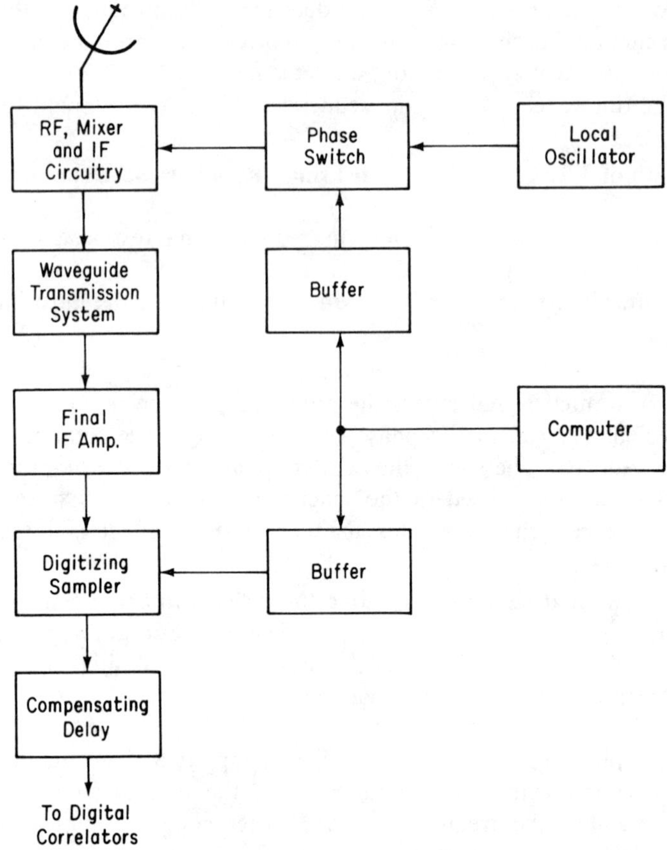

Figure 7.9 Simplified schematic diagram of the receiving channel for one antenna of the VLA. Walsh functions generated by the computer are periodically fed to digital buffers from which they are clocked out to the phase switch and to sign-reversal circuitry at the sampler. From Granlund, Thompson, and Clark (1978); © 1978 IEEE.

circuits do not suffer from unwanted drifts and offsets as analog circuits may do, there is no need to include them between the phase switching and the synchronous detection. Thus, the latter can be performed by reversing the sign bit in the digitized signal data, and need be applied only to n_a signal channels rather than $n_a(n_a - 1)/2$ correlator outputs.

7.6 AUTOMATIC LEVEL CONTROL AND GAIN CALIBRATION

In most synthesis arrays automatic level control (ALC) circuits are used to hold constant the level of the total signal, that is, the cosmic signal plus the system noise, at certain critical points. A fraction of the total signal level is detected, and the resulting voltage compared with a preset value, to generate a control signal that is fed back to some variable-gain element of the signal chain. Points at which the signal level is critical include modulators for transmission of IF signals by microwave carriers and inputs to correlators or digital samplers. For a discussion of level tolerances in samplers see Chapter 8.

The effect of an ALC loop is to hold constant the quantity $|g|^2(T_S + T_A)\,\Delta\nu$, where g is the voltage gain from the antenna output to the point of control, T_S is the system temperature, and T_A is the antenna temperature. Thus $|g|^2$ is made to vary inversely as $(T_S + T_A)$ which can change substantially with the antenna pointing angle as a result of ground radiation in the sidelobes and atmospheric attenuation. To measure such gain changes a signal from a broadband noise source can be injected at the input of the receiving electronics. This noise source is switched on and off, usually at a frequency of a few hertz to a few hundred hertz, and the resulting component sampled and monitored using a synchronous detector at some convenient point in the signal channel. The ALC system can also be switched to respond only to the signal level with the noise source off. When the noise source is on, it adds a component T_c to the overall system temperature, which should not be more than a few percent of T_S or the sensitivity will be significantly degraded. The amplitude of the switched component is a direct measure of the system gain, and for $T_S \gg T_A$ the ratio of the signal levels with the noise source on and off is equal to $1 + T_c/T_S$, which provides a continuous measure of T_S. Of course, this scheme does not measure changes in antenna gain resulting from mechanical deformation, which must be calibrated separately by periodic observation of a radio source.

APPENDIX 7.1   SINGLE-SIDEBAND MIXER

The principle of the single-sideband mixer, or image-rejecting mixer, is shown in Fig. A7.1. The terms $\cos 2\pi\nu_u t$ and $\cos 2\pi\nu_\ell t$ represent frequency components of the input waveform at the upper- and lower-sideband frequencies, respectively. The input is applied to two mixers, for which the local oscillator waveforms at frequency ν_{L0} are in phase quadrature. The mixers generate products of the signal and local oscillator waveforms, and the filters pass only the terms of frequency equal to the difference of ν_{L0} and ν_u or ν_ℓ. The output from the lower mixer also passes through a $\pi/2$ phase lag network. From the resulting terms at points A and B one can see that by applying the waveforms at these points to a summing network the upper-sideband response is obtained. Similarly, by using a differencing network the lower-sideband response is obtained. In either case the accuracy of the suppression of the response to the unwanted sideband depends on the accuracy of the quadrature phase relationships, the matching of the frequency responses of the mixers and filters, and the insertion loss of the phase lag network. In practice, for baseband outputs spanning two to three decades, suppression to a level of -20 dB is routinely

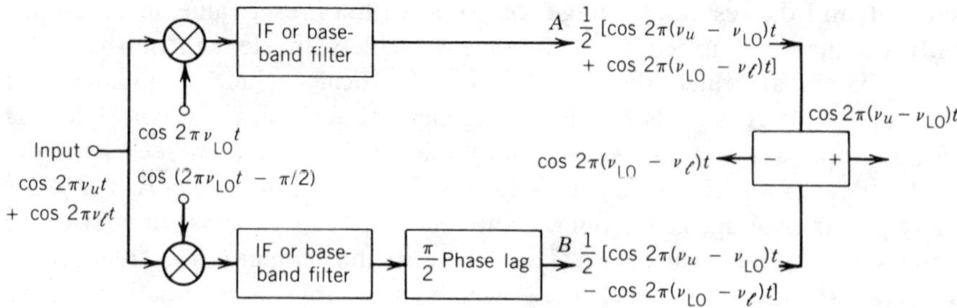

Figure A7.1   Schematic illustration of the principle of the single-sideband (image-rejecting) mixer. The upper-sideband response is obtained from the sum of the outputs A and B, and the lower-sideband response from the difference of these outputs.

achievable. With careful design, suppression to a level below -30 dB can be obtained (Archer, Granlund, and Mauzy 1981).

REFERENCES

Allen, L. R. and R. H. Frater, Wideband Multiplier Correlator, *Proc. IEE*, **117**, 1603–1608, 1970.

Archer, J. W., E. M. Caloccia, and R. Serna, An Evaluation of the Performance of the VLA Circular Waveguide System, *IEEE Trans, Microwave Theory Tech.*, **MTT-28**, 786–791, 1980.

Archer, J. W., J. Granlund, and R. E. Mauzy, A Broadband UHF Mixer Exhibiting High Image Rejection over a Multidecade Baseband Frequency Range, *IEEE J. Solid-State Circuits*, **SC-16**, 385–392, 1981.

Baars, J. W. M., J. F. Van der Brugge, J. L. Casse, J. P. Hamaker, L. H. Sondaar, J. J. Visser, and K. J. Wellington, The Synthesis Radio Telescope at Westerbork, *Proc. IEEE*, **61**, 1258–1266, 1973.

Batty, M. J., D. L. Jauncey, P. T. Rayner, and S. Gulkis, Tidbinbilla Two-Element Interferometer, *Astron. J.*, **87**, 938–944, 1982.

Beauchamp, K. G., *Walsh Functions and Their Applications*, Academic Press, London, 1975.

Bracewell, R. N., R. S. Colvin, L. R. D'Addario, C. J. Grebenkemper, K. M. Price, and A. R. Thompson, The Stanford Five-Element Radio Telescope, *Proc. IEEE*, **61**, 1249–1257, 1973.

Casse, J. L., E. E. M. Woestenburg, and J. J. Visser, Multifrequency Cryogenically Cooled Front-End Receivers for the Westerbork Synthesis Radio Telescope, *IEEE Trans. Microwave Theory Tech.*, **MTT-30**, 201–209, 1982.

Davies, J. G., B. Anderson, and I. Morison, The Jodrell Bank Radio-Linked Interferometer Network, *Nature*, **288**, 64–66, 1980.

Elsmore, B., S. Kenderdine, and M. Ryle, Operation of the Cambridge One-Mile Diameter Radio Telescope, *Mon. Not. R. Astron. Soc.* **134**, 87–95, 1966.

Erickson, W. C., M. J. Mahoney, and K. Erb, The Clark Lake Teepee-Tee Telescope, *Astrophys. J. Suppl.*, **50**, 403–420, 1982.

Frater, R. H., Accurate Wideband Multiplier Square-law Detector, *Rev. Sci. Instrum.*, **35**, 810–813, 1964.

Gardner, F. M., *Phaselock Techniques*, Wiley, New York, 1966 (2nd ed. 1979).

Granlund, J., A. R. Thompson, and B. G. Clark, An Application of Walsh Functions in Radio Astronomy Instrumentation, *IEEE Trans. Electromagn. Compat.*, **EMC-20**, 451–453, 1978.

Harmuth, H. F., Applications of Walsh Functions in Communications, *IEEE Spectrum*, **6**, No.11, 82–91, 1969.

Harmuth, H. F., *Transmission of Information by Orthogonal Functions*, Springer-Verlag, Berlin 1972 (2nd ed.).

Kraus, J. D., *Radio Astronomy*, McGraw-Hill, New York, 1966, pp.261–263.

Legg, T. H., Microwave Phase Comparators for Large Antenna Arrays, *IEEE Trans. Antennas Propag.*, **AP-13**, 428–432, 1965.

Little, A. G., A Phase-Measuring Scheme for a Large Radiotelescope, *IEEE Trans. Antennas Propag.*, **AP-17**, 547–550, 1969.

Morimoto, M., New Calibration Method for Large Aerial Arrays, *Electron. Lett.*, **1**, 192–193, 1965.

Napier, P. J., A. R. Thompson, and R. D. Ekers, The Very Large Array: Design and Performance of a Modern Synthesis Radio Telescope, *Proc. IEEE*, **71**, 1295–1320, 1983.

NRAO, *A Proposal for a Very Large Array Radio Telescope*, Vol. II, National Radio Astronomy Observatory, Green Bank, West Virginia, Ch. 14, 1967.

Phillips, T. G. and D. P. Woody, Millimeter- and Submillimeter-Wave Receivers, *Ann. Rev. Astron. Astrophys.*, **20**, 285–321, 1982.

Read, R. B., Two-Element Interferometer for Accurate Position Determinations at 960 Mc, *IRE Trans. Antennas Propag.*, **AP-9**, 31–35, 1961.

Reid, M. S., R. C. Clauss, D. A. Bathker, and C. T. Stelzried, Low Noise Microwave Receiving Systems in a Worldwide Network of Large Antennas, *Proc. IEEE*, **61**, 1330–1335, 1973.

Swarup, G. and K. S. Yang, Phase Adjustment of Large Antennas, *IRE Trans. Antennas Propag.*, **AP-9**, 75–81, 1961.

Thompson, A. R., B. G. Clark, C. M. Wade, and P. J. Napier, The Very Large Array, *Astrophys. J. Suppl.*, **44**, 151–167, 1980.

Thompson, A. R. and L. R. D'Addario, Frequency Response of a Synthesis Array: Performance Limitations and Design Tolerances, *Radio Sci.*, **17**, 357–369, 1982.

Thompson, M. C., L. E. Wood, D. Smith, and W. B. Grant, Phase Stabilization of Widely Separated Oscillators, *IEEE Trans. Antennas Propag.*, **AP-16**, 683–688, 1968.

USAF, *Handbook of Geophysics and Space Environments*, S. L. Valley, Ed., U.S. Air Force Cambridge Research Laboratories, Bedford, Mass., 1965, pp. 3-20 to 3-22.

Weinreb, S., M. Balister, S. Maas, and P. J. Napier, Multiband Low-Noise Receivers for a Very Large Array, *IEEE Trans. Microwave Theory Tech.*, **MTT-25**, 243–248, 1977a.

Weinreb, S., D. L. Fenstermacher, and R. W. Harris, Ultra-Low-Noise 1.2- to 1.7-GHz Cooled GaAsFET Amplifiers, *IEEE Trans. Microwave Theory Tech.*, **MTT-30**, 849–853, 1982.

Weinreb, S., R. Predmore, M. Ogai, and A. Parrish, Waveguide System for a Very Large Antenna Array, *Microwave J.*, **20**, March, 49–52, 1977b.

Welch, W. J., J. R. Forster, J. Dreher, W. Hoffman, D. D. Thornton, and M. C. H. Wright, An Interferometer for Millimeter Wavelengths, *Aston. Astrophys.*, **59**, 379–385, 1977.

Wright, M. C. H., B. G. Clark, C. H. Moore, and J. Coe, Hydrogen-Line Aperture Synthesis at the National Radio Astronomy Observatory: Techniques and Data Reduction, *Radio Sci.*, **8**, 763–773, 1973.

8

DIGITAL SIGNAL PROCESSING

The use of digital rather than analog instrumentation to introduce the compensating time delays and to measure correlation has important practical advantages. These functions usually occur in the late stages of the signal processing, so in a digital system the output waveform of the final IF amplifier is sampled and digitally encoded. In digital delay circuits the accuracy of the delay depends on the accuracy of the timing pulses in the system, and long delays accurate to a few hundreds of picoseconds are more easily achieved digitally than by using analog delay lines. Furthermore, there is no distortion of the signal by the digital units, other than the calculable effects of quantization. On the other hand, with an analog system it is very difficult to keep the shape of the frequency response within tolerances as delay elements are switched into and out of the signal channels. Correlators with wide dynamic range are easily implemented digitally. If a multichannel output is required for spectral-line observations, this is also readily obtained using digital techniques. Analog implementation of multichannel correlators requires filter banks to divide the signal passband into many narrow channels. Such filters, when subject to temperature variations, can be a source of phase instability. Finally, except at the highest bit rates (frequencies), digital circuits require less adjustment than analog ones and are better suited to replication in large numbers for large arrays.

Digitization of the signal waveforms requires sampling of the voltages at periodic intervals and quantizing the sampled values so that each can be represented by a finite number of bits. Usually only one or two bits are used, so that the digital data rate does not become unmanageably high. The very coarse quantization that is then necessary results in a loss in sensitivity, since modification of the signal levels to the quantized values effectively results in the addition of a component of "quantization noise." In most cases this loss is outweighed by the other advantages. Digital processing of radio astronomical signals was first used in the construction of autocorrelators,[†] which measure the autocorrelation of a signal from a single antenna as a function of time offset, and thereby provide a method of obtaining the power spectrum for studies of spectral lines. The first such system in radio astronomy was constructed by Weinreb (1963). Another early digital autocorrelator was used by Goldstein (1962) to detect radar echoes from Venus.

8.1　BIVARIATE GAUSSIAN PROBABILITY DISTRIBUTION

Before proceeding further it is appropriate to introduce the bivariate Gaussian probability function (see, e.g., Abramowitz and Stegun, 1964, p. 936) which is central to what follows. If x and y are joint Gaussian random variables with zero mean and variance σ^2, the probability that one variable is between x and $x + dx$ and, simultaneously, the other is between y and $y + dy$ is $p(x, y)\,dx\,dy$, where

$$p(x, y) = \frac{1}{2\pi\sigma^2\sqrt{1 - \rho^2}} \exp\left[\frac{-(x^2 + y^2 - 2\rho xy)}{2\sigma^2(1 - \rho^2)}\right]. \qquad (8.1)$$

The form of this function is shown in Fig. 8.1. ρ is the cross-correlation coefficient equal to $\langle xy \rangle / \sqrt{\langle x^2 \rangle \langle y^2 \rangle}$ where $\langle \ \rangle$ denotes the expectation. Note that $-1 \leq \rho \leq 1$. For $\rho \ll 1$, the exponential can be expanded giving

$$p(x, y) \simeq \left[\frac{1}{\sigma\sqrt{2\pi}}\exp\left(\frac{-x^2}{2\sigma^2}\right)\right]\left[\frac{1}{\sigma\sqrt{2\pi}}\exp\left(\frac{-y^2}{2\sigma^2}\right)\right]\left(1 + \frac{\rho xy}{\sigma^2}\right). \qquad (8.2)$$

For $\rho = 0$, the expression is simply the product of two Gaussian functions. Equation (8.1) can also be written

$$p(x, y) = \frac{1}{\sigma\sqrt{2\pi}}\exp\left(\frac{-x^2}{2\sigma^2}\right)\frac{1}{\sigma\sqrt{2\pi(1 - \rho^2)}}\exp\left[\frac{-(y - \rho x)^2}{2\sigma^2(1 - \rho^2)}\right].$$

$$(8.3)$$

[†] The terms *correlator* and *autocorrelator* are commonly used to denote systems containing large numbers of individual correlator circuits, and designed to accommodate many signal products and/or time offsets simultaneously.

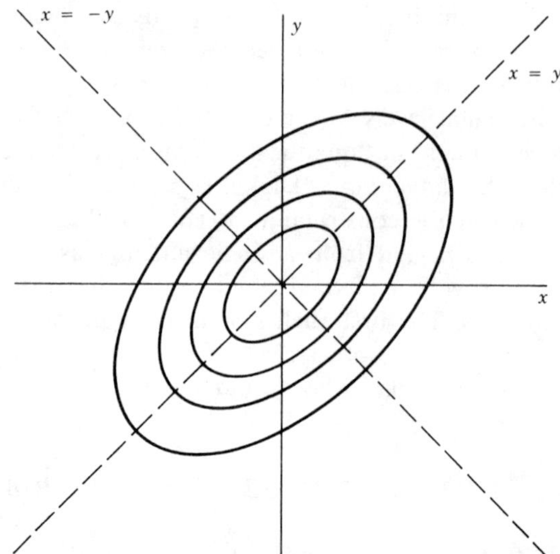

Figure 8.1   Contours of equal probability density from the bivariate Gaussian distribution in Eq. (8.1). The contours are given by $x^2 + y^2 - 2\rho xy = $ const. For $\rho = 0$ they become circles, for $\rho = 1$ they merge into the line $x = y$, and for $\rho = -1$ they merge into $x = -y$.

If this expression is integrated with respect to y from $-\infty$ to $+\infty$ it reduces to a Gaussian function in x. As ρ approaches unity Eq. (8.3) becomes the product of a Gaussian in x and a Gaussian in $y - x$, the latter having a standard deviation $\sigma\sqrt{1 - \rho^2}$ which approaches zero. Equations (8.1) and (8.2) will be used in examining the response of various types of samplers and correlators.

8.2 PERIODIC SAMPLING

We first consider the process of sampling, but without quantization, so that the full accuracy of the signal amplitude is retained.

Nyquist Rate

If the signal is bandlimited, that is, its power spectrum is nonzero only within a finite band of frequencies, no information is lost in the sampling process as long as the sampling rate is high enough. This follows from the sampling theorem discussed in Section 4.11. Here we sample a function of time and wish to avoid aliasing in the frequency domain. For a baseband (lowpass) rectangular spectrum with an upper cutoff frequency $\Delta\nu$, the width of the frequency spectrum, including negative frequencies, is $2\,\Delta\nu$. The function is fully specified by samples spaced in time with an interval no greater than $(2\,\Delta\nu)^{-1}$, that is, a sampling frequency of $2\,\Delta\nu$ or greater. This critical sampling frequency, $2\,\Delta\nu$, is called the Nyquist[†] rate (or Nyquist frequency) for the waveform. For further discussion see, for example, Bracewell (1965) or Oppenheim and Schafer (1975). In most digital systems in radio astronomy the final IF waveform has a baseband spectrum and is sampled at the Nyquist rate. Nyquist sampling can be applied to bandpass spectra also, and if the spectrum is nonzero only within a range of $n\,\Delta\nu$ to $(n+1)\,\Delta\nu$, n being an integer, the Nyquist rate is again $2\,\Delta\nu$. For a rectangular passband of this type, the autocorrelation function, which by the Wiener–Khinchin relation is the Fourier transform of the power spectrum, is

$$R_\infty(\tau) = \frac{\sin(\pi\,\Delta\nu\,\tau)}{\pi\,\Delta\nu\,\tau}\cos\left[2\pi\left(n+\tfrac{1}{2}\right)\Delta\nu\,\tau\right], \tag{8.4}$$

where the subscript ∞ indicates unquantized sampling (i.e., the accuracy is not limited by a finite number of quantization levels). Zeros in this function occur at time intervals τ that are integral multiples of $(2\,\Delta\nu)^{-1}$. Therefore, for a rectangular passband, successive samples at the Nyquist rate are uncorrelated. A baseband spectrum, sometimes referred to as a video spectrum, is the case most often encountered in digital sampling. Then $n=0$ and Eq. (8.4) becomes

$$R_\infty(\tau) = \frac{\sin(2\pi\,\Delta\nu\,\tau)}{2\pi\,\Delta\nu\,\tau}. \tag{8.5}$$

Sampling at frequencies greater than or less than the Nyquist rate is referred to as oversampling or undersampling, respectively.

In the discussion that follows we consider only single-multiplier correlators since complex correlators can be implemented as combinations of these as indicated in Fig. 6.3.

Correlation of Sampled but Unquantized Waveforms

We now investigate the response of a hypothetical correlator for which the input signals are sampled at the Nyquist rate, but are not quantized. This system can be visualized as one in which the samples either remain as analog voltages, or are encoded with a sufficiently large number of bits that quantization errors are negligible. Since no information in lost in sampling or quantization, the signal-to-noise ratio of the correlation measurement may be expected to be the same as would be obtained by applying the waveforms without sampling to an analog correlator. There is probably no reason, in practice, to build a correlator for inputs with unquantized sampling. However, by comparing the results with those for quantized sampling, which we discuss later, the effects of quantization are more easily understood.

[†]Shannon (1949) cites several references relevant to the development of this result, of which the earliest is Nyquist (1928).

The system to be considered is one in which the two bandlimited waveforms $x(t)$ and $y(t)$ are sampled at the Nyquist rate, and for each pair of samples the multiplier produces an output of amplitude proportional to the product of the input amplitudes. The integrator allows the output to be averaged for any required time interval. Now the (normalized) cross-correlation coefficient of $x(t)$ and $y(t)$ for zero time delay between the waveforms is

$$\rho = \frac{\langle x(t)y(t)\rangle}{\sqrt{\langle [x(t)]^2\rangle\langle [y(t)]^2\rangle}}. \tag{8.6}$$

(Note: Do not confuse the cross correlation coefficient ρ with the autocorrelation function of x or y, R_∞.) If x and y have equal variance σ^2, then

$$\langle x(t)y(t)\rangle = \rho\sigma^2. \tag{8.7}$$

The left-hand side is the averaged product of the two waveforms and thus represents the correlator output. The output of the digital correlator after N_q samples is

$$r_\infty = N_q^{-1}\sum_{i=1}^{N_q} x_i y_i, \tag{8.8}$$

where the subscript q denotes the Nyquist rate. Since the samples x_i and y_i obey the same Gaussian statistics as the continuous waveforms $x(t)$ and $y(t)$, we can clearly write

$$\langle r_\infty\rangle = \rho\sigma^2. \tag{8.9}$$

Thus, the output of the correlator is a linear measure of the correlation ρ. The variance of the correlator output is

$$\sigma_\infty^2 = \langle r_\infty^2\rangle - \langle r_\infty\rangle^2, \tag{8.10}$$

and

$$\langle r_\infty^2\rangle = N_q^{-2}\sum_{i=1}^{N_q}\sum_{k=1}^{N_q}\langle x_i y_i x_k y_k\rangle$$

$$= N_q^{-2}\sum_{i=1}^{N_q}\langle x_i^2 y_i^2\rangle + N_q^{-2}\sum_{i=1}^{N_q}\sum_{k\neq i}\langle x_i y_i x_k y_k\rangle, \tag{8.11}$$

where we have separated the terms for which $i = k$ and $i \neq k$. The first summation on the right-hand side of Eq. (8.11) has a value of $\sigma^4(1 + 2\rho^2)N_q^{-1}$:

from Eq. (8.3) it can be shown that

$$\int_{-\infty}^{\infty}\int_{-\infty}^{\infty} x^2 y^2 p(x, y)\, dx\, dy = \sigma^4(1 + 2\rho^2).$$

The second summation term in Eq. (8.11) is readily evaluated by using the fourth-order moment relation in Eq. (6.30). Because successive samples of each signal are uncorrelated (a rectangular passband being assumed), the second summation term has a value of $(1 - N_q^{-1})\rho^2\sigma^4$. Returning to Eq. (8.10) we can write

$$\sigma_\infty^2 = \left(1 + 2\rho^2\right)\sigma^4 N_q^{-1} + \left(1 - N_q^{-1}\right)\rho^2\sigma^4 - \rho^2\sigma^4 = \sigma^4 N_q^{-1}\left(1 + \rho^2\right).$$

$$(8.12)$$

The signal-to-noise ratio with unquantized sampling is

$$\mathscr{R}_{\text{sn}\infty} = \frac{\langle r_\infty \rangle}{\sigma_\infty} = \rho\sqrt{N_q},$$

$$(8.13)$$

where we assume $\rho \ll 1$, and note that the equality in Eq. (8.13) is approximate if ρ is not small. Note that the condition $\rho \ll 1$ is satisfactory for almost all practical purposes. The signal-to-noise ratio at the correlator output, which we are calculating here, is mainly of interest for weak signals. For an averaging time τ the correlation coefficient that represents the threshold of detectability is of order $\sqrt{\Delta\nu\,\tau}$, and the resulting threshold correlation is commonly 10^{-3}–10^{-6}. Thus the condition $\rho \ll 1$ is a safe assumption for small signals. Note also that $N_q = 2\,\Delta\nu\,\tau$, which is of order 10^6–10^{12}. In terms of the signal bandwidth and measurement duration $\mathscr{R}_{\text{sn}\infty} = \rho\sqrt{2\,\Delta\nu\,\tau}$. Now for observations of a point source with identical antennas and receivers, ρ is equal to the ratio of the resulting antenna temperature to the system temperature, T_A/T_S. Thus the present result is identical to that given by Eq. (6.39) for an analog correlator with continuous unsampled inputs and $T_A \ll T_S$.

Before leaving the subject of unquantized sampling we should consider the effect of sampling at rates other than the Nyquist one. Successive sample values from any one signal are then no longer independent. We consider a sampling frequency that is β times the Nyquist rate, and a number of samples $N = \beta N_q$. The sample interval is $\tau_s = (2\beta\,\Delta\nu)^{-1}$ and samples spaced by $q\tau_s$, where q is an integer, have a correlation coefficient which, from Eq. (8.5), is equal to

$$R_\infty(q\tau_s) = \frac{\beta\,\sin(\pi q/\beta)}{\pi q}$$

$$(8.14)$$

for a rectangular baseband response. Since the samples are not independent, we must reconsider the evaluation of the second summation term on the right-hand side of Eq. (8.11). For those terms for which $q = |i - k|$ is small enough that $R_\infty(q\tau_s)$ is significant, there will be an additional contribution given by

$$[\sigma^2 R_\infty(q\tau_s)]^2. \tag{8.15}$$

Now R_∞^2 is very small for all but a very small fraction of the $N(N-1)$ terms in the second summation in Eq. (8.11). From Eq. (8.14) R_∞^2, at its maxima, is equal to $(\beta/\pi q)^2$ and for $q = 10^3$ is of order 10^{-6}. However, as shown above, N is likely to be as high as 10^6–10^{12}. Thus, in the second summation in Eq. (8.11) the contribution made by the terms for which the i and k samples are effectively independent remains essentially unchanged. The products for which R_∞^2 is significant make an additional contribution equal to

$$2\sigma^4 N^{-2} \sum_{q=1}^{N-1} (N - q) R_\infty^2(q\tau_s) \simeq 2\sigma^4 N^{-1} \sum_{q=1}^{\infty} R_\infty^2(q\tau_s). \tag{8.16}$$

The variance of the correlator output now becomes

$$\sigma_\infty^2 = \sigma^4 N^{-1} \left[1 + 2 \sum_{q=1}^{\infty} R_\infty^2(q\tau_s) \right], \tag{8.17}$$

and the signal-to-noise ratio of the correlation measurement is

$$\mathscr{R}_{\text{sn}\infty} = \frac{\rho\sqrt{\beta N_q}}{\sqrt{1 + 2 \sum\limits_{q=1}^{\infty} R_\infty^2(q\tau_s)}}. \tag{8.18}$$

Compare this result with Eq. (8.13) for Nyquist sampling. For values of β of $\frac{1}{2}, \frac{1}{3}, \frac{1}{4}$, and so on, which correspond to undersampling, $R_\infty = 0$ and the denominator in Eq. (8.18) is unity. The sensitivity thus drops as one would expect from the decrease in the data. For oversampling $\beta > 1$, and the summation of $R_\infty^2(q\tau_s)$ in Eq. (8.18) is shown in Appendix 8.1 to be equal to $(\beta - 1)/2$. The denominator in Eq. (8.18) is then equal to $\sqrt{\beta}$, so there is no gain in sensitivity over sampling at the Nyquist rate. This is as expected, since in Nyquist sampling no information is lost. The situation is different for quantized sampling, as will appear in the following sections.

8.3 SAMPLING WITH QUANTIZATION

In some sampling schemes the signal is first quantized and then sampled, and in others it is sampled and then quantized. Ideally, the end result is the same in either case, and in analyzing the processes we can choose the order that is most convenient. Suppose that a bandlimited signal is first quantized and then sampled. Quantization generates new frequency components in the signal waveform, so it is no longer bandlimited. If it is sampled at the Nyquist rate corresponding to the unquantized waveform, as is the usual practice, some information will be lost, and the sensitivity will be less than for unquantized sampling. Also, because quantization is a nonlinear operation, we cannot assume that the measured correlation of the quantized waveforms will be a linear function of ρ, which is what we want to measure. Thus, to utilize digital signal processing there are three main points that should be investigated: (1) the relation between ρ and the measured correlation, (2) the loss in sensitivity, and (3) the extent to which oversampling can restore the lost sensitivity. Investigations of these points can be found in the work of Weinreb (1963), Cole (1968), Burns and Yao (1969), Cooper (1970), Hagen and Farley (1973), and Bowers and Klingler (1974). The discussions given here follow most closely the approach of Hagen and Farley.

Note that in discussing sampling with quantization it is common practice to refer to Nyquist sampling when what is meant is sampling at the Nyquist rate for the unquantized waveform. For brevity, we also follow this usage.

Two-Level Quantization

The quantization characteristic for two-level (one bit) sampling is shown in Fig. 8.2 The quantizing action senses only the sign of the instantaneous signal voltage. In many samplers this is done by first amplifying and strongly clipping the signal voltage. The zero crossings are more sharply defined in the resulting two-state waveform, and errors which might occur if the sampling time coincides with a sign reversal are thereby minimized.

The correlator for two-level signals consists of a multiplying circuit followed by a counter that averages the products of the input samples. The input signals are assigned values of $+1$ or -1 to indicate positive or negative signal voltages, and the products at the multiplier output thus take values of $+1$ or -1 for identical or different input values, respectively. If N samples are fed to

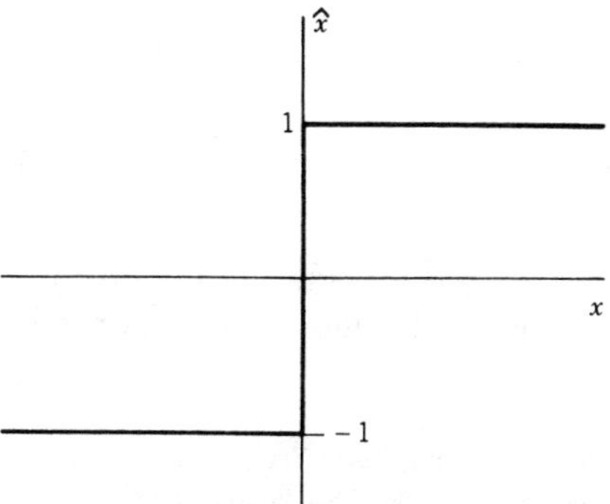

Figure 8.2 Characteristic curve for two-level quantization. The abscissa is the input voltage x and the ordinate is the quantized output $\hat{x}$.

the correlator, the average two-level correlation coefficient is

$$\rho_2 = \frac{(N_{11} + N_{\bar{1}\bar{1}}) - (N_{\bar{1}1} + N_{1\bar{1}})}{N},$$ (8.19)

where N_{11} is the average number of products for which both samples have the value $+1$, $N_{1\bar{1}}$ is the average number of products in which the x sample has the value $+1$ and the y sample -1, and so on. The denominator in Eq. (8.19) is equal to the output that would occur if all pairs of input samples were identical. ρ_2 can be related to the correlation coefficient ρ of the unquantized signals through the bivariate probability distribution Eq. (8.1), from which

$$N_{11} = NP_{11} = \frac{N}{2\pi\sigma^2\sqrt{1-\rho^2}} \int_0^\infty \int_0^\infty \exp\left[\frac{-(x^2 + y^2 - 2\rho xy)}{2\sigma^2(1-\rho^2)}\right] dx\, dy,$$

(8.20)

where P_{11} is the probability of the two unquantized signals being simultaneously greater than zero. The other required probabilities are obtained by changing the limits of the integrals in Eq. (8.20): $\int_{-\infty}^0 \int_{-\infty}^0$ for $P_{\bar{1}\bar{1}}$, $\int_{-\infty}^0 \int_0^\infty$ for $P_{\bar{1}1}$, and $\int_0^\infty \int_{-\infty}^0$ for $P_{1\bar{1}}$. Note that $P_{11} = P_{\bar{1}\bar{1}}$ and $P_{1\bar{1}} = P_{\bar{1}1}$. Thus,

$$\rho_2 = 2(P_{11} - P_{1\bar{1}}).$$ (8.21)

The integral in Eq. (8.20) is evaluated in Appendix 8.2, from which we obtain

$$P_{11} = \frac{1}{4} + \frac{1}{2\pi}\sin^{-1}\rho.$$ (8.22)

Similarly,

$$P_{1\bar{1}} = \frac{1}{4} - \frac{1}{2\pi}\sin^{-1}\rho,$$ (8.23)

so

$$\rho_2 = \frac{2}{\pi}\sin^{-1}\rho.$$ (8.24)

Equation (8.24) is known as the Van Vleck relationship[†] and it allows ρ to be obtained from the measured correlation ρ_2. For small values, ρ is proportional to ρ_2.

[†] This result was first derived by J. H. Van Vleck during the second World War, when studying the power spectrum of strongly clipped noise which was then used for electromagnetic jamming. It was later published by Van Vleck and Middleton (1966).

To determine the signal-to-noise ratio of the correlation measurement we now calculate σ_2^2, the variance of the correlator output r_2:

$$\sigma_2^2 = \langle r_2^2 \rangle - \langle r_2 \rangle^2, \tag{8.25}$$

where

$$r_2 = N^{-1} \sum_{i=1}^{N} \hat{x}_i \hat{y}_i. \tag{8.26}$$

Note that, in this chapter the circumflex ($\hat{\ }$) is used to denote quantized signal waveforms. The expression for $\langle r_2^2 \rangle$ is equivalent to Eq. (8.11) for unquantized waveforms:

$$\langle r_2^2 \rangle = N^{-2} \sum_{i=1}^{N} \langle (\hat{x}_i \hat{y}_i)^2 \rangle + N^{-2} \sum_{i=1}^{N} \sum_{k \neq i} \langle \hat{x}_i \hat{y}_i \hat{x}_k \hat{y}_k \rangle. \tag{8.27}$$

The first summation on the right-hand side of Eq. (8.27) is equal to N^{-1} since the products $\hat{x}_i \hat{y}_i$ takes values of ± 1 for two-level sampling. In evaluating the second summation the situation is similar to that for unquantized sampling, and we assume $\rho_2 \ll 1$. The factor σ^4 in Eq. (8.16) is replaced by the square of the variance of the quantized waveform, which is unity for two-level quantization. For all but a small fraction of the terms, $q = |i - k|$ is large enough that samples i and k from the same waveform are uncorrelated. These terms make a total contribution closely equal to ρ_2^2. Those terms for which samples i and k are correlated make an additional contribution closely equal to

$$2N^{-1} \sum_{q=1}^{\infty} R_2^2(q\tau_s), \tag{8.28}$$

where $R_2(\tau)$ is the autocorrelation coefficient for a signal after two-level quantization. Thus,

$$\sigma_2^2 = N^{-1} + N^{-1}(N-1)\rho_2^2 + 2N^{-1} \sum_{q=1}^{\infty} R_2^2(q\tau_s) - \rho_2^2$$

$$\simeq N^{-1} \left[1 + 2 \sum_{q=1}^{\infty} R_2^2(q\tau_s) \right]. \tag{8.29}$$

Note that again the condition $\rho_2 \ll 1$ is assumed, and the signal-to-noise ratio is

$$\mathcal{R}_{sn2} = \frac{\langle r_2 \rangle}{\sigma_2} = \frac{2\rho\sqrt{N}}{\pi \sqrt{1 + 2 \sum_{q=1}^{\infty} R_2^2(q\tau_s)}}. \tag{8.30}$$

This ratio, relative to that for unquantized sampling at the Nyquist rate given by Eq. (8.13), defines an efficiency factor for the quantized correlation process:

$$\eta_2 = \frac{\mathscr{R}_{sn2}}{\mathscr{R}_{sn\infty}} = \frac{2\sqrt{\beta}}{\pi\sqrt{1 + 2\sum\limits_{q=1}^{\infty} R_2^2(q\tau_s)}}. \tag{8.31}$$

Now Eq. (8.24) gives the relationship between the correlation coefficients for a pair of signals before and after two-level quantization. This result includes the case of autocorrelation where the two signals differ only because of a delay. Thus, we may write

$$R_2(q\tau_s) = \frac{2}{\pi}\sin^{-1}\left[R_\infty(q\tau_s)\right], \tag{8.32}$$

and $R_\infty(q\tau_s)$ is given by Eq. (8.14) for a rectangular baseband signal spectrum sampled at β times the Nyquist rate:

$$R_2(q\tau_s) = \frac{2}{\pi}\sin^{-1}\left[\frac{\beta\sin(\pi q/\beta)}{\pi q}\right]. \tag{8.33}$$

$R_2(q\tau_s)$ thus has zeros at the same values of $q\tau_s$ that $R_\infty(q\tau_s)$ does (the principal value being taken for the inverse sine function), and for $\beta = 1, \frac{1}{2}, \frac{1}{3},$

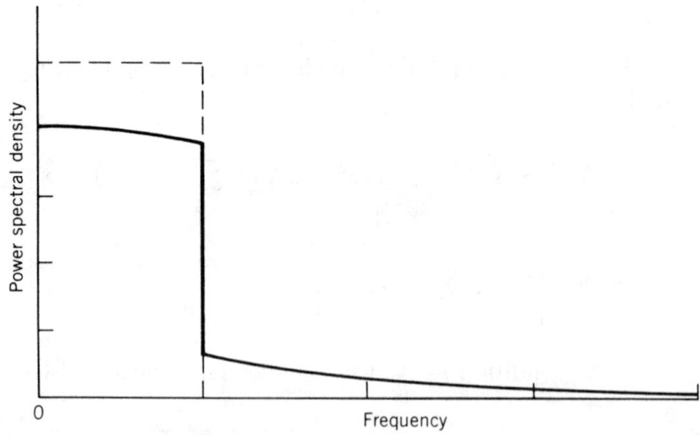

Figure 8.3   Spectra of rectangular bandpass noise before and after two-level quantization. The unquantized spectrum is of lowpass form, as shown by the broken line. The spectrum after quantization is shown by the solid curve. The power levels of the two waveforms (represented by the areas under the curves) are equal, and the Fourier transforms of their spectra are related by Eq. (8.24).

and so on,

$$\sum_{q=1}^{\infty} R_2^2(q\tau_s) = 0. \tag{8.34}$$

In these cases the signal-to-noise ratio is $2/\pi$ ($= 0.64$) times that for unquantized sampling at the same rate. For oversampling with $\beta = 2$ and $\beta = 3$ the signal-to-noise ratio from Eqs. (8.31) and (8.33) is enhanced by factors of 1.17 and 1.21, respectively. Similar results have been obtained by Burns and Yao (1969) and others.

We have mentioned that quantization generates additional spectral components, and we can now compare the power spectra of a signal before and after quantization since these spectra are the Fourier transforms of autocorrelation functions that are related by Eq. (8.24). Figure 8.3 shows the spectrum, after two-level quantization, of noise with an originally rectangular spectrum. A fraction of the original bandlimited spectrum is converted into a broad, low-level skirt which dies away very slowly with frequency.

Four-Level Quantization

The use of two digital bits to represent the amplitude of each sample results in less degradation of the signal-to-noise ratio than is obtained with one-bit quantization. Consideration of two-bit sampling leads naturally to four-level quantization, the performance of which has been investigated by several authors, notably Cooper (1970) and Hagen and Farley (1973). The quantization characteristic is shown in Fig. 8.4 where the quantization thresholds are $-v_0$, 0, and v_0. The four quantization states have designated values, $-n$, -1, $+1$, and $+n$, where n, which is not necessarily an integer, can be chosen to optimize the performance. Products of two samples can take the value ± 1, $\pm n$, or $\pm n^2$. The four-level correlation coefficient ρ_4 can be specified by an expression similar to Eq. (8.19) for the two-level case, that is,

$$\rho_4 = \frac{2n^2 N_{nn} - 2n^2 N_{n\bar{n}} + 4n N_{1n} - 4n N_{1\bar{n}} + 2N_{11} - 2N_{1\bar{1}}}{\left(2n^2 N_{nn} + 2N_{11}\right)_{\rho=1}}. \tag{8.35}$$

Here the numerator is proportional to the correlator output, and reduces to the form in the denominator for $\rho = 1$ when the two input waveforms are identical. The numbers of the various level combinations can be derived from the corresponding joint probabilities. Thus, for example,

$$N_{nn} = NP_{nn}$$

$$= \frac{N}{2\pi\sigma^2\sqrt{1-\rho^2}} \int_{v_0}^{\infty} \int_{v_0}^{\infty} \exp\left[\frac{-(x^2 + y^2 - 2\rho xy)}{2\sigma^2(1-\rho^2)}\right] dx\, dy, \tag{8.36}$$

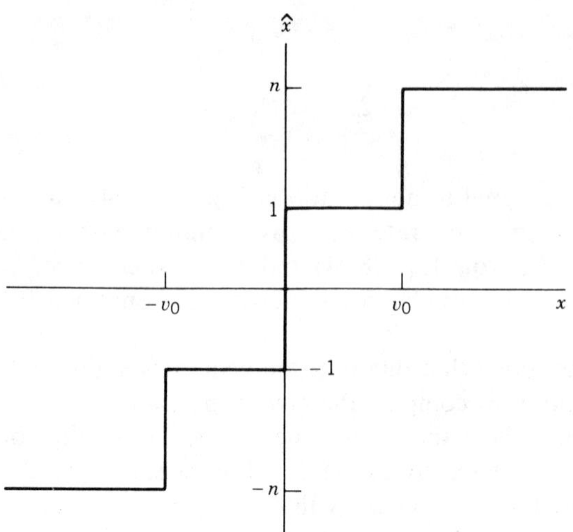

Figure 8.4 Characteristic curve for four-level quantization. The abscissa is the unquantized voltage x and the ordinate is the quantized output $\hat{x}$. v_0 is the threshold voltage.

and, as in the two-level case, the other probabilities are obtained by using the appropriate limits for the integrals. For the case of $\rho \ll 1$, the approximate form of the probability distribution in Eq. (8.2) simplifies the calculation.

Although ρ_4 can be evaluated from Eq. (8.35) in the above manner, an alternative derivation will be used here which provides a more rapid approach to the desired result. This approach follows the treatment of Hagen and Farley (1973) and is based on a theorem by Price (1958). The form of the theorem that we require is

$$\frac{d\langle r_4 \rangle}{d\rho} = \sigma^2 \left\langle \frac{\partial \hat{x}}{\partial x} \frac{\partial \hat{y}}{\partial y} \right\rangle, \tag{8.37}$$

where r_4 is the unnormalized correlator output and $\hat{x}$ and $\hat{y}$ are again the quantized versions of the input signals. For four-level sampling

$$\frac{\partial \hat{x}}{\partial x} = (n-1)\delta(x+v_0) + 2\delta(x) + (n-1)\delta(x-v_0), \tag{8.38}$$

where δ is the delta function, and a similar expression can be written for $\partial \hat{y}/\partial y$. Equation (8.38) is the derivative of the function in Fig. (8.4). To determine the expectation of the product of the two derivatives on the right-hand side of Eq. (8.37), the magnitudes of each of the nine terms in the product must be multiplied by the probability of occurrence. Thus, for example, the term $(n-1)^2\delta(x+v_0)\delta(y+v_0)$ has a magnitude of $(n-1)^2$

and probability

$$\frac{1}{2\pi\sigma^2\sqrt{1-\rho^2}}\exp\left[\frac{-2v_0^2}{2\sigma^2(1+\rho)}\right].$$ (8.39)

By consolidating terms with equal probabilities we obtain

$$\frac{d\langle r_4\rangle}{d\rho} = \frac{1}{\pi\sqrt{1-\rho^2}}\left\{(n-1)^2\left[\exp\left(\frac{-v_0^2}{\sigma^2(1+\rho)}\right) + \exp\left(\frac{-v_0^2}{\sigma^2(1-\rho)}\right)\right]\right.$$

$$\left. +4(n-1)\exp\left(\frac{-v_0^2}{2\sigma^2(1-\rho^2)}\right) + 2\right\},$$ (8.40)

and

$$\langle r_4\rangle = \frac{1}{\pi}\int_0^\rho\frac{1}{\sqrt{1-r^2}}\left\{(n-1)^2\left[\exp\left(\frac{-v_0^2}{\sigma^2(1+r)}\right) + \exp\left(\frac{-v_0^2}{\sigma^2(1-r)}\right)\right]\right.$$

$$\left. +4(n-1)\exp\left(\frac{-v_0^2}{2\sigma^2(1-r^2)}\right) + 2\right\}dr,$$

(8.41)

where r is a dummy variable of integration. To obtain the correlation coefficient ρ_4, $\langle r_4\rangle$ must be divided by the expectation of the correlator output when the inputs are identical four-level waveforms, as in Eq. (8.35):

$$\rho_4 = \frac{\langle r_4\rangle}{\Phi + n^2(1-\Phi)},$$ (8.42)

where Φ is the probability that the unquantized level lies between $\pm v_0$, that is,

$$\Phi = \frac{1}{\sigma\sqrt{2\pi}}\int_{-v_0}^{v_0}\exp\left(\frac{-x^2}{2\sigma^2}\right)dx = \text{erf}\left(\frac{v_0}{\sigma\sqrt{2}}\right).$$ (8.43)

Equations (8.41), (8.42), and (8.43) provide a relationship between ρ_4 and ρ which is equivalent to the Van Vleck relationship for two-level quantization.

The choice of values for n and v_0, which we now investigate, is usually made to maximize the signal-to-noise ratio for weak signals. For $\rho \ll 1$, Eqs. (8.41) and (8.42) reduce to

$$(\rho_4)_{\rho\ll1} = \rho\frac{2[(n-1)E+1]^2}{\pi[\Phi+n^2(1-\Phi)]},$$ (8.44)

224   Digital Signal Processing

where $E = \exp(-v_0^2/2\sigma^2)$. The variance in the measurement of r_4 is

$$\sigma_4^2 = \langle r_4^2 \rangle - \langle r_4 \rangle^2 = \langle r_4^2 \rangle - \rho_4^2 \left[\Phi + n^2(1 - \Phi) \right]^2. \qquad (8.45)$$

The factor $\Phi + n^2(1 - \Phi)$ is the variance of the quantized waveform and here takes the place of σ^2 in the corresponding equations for unquantized sampling. Again we follow the procedure explained for the unquantized case and assume $\rho_4 \ll 1$. Thus, we write

$$\langle r_4^2 \rangle = N^{-2} \sum_{i=1}^{N} \langle (\hat{x}_i \hat{y}_i)^2 \rangle + N^{-2} \sum_{i=1}^{N} \sum_{i \neq k} \langle \hat{x}_i \hat{y}_i \hat{x}_k \hat{y}_k \rangle. \qquad (8.46)$$

To evaluate the first summation note that $(\hat{x}_i \hat{y}_i)^2$ can take values of 1, n^2, or n^4, and the sum of these values multiplied by their probabilities is equal to $[\Phi + n^2(1 - \Phi)]^2$. The contribution of the second summation is

$$(1 - N^{-1})\rho_4^2 \left[\Phi + n^2(1 - \Phi) \right]^2 + 2N^{-1}\left[\Phi + n^2(1 - \Phi) \right]^2 \sum_{q=1}^{\infty} R_4^2(q\tau_s),$$

$$(8.47)$$

where the second term represents the effect of oversampling and is similar to Eq. (8.16), and R_4 is the autocorrelation function after four-level quantization. Thus from Eq. (8.45) we have

$$\sigma_4^2 = N^{-1}\left[\Phi + n^2(1 - \Phi) \right]^2 \left[1 + 2 \sum_{q=1}^{\infty} R_4^2(q\tau_s) \right]. \qquad (8.48)$$

The signal-to-noise ratio for the four-level correlation measurement is

$$\mathcal{R}_{sn4} = \frac{\langle r_4 \rangle}{\sigma_4} = \frac{2\rho\left[(n - 1)E + 1\right]^2 \sqrt{N}}{\pi\left[\Phi + n^2(1 - \Phi) \right] \sqrt{1 + 2 \sum_{q=1}^{\infty} R_4^2(q\tau_s)}}. \qquad (8.49)$$

The signal-to-noise ratio relative to that for unquantized Nyquist sampling is, from Eq. (8.13),

$$\eta_4 = \frac{\mathcal{R}_{sn4}}{\mathcal{R}_{sn\infty}} = \frac{2\left[(n - 1)E + 1\right]^2 \sqrt{\beta}}{\pi\left[\Phi + n^2(1 - \Phi) \right] \sqrt{1 + 2 \sum_{q=1}^{\infty} R_4^2(q\tau_s)}}. \qquad (8.50)$$

For sampling at the Nyquist rate $\beta = 1$ and

$$\eta_4 = \frac{\mathscr{R}_{\mathrm{sn4}}}{\mathscr{R}_{\mathrm{sn\infty}}} = \frac{2\left[(n-1)E+1\right]^2}{\pi\left[\Phi + n^2(1-\Phi)\right]}. \tag{8.51}$$

Cooper (1970) has examined the values of n and v_0 that maximize Eq. (8.51) and shown that values of $n = 3$, $v_0 = \sigma$, and $n = 4$, $v_0 = 0.95\sigma$ both result in a maximum value of 0.88 for the signal-to-noise ratio relative to that for unquantized sampling. Curves of the relative sensitivity (signal-to-noise ratio) as function of v_0/σ for $n = 2$, 3, and 4 are shown in Fig. 8.5. Similar conclusions are derived by Hagen and Farley (1973) and Bowers and Klingler (1974).

Having chosen values for n and v_0 we can now return to Eqs. (8.41) and (8.42) to examine the relationship of ρ and ρ_4. Curve 1 of Fig. 8.6 shows a plot of ρ and ρ_4. Note that for $\rho \ll 0.8$, the relationship can be considered linear, the slope being obtainable from Eq. (8.44). This linearity is an important feature since it simplifies the derivation of ρ. It also simplifies the final step that we require in discussing four-level sampling, namely, calculation of the improvement in sensitivity resulting from oversampling.

The relationship between the autocorrelation function for unquantized noise R_∞ and that for the same waveform after four-level quantization is the same as

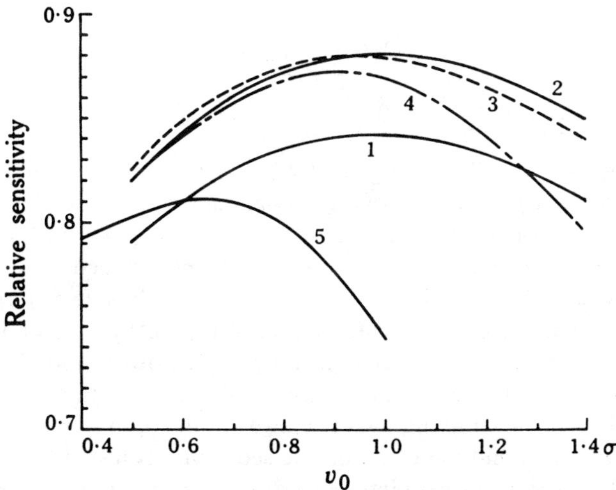

Figure 8.5   Signal-to-noise ratio relative to that for unquantized correlation for the four-level system and several modifications of it. (From Cooper 1970.) The abscissa is the quantization threshold v_0 in units of the rms level of the waveforms at the quantizer input. The ordinate is sensitivity (signal-to-noise ratio) relative to an unquantized system. The curves are for: (1) full four-level system with $n = 2$; (2) full four-level system with $n = 3$; (3) full four-level system with $n = 4$; (4) four-level system with $n = 3$ and low-level products omitted; (5) three-level system.

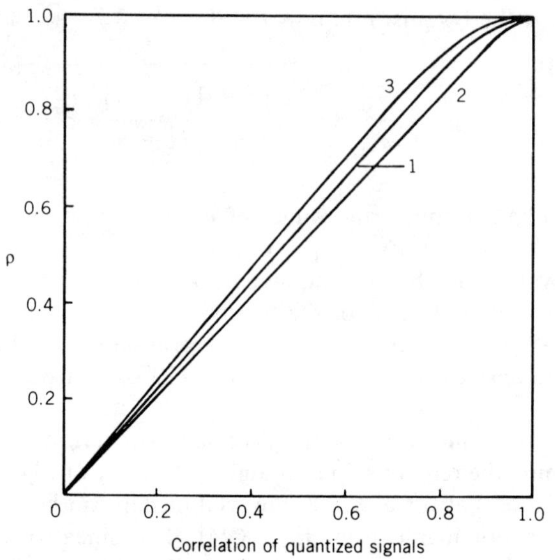

Figure 8.6 Correlation coefficient ρ for unquantized signals plotted as a function of the correlation that would be measured after quantization. (From Cooper 1970.) The curves are for: (1) full four-level system with $n = 3$ and $v_0 = \sigma$, or $n = 4$ and $v_0 = 0.95\sigma$; (2) four-level system with low-level products omitted, $n = 4$ and $v_0 = 0.9\sigma$; (3) three-level system with $v_0 = 0.6\sigma$.

for the cross-correlation functions in Eq. (8.44), so

$$R_4 = \frac{2\big[(n-1)E+1\big]^2 R_\infty}{\pi\big[\Phi + n^2(1-\Phi)\big]}, \tag{8.52}$$

provided that $R_\infty < 0.8$. R_∞ is given by Eq. (8.14) and does not exceed 0.8 for adjacent samples ($q = 1$) if the oversampling factor, β, is less than 2.8. For the optimum values of $v_0 = \sigma$ and $n = 3$, $E = 0.607$, $\Phi = 0.683$, and $R_4 = 0.88 R_\infty$. The signal-to-noise ratio relative to an unquantized system is given by Eq. (8.51) and is 0.88 for $\beta = 1$. For $\beta = 2$ we use Eqs. (8.52) and (A8.5) to evaluate the summation in the denominator of Eq. (8.50) and obtain $\eta_4 = 0.94$, which is a factor of 1.06 greater than for $\beta = 1$. Bowers and Klingler (1974) have pointed out that the optimum value of the quantization level v_0 changes slightly with the oversampling factor. However, the optimum values are rather broad (see Fig. 8.5) and the effect on the sensitivity is insignificant.

In his discussion of two-bit quantization, Cooper (1970) considered the effect of omitting certain products in the multiplication process. For example, if all products of the two low-level bits are counted as zero instead of ± 1, the loss in signal-to-noise ratio is approximately 1% as shown in curve 4 of Fig. 8.5. The products to be accumulated are only those counted as $\pm n$ and $\pm n^2$ in the full four-level system described above, and in the modified system they can

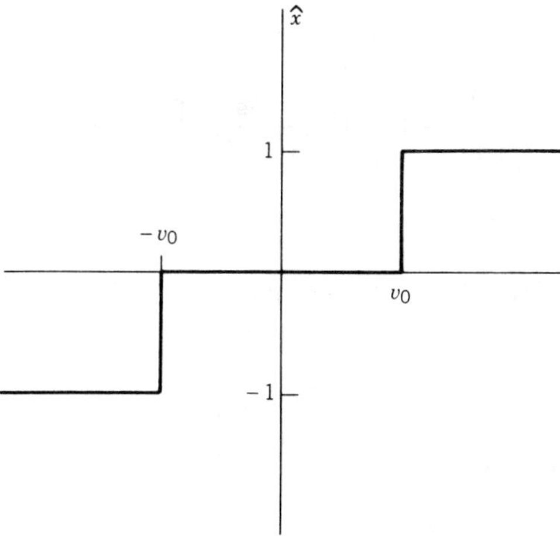

Figure 8.7 Characteristic curve for three-level quantization. The abscissa is the unquantized voltage x and the ordinate is the quantized output $\hat{x}$. v_0 is the threshold voltage.

be assigned values of ± 1 and $\pm n$, respectively, thereby simplifying the counter circuitry of the integrator. An even greater simplification can be accomplished by omitting the intermediate level products also and assigning values ± 1 to the high-level products. This last type of modification yields 92% of the sensitivity of a full four-level correlator. We shall not analyze the case where only the low-level products are omitted, but we note that to derive the correlation coefficient as a function of ρ one can express the action of the correlator in terms of two different quantization characteristics (Hagen and Farley 1973) or else return to Eq. (8.35) and omit the appropriate terms. If both the low- and intermediate-level products are omitted, however, the action can be described more simply in terms of a new quantization characteristic, known as three-level quantization, without arbitrary omission of product terms.

Three-Level Quantization

Three-level quantization has proved to be an important practical technique, and the quantization characteristic is shown in Fig. 8.7 In this case the approach using Price's theorem will again be followed.

The expressions for the operating characteristics of a three-level correlator can be obtained from those in the preceding section by omitting the terms that refer to low- and intermediate-level products and adjusting the weighting factors as appropriate. Thus, the equivalent derivative needed in Price's theorem is

$$\frac{\partial \hat{x}}{\partial x} = \delta(x - v_0) + \delta(x + v_0), \tag{8.53}$$

and the expectation of the correlator output $\langle r_3 \rangle$ is, from Price's theorem,

$$\langle r_3 \rangle = \frac{1}{\pi} \int_0^\rho \frac{1}{\sqrt{1 - r^2}} \left[\exp\left(\frac{-v_0^2}{\sigma^2(1 + r)} \right) + \exp\left(\frac{-v_0^2}{\sigma^2(1 - r)} \right) \right] dr.$$

(8.54)

The normalized correlation coefficient is

$$\rho_3 = \frac{\langle r_3 \rangle}{1 - \Phi},$$

(8.55)

and for small correlation

$$(\rho_3)_{\rho \ll 1} = \rho \frac{2E^2}{\pi(1 - \Phi)}.$$

(8.56)

For $\rho_3 \ll 1$ the variance of r_3 is

$$\sigma_3^2 = N^{-1}(1 - \Phi)^2 \left[1 + 2 \sum_{q=1}^{\infty} R_3^2(q\tau_s) \right],$$

(8.57)

where R_3 is the autocorrelation coefficient after three-level quantization. The signal-to-noise ratio relative to a nonquantizing correlator is

$$\eta_3 = \frac{\mathscr{R}_{sn3}}{\mathscr{R}_{sn\infty}} = \frac{\langle r_3 \rangle}{\sigma_3 \mathscr{R}_{sn\infty}} = \frac{2\sqrt{\beta}\,E^2}{\pi(1 - \Phi)\sqrt{1 + 2 \sum_{q=1}^{\infty} R_3^2(q\tau_s)}}.$$

(8.58)

For Nyquist sampling the maximum sensitivity relative to the nonquantizing case is 0.81 which is obtained with $v_0 = 0.612\sigma$ as shown in curve 5 of Fig. 8.5. With this optimized threshold value $\Phi = 0.459$, $E = 0.829$, and for $\rho < 0.8$ we can write $R_3(q\tau_s) = 0.81 R_\infty(q\tau_s)$. Then from Eqs. (8.58) and (A8.5) we find that for a rectangular baseband spectrum, increasing the oversampling factor β from 1 to 2 increases the signal-to-noise ratio by a factor of 1.10.

8.4   COMPARISON OF QUANTIZATION SCHEMES

At this point it is necessary to put into perspective the characteristics derived for sampling with different quantization schemes discussed above. The results, which are summarized in Table 8.1, are entirely theoretical, but it seems to be the general experience that performance agrees with theoretical prediction to within a few percent.

TABLE 8.1   Efficiency Factor η_Q for Various Quantization Schemes

| Number of Quantization | Sensitivity Relative to Unquantized Case, η_Q | |
Levels (Q)	$\beta = 1$	$\beta = 2$
2	0.64	0.74
3	0.81	0.89
4	0.88	0.94

In considering the relative advantages of different quantization schemes we note first that both the efficiency factor η_Q and the receiving bandwidth $\Delta\nu$ may be limited by the size and speed of the correlator system. The overall sensitivity is proportional to $\eta_Q\sqrt{\Delta\nu}$. Consider two conditions. In the first, the observing bandwidth is limited by factors other than the capacity of the digital system. This occurs in spectral-line observing or when the interference-free band is of limited width. The sensitivity limitation imposed by the correlator system then involves only the efficiency factor η_Q in Table 8.1, and the choice of quantization scheme is one between simplicity and sensitivity. In the second case the observing bandwidth is set by the maximum bit rate that the digital system can handle, as may occur in continuum observation in the higher-frequency bands. For a fixed bit rate ν_b the sample rate is ν_b/N_b where N_b is the number of bits per sample, and the maximum signal bandwidth $\Delta\nu$ is $\nu_b/2\beta N_b$. Thus the sensitivity is proportional to $\eta_Q/\sqrt{\beta N_b}$, and this factor is listed for various systems in Table 8.2, in which $N_b = 1$ (Q = number of quantization levels = 2) or $N_b = 2$ ($Q = 3$ or 4). Note that oversampling always reduces the performance under these conditions. For those situations in which the capacity of the data processing is limited by the correlator system, the value of 0.64 for Nyquist sampling with two-level quantization results in the highest overall performance. Encoding schemes involving nonintegral values of N_b are also of interest, for example, in tape recording of data. In that

TABLE 8.2   Sensitivity Factor $\eta_Q/\sqrt{\beta N_b}$ for a Correlator-Limited System

| Number of Quantization Levels Q | $\dfrac{\eta_Q}{\sqrt{\beta N_b}}$ | |
	$\beta = 1$	$\beta = 2$
2	0.64	0.52
3	0.57	0.45
4	0.62	0.47

case the amount of information stored per bit is a prime consideration as is discussed in Chapter 9.

In some circumstances quantization with more than four levels may be worthwhile. It has been reported that 16-level (four-bit) quantization results in a value of η_Q equal to 0.97 (D. T. Emerson, private communication). There is sometimes a loss in sensitivity resulting from truncation as data are accumulated. This problem is exacerbated as the number of quantization levels is increased.

An interesting development that has been applied to autocorrelators is the use of different numbers of quantization levels for the signals at the two inputs of the correlator. Autocorrelators are used for spectral observations with single antennas, and the signal is correlated with a delayed version of itself. Minimization of the number of digital circuits and optimization of sensitivity lead to the use of fewer bits to represent the version of the signal that is made to pass through the delay circuits than for the one that is applied directly to the correlators. A three-level by five-level correlator, for which the quantization efficiency factor η_Q is 0.86, has been constructed by Bowers et al. (1973) for spectral line mapping with a two-element interferometer. The use of different numbers of bits at the correlator inputs offers some simplification for a two-antenna system if the instrumental delay is applied to the signal with the fewer quantization levels. Another practical example is one in which a remote antenna is to be linked by microwave transmission to an array that uses three-level data processing. If the bit rate of the link sets the limit on the system bandwidth, the sensitivity is maximized by transmitting two-level data and performing two-level by three-level correlation, for which the quantization efficiency is 0.72.

8.5   ACCURACY IN DIGITAL SAMPLING

Principal Causes of Errors

Deviations from ideal performance in practical samplers result in errors that, if not corrected for, can limit the accuracy of maps synthesized from the data. Once the signal is in digital form, however, the rate at which errors are introduced is usually negligibly small.

Two-level samplers, which sense only the sign of the signal voltages, are the simplest samplers to construct. The most serious error that is likely to occur is in the definition of the zero level, in which a small voltage offset may occur. The effect of offsets in the samplers is to produce small offsets of positive or negative polarity in the correlator outputs, which can be largely eliminated by phase switching, as described in Chapter 7. Alternatively, the offsets in the samplers can be measured by incorporating counters to compare the numbers of positive and negative samples produced. Correction for the offsets can then be applied to the correlator output data (e.g., Davis 1974).

In samplers with three or more quantization levels the performance depends on the specification of the levels with respect to the rms signal level, σ. An automatic level control (ALC) circuit is therefore commonly used at the sampler input. Errors resulting from incorrect signal amplitude become less important as the number of quantization levels is increased; with many levels the signal amplitude becomes simply a linear factor in the correlator output. In systems using complex correlators two samplers are usually required for each signal, one at each output of the quadrature network. The accuracy of the quadrature network, and the possible errors in the relative timing of the two sample pulses, can also introduce errors in the data.

Tolerances in Three-Level Sampling

The results in this section are based largely on a study of accuracy requirements in three-level sampling by D'Addario et al. (1984). We start by considering the diagram in Fig. 8.8 which shows the sampling thresholds for a correlated pair of signals. Thresholds v_1 and $-v_2$ apply to the signal waveform $x(t)$ and v_3 and $-v_4$ to $y(t)$. The probability distribution of x and y is given by Eq. (8.1), and the correlator output is proportional to this probability integrated over the (x, y) plane with the weighting factors ± 1 and zero indicated in the figure. This approach enables us to investigate the effect of deviations of the sampler thresholds from the optimum, $v_0 = 0.612\sigma$. For three-level sampling the correlator output can be written

$$\langle r_3(\alpha, \rho) \rangle = \left[L(\alpha_1, \alpha_3, \rho) + L(\alpha_2, \alpha_4, \rho) \right.$$

$$\left. - L(\alpha_1, \alpha_4, -\rho) - L(\alpha_2, \alpha_3, -\rho) \right], \qquad (8.59)$$

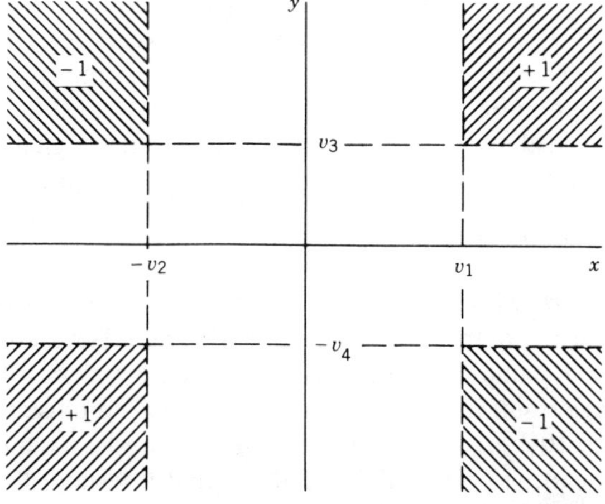

Figure 8.8 Threshold diagram for a correlator, the inputs of which are three-level quantized signals. x and y represent the unquantized signals, and the shaded areas show the combinations of input levels for which the output is nonzero.

where $\alpha_i = v_i/\sigma$, and so on, and

$$L(\alpha_i, \alpha_k, \rho) = \int_{\alpha_i}^{\infty} \int_{\alpha_k}^{\infty} \frac{1}{2\pi\sqrt{1-\rho^2}} \exp\left[\frac{-(X^2 + Y^2 - 2\rho XY)}{2(1-\rho^2)}\right] dX\, dY.$$

$$(8.60)$$

Note that the integral in Eq. (8.60) is simply Eq. (8.1) with the variables measured in units of σ. Inversion formulas for determination of ρ from r_3 and the threshold values are given by Kulkarni and Heiles (1980) and D'Addario et al. (1984). The latter authors point out that since less than 5% loss in signal-to-noise ratio occurs for threshold departures of $\pm 40\%$ from optimum, the required accuracy of the threshold settings, in practice, depends mainly on the correction algorithm.

Suppose that the thresholds are kept close to, but not exactly equal to, the optimum value. For the x sampler in Fig. 8.8 the deviations from the ideal threshold value α_0 can be expressed in terms of an even part

$$\Delta_{gx} = \tfrac{1}{2}(\alpha_1 + \alpha_2) - \alpha_0, \qquad (8.61)$$

and an odd part

$$\Delta_{ox} = \tfrac{1}{2}(\alpha_1 - \alpha_2), \qquad (8.62)$$

and Δ_{gy} and Δ_{oy} are similarly defined for the y sampler. The Δ_g terms produce gain errors. They are equivalent to an error in the level of the signal at the sampler, and they have the effect of introducing a multiplicative error in the measured cross correlation. The Δ_o terms produce offset errors in the correlator output and are potentially more damaging since such errors can be large compared with the low levels of cross correlation resulting from weak sources. The offset errors, however, can be removed with high precision by phase switching. The cancellation of the offset results from the sign reversal of the digital samples, or of the correlator output, as described in Section 7.5. The synchronous reversal of a local oscillator allows the signal to remain unaffected. The correlator output of a phase-switched system is of the form

$$r_{3s}(\alpha, \rho) = \tfrac{1}{2}\left[r_3(\alpha, \rho) - r_3(\alpha, -\rho)\right]. \qquad (8.63)$$

If all α values are within $\pm 10\%$ of α_0, the output is always within 10^{-3} (relative error) of the output of a correlator with no offset errors and the same gain errors in the samplers. Note also from Fig. 8.5 that 10% errors in the threshold settings result in less than 1% loss in signal-to-noise ratio. Thus, with phase switching, 10% errors are tolerable in the thresholds, and the effects of the gain errors can be corrected for if the actual threshold levels are known. Since the probability density distribution of the signal amplitudes can be assumed to be Gaussian, the threshold levels can be determined by counting the relative numbers of $+1, 0$, and -1 outputs from each sampler. When ρ is small (a few percent), a simple correction for the gain error can be obtained by dividing the correlator output by the arithmetic mean of the numbers of high-level (± 1) samples for the two signals. Ten percent errors in the threshold settings then result in errors of less than 1% in ρ. More accurate formulas for determining ρ from r_3 are given by Kulkarni and Heiles (1980) and D'Addario et al. (1984), and are not restricted to small ρ.

Another nonideal aspect of the behavior of the sampler and quantizer is that the threshold level may not be precisely defined but may be influenced by the direction and rate of change of the signal voltage, the previous sample value (hysteresis), and other effects. The result can be modeled by including an indecision region in the sampler response extending from $\alpha_k - \Delta$ to $\alpha_k + \Delta$. A signal that falls within this region results in an output that takes either of the two values associated with the threshold randomly and with equal probability. The threshold diagram with indecision regions included is shown in Fig. 8.9. The weighting in the indecision regions depends on the probability of the random sample values and is $\frac{1}{4}$ when both signals fall within indecision regions, and $\frac{1}{2}$ when one signal is within an indecision region and the other produces a nonzero output. As before, the correlator output can be obtained by integrating the weighted probability of the signal values over the (X, Y) plane. Figure 8.10 shows the decrease in the correlator output as a function of

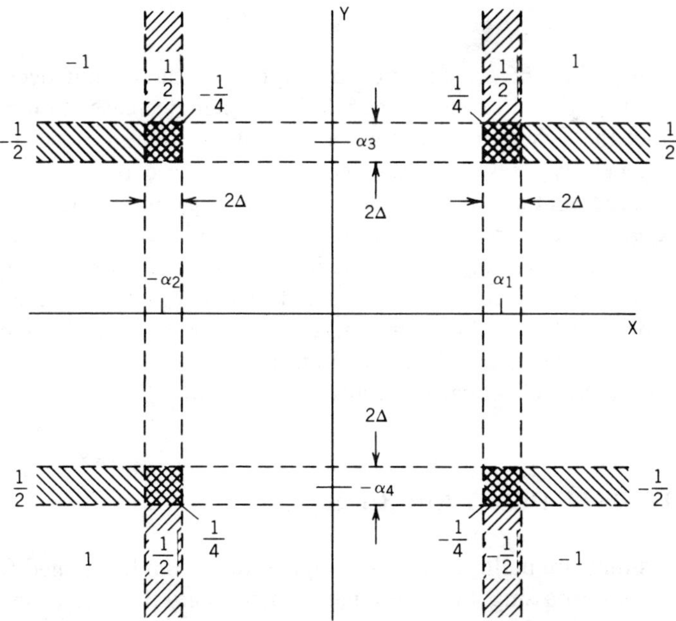

Figure 8.9   Threshold diagram for a three-level correlator showing indecision regions and the shaded areas within them for which the response is nonzero. The figures ± 1, $\pm \frac{1}{2}$, and $\pm \frac{1}{4}$ indicate the correlator response. The diagram shows the (X, Y) plane in which the signals are normalized to the rms value σ.

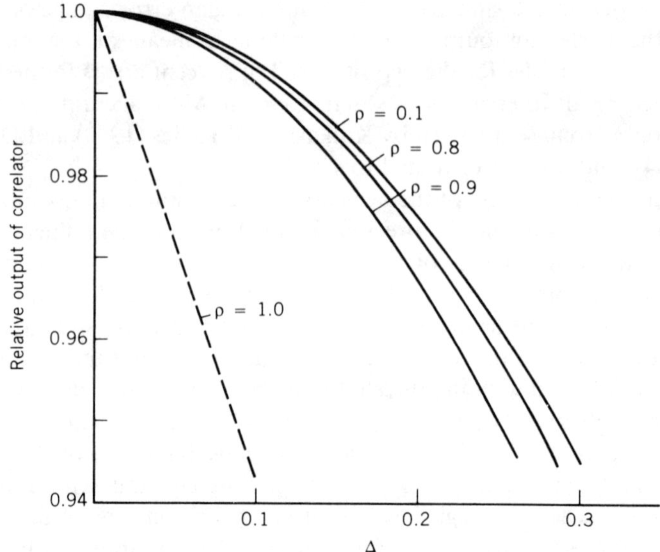

Figure 8.10   Effect of indecision regions on the output of a three-level correlator. The thresholds are assumed to be set to the optimum value 0.612σ, and the widths of the indecision regions are $2\sigma\Delta$. The output is given as a fraction of the output for $\Delta = 0$.

Δ for several values of ρ, computed by expressing the output decrease as a Maclaurin series in Δ. For $\rho < 1$, the relatively small decrease in output results from the fact that when one input waveform falls within an indecision region, the other generally does not. For the particular case of $\rho = 1$, the input waveforms are identical and fall within these regions simultaneously. The output decrease is then proportional to Δ as shown by the broken line in Fig. 8.10: however, this case is not of practical importance. For a 1% maximum error, Δ must not exceed 0.11σ so the indecision region can be as large as $\pm18\%$ of the threshold value. For a maximum error of 0.1% the above limits must be divided by $\sqrt{10}$. Thus the indecision regions have large enough tolerances that their effect can generally be neglected.

8.6   DIGITAL DELAY CIRCUITS

Time delays that are multiples of the sample interval can be applied to streams of digital bits by passing them through shift registers that are clocked at the sampling frequency. Registers with different numbers of stages thus provide different fixed delays. A series of shift registers with delays of 1, 2, 4, 8, and so on, times the clock interval can be used in a similar way to a binary series of cable lengths in an analog delay system. A method of using two shift registers to obtain a delay that is variable in increments of the clock pulse interval is described by Napier, Thompson, and Ekers (1983). Random access memories can also be used to store data for variable delay periods.

Another useful technique is serial-to-parallel conversion, that is, the division of a bit stream at frequency ν into 2^n parallel streams at frequency $\nu/2^n$. This allows the use of slower and more economical types of digital circuits for delay, correlation, and other processes.

The precision required in setting a delay has been discussed in Chapter 7, and is usually some fraction of the reciprocal analog bandwidth. In the schemes discussed above the smallest delay increment is the reciprocal of the sampling frequency, that is, half the reciprocal bandwidth for the Nyquist rate. A finer delay step can be obtained digitally by varying the timing of the sample pulse in a number of steps, for example, 16, between the basic timing pulses. Thus, if an extra delay of, say, 3/16 of a clock interval is required the sampler is activated 13/16 of a clock interval after the clock pulse, and the data are held for 3/16 of an interval to bring them into phase with the clock pulse timing. Alternatively, the fine delay steps can be obtained by using small analog delay lines, since the line lengths required are then short enough to avoid the problems of mismatch and attenuation typical of long cable delays.

8.7 QUADRATURE PHASE SHIFT OF A DIGITAL SIGNAL

We have mentioned that complex correlators for digital signals can be implemented by introducing the quadrature phase shift in the analog signal as in Fig. 6.3, and then using separate samplers for the signal and its phase-shifted version. The Hilbert transformation that the phase shift represents can also be performed on the digital signal, thus eliminating the quadrature network and saving samplers and delay lines, but the accuracy is limited. Hilbert transformation is mathematically equivalent to convolution with the function $(-\pi\tau)^{-1}$, which extends to infinity in both directions (e. g., Bracewell 1965). A truncated sequence of the same form, for example, $\frac{1}{3}$, 0, 1, 0, -1, 0, $-\frac{1}{3}$, provides a convolving function for the digital data which introduces the required phase shift. However, the truncation results in convolution of the resulting signal spectrum with the Fourier transform of the truncating function, that is, a sinc function. This introduces ripples and degrades the signal-to-noise ratio by a few percent. Also, the summation process in the digital convolution increases the number of bits in the data samples, and the low-order bits must be discarded to avoid a major increase in the complexity of the correlator. There is thus a further loss of information. The overall result is that the imaginary output of the correlator suffers spectral distortion and some loss in signal-to-noise ratio relative to the real output. These effects are most serious, of course, in broad bandwidth systems in which the high data rate permits only simple processing. Lo et al. (1984) have described a system in which the real part of the correlation is measured as a function of time offset, as described below for the spectral correlator, and the imaginary part is then computed by Hilbert transformation.

8.8 DIGITAL CORRELATORS FOR CONTINUUM OBSERVATIONS

In continuum observations the average correlation over the signal bandwidth is measured, and no spectral data on a finer frequency scale are obtained. Thus, the correlation of the signals is measured only for zero time-delay offset. A complex correlator, as described in section 6.1, is used to obtain the real and imaginary products.

Digital correlators can run at the sampling frequency of the signals, or at a submultiple resulting from dividing the bit stream from the sampler into a number of parallel streams. In the latter case the number of correlator units must be proportionally increased, and their outputs can be additively combined in the control computer. The choice of input frequency for the correlator depends on the complexity of the system and the type of logic circuitry, both of which affect the maximum frequency at which it can be designed to operate. Two-level and three-level correlators, for which the products are represented by values of -1, 0, and $+1$, are the simplest to construct. Correlators in which one of the inputs is a two-level or three-level signal and the other input is more highly quantized also have a degree of simplicity. In this case, the correlator is essentially an accumulating register into which the higher-quantization value is entered. The two-level or three-level value is used to specify whether the other number is to be added, subtracted, or ignored. In correlators in which both inputs have more than three levels of quantization the multiplier output for any single product can be one of a range of numbers. One method of implementing such a multiplier is to use a read-only memory unit in which the possible product values are stored. The input bits to be multiplied are used to specify the address of the required product in the memory.

The output of a multiplier can take both positive and negative values, and, ideally, an up-down counter is required as an integrator. Since such counters are usually slower than simple adding counters, two of the latter are sometimes used to accumulate the positive and negative counts independently. Another technique is to count, for example, -1, 0, and $+1$ as 0, 1, or 2, and then subtract the excess values, in this case equal to the number of products, in the subsequent processing.

8.9 DIGITAL CORRELATORS FOR SPECTRAL-LINE OBSERVATIONS

Principles of Digital Spectral Measurements

In spectral-line observations, measurements at different frequencies across the signal band are required. These measurements can be obtained by digital techniques using a spectral correlator system, which is most commonly implemented by measuring the correlation of the signals as a function of time offset (also referred to as time lag). The Fourier transform of this quantity is the cross power spectrum, which can be interpreted as the complex visibility as a function of frequency. The Fourier transform relationship is discussed in Chapter 3. Note that in an autocorrelator the two input signals are the same waveform with a time offset, so the autocorrelation function is symmetrical and the power spectrum is entirely real. However, the cross power spectrum of the signals from two different antennas is complex, and the cross-correlation function has odd as well as even parts.

The output of a spectral correlator system provides values of the visibility at N frequency intervals across the signal band. These intervals are sometimes spoken of as frequency channels, and their spacing as the channel bandwidth, by analogy with the analog type of spectral correlator in which the signal band is broken up into channels by a bank of N filters with separate analog correlators for each filter channel. To explain the action of a digital spectral correlator we consider the cross power spectrum $\mathscr{S}(\nu)$ of the signals from two antennas, as shown in idealized form in Fig. 8.11. Here it is assumed that the source under observation has a flat spectrum with no line features, and the final IF amplifier before the sampler has a rectangular baseband response. In Fig. 8.11 we have included the negative frequencies since they are necessary in discussing the Fourier transform relationships. For $-\Delta\nu \le \nu \le \Delta\nu$ the real and imaginary parts of $\mathscr{S}(\nu)$ have magnitudes a and b, respectively, and the corresponding visibility phase is $\tan^{-1}(b/a)$, The cross-correlation function $\rho(\tau)$ is the Fourier transform of $\mathscr{S}(\nu)$, τ being the time offset:

$$\rho(\tau) = (a - jb)\int_{-\Delta\nu}^{0} e^{j2\pi\nu\tau}d\nu + (a + jb)\int_{0}^{\Delta\nu} e^{j2\pi\nu\tau}d\nu$$

$$= 2\,\Delta\nu\left[a\frac{\sin(2\pi\,\Delta\nu\,\tau)}{2\pi\,\Delta\nu\,\tau} - b\frac{1 - \cos(2\pi\,\Delta\nu\,\tau)}{2\pi\,\Delta\nu\,\tau}\right]. \tag{8.64}$$

Thus $\rho(\tau)$ has an even component of the form $\sin x/x$ which is related to the real part of $\mathscr{S}(\nu)$, and an odd component of the form $(1 - \cos x)/x$ which is related to the imaginary part. The spectral correlator measures $\rho(\tau)$ for integral values of the sampling interval τ_s. We consider the case of Nyquist sampling for which $\tau_s = 1/2\,\Delta\nu$. The measured cross correlation refers, of course, to the quantized waveforms, and the results in Section 8.3 show how this is related to the cross correlation of the unquantized waveforms. For correlation levels that are not too large, the two quantities are closely proportional, so for simplicity we assume that Eq. (8.64) represents the behavior of the measured cross correlation. The measurements are made with $2N$ time offsets from $-N\tau_s$ to $(N - 1)\tau_s$ between the signals, and Fourier transformation of these discrete values yields the cross power spectrum at intervals of $(2N\tau_s)^{-1} = \Delta\nu/N$ in frequency for Nyquist sampling. The N complex values of the positive frequency spectrum are the data required. Of these, the imaginary part comes from the odd component of $r(\tau)$. Thus, in the correla-

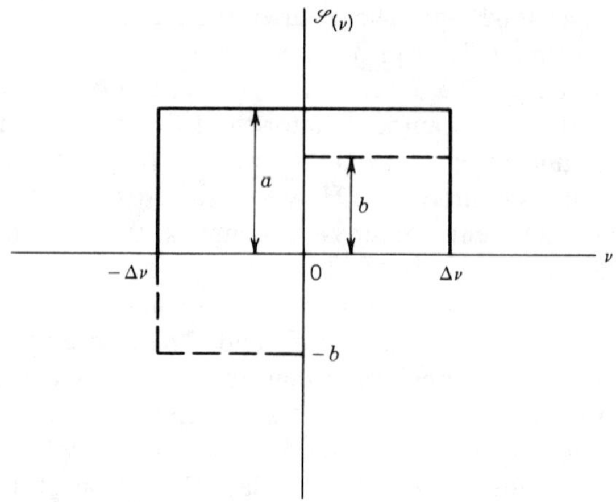

Figure 8.11   Cross power spectrum $\mathscr{S}(\nu)$ of two signals for which the power spectra are rectangular bands extending in frequency from zero to $\Delta\nu$. Negative frequencies are included. The full line represents the real part of $\mathscr{S}(\nu)$ and the broken line the imaginary part. The corresponding correlation function is derived in Eq. (8.64).

tion measurement it suffices to use single-multiplier correlators to measure $2N$ real values of $r(\tau)$ over both positive and negative values of τ for one antenna with respect to the other. Note that, if required, the imaginary part of the correlation can be computed by Hilbert transformation of the real part as a function of τ. The Hilbert transform relationships between the real and imaginary parts is easily demonstrated (e.g., Lo et al. 1984) and follows, for example, from the associative property of convolution. As an alternative to measuring only the real part of the correlation, we could use complex correlators to measure both the real and imaginary parts, as in Fig. 6.3, for a range of time offsets from zero to $(N-1)\tau_s$. In either case the same information would be obtained. From a practical viewpoint, it is generally preferable to use single-multiplier correlators to avoid constructing the broadband quadrature networks used in complex correlators.

Measurement of the cross correlation over the limited time offset range is equivalent to measuring $r(\tau)$ multiplied by a rectangular function of width $2N\tau_s$. The cross power spectrum derived from the limited measurements is therefore equal to the true cross power spectrum convolved with the Fourier transform of the rectangular function, that is, with the sinc function

$$\frac{\sin(\pi\nu N/\Delta\nu)}{\pi\nu}, \tag{8.65}$$

which is normalized to unit area with respect to ν. Any line feature within the spectrum is broadened by the sinc function (8.65) and, depending on its frequency profile, may show the characteristic oscillating skirts. The width of

TABLE 8.3   Commonly Used Smoothing Functions

Weighting Function $w(\tau)$. $[w(\tau) = 0, |\tau| > \tau_1 = N\tau_s]$	Half-Amplitude Width Unit $= \Delta\nu/N$	Peak Sidelobe
Uniform, $w(\tau) = 1$	1.21	0.22
Bartlett, $w(\tau) = 1 - (|\tau|/\tau_1)$	1.77	0.047
Hanning, $w(\tau) = \frac{1}{2} + \frac{1}{2}\cos(\pi\tau/\tau_1)$	2.00	0.027
Hamming, $w(\tau) = 0.54 + 0.46\cos(\pi\tau/\tau_1)$	1.82	0.0073
Blackman, $w(\tau) = 0.42 + 0.50\cos(\pi\tau/\tau_1)$ $\qquad\qquad\quad + 0.08\cos(2\pi\tau/\tau_1)$	2.30	0.0012

the sinc function at the half-maximum level is 1.2 $\Delta\nu/N$, that is, 1.2 times the channel separation, and this width defines the effective frequency resolution.

The oscillations of the sinc function introduce structure in the frequency spectrum similar to the sidelobe responses of an antenna beam. They result from the sharp edges of the rectangular function that multiplies the correlation function. Such sidelobes are undesirable and can be reduced by choosing weighting functions, other than the rectangular truncation, that are constrained to be zero outside the measurement range. Weighting functions are generally chosen to taper smoothly to zero at $|\tau| = N\tau_s$, thereby avoiding unwanted ripples in the smoothing (convolving) function, but to be as wide as possible in order to keep the width of the smoothing function as narrow as possible. These requirements are not entirely compatible, so that weighting functions that produce smoothing functions with very low sidelobes have poor frequency resolution. Some commonly used weighting functions are listed in Table 8.3. Hanning weighting, also known as raised cosine weighting, reduces the first sidelobe by a factor of 9, and the resolution by 1.67, compared to uniform weighting. For discretely sampled spectral data, as obtained from a digital correlator, the value of any point in a sequence is replaced by the sum of one-half the original value and one-quarter the original values at the two adjacent points. Then the smoothed value of the cross power spectrum for frequency interval n is given by

$$\mathscr{S}'\left(\frac{n\,\Delta\nu}{N}\right) = \frac{1}{4}\mathscr{S}\left[\frac{(n-1)\Delta\nu}{N}\right] + \frac{1}{2}\mathscr{S}\left(\frac{n\,\Delta\nu}{N}\right) + \frac{1}{4}\mathscr{S}\left[\frac{(n+1)\Delta\nu}{N}\right].$$

$$(8.66)$$

The above expression is the convolution of the power spectrum $\mathscr{S}(\nu)$ with three delta functions, and the Hanning-smoothed curve in Fig. 8.12 is thus the sum of three sinc functions. The Fourier transform of the delta functions is equal to the raised-cosine function $\frac{1}{2}[1 + \cos(2\pi\,\Delta\nu\,\tau/N)]$. In the time domain the effect of the smoothing is to multiply the cross correlation by this raised-cosine weighting, which falls to zero at $\tau = \pm N\tau_s$. The Hamming weighting function is very similar to the Hanning function and would appear to be superior because it produces a better resolution and a lower peak sidelobe level. However, the sidelobes of the Hamming smoothing function do not decrease in amplitude as rapidly as those of the Hanning smoothing function. Weighting functions are discussed in detail by Blackman and Tukey (1959) and Harris (1978).

A further effect of the finite time offset range complicates the calibration of the instrumental frequency response in the following way (Willis and Bregman 1981). The frequency responses of the amplifiers associated with the different antennas in general are not identical, as discussed in Chapter 7. To calibrate the response of each antenna pair over the spectral channels, it is usual to measure the cross power spectrum of an unresolved source for which the actual radiated spectrum is known to be flat across the receiving passband. We can consider the result in terms of the idealized power spectra in Fig. 8.11. If no special weighting function is used the real and imaginary parts are both convolved with Expression (8.65). When a function with a sharp edge is convolved with a sinc function the result is the appearance of oscillations, sometimes called the Gibbs' phenomenon, near the edge, as shown in Fig. 8.13. The point here is that the real component of $\mathscr{S}(\nu)$ in Fig. 8.11 is continuous through zero frequency, but the imaginary part shows a sharp sign reversal. Thus, near zero frequency the observed imaginary part of $\mathscr{S}(\nu)$ will show

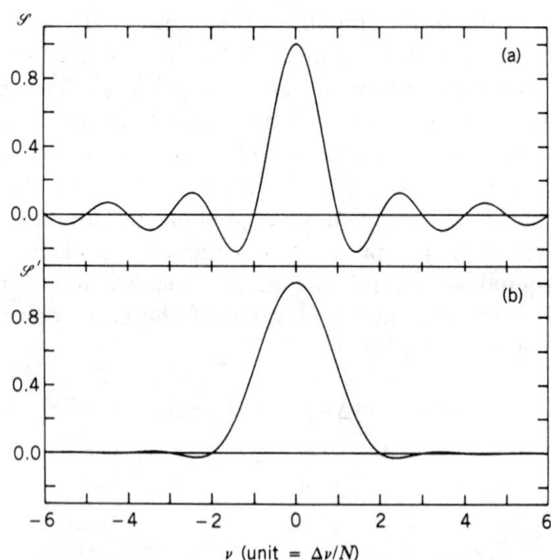

Figure 8.12   (a) The sinc function $\sin(\pi\nu N/\Delta\nu)/(\pi\nu N/\Delta\nu)$ which represents the frequency response of a spectral correlator with channels of width $\Delta\nu/N$ to a narrow line at $\nu = 0$. Here ν is measured with respect to the center of the received signal band. (b) The same curve after the application of Hanning smoothing as in Eq. (8.66).

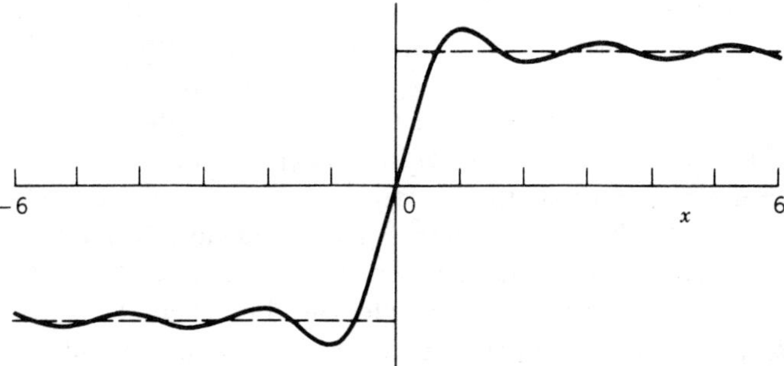

Figure 8.13   Convolution of a step function at the origin (broken line) with the sinc function $\sin(\pi x)/\pi x$.

oscillations that may be as high as 18% in peak amplitude, whereas the real component will show relatively small oscillations at that point. As a result the magnitude and phase measured for $\mathscr{S}(\nu)$ will show oscillations or ripples, the amplitude of which will depend on the relative amplitudes of the real and imaginary parts, that is, on the phase of the uncalibrated visibility. The uncalibrated phase measured for any source depends on instrumental factors such as the lengths of cables as well as the source position which may not be known. In general, the phase will not be the same for the source under investigation and the calibrator. Hence, near zero frequency some precautions must be taken in applying the calibration. Possible solutions to the problem include (1) calibrating the real and imaginary parts separately, (2) observing over a wide enough band that the end channels in which the ripples are strongest can be discarded, or (3) applying smoothing in frequency to reduce the ripples. The problem is most serious in the study of weak spectral lines in the presence of unresolved continuum emission. In certain VLBI systems it is avoided because of filtering applied in the fringe rotation.

Another problem encountered when observing a spectral line in the presence of a continuum background is caused by reflections in the antenna structure. These reflections cause a sinusoidal gain variation across the passband, the period of which is equal to the reciprocal of the delay of the reflected signal. In a correlation interferometer the magnitude of the ripple is a nearly constant fraction of the correlated continuum flux density, and is removed when the spectrum of the source under investigation is divided by the spectrum of the calibration source.

Spectral Correlator Systems

A simplified schematic diagram of a system that performs the basic measurements described in the last section is shown in Fig. 8.14. Practical systems are often more complicated and are designed to take full advantage of the flexibility of digital processing techniques. The numbers and bandwidths of channels required for spectral-line studies vary greatly, from a few hundred hertz to tens of megahertz. This versatility is necessary because the widths of spectral features depend mainly on Doppler shifts, which are proportional to the rest frequencies of the lines and the velocities of the emitting molecules. For this reason many digital spectral-line systems incorporate a series of filters in the IF amplifiers so that the overall signal bandwidth can be reduced by factors of $\frac{1}{2}$, $\frac{1}{4}$, $\frac{1}{8}$, and so on. When the signal bandwidth is halved, the Nyquist frequency is halved, and the samplers can be run at half the maximum frequency (or else every other sample can be deleted). However, if the correlators are run at the frequency used for the maximum bandwidth, the data samples can be processed by the correlators twice, and the range of time offsets can thereby be doubled. As a result, the number of channels is doubled and the channel bandwidth decreased by a factor of 4. The use of this principle allows the signal bandwidths to be further decreased, and the number of channels increased, as required. Usually the number of correlators is an integral power

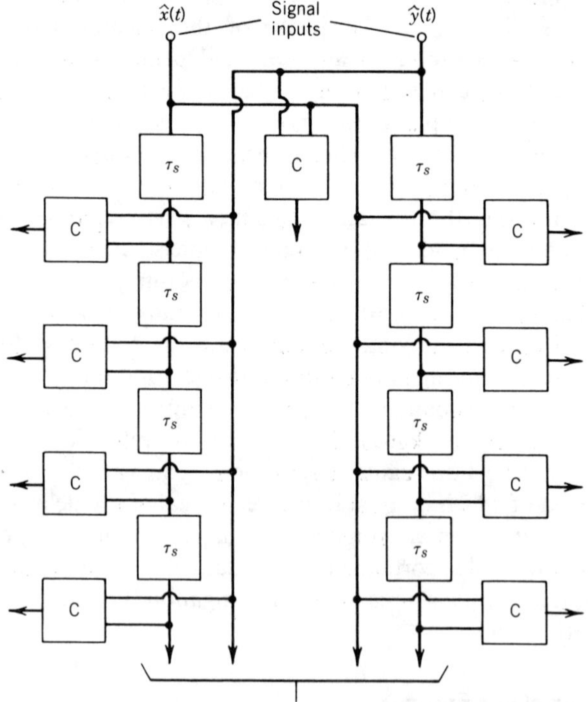

Figure 8.14 Simplified schematic diagram of a spectral correlator for two sampled signals. τ_s indicates a time delay equal to the sample interval and C indicates a correlator. The correlation is measured for zero delay, for the $\hat{x}$ input delayed with respect to the $\hat{y}$ input (left-hand correlator bank), and for $\hat{y}$ delayed with respect to $\hat{x}$ (right-hand correlator bank). The delays are integral multiples of τ_s.

of 2, and the signal bandwidths are decreased by powers of 2 to be compatible with digital computing techniques. To implement the above scheme recirculator units are required, which are basically memories that store blocks of input samples and allow them to be read out at the correlator input rate. These memory units are required in pairs, so that one is filled with data at the Nyquist rate appropriate to the chosen signal bandwidth, while the other is being read repeatedly at the maximum data rate. One memory becomes filled in the time that the other is read for the required number of times, and the two are then interchanged. Correlators that incorporate the above principles are described as recirculating correlators (Ball 1973).

Another approach to spectral correlator design involves Fourier transformation of the signals before cross correlation. One method of implementing this scheme is discussed by Chicada et al. (1984). For each antenna, both in-phase and quadrature components of each signal are sampled, as in sampling for a complex correlator. The samples then go to a special processor that performs a fast Fourier transform (FFT) in which sequences of N samples emerge as N values of complex signal amplitude for N channels of the input band. The phase in these output values is measured relative to the sampler clock. The complex amplitudes are then cross correlated by antenna pairs and the resulting visibility data processed in the usual way. Writing down the expressions for the transformation and correlation steps in this process, and in that discussed earlier in which correlation precedes transformation, shows that the result obtained is of the same form in both cases (Moran 1976). To compare the number of correlators required in the two systems, we note that for correlation before transformation $4\,\Delta\nu\,N$ real correlations per second are required for each antenna pair for Nyquist sampling. For an equivalent system using correlation after transformation $2\,\Delta\nu$ samples per second from each antenna are fed to each Fourier transformer. Then for each antenna pair there are $2\,\Delta\nu/N$ complex correlations per second for each of the N channels, that is, a total rate of $2\,\Delta\nu$ complex correlations per second. Each signal datum from the Fourier transformer is represented by substantially more bits than a datum from the sampler. Thus, overall, the correlation-after-transformation scheme requires fewer (but higher level) multiplications in the correlators by a factor of order N. The number of correlations does not increase with the number of channels, which is helpful in making a wide bandwidth system with many channels. The advantages are gained at the expense of providing the real-time FFT processors. However, the number of such processors is proportional to the number of antennas, rather than the number of antenna pairs, and thus with increasing array size they constitute a decreasing fraction of the entire system. For a more detailed discussion see Chicada et al. (1984).

The flexibility of digital correlators and the reliability of digital circuitry has led to the gradual displacement of analog filter-bank systems, although this trend has been slower for millimeter-wavelength instruments in which very wide signal bandwidths are often used. Hybrid systems in which the signal band is subdivided by analog filters before being digitally sampled and correlated offer some advantages for very wide signal bandwidths. In general, the speed of the sampler is the limiting factor, since in the later stages it is always possible to use serial-to-parallel conversion and increase the number of circuits to accommodate any sampling rate. The systems described in this chapter illustrate the principles involved, but methods of implementation can be expected to change as digital techniques develop.

APPENDIX 8.1   EVALUATION OF $\sum_{q=1}^{\infty} R_{\infty}^2(q\tau_s)$

The periodic function $f(t)$ can be expressed as a Fourier series as follows:

$$f(t) = \frac{a_0}{2} + \sum_{q=1}^{\infty} \left[a_q \cos\left(\frac{2\pi qt}{\beta}\right) + b_q \sin\left(\frac{2\pi qt}{\beta}\right) \right], \qquad (A8.1)$$

where β is the period and

$$\left. \begin{matrix} a_q \\ b_q \end{matrix} \right\} = \frac{2}{\beta} \int_0^{\beta} f(t) \left\{ \begin{matrix} \cos \\ \sin \end{matrix} \right\} \left(\frac{2\pi qt}{\beta}\right) dt. \qquad (A8.2)$$

Parseval's theorem for Eq. (A8.1) takes the form

$$\frac{2}{\beta} \int_0^{\beta} f^2(t)\, dt = \frac{a_0^2}{2} + \sum_{q=1}^{\infty} \left(a_q^2 + b_q^2 \right). \qquad (A8.3)$$

Now let $f(t)$ be a series of rectangular functions of unit height and width, one centered on $t = 0$ and the others centered on integral multiples of $\pm \beta$. Then one obtains

$$\left. \begin{matrix} a_0 = \dfrac{2}{\beta}, \qquad a_q = \dfrac{2}{\beta}\dfrac{\sin(\pi q/\beta)}{\pi q/\beta}, \\[2ex] b_q = 0, \qquad \int_0^{\beta} f^2(t)\, dt = 1. \end{matrix} \right\} \qquad (A8.4)$$

From Eqs. (A8.3) and (A8.4)

$$\sum_{q=1}^{\infty} \left[\frac{\sin(\pi q/\beta)}{\pi q/\beta} \right]^2 = \frac{\beta - 1}{2}, \qquad (A8.5)$$

which, from Eq. (8.14), is the required summation $\sum_{q=1}^{\infty} R_{\infty}^2(q\tau_s)$.

APPENDIX 8.2 EVALUATION OF THE PROBABILITY INTEGRAL FOR TWO-LEVEL QUANTIZATION

The probability integration required in Eq. (8.20) can be performed as follows. The integral is

$$P_{11} = \frac{1}{2\pi\sigma^2\sqrt{1-\rho^2}} \int_0^\infty \int_0^\infty \exp\left[\frac{-(x^2 + y^2 - 2\rho xy)}{2\sigma^2(1-\rho^2)} \right] dx\, dy.$$

$$(A8.6)$$

Restore circular symmetry in the integral by the substitutions

$$z = \frac{y - \rho x}{\sqrt{1-\rho^2}}, \qquad dy = \sqrt{1-\rho^2}\, dz. \qquad (A8.7)$$

Then,

$$P_{11} = \frac{1}{2\pi\sigma^2} \int_0^\infty dx \int_{\frac{-\rho x}{\sqrt{1-\rho^2}}}^\infty \exp\left[\frac{-(x^2 + z^2)}{2\sigma^2} \right] dz. \qquad (A8.8)$$

Next substitute $x = r\cos\theta$ and $z = r\sin\theta$. The lower limit of the z integral in Eq. (A8.8) represents the line $z = -\rho x/\sqrt{1-\rho^2}$ which makes an angle θ with the x axis given by $\theta = -\sin^{-1}\rho$. The integral covers an area of the (x, z) plane between this line and the z axis ($\theta = \pi/2$). Thus,

$$P_{11} = \frac{1}{2\pi\sigma^2} \int_0^\infty dr \int_{-\sin^{-1}\rho}^{\pi/2} r\exp\left(\frac{-r^2}{2\sigma^2} \right) d\theta. \qquad (A8.9)$$

Finally, substitute $u = r^2/2\sigma^2$:

$$P_{11} = \frac{1}{2\pi} \int_0^\infty du \int_{-\sin^{-1}\rho}^{\pi/2} e^{-u}\, d\theta. \qquad (A8.10)$$

Equation (A8.10) can be integrated directly to give

$$P_{11} = \frac{1}{4} + \frac{1}{2\pi}\sin^{-1}\rho. \qquad (A8.11)$$

REFERENCES

Abramowitz, M. and I. A. Stegun, *Handbook of Mathematical Functions*, National Bureau of Standards, Washington, DC, 1964, reprinted by Dover, New York, 1965.

Ball, J. A., The Harvard Minicorrelator, *IEEE Trans. Instrum. Meas.*, **IM-22**, 193, 1973.

Blackman, R. B. and J. W. Tukey, *The Measurement of Power Spectra*, Dover, New York, 1959.

Bowers, F. K. and R. J. Klingler, Quantization Noise of Correlation Spectrometers, *Astron. Astrophys. Suppl.* **15**, 373–380, 1974.

Bowers, F. K., D. A. Whyte, T. L. Landecker, and R. J. Klingler, A Digital Correlation Spectrometer Employing Multiple-Level Quantization, *Proc. IEEE*, **61**, 1339–1343, 1973.

Bracewell, R. N., *The Fourier Transform and its Applications*, McGraw-Hill, New York, 1965 (2nd ed. 1978).

Burns, W. R. and S. S. Yao, Clipping Noise Loss in the One-Bit Autocorrelation Spectral Line Receiver, *Radio Sci.*, **4**, 431–436, 1969.

Chicada, Y., M. Ishiguro, H. Hirabayashi, M. Morimoto, K. Morita, K. Miyazawa, K. Nagane, K. Murata, A. Tojo, S. Inoue, T. Kanzawa, and H. Iwashita, A Digital FFT Spectro-Correlator for Radio Astronomy, in *Indirect Imaging*, J. A. Roberts, Ed., Cambridge University Press, Cambridge, England, 1984, pp 387–404.

Cole, T., Finite sample Correlations of Quantized Gaussians, *Aust. J. Phys.*, **21**, 273–282, 1968.

Cooper, B. F. C., Correlators with Two-Bit Quantization, *Aust. J. Phys.*, **23**, 521–527, 1970.

D'Addario, L. R., A. R. Thompson, F. R. Schwab, and J. Granlund, Complex Cross Correlators with Three-Level Quantization: Design Tolerances, *Radio Sci.*, **19**, 931–945, 1984.

Davis, W. F., Real-Time Compensation for Autocorrelation Clipper Bias, *Astron. Astrophys. Suppl.* **15**, 381–382, 1974.

Goldstein, R. M., A Technique for the Measurement of the Power Spectra of Very Weak Signals, *IRE Trans. Space Electron. Telem.*, **8**, 170–173, 1962.

Hagen, J. B. and D. T. Farley, Digital Correlation Techniques in Radio Science, *Radio Sci.*, **8**, 775–784, 1973.

Harris, F. J., The Use of Windows for Harmonic Analysis with the Discrete Fourier Transform, *Proc. IEEE*, **66**, 51–83, 1978.

Kulkarni, S. R. and C. Heiles, How to Obtain the True Correlation From a Three-Level Digital Correlator, *Astron. J.*, **85**, 1413–1420, 1980.

Lo, W. F., P. E. Dewdney, T. L. Landecker, D. Routledge, and J. F. Vaneldik, A Cross-Correlation Receiver for Radio Astronomy Employing Quadrature Channel Generation by Computed Hilbert Transform, *Radio Sci.*, **19**, 1413–1421, 1984.

Moran, J. M., Very Long Baseline Interferometer Systems, in *Methods of Experimental Physics*, Vol. 12C, M. L. Meeks, Ed., Academic Press, New York, 1976, pp. 174–197.

Napier, P. J., A. R. Thompson, and R. D. Ekers, The Very Large Array; Design and Performance of a Modern Synthesis Radio Telescope, *Proc. IEEE*, **71**, 1295–1320, 1983.

Nyquist, H., Certain Topics in Telegraph Transmission Theory, *Trans. Am. Inst. Electr. Eng.*, **47**, 617–644, 1928.

Oppenheim, A. V. and R. W. Schafer, *Digital Signal Processing*, Prentice-Hall, Englewood Cliffs, New Jersey, 1975.

Price, R., A Useful Theorem for Nonlinear Devices Having Gaussian Inputs, *IRE Trans. Inf. Theory*, **IT-4**, 69–72, 1958.

Shannon, C. E., Communication in the Presence of Noise, *Proc. IRE*, **37**, 10–21, 1949.

Van Vleck, J. H. and D. Middleton, The Spectrum of Clipped Noise, *Proc. IEEE*, **54**, 2–19, 1966.

Weinreb, S., A Digital Spectral Analysis Technique and Its Application to Radio Astronomy, Technical Report No. 412, Res. Lab. for Electronics, Massachusetts Institute of Technology, Cambridge, Mass., 1963.

Willis A. G. and J. D. Bregman, Effects in Fourier Transformed Spectra, *User's Manual for Westerbork Synthesis Radio Telescope*, Netherlands Foundation for Radio Astronomy, Westerbork, Netherlands, Ch. 2, Appendix 2, 1981.

9

VERY-LONG-BASELINE INTERFEROMETRY

In 1967 a new technique of interferometry was developed in which the receiving elements were separated by such a large distance that it was expedient to operate them independently with no real-time communication link. This was accomplished by recording the data on magnetic tape for cross correlation at a later time at a central processing station. The technique was called very-long-baseline interferometry (VLBI), a term recalling the earlier long-baseline interferometers at Jodrell Bank, in which the elements were connected by microwave links that reached 127 km in length. The principles involved in VLBI are fundamentally the same as those involved in interferometers with connected elements. The tape recorder can be considered as an IF delay line of limited capacity with an unusually long propagation time, weeks instead of microseconds. The use of tape recorders is motivated entirely by economics and places substantial limitations upon the system. Satellite links would be superior and have been demonstrated (Yen et al. 1977), but their high cost discourages their use.

9.1 EARLY DEVELOPMENT

The motivation to develop VLBI came from the realization that many radio sources have structures that cannot be resolved by interferometers with baselines of a few hundred kilometers. By the mid-1960s it was well known that scintillation, discussed in Chapter 13, and time variability of the radiation from quasars implied angular sizes of < 0.01 arcsec. Maser emission from OH molecules at 18-cm wavelength was unresolved at 0.1 arcsec. Low frequency burst radiation from Jupiter was believed to emanate from regions of small angular size. The aim of the first VLBI experiments was to measure the angular sizes of these radio sources. It is instructive to consider the operation of these early VLBI experiments in their most primitive form. Consider two telescopes with system temperatures T_{S1} and T_{S2}, which are pointed at a compact source giving antenna temperatures T_{A1} and T_{A2}. Each station records N data samples within the *coherence time*, that is, the interval during which the independent oscillators remain sufficiently stable that fringes can be averaged. In the subsequent processing these data streams are aligned, cross correlated, and time averaged after removing the quasi-sinusoidal fringes. The expected correlation for a point source is

$$\rho_0 \simeq \eta \sqrt{\frac{T_{A1}T_{A2}}{(T_{S1} + T_{A1})(T_{S2} + T_{A2})}} , \tag{9.1}$$

where η is a factor of value $\simeq 0.5$ to account for losses due to quantization and processing (see Section 9.7). Here it is convenient to consider a normalized form of the visibility $\mathcal{V}_N = \rho/\rho_0$:

$$\mathcal{V}_N = \frac{\rho}{\rho_0} = \frac{\rho}{\eta} \sqrt{\frac{T_{S1}T_{S2}}{T_{A1}T_{A2}}} , \tag{9.2}$$

where ρ is the measured correlation and we assume $T_A \ll T_S$. The rms noise level is

$$\Delta\rho \simeq \frac{1}{\sqrt{N}} \simeq \frac{1}{\sqrt{2 \, \Delta\nu \, \tau_c}} , \tag{9.3}$$

where $\Delta\nu$ is the IF bandwidth and τ_c is the coherent integration time. Hence from Eqs. (9.1)–(9.3) the signal-to-noise ratio is

$$\frac{\rho}{\Delta\rho} = \eta \mathcal{V}_N \sqrt{\frac{T_{A1}T_{A2}}{T_{S1}T_{S2}}} (2 \, \Delta\nu \, \tau_c). \tag{9.4}$$

If the minimum useful signal-to-noise ratio is 4, the smallest detectable flux density is, from Eqs. (1.3), (1.5), and (9.4),

$$S_{\min} \simeq \frac{8k}{\mathcal{V}_N \eta} \sqrt{\frac{T_{S1}T_{S2}}{A_1 A_2}} \frac{1}{\sqrt{2 \, \Delta\nu \, \tau_c}} , \tag{9.5}$$

where k is Boltzmann's constant, and A_1 and A_2 are the antenna collecting areas. Typical parameters in 1967 were $A \simeq 250$ m^2 (25-m diameter telescope), $T_S \simeq 100$ K, $\eta \simeq 0.5$, and $N = 1.4 \times 10^8$ bits (one bit per sample), the capacity of a tape at a standard density of 800 bpi used in the NRAO Mark I system. For an unresolved source, $S_{min} \simeq 2$ Jy. The development after two decades is indicated by the following parameter values: $A \simeq 1600$ m^2 (64-m diameter telescope), $T_S \simeq 30$ K, and $N = 10^{11}$ bits, the capacity of an instrumentation tape operated at 50 MHz bandwidth. For $\mathcal{V}_N = 1$, Eq. (9.5) gives $S_{min} \simeq 4$ mJy. In both examples, the coherence time is assumed to be greater than the running time of the tape. The source size can be estimated from a single measurement of $\mathcal{V}_N$ by comparison with the visibility expected for a symmetric Gaussian model. Hence, as in Fig. 1.5, the full width at half maximum, a, is given by

$$a = \frac{2\sqrt{\ln 2}}{\pi u}\sqrt{-\ln \mathcal{V}_N}, \qquad (9.6)$$

where u is the projected baseline (in wavelengths).

VLBI can be used only to study objects of exceedingly high brightness. Thus, the emission processes must normally be of nonthermal origin. To be detected on a baseline of length D, the source must be smaller than the fringe spacing. Since the flux density S is $2kT_B\Omega/\lambda^2$, where T_B is the brightness temperature, λ is the wavelength, and Ω is the source solid angle, the minimum detectable brightness temperature is

$$(T_B)_{min} \simeq \frac{2}{\pi k}D^2 S_{min}, \qquad (9.7)$$

since $\Omega \simeq \pi(\lambda/2D)^2$. If $D = 10^4$ km and $S_{min} = 2$ mJy, then $(T_B)_{min} = 10^8$ K. Therefore, observations of thermal phenomena occurring in molecular clouds and compact HII regions are not possible. On the other hand, synchrotron sources such as supernova remnants, radio galaxies, and quasars, which are limited to 10^{12} K by Compton losses; masers in which $T_B \simeq 10^{15}$ K; and pulsars can be readily studied.

Three things were accomplished by early VLBI measurements:

1. Simple brightness distributions were derived by comparing measured visibilities with source models.
2. Masers were mapped by comparing fringe frequencies for different spectral features.
3. Source positions were measured to an accuracy of $\simeq 1''$, and baselines to a few meters.

For a review of early techniques see Klemperer (1972). Since then the technique has moved steadily toward the mainstream of interferometry in terms of being able to produce reliable images of complex radio sources. The principal reason for this is the use of phase closure (see Chapter 11), which provides most of the phase information when a large enough number of antennas is available in the VLBI network.

9.2 DIFFERENCES BETWEEN VLBI AND CONVENTIONAL INTERFEROMETRY

In this section we discuss briefly the differences between VLBI and connected-element interferometry. Later sections in this chapter contain elaborations on these differences. Before beginning we emphasize the theoretical unity of interferometry. The fundamental aim of all interferometry is to measure the coherence properties of the electromagnetic field. Thus the principles of connected-element interferometry and VLBI are basically identical. However, there are various techniques used in VLBI that are needed because of the particular observational constraints. We can imagine that sometime in the future, when (u, v) coverage is continuous from a few meters to more than 10^5 km, with the largest spacing achieved by elements on distant satellites, and advanced communication systems making recording unnecessary, the concept of VLBI as a distinct technique will be a matter of history. Until that time we must deal with certain limitations that make VLBI practices somewhat distinct from those of connected-element interferometry.

Early VLBI experiments were conducted by organizing a diverse group of observatories that had been constructed for general radio astronomical research. Each telescope had its own limitations, calibration procedures, and management personnel. Various networks were formed (see, e.g., Table 5.1) to standardize procedures and automate the execution of VLBI experiments. Such VLBI networks operate on an intermittent basis, and during observations communication between elements to verify proper operation is limited. Small amounts of data from strong sources are transmitted from the antennas to the correlator over telephone lines and are cross correlated to determine the instrumental delays and to check that the equipment is working properly.

In VLBI one has less control over the system stability because of the use of an independent frequency standard at each element (Section 9.4). There are instrumental timing errors caused by frequency offsets in the standards. These errors usually include an epoch error of a few microseconds and a drift of a few tenths of a microsecond per day (Section 9.5). Therefore, the correlation function of the received signals must be measured in order to determine and track the instrumental delay. In contrast, delay errors in connected-element interferometers, due mainly to baseline errors and atmospheric propagation delays, are usually less than 30 psec, corresponding to 1 cm of path length. These errors are negligible for bandwidths less than 1 GHz. Thus, the response in connected-element interferometers is always centered on the "white light fringe." The only contexts in which delay becomes important are those in which the field of view becomes too large for the bandwidth (see Chapter 6) or when spectral-line measurements are made by introducing time offsets. In VLBI it is necessary to search a range of delay values to find the correct time relationship that maximizes the correlation. Correlations for a number of delay offsets are usually formed simultaneously, so a VLBI correlator closely resembles a digital spectral correlator. The frequency offsets in the standards, which cause drifts with time in the instrumental delay, also introduce offsets in the fringe frequency. Thus, a VLBI experiment must begin with a two-dimensional search in delay and fringe frequency (delay rate) to find the peak of the correlation function.

The concept of coherence has different implications in VLBI and connected-element interferometry. In connected-element interferometry there is generally a suitable calibration source within a few degrees of the source of interest that can be observed every few minutes. Thus, even if the instrumental phase drifts, there is no fundamental limit on integration time and the concept of a coherence time is replaced by that of the interval between calibrations. In VLBI, the short-term phase stability ($t < 10^3$ sec) is worse. The atmospheric fluctuations above the stations are generally completely uncorrelated, and the frequency standards and frequency multipliers introduce phase noise in the fringes. However, the fundamental difference between connected-element inter-ferometry and VLBI comes from the fact that, at current sensitivity levels, there are very few unresolved sources that can be used as calibrators. They are usually so far apart in angle from the source under investigation that they cannot be used as phase references because of the time required to repoint the antennas, and because of the decorrelation introduced by the atmosphere which increases with angle. Thus, VLBI is subject to a fundamental coherence time that limits its sensitivity. For integration beyond the coherence time, it is necessary to average the fringe *amplitudes*, for which sensitivity improves only as the fourth root of the integration time (Section 9.3). It is more difficult to calibrate phase in VLBI systems. Phase information is used primarily in phase closure analysis and in differential measurements. In measuring positions, fringe frequency and group delay (the delay pattern effect discussed in Chapter 2) are useful as measurement quantities.

The storage of the undetected signals before correlation presents VLBI with several problems. The average IF bandwidth is limited by the recording medium, which therefore limits the sensitivity of VLBI. The data must be stored as efficiently as possible, which requires a coarsely quantized representation of the signal, sampled at the Nyquist rate. With such a representation the basic operations of fringe rotation and delay tracking, when performed on the recorded data, introduce significant effects that must be allowed for in deriving the visibility (Section 9.7).

9.3 BASIC PERFORMANCE OF A VLBI SYSTEM

Time and Frequency Errors

A block diagram of a basic VLBI system and a possible processor configuration is shown in Fig. 9.1. The atomic frequency standards control the phases of the local oscillators and the sampling time for the tape recorders. In many VLBI applications, such as spectral-line observations or astrometric programs,

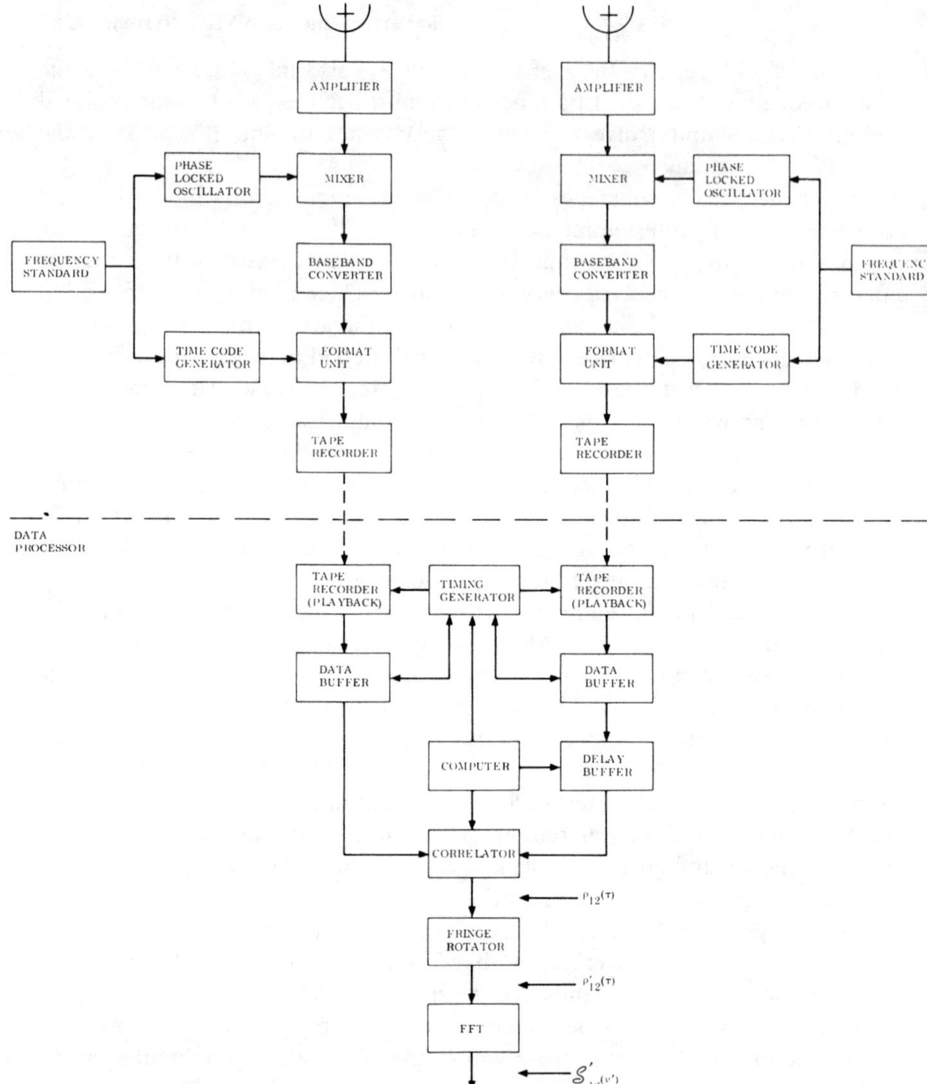

Figure 9.1 Block diagram of the essential elements of a VLBI system including data acquisition and processing. Quantization and sampling of the signals occur in the format units. The processor system shown illustrates the configuration described analytically by Eqs. (9.16)–(9.21). Major variations in the processing system relate to the position of the fringe rotator, which can also be located before the correlator (see Fig. 9.17).

frequency-dependent effects must be accounted for precisely. To obtain a spectral analysis of the system, we consider the phase shifts encountered by a single frequency component. The signals received from a plane wave are $e^{j2\pi\nu t}$ at antenna 1, which we also designate as the time-reference antenna, and $e^{j2\pi\nu(t-\tau_g)}$ at antenna 2. The local oscillators have phases $2\pi\nu_{\mathrm{LO}}t + \theta_1$ and $2\pi\nu_{\mathrm{LO}}t + \theta_2$, where ν_{LO} is the local oscillator frequency and θ_1 and θ_2 are the slowly varying terms that represent the phase noise due to the frequency standards. We assume that the system passes only one sideband and the local oscillator frequency is below the signal frequency. Thus, the phases after mixing are

$$\phi_1^{(1)} = 2\pi(\nu - \nu_{\mathrm{LO}})t - \theta_1,$$

$$\phi_2^{(1)} = 2\pi(\nu - \nu_{\mathrm{LO}})t - 2\pi\nu\tau_g - \theta_2. \qquad (9.8)$$

The recorded signals each have clock errors τ_1 and τ_2 so that the phases of the recorded signals are

$$\phi_1^{(2)} = 2\pi(\nu - \nu_{\mathrm{LO}})(t - \tau_1) - \theta_1,$$

$$\phi_2^{(2)} = 2\pi(\nu - \nu_{\mathrm{LO}})(t - \tau_2) - 2\pi\nu\tau_g - \theta_2. \qquad (9.9)$$

During processing, the time series of signal samples from antenna 2 is advanced by τ_g', the estimate of τ_g, so

$$\phi_2^{(3)} = 2\pi(\nu - \nu_{\mathrm{LO}})(t - \tau_2 + \tau_g') - 2\pi\nu\tau_g - \theta_2. \qquad (9.10)$$

The output of the multidelay correlator and Fourier transform processor is the cross power spectrum. The phase at the output of the processor for the signal component at frequency ν is

$$\phi_{12} = \phi_1^{(2)} - \phi_2^{(3)}$$

$$= 2\pi(\nu - \nu_{\mathrm{LO}})(\tau_2 - \tau_1) + (\theta_2 - \theta_1) + 2\pi(\nu\Delta\tau_g + \nu_{\mathrm{LO}}\tau_g'),$$

$$(9.11)$$

where $\Delta\tau_g = \tau_g - \tau_g'$. Note that if $\tau_1 = \tau_2$, $\theta_1 = \theta_2$, and $\tau_g = \tau_g'$, then

$$\phi_{12} = 2\pi\nu_{\mathrm{LO}}\tau_g, \qquad (9.12)$$

which is independent of the signal frequency because the delay tracking removes the delay-induced phase variation across the band. The correlation function is real, but not even; thus, the cross power spectrum for a source of continuum radiation has the property

$$\mathcal{S}_{12}(\nu') = \mathcal{S}_{12}^{*}(-\nu'),\qquad(9.13)$$

where ν' is the intermediate frequency, $\nu - \nu_{\text{LO}}$. We assume that the filters in the electronics have identical responses and therefore do not introduce any net phase shifts. The filters are characterized by a real power response function $\mathcal{S}(\nu')$. By combining the phase from Eq. (9.11) and the amplitude response, the cross power spectrum can be written

$$\mathcal{S}_{12}(\nu') = \mathcal{S}(\nu')\exp\left\{ j\left[2\pi\nu'(\tau_e + \Delta\tau_g) + 2\pi\nu_{\text{LO}}\tau_g + \theta_{21}\right]\right\},\qquad \nu' > 0$$

$$(9.14)$$

where $\tau_e = \tau_2 - \tau_1$ is the clock error, and $\theta_{21} = \theta_2 - \theta_1$. The cross-correlation function can be calculated from Eqs. (9.13) and (9.14) as

$$\rho_{12}(\tau) = \int_{-\infty}^{\infty} \mathcal{S}_{12}(\nu')e^{j2\pi\nu'\tau}d\nu',\qquad(9.15)$$

which is

$$\rho_{12}(\tau) = 2F_1(\tau')\cos\left(2\pi\nu_{\text{LO}}\tau_g + \theta_{21}\right) - 2F_2(\tau')\sin\left(2\pi\nu_{\text{LO}}\tau_g + \theta_{21}\right),$$

$$(9.16)$$

where $\tau' = \tau + \tau_e + \Delta\tau_g$ and

$$F_1(\tau) = \int_0^{\infty} \mathcal{S}(\nu')\cos(2\pi\nu'\tau)d\nu',$$

$$(9.17)$$

$$F_2(\tau) = \int_0^{\infty} \mathcal{S}(\nu')\sin(2\pi\nu'\tau)d\nu'.$$

If $\mathcal{S}(\nu')$ is a rectangular lowpass spectrum with bandwidth $\Delta\nu$, then

$$F_1(\tau) = \Delta\nu\,\frac{\sin 2\pi\Delta\nu\tau}{2\pi\Delta\nu\tau},$$

$$(9.18)$$

$$F_2(\tau) = \Delta\nu\,\frac{\sin^2\pi\Delta\nu\tau}{\pi\Delta\nu\tau}.$$

These functions are shown in Fig. 9.2. By substituting Eq. (9.18) into Eq.

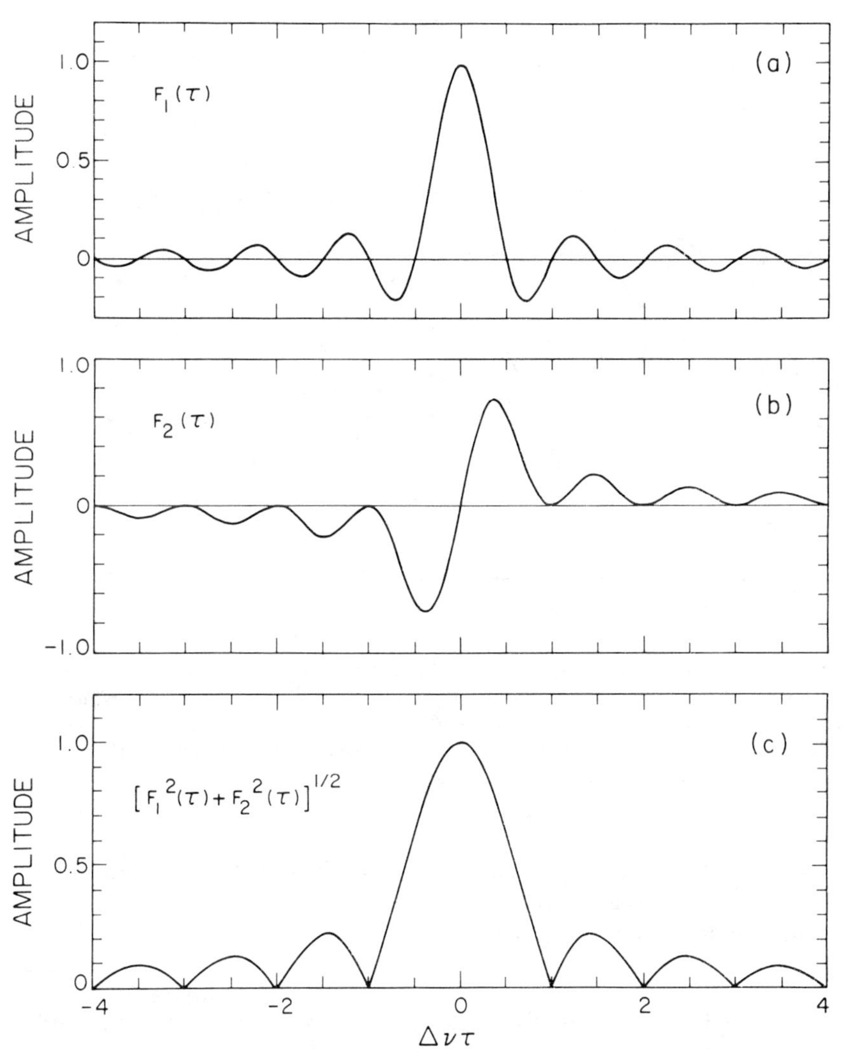

Figure 9.2   Functions $F_1(\tau)$ and $F_2(\tau)$, defined in Eqs. (9.18), and the quantity $\sqrt{F_1^2(\tau) + F_2^2(\tau)}$.

(9.16), the cross-correlation function can be written

$$\rho_{12}(\tau) = 2\Delta\nu \cos(2\pi\nu_{\mathrm{LO}}\tau_g + \theta_{21} + \pi\Delta\nu\tau') \frac{\sin \pi\Delta\nu\tau'}{\pi\Delta\nu\tau'}. \qquad (9.19)$$

The variation of τ_g as the earth rotates results in the usual fringe oscillations. The fringe frequency is reduced to approximately zero by a process that
usually involves applying a phase shift to the quantized signals at the correlator input or output (see Section 9.7). The effect on the cross-correlation

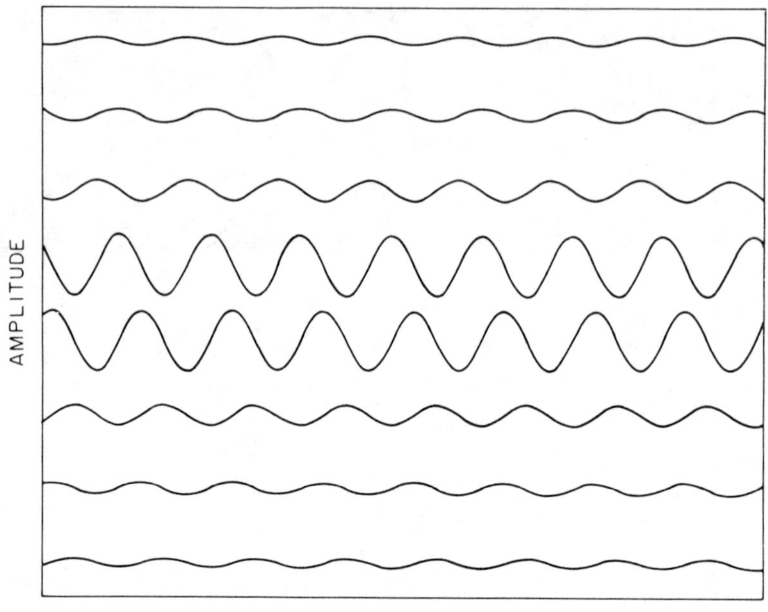

Figure 9.3   Each sinusoid represents the correlation function [the real part of Eq. (9.20)] versus time for a particular delay offset (from the top: $\frac{7}{2}, \frac{5}{2}, \frac{3}{2}, \frac{1}{2}, -\frac{1}{2}, -\frac{3}{2}, -\frac{5}{2}, -\frac{7}{2}$ times the Nyquist interval). The oscillations result from the residual fringe frequency including any offsets in the frequency standards at the two antennas. Note the progressive phase shift of 90° between values of the correlation function at successive delay offsets at the same time.

function or the cross power spectrum can be described as multiplication by $e^{-j2\pi\nu_{LO}\tau'_g}$ and filtering to select the low-frequency term. This process results in a complex correlation function

$$\rho'_{12}(\tau) = \Delta\nu \, \exp\left[j\left(2\pi\nu_{LO}\Delta\tau_g + \theta_{21} + \pi\Delta\nu\tau'\right)\right]\frac{\sin \pi\Delta\nu\tau'}{\pi\Delta\nu\tau'}, \qquad (9.20)$$

and thus a cross power spectrum

$$\mathcal{S}'_{12}(\nu') = \mathcal{S}(\nu')\exp\left\{j\left[2\pi\nu'(\tau_e + \Delta\tau_g) + 2\pi\nu_{LO}\Delta\tau_g + \theta_{21}\right]\right\}, \quad \nu' > 0$$

$$= 0, \qquad \nu' < 0. \qquad (9.21)$$

An example of $\rho'_{12}(\tau)$ for several values of τ is shown in Fig. 9.3. Note that there is a phase shift of $\pi/2$ between adjacent delay steps. The fringe phase can be recovered by a proper interpolation (see Section 9.7) to the peak of the correlation function or from the phase of the cross power spectrum at $\nu' = 0$. The group delay can be derived from the position of the correlation peak or the slope of the phase of the cross power spectrum. Note that the measured delay is $(1/2\pi)\,d\phi/d\nu'$ and is therefore a group delay, not a phase delay.

The actual local oscillator frequencies may differ from the nominal value ν_{LO} due to an intentional offset from the nominal frequency or due to an offset error in the frequency standard. We can expand the phase terms θ_1 and θ_2 to include these frequency offsets, $\Delta\nu_1$ and $\Delta\nu_2$, and zero-mean phase components, θ_1' and θ_2':

$$\theta_1 = 2\pi\Delta\nu_1 t + \theta_1',$$

$$\theta_2 = 2\pi\Delta\nu_2 t + \theta_2'. \tag{9.22}$$

Thus the fringe phase from Eq. (9.21) becomes

$$\phi_{12}(\nu') = 2\pi\left[\nu'(\tau_e + \Delta\tau_g) + \nu_{LO}\Delta\tau_g + \Delta\nu_{LO}t\right] + \theta_{21}', \tag{9.23}$$

where $\Delta\nu_{LO} = \Delta\nu_2 - \Delta\nu_1$, the difference in the local oscillator frequencies, and $\theta_{21}' = \theta_2' - \theta_1'$. The fringe frequency $(1/2\pi)d\phi_{12}/dt$ contains this local oscillator difference term. If $\Delta\nu_1$ is due to a frequency standard offset and is not zero, the measured fringe phase is actually more complicated than shown in Eq. (9.23). The clock error changes with time because of the frequency standard offset and is

$$\tau_1 = (\tau_1)_{t=0} + \frac{\Delta\nu_1}{\nu_{LO}}t. \tag{9.24}$$

The recovered time in the processor, based on the time of station 1, is related to the "true" time t by

$$t_1 = (\tau_1)_{t=0} + \left(1 + \frac{\Delta\nu_1}{\nu_{LO}}\right)t, \tag{9.25}$$

so that there is a slight shift in all measured frequencies and phases. Thus, there is a fundamental asymmetry in the processing between the reference station from which time is derived and the other stations (Whitney et al. 1976).

For spectral line observations the quantity $\mathscr{S}(\nu')$ in Eq. (9.21) is the (temporal frequency) spectrum of the visibility of the source multiplied by the bandpass response of the interferometer. The bandpass response can be obtained by observation of the cross power spectrum of a continuum source with a flat spectrum. Alternatively, if the phase responses of the interferometer elements are identical, the bandpass response can be obtained from the geometric mean of the power spectra from the individual elements. These power spectra are obtained by observing a continuum source or blank sky, and measuring the autocorrelation of the waveform from each element. The frequency spectrum of the normalized visibility can be obtained by dividing the visibility spectrum by the geometric mean of the power spectra of the source as measured with each interferometer element. Details of calibration procedures in VLBI spectral line observations are given by Moran (1973) and Reid et al. (1980).

Retarded Baselines

The estimate of delay τ_g must be accurate enough to ensure that the signal is within the delay and fringe-frequency ranges of the processor. The simplest approximation is

$$\tau_g = \frac{1}{c}\mathbf{D} \cdot \mathbf{s}_0, \tag{9.26}$$

where $\mathbf{D} = \mathbf{r}_1 - \mathbf{r}_2$, $\mathbf{r}_1$ and $\mathbf{r}_2$ are vectors from the center of the earth to each station, and $\mathbf{s}_0$ is the unit vector to the center of the field. Account must be taken of the fact that the earth moves in the time between the arrival of a wave crest at one station and at another. Therefore, in calculating the delay we should not use the instantaneous baseline but the "retarded" baseline (Cohen and Shaffer 1971). A plane wave reaches the first station at time t_1 and the second station at a time t_2, which satisfies the equation

$$\mathbf{k} \cdot \mathbf{r}_1(t_1) - 2\pi\nu t_1 = \mathbf{k} \cdot \mathbf{r}_2(t_2) - 2\pi\nu t_2, \tag{9.27}$$

where $\mathbf{k} = (2\pi/\lambda)\mathbf{s}_0$. Now $t_2 - t_1 = \tau_g$, so

$$2\pi\nu\tau_g = \mathbf{k} \cdot \left[\mathbf{r}_2(t_1 + \tau_g) - \mathbf{r}_1(t_1)\right]. \tag{9.28}$$

Expansion of $\mathbf{r}_2$ in a Taylor series gives

$$\mathbf{r}_2(t_1 + \tau_g) \simeq \mathbf{r}_2(t_1) + \dot{\mathbf{r}}_2(t_1)\tau_g \tag{9.29}$$

and

$$2\pi\nu\tau_g \simeq \mathbf{k} \cdot \left[\mathbf{D}(t_1) + \dot{\mathbf{r}}_2(t_1)\tau_g\right]. \tag{9.30}$$

Solving for τ_g yields

$$\tau_g = \frac{\mathbf{s}_0 \cdot \mathbf{D}}{c}\left[1 - \frac{\mathbf{s}_0 \cdot \dot{\mathbf{r}}_2}{c}\right]^{-1}, \tag{9.31}$$

where all quantities are evaluated at t_1. Since $\dot{\mathbf{r}} = \omega_e \times \mathbf{r}$, where ω_e is the angular velocity vector of the earth and $\times$ indicates the vector cross product, we can rewrite Eq. (9.31) as

$$\tau_g \simeq \frac{\mathbf{s}_0 \cdot \mathbf{D}}{c}\left[1 - \frac{\mathbf{s}_0 \cdot (\omega_e \times \mathbf{r}_2)}{c}\right]^{-1}, \tag{9.32}$$

or

$$\tau_g \simeq \tau_{g0}(1 + \Delta),\tag{9.33}$$

where $1 + \Delta$ is the term in brackets on the right-hand side in Eq. (9.32) and, from the w term in Eq. (4.17),

$$\tau_{g0} = \frac{D}{c}\left[\sin d \sin \delta + \cos d \cos \delta \cos(H - h)\right].\tag{9.34}$$

Here (H, δ) and (h, d) are the hour angle and declination coordinates of the source and baseline, respectively, the hour angles usually being specified with respect to the Greenwich meridian in VLBI practice. Also, we have

$$\Delta = \frac{\omega_e r_2}{c}\cos \mathscr{L}_2 \cos \delta \sin(h_2 - H),\tag{9.35}$$

where $\mathscr{L}_2$, h_2, and r_2 are the latitude, hour angle, and magnitude of $\mathbf{r}_2$. The function Δ has a maximum value of 1.5×10^{-6}, and τ_g can differ from τ_{g0} by a maximum of about 0.05 μsec. It is important to note that the appropriate coordinates in Eq. (9.34) are those that are uncorrected for refraction or diurnal aberration. An equivalent way of accounting for the retarded baseline is to use Eq. (9.26) for the delay but correct h and δ for the diurnal aberration at the remote site.

There are different ways to formulate VLBI observables. One system that may be described as station-oriented is to refer the measurements to the center of the earth so that if tapes from two antennas are processed once, and then interchanged and reprocessed, the phase obtained on the second pass will be the negative of that obtained on the first pass. This method presupposes an earth model, since the radius vectors must be known. For applications to astrometry or geodesy, a baseline-oriented system is usually preferred in which the observables have no dependence on a priori values of earth parameters. A more precise discussion of VLBI observables can be found in Shapiro (1976) and Cannon (1978).

Noise in VLBI Observations

We begin the discussion of noise by reviewing the statistical properties of the fringe amplitude and phase, which were introduced in Chapter 6 (see also Moran 1976). The measured visibility is represented by a vector $\mathbf{Z} = \mathscr{V} + \boldsymbol{\varepsilon}$, where $\mathscr{V}$ and ε represent the true visibility (the signal) and noise components, respectively. We then select coordinates with x (real) and y (imaginary) so that $\mathscr{V}$ lies along the x axis, as shown in Fig. 6.7. The phase of the measured visibility resulting from the noise is a random variable denoted by ϕ. The components of ε have independent zero-mean Gaussian probability distributions in the x and y coordinates with an rms deviation σ given by Eq. (6.43).

In polar coordinates the magnitude of ε has a Rayleigh probability distribution, and the phase of ε has a uniform probability distribution (Papoulis 1965). $\mathbf{Z}$ is therefore a random variable whose x and y components, Z_x and Z_y, have a probability distribution given by

$$p(Z_x, Z_y) = \frac{1}{2\pi\sigma^2}\exp\left[-\frac{(Z_x - |\mathscr{V}|)^2 + Z_y^2}{2\sigma^2}\right]. \tag{9.36}$$

It is often necessary to deal with the magnitude and phase of the visibility, denoted by Z and ϕ, respectively, whose probability distributions are [Eqs. (6.56a) and (6.56b)]:

$$p(Z) = \frac{Z}{\sigma^2}\exp\left(-\frac{Z^2 + |\mathscr{V}|^2}{2\sigma^2}\right)I_0\left(\frac{Z|\mathscr{V}|}{\sigma^2}\right), \qquad Z > 0 \tag{9.37}$$

where $Z = \sqrt{Z_x^2 + Z_y^2}$, and

$$p(\phi) = \frac{1}{2\pi}\exp\left(-\frac{|\mathscr{V}|^2}{2\sigma^2}\right)\left\{1 + \sqrt{\frac{\pi}{2}}\frac{|\mathscr{V}|\cos\phi}{\sigma}\exp\left(\frac{|\mathscr{V}|^2\cos^2\phi}{2\sigma^2}\right)\right.$$

$$\left.\times\left[1 + \operatorname{erf}\left(\frac{|\mathscr{V}|\cos\phi}{\sqrt{2}\,\sigma}\right)\right]\right\},$$

$$\tag{9.38}$$

where I_0 is the modified Bessel function of order zero, and erf is the error function. $p(Z)$ is known as the Rice distribution. Note that $\langle\phi\rangle = 0$, as expected, since the phase of $\mathscr{V}$ was set to zero. These probability distributions are plotted in Fig. 6.8. The expectations of Z, Z^2, and Z^4 are

$$\langle Z\rangle = \sqrt{\frac{\pi}{2}}\,\sigma\exp\left(\frac{-|\mathscr{V}|^2}{4\sigma^2}\right)\left[\left(1 + \frac{|\mathscr{V}|^2}{2\sigma^2}\right)I_0\left(\frac{|\mathscr{V}|^2}{4\sigma^2}\right) + \frac{|\mathscr{V}|^2}{2\sigma^2}I_1\left(\frac{|\mathscr{V}|^2}{4\sigma^2}\right)\right],$$

$$\tag{9.39}$$

$$\langle Z^2\rangle = |\mathscr{V}|^2 + 2\sigma^2, \tag{9.40}$$

and

$$\langle Z^4\rangle = |\mathscr{V}|^4 + 8\sigma^2|\mathscr{V}|^2 + 8\sigma^4, \tag{9.41}$$

where I_1 is the modified Bessel function of order one. Higher even-order moments of Z can be readily calculated using the moment theorem for a Gaussian random distribution. When no signal is present, $I_0(0) = 1$, and the probability distributions of Z and ϕ are those of the noise, which are Rayleigh and uniform distributions, respectively:

$$p(Z) = \frac{Z}{\sigma^2}\exp\left(-\frac{Z^2}{2\sigma^2}\right), \qquad Z > 0. \tag{9.42}$$

and

$$p(\phi) = \frac{1}{2\pi}, \qquad 0 \le \phi < 2\pi. \tag{9.43}$$

For the no-signal case $\langle Z \rangle = \sqrt{\pi/2}\,\sigma$, $\sigma_Z = \sqrt{\langle Z^2 \rangle - \langle Z \rangle^2} = \sigma\sqrt{2 - \pi/2}$, and $\sigma_\phi = \pi/\sqrt{3}$.

For the weak-signal case, defined as $|\mathscr{V}| \ll \sigma$, the probability distributions of Z and ϕ are

$$p(Z) \simeq \frac{Z}{\sigma^2}\exp\left(-\frac{Z^2}{2\sigma^2}\right)\left[1 - \frac{1}{2}\frac{|\mathscr{V}|^2}{\sigma^2} + \frac{1}{4}\left(\frac{Z|\mathscr{V}|}{\sigma^2}\right)^2\right] \tag{9.44}$$

and

$$p(\phi) \simeq \frac{1}{2\pi} + \frac{1}{\sqrt{8\pi}}\frac{|\mathscr{V}|}{\sigma}\cos\phi, \tag{9.45}$$

to first order in $|\mathscr{V}|/\sigma$. Thus,

$$\langle Z \rangle \simeq \sigma\sqrt{\frac{\pi}{2}}\left(1 + \frac{|\mathscr{V}|^2}{4\sigma^2}\right), \tag{9.46}$$

$$\sigma_Z \simeq \sigma\sqrt{2 - \frac{\pi}{2}}\left(1 + \frac{|\mathscr{V}|^2}{4\sigma^2}\right), \tag{9.47}$$

and

$$\sigma_\phi \simeq \frac{\pi}{\sqrt{3}}\left(1 - \sqrt{\frac{9}{2\pi^3}}\frac{|\mathscr{V}|}{\sigma}\right). \tag{9.48}$$

For the strong-signal case, $|\mathscr{V}| \gg \sigma$, the probability functions for Z and ϕ are approximately Gaussian distributions and are given by

$$p(Z) \simeq \frac{1}{\sqrt{2\pi}\,\sigma}\sqrt{\frac{Z}{|\mathscr{V}|}}\exp\left[-\frac{(Z - |\mathscr{V}|)^2}{2\sigma^2}\right] \tag{9.49}$$

and

$$p(\phi) \simeq \frac{1}{\sqrt{2\pi}}\frac{|\mathscr{V}|}{\sigma}\exp\left(-\frac{|\mathscr{V}|^2\phi^2}{2\sigma^2}\right). \tag{9.50}$$

For this case,

$$\langle Z \rangle \simeq |\mathscr{V}|\left(1 + \frac{\sigma^2}{2|\mathscr{V}|^2}\right), \tag{9.51}$$

$$\sigma_Z \simeq \sigma\left(1 - \frac{\sigma^2}{8|\mathscr{V}|^2}\right), \tag{9.52}$$

and

$$\sigma_\phi \simeq \frac{\sigma}{|\mathscr{V}|}. \tag{9.53}$$

Hence, in the strong-signal case, the statistics of Z are approximately Gaussian (see Fig. 6.8) and $\langle Z \rangle$ approaches $|\mathscr{V}|$. In this case, N samples of Z can be averaged and the signal-to-noise ratio improves with $\sqrt{N}$. In the weak-signal case the perturbation of the Rayleigh noise distribution by the signal is small and, as we shall discuss later in this section, it is difficult to improve the signal-to-noise ratio by averaging beyond the coherence time of the system.

Equations (9.46) and (9.51) show that $\langle Z \rangle$ is a biased estimate of $|\mathscr{V}|$. If only one measurement of Z is available, the most likely value of $|\mathscr{V}|$ is the one for which $p(Z)$, given by Eq. (9.37), is a maximum. This maximum is closely approximated by the equation $Z_{\mathrm{max}} = \sqrt{|\mathscr{V}|^2 + \sigma^2}$, which is accurate to better than 8 percent for all values of $|\mathscr{V}|$ and to better than 1 percent for $|\mathscr{V}|/\sigma > 2$. Hence when one measurement of Z is available, the most likely value of $|\mathscr{V}|$ is approximately $\sqrt{Z^2 - \sigma^2}$.

Probability of Error in the Signal Search

The first task in the processing of any VLBI experiment is to search for fringes, which is necessary because of the uncertainties in the station clocks and their drift rates. This means that the instrumental delay and fringe frequency must be found. The search must be carried out on a large two-dimensional grid, as shown in Fig. 9.4. For example, consider an experiment where $\Delta\nu = 50$ MHz at an observing frequency of 10^{11} Hz. The delay increments are equal to the sampling interval of 0.01 μsec. An instrumental delay uncertainty of ± 1 μsec requires a search of 200 delay intervals. If the coherent integration time is 200 sec and the frequency standards are only set to a fractional accuracy of 10^{-11}, then ± 1 Hz must be searched, which, at an interval size of 0.005 Hz, is 400 discrete frequencies. The total number of cells to be searched is 80,000. If there is no signal present, then $p(Z)$ will be given by Eq. (9.42). The cumulative probability distribution, that is, the probability that Z is less than Z_0, is in this

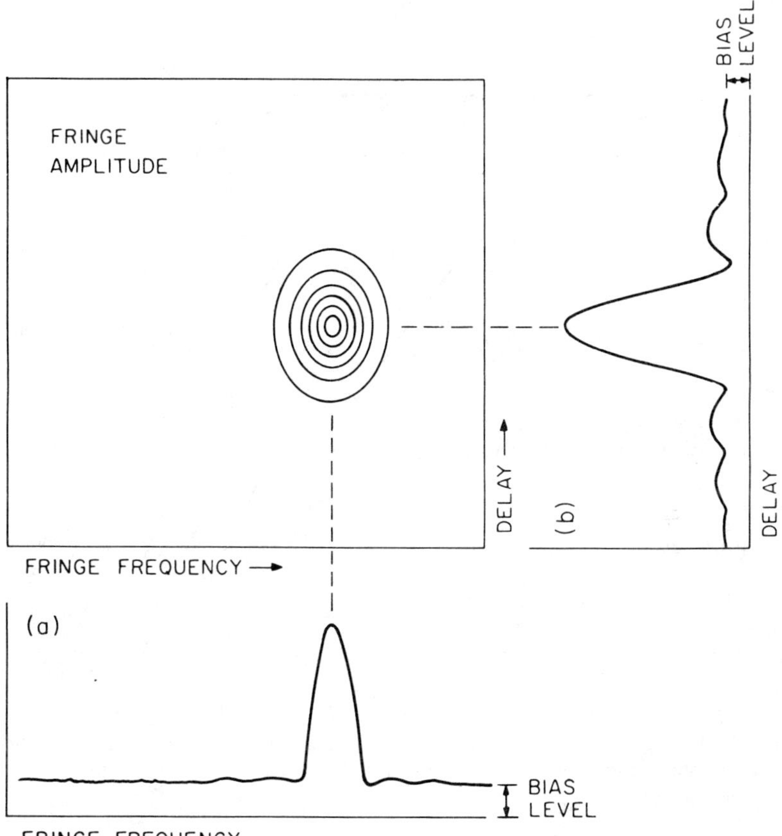

Figure 9.4 Fringe amplitude as a function of residual fringe frequency and delay. The one-dimensional plots are the peak fringe amplitude versus delay and fringe frequency. The probability distribution of the noise in these plots is given by Eq. (9.57) and the bias level by Eq. (9.58).

case the integral of Eq. (9.42) from zero to Z_0, or

$$P(Z_0) = 1 - \exp\left(-\frac{Z_0^2}{2\sigma^2}\right). \tag{9.54}$$

The cumulative probability distribution for the maximum of n independent samples $Z_m = \max\{Z_1, Z_2, \ldots, Z_n\}$ is

$$P(Z_m) = \left[1 - \exp\left(-\frac{Z_m^2}{2\sigma^2}\right)\right]^n. \tag{9.55}$$

Thus, the probability of one or more samples exceeding Z_m, which we call the

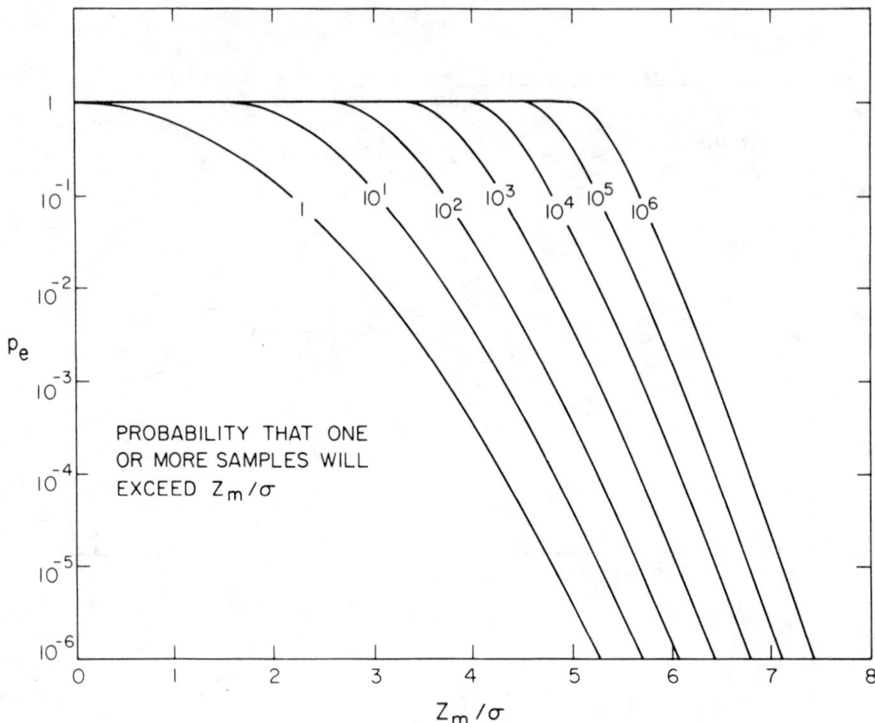

Figure 9.5 Probability that one or more samples of the fringe amplitude will exceed the value Z_m/σ in the absence of a signal, as given by Eq. (9.56). The curves are labeled by the number of samples measured.

probability of error, p_e, is

$$p_e = 1 - \left[1 - \exp\left(-\frac{Z_m^2}{2\sigma^2}\right)\right]^n. \tag{9.56}$$

This function is shown in Fig. 9.5. The probability distribution of Z_m is obtained by differentiating Eq. (9.55),

$$p(Z_m) = \frac{nZ_m}{\sigma^2}\exp\left(-\frac{Z_m^2}{2\sigma^2}\right)\left[1 - \exp\left(-\frac{Z_m^2}{2\sigma^2}\right)\right]^{n-1}. \tag{9.57}$$

For large n, this probability distribution is nearly Gaussian with mean value and standard deviation given by

$$\langle Z_m \rangle \simeq \sigma\sqrt{2\ln n}, \tag{9.58}$$

$$\sigma_m \simeq \frac{0.77\sigma}{\sqrt{\ln n}}. \tag{9.59}$$

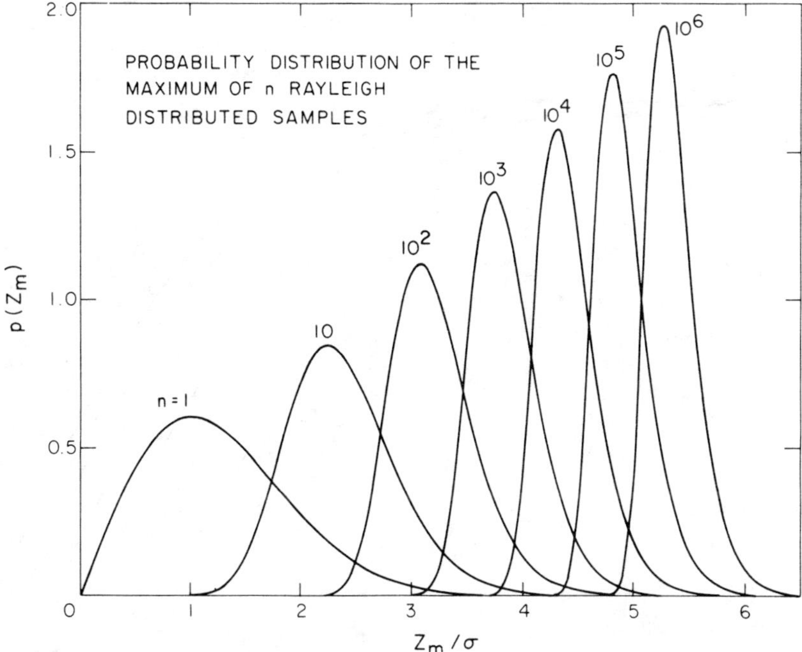

Figure 9.6   Probability distribution of the maximum of n random variables that have Rayleigh distributions, as given by Eq. (9.57).

Examples of $p(Z_m)$ for various values of n are shown in Fig. 9.6. It is frequently useful to reduce a two-dimensional function, such as the one shown in Fig. 9.4 of fringe amplitude versus fringe frequency and delay, to a one-dimensional one by searching for the maximum value of the function over one variable. This search process introduces a bias or mean level $\langle Z_m \rangle$ into the one-dimensional function, which increases with the number of samples and obscures weak signals.

We can also calculate the probability of misidentifying a signal. Suppose we have measurements of fringe amplitude at two values of delay or fringe frequency with the signal present at one value. The probability that the amplitude in the channel with the signal (Z_1) is larger than the amplitude in the channel with only the noise (Z_2) is

$$p(Z_1 > Z_2) = \int_0^\infty p(Z_1)\left[\int_0^{Z_1} p(Z_2)\,dZ_2\right]dZ_1. \qquad (9.60)$$

$p(Z_1)$ is given by Eq. (9.37), and $p(Z_2)$ is given by Eq. (9.42). We can generalize this result for a search over n channels where the signal channel amplitude is Z_s. The probability that Z_s will exceed the values of Z in the

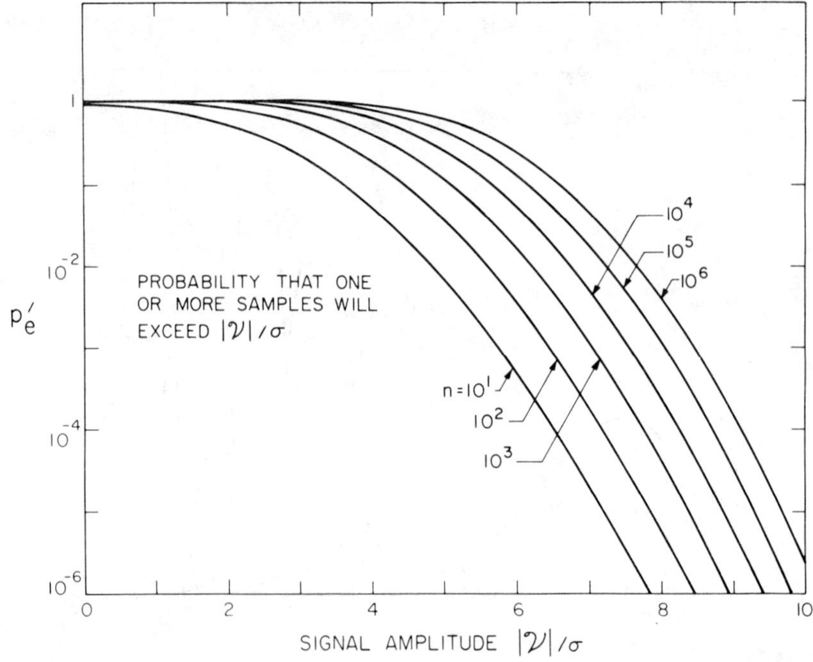

Figure 9.7   Probability that one or more samples of fringe amplitude among the samples with no signal will exceed the fringe amplitude of the sample with the signal, versus the signal amplitude, $|\mathcal{V}|$, as given in Eq. (9.62). The curves are labeled according to the total number of samples n. The asymptotic value of p'_e as $|\mathcal{V}|/\sigma$ goes to zero is $1 - 1/n$.

other channels is, from Eqs. (9.54) and (9.60),

$$p(Z_s > Z_1, \ldots, Z_n) = \int_0^\infty p(Z)\left[1 - \exp\left(-\frac{Z^2}{2\sigma^2}\right)\right]^{n-1} dZ, \quad (9.61)$$

where $p(Z)$ is given by Eq. (9.37). Thus, the probability of one or more samples exceeding the amplitude of the signal is

$$p'_e = 1 - \int_0^\infty p(Z)\left[1 - \exp\left(-\frac{Z^2}{2\sigma^2}\right)\right]^{n-1} dZ. \quad (9.62)$$

p'_e is plotted in Fig. 9.7. For example, if the search is over 100 channels, a probability of misidentification of less than 0.1% requires $|\mathcal{V}|/\sigma > 6.5$.

Coherent and Incoherent Averaging

We wish to estimate the amplitude of a barely detectable signal. We examine a time series of the correlator output values in which $\phi(t)$ is the instrumental phase due to receiver noise, or to fluctuations in the frequency standards or in

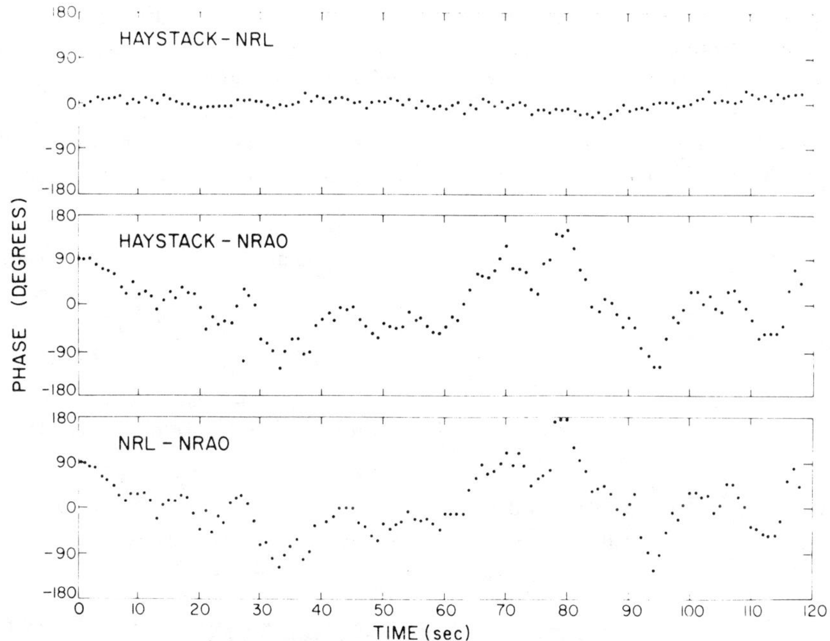

Figure 9.8   Fringe phase versus time from an observation of a strong source [the water vapor maser in W3 (OH)] on a three-baseline VLBI experiment at 22 GHz. Two of the stations, Haystack Observatory and the Naval Research Laboratory (Maryland Point Observatory), were equipped with hydrogen-maser frequency standards, while the National Radio Astronomy Observatory used a rubidium vapor frequency standard. The phase noise on the top plot is dominated by contributions from the receivers and the atmosphere, while the phase noise on the bottom two plots is dominated by the phase noise in the rubidium frequency standard. These data were obtained in 1971 with the Mark I VLBI system.

the atmosphere. An example of the phase versus time from a VLBI measurement is shown in Fig. 9.8. The correlator output is

$$r(t) = Z(t)e^{j\phi(t)}. \tag{9.63}$$

How do we estimate $|\mathscr{V}|$ when the time range of the data exceeds the coherence time? There are two useful procedures. First, note that $r(t)$ has a Fourier transform $R(\nu)$ for a time segment of length τ. Using Parseval's theorem,

$$\int_0^\tau r(t)r^*(t)\,dt = \int_{-\infty}^\infty R(\nu)R^*(\nu)\,d\nu, \tag{9.64}$$

and Eq. (9.40), we obtain an estimate $|\mathscr{V}|_e$ of the amplitude:

$$|\mathscr{V}|_e^2 = \frac{1}{\tau}\int_{-\infty}^\infty |R(\nu)|^2\,d\nu - 2\sigma^2. \tag{9.65}$$

Equation (9.65) shows that one can take a time series of duration greater than τ_c and estimate the visibility amplitude from the fringe-frequency spectrum, $|R(\nu)|^2$. This procedure is a form of incoherent averaging because squares of spectral components in the fringe-frequency domain are summed.

Another method of incoherent averaging involves the averaging of the squares of the time-series samples. An unbiased estimate of the amplitude is

$$|\mathscr{V}|_e^2 = \left(\frac{1}{N}\sum_{i=1}^{N} Z_i^2\right) - 2\sigma^2,$$

(9.66)

where

$$Z_i e^{j\phi_i} = \frac{1}{\tau_c}\int_{t_i}^{t_i+\tau_c} Z(t) e^{j\phi(t)}\, dt.$$

(9.67)

From Eqs. (9.40), (9.41), and (9.66) we have $\langle|\mathscr{V}|_e^2\rangle = |\mathscr{V}|^2$ and $\langle|\mathscr{V}|_e^4\rangle = |\mathscr{V}|^4 + 4\sigma^2(|\mathscr{V}|^2 + \sigma^2)/N$, so that the signal-to-noise ratio is

$$\mathscr{R}_{sn} = \frac{\langle|\mathscr{V}|_e^2\rangle}{\sqrt{\langle|\mathscr{V}|_e^4\rangle - \langle|\mathscr{V}|_e^2\rangle^2}} = \frac{\sqrt{N}}{2\sigma^2}|\mathscr{V}|^2\frac{1}{\sqrt{1 + |\mathscr{V}|^2/\sigma^2}}.$$

(9.68)

$|\mathscr{V}|/\sigma$ is equal to the signal-to-noise ratio at the output of a single-multiplier correlator as given by Eq. (6.42). For VLBI observations, η_Q, the quantization loss described in Chapter 8, is replaced by the general loss factor η described in Section 9.7, and from Eq. (6.57) we obtain $|\mathscr{V}|/\sigma = (T_A\eta/T_S)\sqrt{2\Delta\nu\tau_c}$. Equation (9.68) then becomes

$$\mathscr{R}_{sn} = \frac{T_A^2\eta^2}{T_S^2}\sqrt{\frac{\Delta\nu^2\tau\tau_c}{\left(1 + 2T_A^2\eta^2\Delta\nu\tau_c/T_S^2\right)}},$$

(9.69)

where $\tau = N\tau_c$ is the total integrating time. The two limiting cases of Eq. (9.69) are

$$\mathscr{R}_{sn} \simeq \frac{\eta}{\sqrt{2}}\frac{T_A}{T_S}\sqrt{\Delta\nu\tau}, \qquad T_A \gg \frac{T_S}{\sqrt{2\Delta\nu\tau_c}},$$

(9.70)

$$\mathscr{R}_{sn} \simeq \left(\frac{T_A\eta}{T_S}\right)^2\Delta\nu\sqrt{\tau\tau_c}, \qquad T_A \ll \frac{T_S}{\sqrt{2\Delta\nu\tau_c}}.$$

(9.71)

Note that in the strong-signal case, the incoherent averaging is not needed. When incoherent averaging is used, the coherent averaging time should be as long as possible without decreasing the fringe amplitude. If we assume that

$\mathcal{R}_{sn} = 4$ for detection, and recall that $\tau = N\tau_c$, then for the weak-signal case the minimum detectable antenna temperature can be found from Eq. (9.71) to be

$$(T_A)_{min} = \frac{2T_S}{\eta N^{1/4}\sqrt{\Delta\nu\,\tau_c}}.$$ (9.72)

Thus, because of the $N^{1/4}$ dependence in Eq. (9.72), incoherent averaging is effective only if N is very large.

If the coherence time is of the order of $1/\Delta\nu$, then the observing system reduces to a form of incoherent or intensity interferometer (see Chapter 15 and Clark 1968). For the weak-signal case, Eq. (9.71) then becomes

$$\mathcal{R}_{sn} \simeq \left(\frac{T_A\eta}{T_S}\right)^2 \sqrt{\Delta\nu\,\tau}.$$ (9.73)

9.4   PHASE STABILITY AND ATOMIC FREQUENCY STANDARDS

Precision oscillators have been steadily improved since the 1920s when the invention of the quartz oscillator had immediate application to the problem of precise timekeeping. In the early 1950s cesium-beam clocks allowed better timekeeping than could be obtained from astronomical observations. This development led to an atomic definition of time that differs from the astronomical one, and to the establishment of the definition of the second of time based on a particular transition frequency of cesium.

The mathematical theory of the interpretation of measurements of oscillator phase was systematized by an IEEE committee (Barnes et al. 1971). This paper helped to standardize the approach to handling low-frequency divergence in the noise of oscillators. The physical theory of noise in oscillators was treated by Edson (1960). In this section we develop relevant aspects of the theory and describe the operation of atomic frequency standards with particular emphasis on the hydrogen maser. These topics are discussed in more detail by Blair (1974) and Rutman (1978).

Analysis of Phase Fluctuations

The desired signal from an oscillator is a pure sine wave,

$$V(t) = V_0\cos 2\pi\nu_0 t.$$ (9.74)

This is unobtainable since all devices have some phase noise. A more realistic model is given by

$$V(t) = V_0\cos[2\pi\nu_0 t + \phi(t)],$$ (9.75)

where $\phi(t)$ is a random process characterizing the phase departure from a pure sine wave. We ignore amplitude fluctuations since they do not directly affect performance in VLBI applications. The instantaneous frequency $\nu(t)$ is the derivative of the argument of Eq. (9.75) divided by 2π, that is,

$$\nu(t) = \nu_0 + \delta\nu(t), \tag{9.76}$$

where

$$\delta\nu(t) = \frac{1}{2\pi}\frac{d\phi(t)}{dt}. \tag{9.77}$$

The instantaneous fractional frequency deviation is defined as

$$y(t) = \frac{\delta\nu(t)}{\nu_0} = \frac{1}{2\pi\nu_0}\frac{d\phi}{dt}. \tag{9.78}$$

This definition allows the performance of oscillators at different frequencies to be compared.

We assume that the random processes $\phi(t)$ and $y(t)$ are statistically stationary, so that correlation functions can be defined. This assumption is not always valid and can cause difficulty (Rutman 1978). The autocorrelation function of $y(t)$ is

$$R_y(\tau) = \langle y(t)y(t+\tau)\rangle. \tag{9.79}$$

$R_y(\tau)$ is a real and even function so that its power spectrum $\mathscr{S}_y'(f)$, is a real and even function of frequency f. In order to prevent confusion between $\nu(t)$ and its frequency components, we use the symbol f for the frequency variable in the following spectral analysis. Following the somewhat nonstandard convention that is used in most of the literature on phase stability (Barnes et al. 1971), we replace the double-sided spectrum $\mathscr{S}_y'(f)$ with a single-sided spectrum $\mathscr{S}_y(f)$ where $\mathscr{S}_y(f) = 2\mathscr{S}_y'(f)$ for $f \geq 0$, and $\mathscr{S}_y(f) = 0$ for $f < 0$. Thus we can write the Fourier transform relation $R_y(\tau) \rightleftharpoons \mathscr{S}_y'(f)$ as

$$\mathscr{S}_y(f) = 4\int_0^\infty R_y(\tau)\cos(2\pi f\tau)\,d\tau,$$

$$\tag{9.80}$$

$$R_y(\tau) = \int_0^\infty \mathscr{S}_y(f)\cos(2\pi f\tau)\,df.$$

Similarly, the autocorrelation function of the phase is

$$R_\phi(\tau) = \langle \phi(t)\phi(t+\tau)\rangle. \tag{9.81}$$

$\mathscr{S}_\phi(f)$, the power spectrum of ϕ, and $R_\phi(\tau)$ are related by a Fourier transform. From the derivative property of Fourier transforms, the relationship between $\mathscr{S}_y(f)$ and $\mathscr{S}_\phi(f)$ can be shown to be

$$\mathscr{S}_y(f) = \frac{f^2}{\nu_0^2}\mathscr{S}_\phi(f). \tag{9.82}$$

$\mathscr{S}_y(f)$ and $\mathscr{S}_\phi(f)$ serve as primary measures of frequency stability.

A second measure of frequency stability is based on time domain measurements. The average fractional frequency deviation is

$$\bar{y}_k = \frac{1}{\tau}\int_{t_k}^{t_k+\tau} y(t)\,dt, \tag{9.83}$$

which, from Eq. (9.78), becomes

$$\bar{y}_k = \frac{\phi(t_k + \tau) - \phi(t_k)}{2\pi\nu_0\tau}, \tag{9.84}$$

where the measurements of $\bar{y}_k$ are made with a repetition interval T $(T \geq \tau)$ such that $t_{k+1} = t_k + T$ (see Fig. 9.9a). Measurements of $\bar{y}_k$ are directly

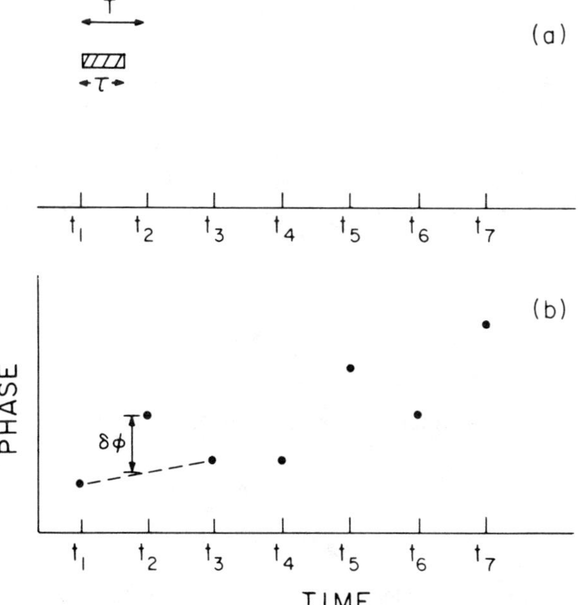

Figure 9.9   (a) Time intervals involved in the measurement of $\bar{y}_k$ as defined in Eq. (9.84). (b) Plot of a series of phase samples versus time. The Allan variance, defined in Eq. (9.86), is the average of the square of the deviation, $(\delta\phi)^2$, of each sample from the mean of its two adjacent samples.

obtainable with conventional frequency counters. The measure of frequency stability is the sample variance of $\bar{y}_k$ given by

$$\langle \sigma_y^2(N, T, \tau) \rangle = \frac{1}{N-1} \left\langle \sum_{n=1}^{N} \left(\bar{y}_n - \frac{1}{N} \sum_{k=1}^{N} \bar{y}_k \right)^2 \right\rangle. \qquad (9.85)$$

In the limit as $N \to \infty$ the above quantity is the true variance, $I^2(\tau)$. However, in many cases Eq. (9.85) does not converge because of the low-frequency behavior of $\mathscr{S}_y(f)$, and $I^2(\tau)$ is then not defined. To avoid some of the convergence problems, a particular case of Eq. (9.85), the two-sample or Allan variance, $\sigma_y^2(\tau)$, has gained wide acceptance (Allan 1966). It is defined by setting $T = \tau$ (no dead time between measurements) and $N = 2$. Therefore, the Allan variance is

$$\sigma_y^2(\tau) = \frac{\langle (\bar{y}_{k+1} - \bar{y}_k)^2 \rangle}{2}, \qquad (9.86)$$

or, from Eq. (9.84),

$$\sigma_y^2(\tau) = \frac{\langle [\phi(t + 2\tau) - 2\phi(t + \tau) + \phi(t)]^2 \rangle}{8\pi^2 v_0^2 \tau^2}. \qquad (9.87)$$

The procedure for estimating the Allan variance can be understood as follows. Take a series of phase measurements at interval T as shown in Fig. 9.9b. For each set of three independent points, draw a straight line between the outer two and determine the deviation of the center point from the line. With m samples of $\bar{y}$, the average of the squared deviations divided by $(2\pi v_0 \tau)^2$ is an estimate of $\sigma_y^2(\tau)$, denoted $\sigma_{ye}^2(\tau)$, where

$$\sigma_{ye}^2(\tau) = \frac{1}{2(m-1)} \sum_{k=1}^{m-1} (\bar{y}_{k+1} - \bar{y}_k)^2. \qquad (9.88)$$

The accuracy of this estimate is (Lesage and Audoin 1979)

$$\sigma(\sigma_{ye}) \simeq \frac{K}{\sqrt{m}} \sigma_y, \qquad (9.89)$$

where K is a constant of order unity, whose exact value depends on the power spectrum of y.

We can now relate the true variance and the Allan variance to the power spectrum of y or ϕ. From Eq. (9.84) the true variance is $I^2(\tau) = \langle \bar{y}_k^2 \rangle$, given by

$$I^2(\tau) = \frac{1}{(2\pi\nu_0\tau)^2}\left[\langle\phi^2(t+\tau)\rangle - 2\langle\phi(t+\tau)\phi(t)\rangle + \langle\phi^2(t)\rangle\right],$$

$$(9.90)$$

which, from Eq. (9.81), is

$$I^2(\tau) = \frac{1}{2(\pi\nu_0\tau)^2}\left[R_\phi(0) - R_\phi(\tau)\right]. \tag{9.91}$$

Then, since $R_\phi(\tau)$ is the Fourier transform of $\mathcal{S}_\phi(f)$, by using Eq. (9.82), we obtain from Eq. (9.91) the result

$$I^2(\tau) = \int_0^\infty \mathcal{S}_y(f)\left(\frac{\sin \pi f\tau}{\pi f\tau}\right)^2 df. \tag{9.92}$$

Similarly, from Eq. (9.87),

$$\sigma_y^2(\tau) = \frac{1}{(2\pi\nu_0\tau)^2}\left[3R_\phi(0) - 4R_\phi(\tau) + R_\phi(2\tau)\right], \tag{9.93}$$

and, therefore,

$$\sigma_y^2(\tau) = 2\int_0^\infty \mathcal{S}_y(f)\left[\frac{\sin^4\pi f\tau}{(\pi f\tau)^2}\right] df. \tag{9.94}$$

We can think of $I^2(\tau)$ and $\sigma_y^2(\tau)$ as the power obtained after filtering $y(t)$ with two different frequency responses $H_I^2(f)$ and $H_A^2(f)$, respectively, given by

$$H_I^2(f) = \left(\frac{\sin \pi f\tau}{\pi f\tau}\right)^2 \tag{9.95}$$

and

$$H_A^2(f) = \frac{2\sin^4\pi f\tau}{(\pi f\tau)^2}. \tag{9.96}$$

The functions $H_I^2(f)$ and $H_A^2(f)$ and corresponding impulse responses $h_I(t)$ and $h_A(t)$ are shown in Fig. 9.10. Note that $I^2(\tau)$ can be estimated from a series of measurements $\bar{y}_k$ as the average of the square of $h_I(t_k) * \bar{y}_k$.

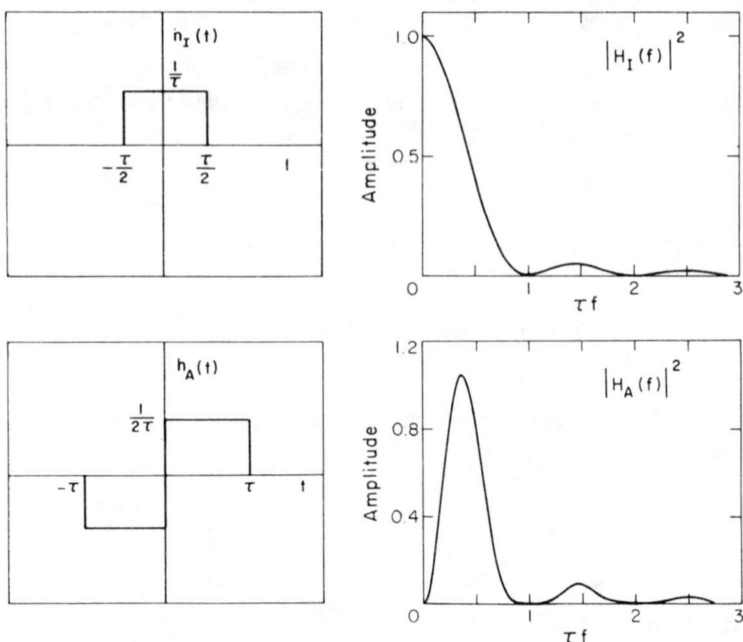

Figure 9.10   (*Top*)   The impulse function $h_I(t)$ and the square of its Fourier transform $|H_I(f)|^2$, given by Eq. (9.95), which is used to relate the power spectrum $\mathscr{S}_y(f)$ to the true variance $I^2(\tau)$, as defined in Eq. (9.92). (*Bottom*)   The impulse response $h_A(t)$ and the square of its Fourier transform $|H_A(f)|^2$, given by Eq. (9.96), which is used to relate the power spectrum $\mathscr{S}_y(f)$ to the Allan variance $\sigma_y^2(\tau)$, as defined in Eq. (9.94).

Similarly, $\sigma_y^2(\tau)$ can be estimated as the average of the square of $h_A(t_k) * \bar{y}_k$. Other transfer functions could be chosen. In time domain measurements, additional filtering with high- and low-frequency cutoffs can be performed. For example, removing a long-term trend from the frequency data is a form of highpass filtering. Clearly, measurements of $\mathscr{S}_y(f)$ are preferable to those of $\sigma_y^2(\tau)$ because σ_y^2 can be calculated from $\mathscr{S}_y$ using Eq. (9.94), but $\mathscr{S}_y$ cannot be calculated from σ_y^2. However, in many cases of interest, as in power-law spectra discussed below, the form of σ_y^2 is indicative of the behavior of $\mathscr{S}_y$. Traditionally, it has been easier to make time domain measurements, and most published results are given in terms of the Allan variance, σ_y^2.

Laboratory measurements show that $\mathscr{S}_y(f)$ is often a combination of power-law components. A useful model, shown in Fig. 9.11, is

$$\mathscr{S}_y(f) = \sum_{\alpha = -2}^{2} h_\alpha f^\alpha, \qquad 0 < f < f_h, \qquad (9.97)$$

where α is the power-law exponent that has integer values between -2 and 2, and f_h is the cutoff frequency of a lowpass filter. An equation similar to Eq.

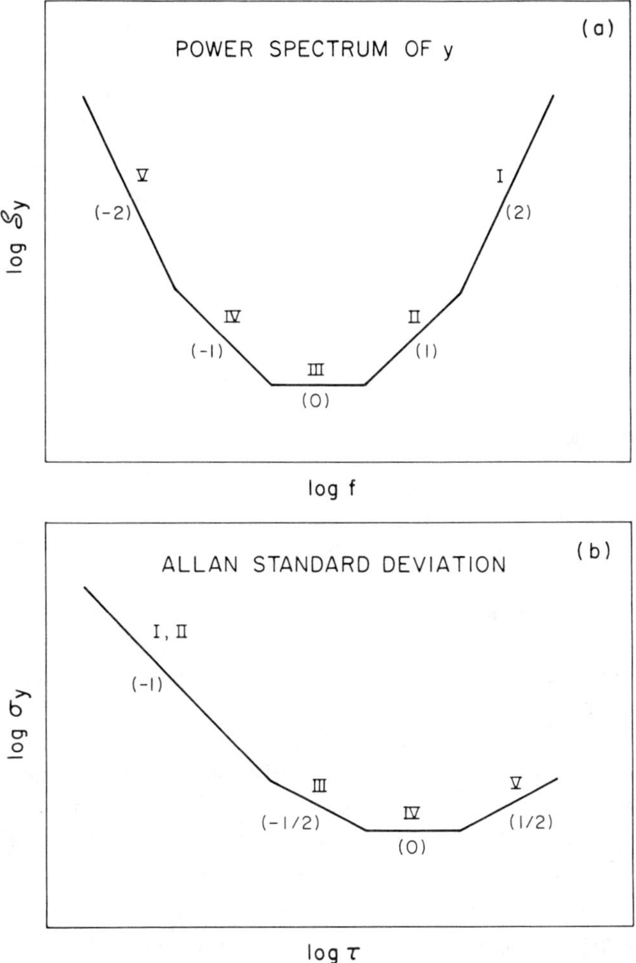

Figure 9.11   (a) Idealized power spectrum of $y(t)$, the fractional frequency deviation [see Eq. (9.97)]. The various spectral regimes are marked by Roman numerals, and the power-law coefficients are given in parentheses. The regimes are: I, white phase noise; II, flicker-phase noise; III, white-frequency noise; IV, flicker-frequency noise; and V, random-walk-of-frequency noise. (b) Two-point rms deviation or Allan standard deviation versus the time between samples. The spectral regimes are marked by the Roman numerals, and the power-law coefficients are given in parentheses.

(9.97) can be written for $\mathcal{S}_\phi(f)$ using Eq. (9.82). Each of the terms in Eq. (9.97) or the equivalent equation for $\mathcal{S}_\phi(f)$ has a name based on traditional terminology (see Table 9.1). Noise with a power-law dependence f^0, independent of frequency, is called "white noise;" f^{-1} is called "flicker noise," or colloquially, "one-over-f noise;" and f^{-2} is called "random-walk noise." There are well-known origins for some of these processes that we discuss

TABLE 9.1   Characteristics of Noise in Oscillators[a]

Noise Type	$\mathcal{S}_y(f)$	$\mathcal{S}_\phi(f)$	$\sigma_y^2(\tau)$	[b]μ	$I^2(\tau)$
White phase	$h_2 f^2$	$\nu_0^2 h_2$	$\dfrac{3h_2 f_h}{4\pi^2\tau^2}$[c]	-2	$\dfrac{h_2 f_h}{2\pi^2\tau^2}$
Flicker phase	$h_1 f$	$\nu_0^2 \dfrac{h_1}{f}$	$\dfrac{3h_1}{4\pi^2\tau^2}\ln(2\pi f_h\tau)$	~ -2	—
White frequency or random walk of phase	h_0	$\nu_0^2 \dfrac{h_0}{f^2}$	$\dfrac{h_0}{2\tau}$	-1	$\dfrac{h_0}{2\tau}$
Flicker frequency	$\dfrac{h_{-1}}{f}$	$\nu_0^2 \dfrac{h_{-1}}{f^3}$	$(2\ln 2)h_{-1}$	0	—
Random walk of frequency	$\dfrac{h_{-2}}{f^2}$	$\nu_0^2 \dfrac{h_{-2}}{f^4}$	$\dfrac{2\pi^2\tau}{3}h_{-2}$	1	—

[a]Adapted from Barnes et al. (1971).

[b]Power-law exponent of Allan variance: $\sigma_y^2(\tau) \propto \tau^\mu$.

[c]$2\pi f_h\tau \gg 1$.

briefly (see also Vessot 1976). The frequency dependence given in parentheses below is for $\mathcal{S}_y$.

1. *White-phase noise* (f^2) is usually due to additive noise outside the oscillator, introduced, for example, by amplifiers. This process dominates at large values of f, corresponding to short averaging times.

2. *Flicker-phase noise* (f^1) is seen in transistors and may be due to diffusion processes across junctions.

3. *White-frequency* or *random-walk-of-phase noise* (f^0) is due to the internal additive noise within the oscillator such as the thermal noise inside the resonant cavity. Shot noise also has this spectral dependence.

4. *Flicker-frequency noise* (f^{-1}) and *random-walk-of-frequency noise* (f^{-2}) are the processes that limit the long-term stability of oscillators. They are due to random changes in temperature, pressure, and magnetic field in the oscillator environment. This noise is associated with long-term drift. There is a large body of literature on flicker-frequency noise that is encountered in many situations (see Keshner 1982 for a general discussion; Dutta and Horn 1981 for applications in solid-state physics; and Press 1978 for applications in astrophysics).

The variances $I^2(\tau)$ and $\sigma_y^2(\tau)$ can be calculated for the various types of noise described above. For $\alpha = 1$ and 2, the variances converge only if a high-frequency cutoff f_h is specified. With this restriction, σ_y^2 converges for all cases. $I(\tau)$ converges only for $\alpha \geq 0$. These functions are listed in Table 9.1. Except for the logarithmic dependence in flicker phase noise, each noise component maps into a component of Allan variance of the form τ^{μ}. From Table 9.1 we can write the total Allan variance as

$$\sigma_y^2(\tau) = \left[K_2^2 + K_1^2 \ln(2\pi f_h \tau) \right] \tau^{-2} + K_0^2 \tau^{-1} + K_{-1}^2 + K_{-2}^2 \tau, \quad (9.98)$$

where the K's are constants. The subscripts correspond to the subscripts of h (see Table 9.1). White-phase and flicker-phase noise both result in $\mu \simeq -2$, but these two processes can be distinguished by varying f_h. Note that for white-phase and white-frequency noise the following relations hold [see Eqs. (9.92) and (9.94)]:

$$\sigma_y^2(\tau) = \tfrac{3}{2} I^2(\tau), \qquad \alpha = 2, \quad (9.99)$$

$$\sigma_y^2(\tau) = I^2(\tau), \qquad \alpha = 0. \quad (9.100)$$

In general, when $I^2(\tau)$ is defined, we see from Eqs. (9.91) and (9.93) that

$$\sigma_y^2(\tau) = 2\left[I^2(\tau) - I^2(2\tau) \right]. \quad (9.101)$$

Oscillator Coherence Time

A quantity of special interest in VLBI is the coherence time. The approximate coherence time is that time τ_c for which the rms phase error is 1 radian,

$$2\pi \nu_0 \tau_c \sigma_y(\tau_c) \simeq 1. \quad (9.102)$$

Rogers and Moran (1981) calculated a more exact expression for the coherence time that they defined in terms of the coherence function

$$C(T) = \left| \frac{1}{T} \int_0^T e^{j\phi(t)} dt \right|, \quad (9.103)$$

where $\phi(t)$ is the component of fringe phase of instrumental origin, and T is an arbitrary integration time. $\phi(t)$ includes effects that cause the fringe phase to wander, such as atmospheric irregularities and noise in frequency standards. The rms value of $C(T)$ is a monotonically decreasing function of time with the range 1–0. The coherence time is defined as the value of T for which $\langle C^2(T) \rangle$ drops to some specified value, say, 0.5. The mean-square value of C is

$$\langle C^2(T) \rangle = \frac{1}{T^2} \int_0^T \int_0^T \langle \exp\{ j[\phi(t) - \phi(t')] \} \rangle \, dt \, dt'. \quad (9.104)$$

If ϕ is a Gaussian random variable then

$$\langle C^2(T)\rangle = \frac{1}{T^2}\int_0^T\int_0^T \exp\left[-\frac{\sigma^2(t,t')}{2}\right]dt\,dt', \qquad (9.105)$$

where $\sigma^2(t,t')$ is the variance $\langle[\phi(t)-\phi(t')]^2\rangle$, which we assume depends only on $\tau = t' - t$. Then from Eq. (9.81),

$$\sigma^2(t',t) = \sigma^2(\tau)$$

$$= \langle[\phi(t)-\phi(t')]^2\rangle = 2\left[R_\phi(0) - R_\phi(\tau)\right]. \qquad (9.106)$$

Note that $\sigma^2(\tau)$ is the structure function of phase and is related to $I^2(\tau)$ by Eq. (9.91):

$$\sigma^2(\tau) = 4\pi^2\tau^2\nu_0^2 I^2(\tau). \qquad (9.107)$$

The integral in Eq. (9.105) can be simplified by noting that the integrand is constant along diagonal lines in (t, t') space for which $t' - t = \tau$. These lines have length $\sqrt{2}\,(T - \tau)$ so that

$$\langle C^2(T)\rangle = \frac{2}{T}\int_0^T\left(1 - \frac{\tau}{T}\right)\exp\left[-\frac{\sigma^2(\tau)}{2}\right]d\tau. \qquad (9.108)$$

Thus, from Eqs. (9.92) and (9.107)

$$\langle C^2(T)\rangle = \frac{2}{T}\int_0^T\left(1 - \frac{\tau}{T}\right)\exp\left[-2(\pi\nu_0\tau)^2\int_0^\infty \mathscr{S}_y(f)H_I^2(f)\,df\right]d\tau,$$

$$(9.109)$$

where $H_I^2(f)$ is defined in Eq. (9.95). Since $\mathscr{S}_y(f)$ is often not available, it is useful to relate $\langle C^2(T)\rangle$ to $\sigma_y^2(\tau)$. We can solve Eq. (9.101) for $I(\tau)$ by series expansion, obtaining

$$2I^2(\tau) = \sigma_y^2(\tau) + \sigma_y^2(2\tau) + \sigma_y^2(4\tau) + \sigma_y^2(8\tau) + \cdots, \qquad (9.110)$$

provided that the series converges. Therefore, from Eqs. (9.107), (9.108), and (9.110),

$$\langle C^2(T)\rangle = \frac{2}{T}\int_0^T\left(1 - \frac{\tau}{T}\right)\exp\left\{-\pi^2\nu_0^2\tau^2\left[\sigma_y^2(\tau) + \sigma_y^2(2\tau) + \cdots\right]\right\}d\tau.$$

$$(9.111)$$

This integral is readily calculable for the cases where $I^2(\tau)$ is defined. We now consider white-phase noise and white-frequency noise, which are important processes in frequency standards on short time scales.

For the case of white-phase noise, $\sigma_y^2 = K_2^2 \tau^{-2}$, where $K_2^2 = 3h_2 f_h / 4\pi^2$ is the Allan variance in 1 sec (Table 9.1), and the coherence function can be evaluated from Eq. (9.109) or Eq. (9.111):

$$\langle C^2(T) \rangle = \exp\left(\frac{-4\pi^2 \nu_0^2 K_2^2}{3} \right) = \exp\left(-h_2 f_h \nu_0^2 \right). \tag{9.112}$$

For white-frequency noise, $\sigma_y^2 = K_0^2 \tau^{-1}$, where $K_0^2 = h_0/2$, and we obtain

$$\langle C^2(T) \rangle = \frac{2(e^{-aT} + aT - 1)}{a^2 T^2}. \tag{9.113}$$

Here, $a = 2\pi^2 \nu_0^2 K_0^2 = \pi^2 h_0 \nu_0^2$. The limiting cases for white-frequency noise are

$$\langle C^2(T) \rangle = 1 - \frac{2\pi^2 \nu_0^2 K_0^2 T}{3}, \qquad 2\pi^2 \nu_0^2 K_0^2 T \ll 1,$$

$$= \frac{1}{\pi^2 \nu_0^2 K_0^2 T}, \qquad 2\pi^2 \nu_0^2 K_0^2 T \gg 1. \tag{9.114}$$

The approximate relation for coherence time in Eq. (9.102) corresponds to rms values of the coherence function of 0.85 and 0.92 for white-phase noise and white-frequency noise, respectively. These calculations assume that one station has a perfect frequency standard. In practice, the effective Allan variance is the sum of the Allan variances of the two oscillators,

$$\sigma_y^2 = \sigma_{y1}^2 + \sigma_{y2}^2. \tag{9.115}$$

Thus, if two stations have similar standards the coherence loss is doubled, if the loss is small. If the short-term stability is dominated by white-phase noise, which is usually the case for hydrogen masers, the coherence function is independent of time. This means that there is a maximum frequency above which a particular standard will not be useable for VLBI, regardless of the integration time. This frequency is approximately $1/2\pi K_2$ Hz, which for hydrogen masers is about 1000 GHz.

In practice, the coherence $C(T)$ is measured at the peak amplitude of the correlator output, which varies as a function of fringe frequency. This operation is equivalent to removing a constant frequency drift from the phase data and can be considered as highpass filtering of the data with a cutoff frequency of $1/T$. Modeling this operation as the response of a single-pole, highpass

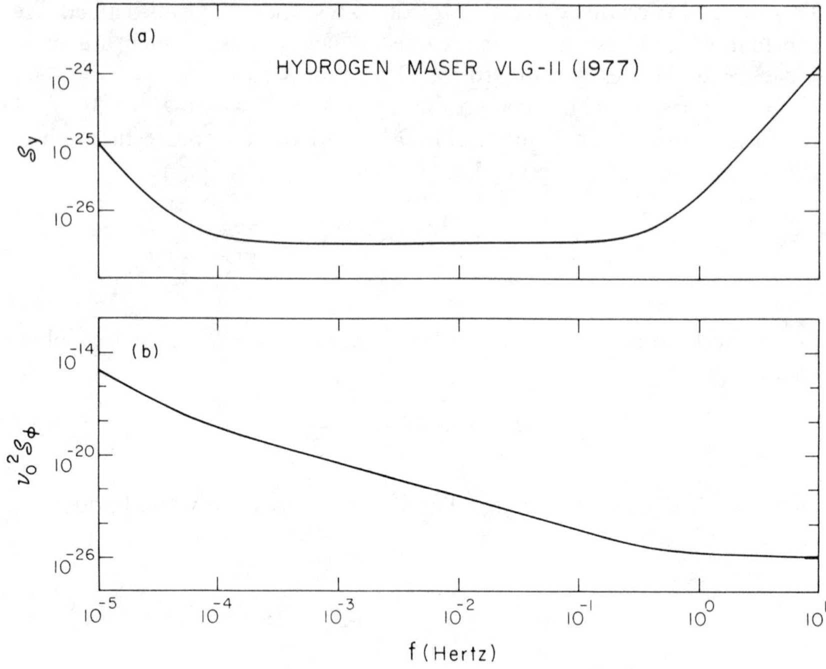

Figure 9.12   (a) Power spectrum of the fractional frequency deviation $\mathscr{S}_y(f)$ for a hydrogen-maser frequency standard, and (b) the normalized power spectrum of the phase noise $v_0^2\mathscr{S}_\phi(f)$. $\mathscr{S}_y(f)$ and $\mathscr{S}_\phi(f)$ are related by Eq. (9.82). For frequencies above 10 Hz, $\mathscr{S}_\phi(f)$ approaches the spectrum of the crystal oscillator to which the maser is locked, which declines as f^{-3}. The data were adapted from the measurements of Vessot (1979).

filter, one can show that it ensures the convergence of Eq. (9.111) for all processes for which the Allan variance exponent $\mu < 1$. To compare the various representations of frequency stability, we show in Figs. 9.12 and 9.13 an example of the performance of a hydrogen maser given by the functions σ_y^2, $\mathscr{S}_y(f)$, and $\langle C^2(T)\rangle^{1/2}$.

Precise Frequency Standards

Precise frequency standards of interest for VLBI include crystal oscillators and atomic frequency standards such as rubidium vapor cells, cesium-beam resonators, and hydrogen masers. Atomic frequency standards incorporate crystal oscillators that are phase locked or frequency locked to the atomic process, using loops with time constants in the range 0.1–1 sec, so that short-term performance becomes that of the crystal oscillator. Details of how these loops are implemented are given by Vanier, Têtu, and Bernier (1979). The performance of the crystal oscillator is very important because unless it has high spectral purity the phase-locked loops involved in generating the local oscilla-

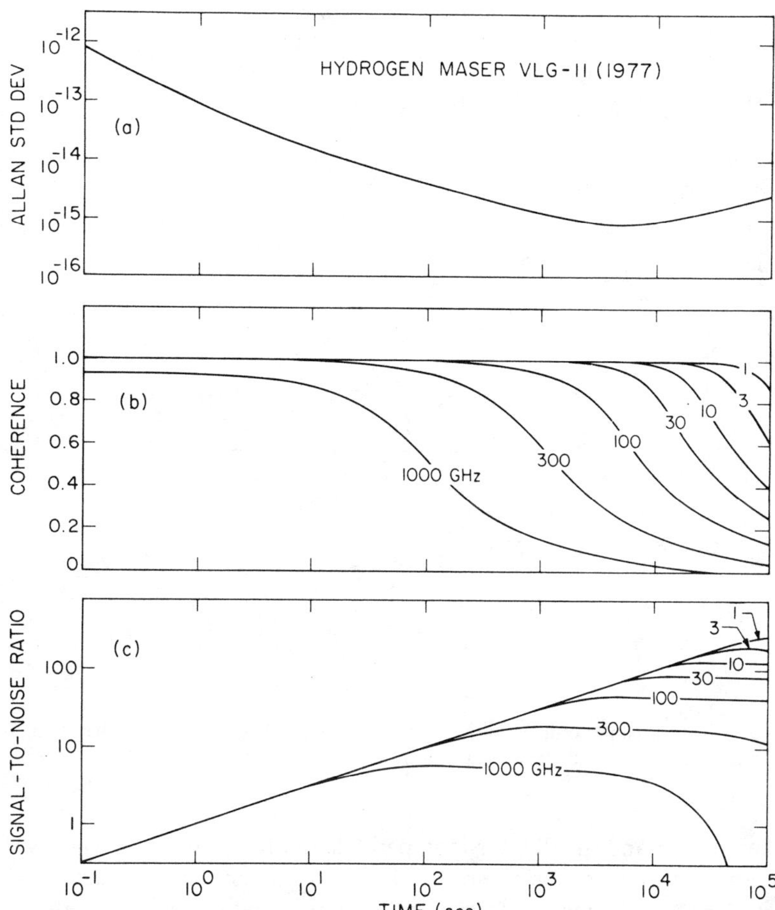

Figure 9.13 (a) Allan standard deviation versus sample time for a hydrogen-maser frequency standard. Data from Vessot (1979). (b) Coherence, $\sqrt{\langle C^2(T)\rangle}$, defined by Eq. (9.108) for various radio frequencies based on two frequency standards with Allan standard deviations given in the top panel. (c) Signal-to-noise ratio, normalized to unity at one second, of the measured visibility versus integration time for various frequencies. In practice, the coherence and signal-to-noise ratios will be less due to atmospheric fluctuations.

tor signal from the frequency standard will not operate properly (Vessot 1976).

We first consider a frequency standard as a "black box" that puts out a stable sinusoid at a convenient frequency such as 5 MHz or some higher frequency included in the phase-locked loop. The performance of various devices is shown in Fig. 9.14. These somewhat idealized plots show that the Allan variances of the standards have three regions: short-term noise dominated by either white-phase or white-frequency noise; flicker-frequency noise, which gives the lowest value of Allan variance and is therefore referred to as the

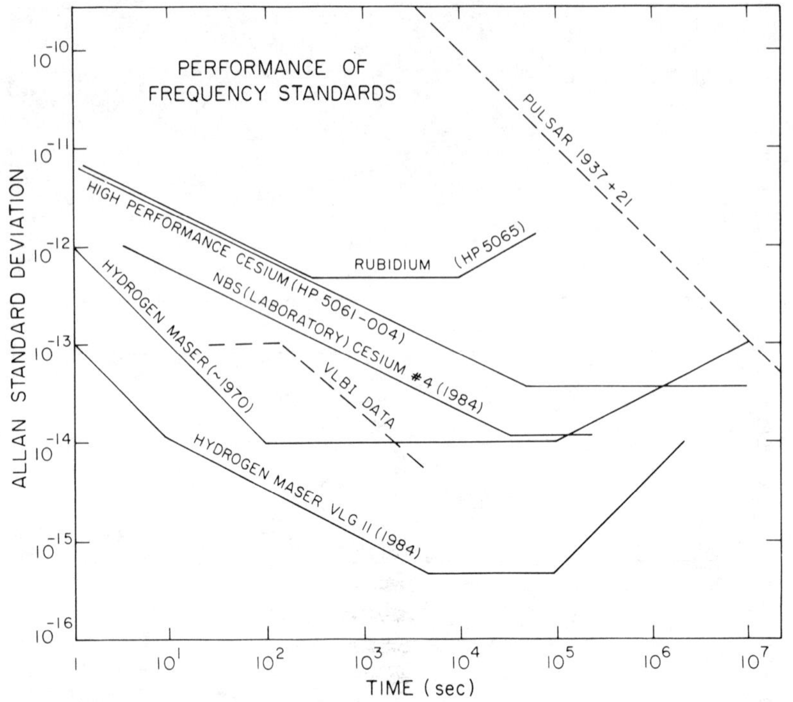

Figure 9.14   Idealized performance of various frequency standards and other systems. Pulsar data from Davis et al. (1985) and VLBI data from Rogers and Moran (1981).

"flicker floor"; and, finally, for long periods, random-walk-of-frequency noise. Two other parameters can be specified, a drift rate and an accuracy. The drift rate is the linear change in frequency per unit time interval. Note that if the standard drives a clock, then a constant drift rate results in a clock error that accumulates as time squared. The accuracy refers to how well the standard can be set to its nominal frequency. The performance parameters are summarized in Table 9.2.

Atomic frequency standards are based on the detection of an atomic or molecular resonance. There are three parts to any frequency standard (e.g., Kartashoff and Barnes 1972). These are (1) particle preparation, (2) particle confinement, and (3) particle interrogation. Particle preparation involves enhancing the population difference in the desired transition. This is necessary for radio transitions in a gas with temperature T_g for which $h\nu/kT_g \ll 1$, so that the level populations are nearly equal. Preparation is usually done by state selection in a beam passing through a magnetic or electric field, or by optical pumping. Particle confinement makes it possible to obtain narrow resonance lines from long interaction times since, according to the Heisenberg uncertainty principle, the linewidth is equal to the reciprocal of the interaction time. Particles can be confined in beams or storage cells. Storage cells either contain

TABLE 9.2   Typical Performance[a] Data on Available Frequency Standards[b]

Type	K_2 (10^{-12} sec)	K_0 (10^{-12} sec$^{1/2}$)	K_{-1} (10^{-15})	K_{-2} (10^{-17} sec$^{-1/2}$)	Drift Rate[c] (10^{-15})	Fractional Accuracy (10^{-12})
H (active)	0.1	0.03	0.4	0.1	< 1	1
Cs	—	50	100	3	1	5
Cs (high performance)	—	7	40	3	1	2
Rb	—	7	500	300	10^2	10^2
Crystal	1	—	500	300	10^3	—

Two-point Allan standard deviation; coefficients defined by Eq. (9.98).

Updated from Hellwig (1979).

Fractional frequency change per day.

a buffer gas or have specially coated walls so that particle collisions do not result in phase changes. Finally, particle interrogation is the process of sensing the interaction of particles and radiation fields. Frequency standards can be either active or passive. An example of an active standard is a maser oscillator. Passive standards require an external radiation field, and transitions are observed by (1) absorption, (2) re-emission, (3) detection of particles having made the transition, or (4) indirectly by detection of a quantity such as a variation in the rate of optical pumping. To show how some principles are implemented in practice, we give brief descriptions of the operation of several types of standard. Other types of frequency standards are under development (Wineland 1984).

Rubidium Gas-Cell Resonator

Rubidium is an alkali metal with a single valence electron and thus a hydrogen-like spectrum. The electronic ground state is split into two levels, with a transition frequency of 6835 MHz. These levels correspond to the spin of the unpaired electron being parallel or antiparallel to the nuclear spin vector. A schematic diagram of the oscillator system is shown in Fig. 9.15. An RF plasma discharge in a tube containing ^{87}Rb excites the gas to an electronic level about 0.8 μm above the ground state. The light from this discharge passes through a filter that removes the components involving the $F = 2$ level and passes the light at 0.7948 μm. This filter consists of a cell of ^{85}Rb atoms whose energy levels are slightly shifted from those of the ^{87}Rb atoms, such that both gases have transitions near 0.7800 μm. The filtered light passes through another cell of ^{87}Rb gas inside a microwave cavity resonant at the transition frequency between the $F = 2$ and $F = 1$ levels. With no RF signal applied to the cavity, the gas is nearly transparent and the discharge beam is

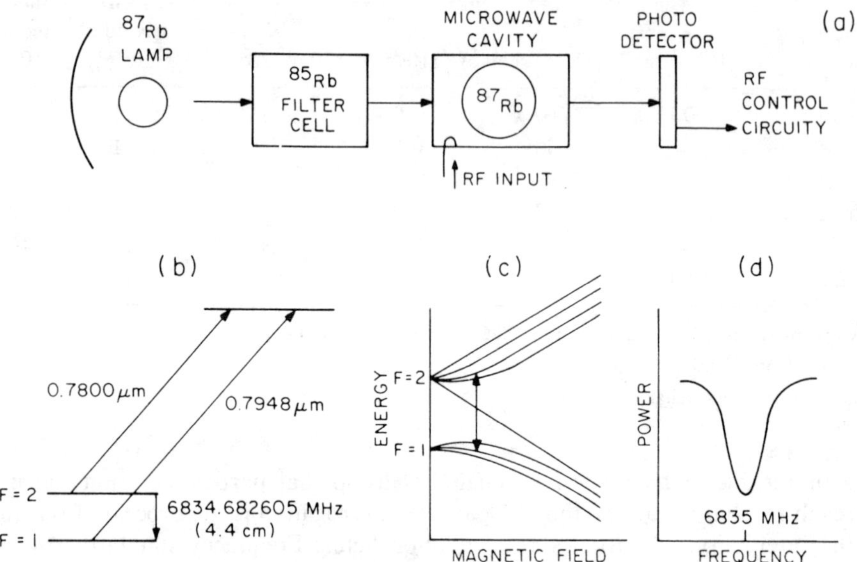

Figure 9.15   (a) Schematic diagram of a rubidium gas-cell frequency standard; (b) pump and microwave transitions; (c) magnetic sublevels of microwave transition versus magnetic field; and (d) absorption of ^{87}Rb light versus microwave frequency. Adapted from Vessot (1976).

unattenuated as it reaches the photodetector. The application of an RF signal at 6835 MHz stimulates transitions from the $F = 2$ to $F = 1$ level. The atoms reaching the lower level are then pumped to the excited state by the light from the filtered ^{87}Rb lamp. The ^{87}Rb light therefore suffers absorption. A buffer gas, consisting of inert atoms that collide elastically with the ^{87}Rb atoms in the resonance cell, extends the interaction time to about 10^{-2} sec, the mean collision time with the cell walls, and gives an absorption resonance with a linewidth of about 10^2 Hz. The cavity is magnetically shielded to minimize external fields. A weak homogeneous field is applied so that only $\Delta M_F = 0$ transitions, which have zero first-order Doppler shift, are obtained. The absorption resonance has a width of 10^2–10^3 Hz. The shot noise of individual arriving photons leads to white-frequency noise.

The RF signal is frequency or phase modulated so that the resonance line is continuously scanned. A control voltage is generated by comparing the modulation signal and the detector signal, and is fed back to the slave oscillator driving the cavity to correct its frequency to the peak of the resonance.

Rubidium standards have the advantage of being small, inexpensive, and readily transportable. They are used in VLBI below 1 GHz, where the ionosphere dominates the system stability. At higher frequencies the use of rubidium standards results in degraded performance. They are also useful in transferring time between stations and in maintaining timing at a station during servicing of a primary standard.

Cesium-Beam Resonator

The cesium-beam resonator originated in the earliest beam experiments in the mid-1940s. Cesium, like rubidium, is an alkali metal with a single valence electron. The ground electronic state is split into two levels depending on whether the spin of the unpaired electron is aligned parallel or antiparallel to the nuclear spin. The total angular momentum quantum numbers for these states are $F = 4$ and $F = 3$, respectively. These states have magnetic sublevels denoted by M_F. The frequency of the ground-state, spin-flip transition is exactly 9192.631770 MHz, which is used to define the second of atomic time. A ribbon-shaped beam of cesium gas is passed through a state selector magnet that passes the atoms in the $F = 3$ level into a resonator. The resonator consists of a long tube with cavities at each end that are excited in phase. The RF fields, in addition to a weak homogeneous magnetic field, excite transitions to the $F = 4$, $M_F = 0$, level. The cesium atoms then pass through another state-selecting magnetic field that removes the atoms in the $F = 3$ level. The remaining atoms, those that were stimulated to make the $F = 3$, $M_F = 0$ to $F = 4$, $M_F = 0$ transition, fall onto a detector, which controls the oscillator that drives the resonator. The dual cavities produce an interference phenomenon that narrows the resonance to about 500 Hz. The narrow resonance envelope is achieved because the particles are non-interacting and because the RF magnetic field is perpendicular to the beam direction so that the first-order Doppler shift is nearly eliminated.

Cesium frequency standards are larger and substantially more expensive than rubidium standards. Because of their low signal-to-noise ratio, their short-term stability is poor. Thus, they are not used in VLBI for controlling local oscillators. However, they provide excellent long-term stability and are used by the U.S. Naval Observatory to monitor time. They have also been used to verify the capability of transferring time via VLBI (Clark et al. 1979). The historical development of the cesium-beam resonator is described by Forman (1985).

Hydrogen-Maser Frequency Standard

The hydrogen maser is the usual VLBI standard, and we discuss its operating principles in some detail. The quantum mechanical analysis of the hydrogen maser is presented in a classic paper by Kleppner, Goldenberg, and Ramsey (1962). Fundamental principles of masers are given by Shimoda, Wang, and Townes (1956) and details of maser construction are given by Kleppner et al. (1965) and Vessot et al. (1976).

The hydrogen-maser oscillator uses the ground-state spin-flip transition at 1420.405 MHz, the well-known 21-cm line in radio astronomy. A schematic

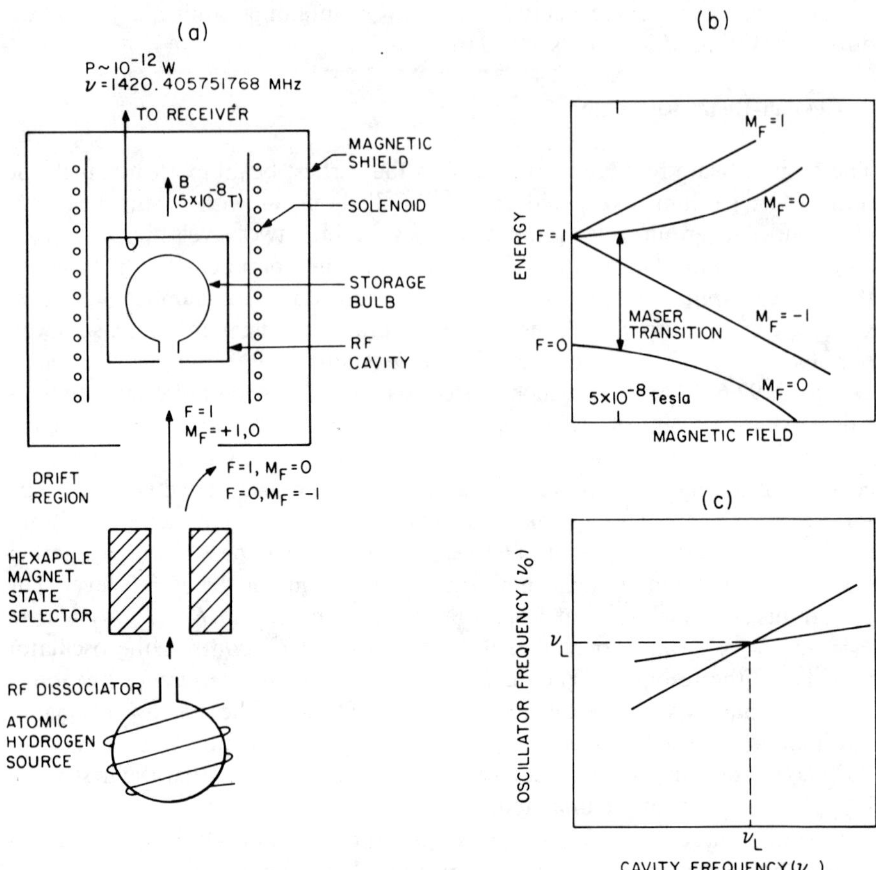

Figure 9.16 (a) Schematic diagram of a hydrogen-maser frequency standard. The line frequency given is the rest frequency of the transition in free space. The actual frequency will differ because of cavity pulling, second order Doppler and the wall shift. A typical frequency is 1420.40575168 MHz. (b) Energies of magnetic sublevels versus magnetic field for the 21-cm transition. Adapted from Vessot (1976). (c) Curves of resonance frequency, ν_0, versus cavity frequency, ν_C, for two values of linewidth [see Eq. (9.121)]. The intersection of the curves, which can be found empirically, gives the best operating frequency.

diagram of the oscillator is shown in Fig. 9.16. The hydrogen for the maser comes from a tank of molecular hydrogen gas that is dissociated in an RF discharge. The gas in the discharge is ionized and emits the reddish glow of the Balmer lines as the hydrogen atoms recombine and cascade to the ground state. The atomic gas flows out of the dissociator through a hexapole-magnet state selector. The inhomogeneous magnetic field separates the two upper states, $F = 1$, $M_F = 1$, and $F = 1$, $M_F = 0$, from the lower states, $F = 1$, $M_F = -1$, and $F = 0$, $M_F = 0$. The beam of atoms in the two upper states is directed into the storage bulb that is located inside a microwave cavity resonant in the TE_{011} or TE_{111} mode at 1420.405 MHz. The atoms bounce around the inside of the bulb about 10^5 times before escaping through the entrance hole. The spent atoms are evacuated from the system, which operates at low pressure, by an ion pump. The cavity is surrounded by several layers of material with high magnetic permeability that shields it from ambient magnetic fields. Inside the shield is a solenoid that creates a weak homogeneous field. This field allows the $(F = 1, M_F = 0)$-to-$(F = 0, M_F = 0)$ transition to radiate and minimizes transitions from the $F = 1$, $M_F = 1$ level. There is no first-order Zeeman effect for the $\Delta M_F = 0$ transition (see Fig. 9.14). The maser will oscillate if the cavity is tuned close to the transition frequency and the losses are small enough. In the active maser, the 1420-MHz signal is picked up by a cavity probe and used to phase lock a crystal oscillator from which a signal at the hydrogen-line frequency has been synthesized.

The interaction lifetime of an atom in the bulb can be described by an exponential probability function

$$f(t) = \gamma e^{-\gamma t}, \tag{9.116}$$

where γ is the total relaxation rate. The line has an approximately Lorentzian profile with a linewidth (full width at half maximum) $\Delta \nu_0$ of γ/π. The most important contribution to γ is the rate at which atoms escape through the entrance hole. This rate is

$$\gamma_e = \frac{v_0 A_h}{6V}, \tag{9.117}$$

where $v_0 = \sqrt{8kT_g/m}$ is the average particle speed, m is the mass of a hydrogen atom, A_h is the area of the entrance hole, and V is the volume of the bulb. γ_e is about 1 sec^{-1}. The atoms lose coherence after many wall collisions, and this leads to a loss rate $\gamma_w \simeq 10^{-4}$ sec^{-1}. Collisions between hydrogen atoms cause spin-exchange relaxation at a rate γ_{se} that is proportional to the gas density and to v_0. The net relaxation rate is approximately the sum of the three most important terms,

$$\gamma = \gamma_e + \gamma_w + \gamma_{se} = \pi \Delta \nu_0. \tag{9.118}$$

All three terms are proportional to v_0 and thus also to $\sqrt{T_g}$. Note that the random thermal motions of the atoms do not give rise to a first order Doppler broadening of the line because the interaction between the atoms and the RF field takes place in a resonant cavity (see Kleppner, Goldenberg and Ramsey 1962).

The maser oscillator has two resonant frequencies, the line frequency ν_L and the electromagnetic cavity resonance frequency ν_C defined by the cavity's dimensions. In classical oscillators the frequency is the mean of these two weighted by the respective Q factors, Q_L for the line and Q_C for the cavity.

That is,

$$\nu_0 = \frac{\nu_L Q_L + \nu_C Q_C}{Q_L + Q_C}. \tag{9.119}$$

The Q-factor is defined as π times the reciprocal of the fractional loss in energy per cycle of the resonant frequency. Hence, from Eq. (9.116), Q_L is given by (e.g., Siegman 1971)

$$Q_L \simeq \frac{\pi \nu_0}{\gamma} = \frac{\nu_0}{\Delta \nu_0}. \tag{9.120}$$

A typical value of Q_L is about 10^9. The practical value of Q_C for a silver plated cavity is about 5×10^4. Since $Q_L \gg Q_C$, the resonance frequency is approximately

$$\nu_0 \simeq \nu_L + \frac{Q_C}{Q_L}(\nu_C - \nu_L). \tag{9.121}$$

Equation (9.121) describes the effect of "cavity pulling" on the resonance frequency. Temperature changes cause the size, and thus the resonant frequency, of the cavity to change. Hence, a fractional frequency stability of 10^{-15} for the maser requires a fractional mechanical stability of about 5×10^{-10} for the cavity. The cavity dimensions therefore must be stable to about 10^{-8} cm. The cavity must be made from material with a small thermal expansion coefficient or the temperature must be carefully controlled. Extreme mechanical stability is also required so that atmospheric pressure changes do not affect the frequency. The TE_{011} cavity is a cylinder about 27 cm in length and diameter, appreciably larger than the free space wavelength because of the loading by the storage bulb. Coarse tuning is accomplished by moving the end plate of the cavity and fine tuning by a varactor diode. From Eq. (9.121) it is clear that the maser frequency is most stable when ν_C is set to ν_L so that ν_0 equals ν_L regardless of the values of Q_C and Q_L. This optimal tuning point of the maser can be found by making a plot of ν_0 versus ν_C, which is a straight line with slope Q_C/Q_L according to Eq. (9.121). By varying Q_L, for example, by varying the gas pressure and thereby changing γ, a family of straight lines can be generated that intersect at the desired frequency $\nu_0 = \nu_L = \nu_C$ (see Fig. 9.14). Servomechanisms are used in some systems to keep the maser cavity continuously tuned.

Performance of hydrogen masers is shown in Figs. 9.13 and 9.14. For periods less than 10^3 sec the performance is limited by two fundamental processes, white-frequency noise due to thermal noise generated inside the cavity and white-phase noise due to thermal noise in the external amplifier. The thermal noise generated inside the cavity produces a fractional frequency variance (Allan variance) of (Edson 1960; Kleppner, Goldenberg, and Ramsey

1962)

$$\sigma_{yf}^2 = \frac{1}{Q_L^2} \frac{kT_g}{P_0 \tau},$$

(9.122)

where P_0 is the power delivered by the atoms. There is also shot noise in the cavity due to the discrete radiation of photons. However, this process, described by the Allan variance σ_{ys}^2, is smaller than σ_{yf}^2 by the ratio $h\nu/kT_g$, which is 2×10^{-4} at room temperature. Spontaneous emission also contributes a small amount of noise equivalent to increasing T_g by $h\nu/k \simeq 0.07$ K. Finally, the maser receiver adds a noise power $kT_R\Delta\nu$ to the signal coupled out of the cavity, where T_R is the receiver noise temperature and $\Delta\nu$ is the receiver bandwidth. This noise causes an Allan variance of (Cutler and Searle 1966)

$$\sigma_{yR}^2 = \frac{1}{(2\pi\nu_0\tau)^2} \frac{kT_R\Delta\nu}{P_0}.$$

(9.123)

These two processes are independent, so the net Allan variance is $\sigma_y^2 = \sigma_{yf}^2 + \sigma_{yR}^2$. The effects of both processes are clearly evident in the data in Fig. 9.13. Note that a flicker floor is not reached because of long-term drifts. The short-term performance can be improved by increasing the atomic flux level, which increases P_0. However, increasing the flux increases the spin-exchange rate, which decreases Q_L, thereby making the oscillator more susceptible to the long-term effects of cavity pulling.

The frequency of a maser is not exactly equal to the atomic transition frequency because of several effects. These effects limit the accuracy to which the frequency can be set, and because most of them are temperature dependent, they probably contribute to flicker-frequency and random-walk-of-frequency noise. Cavity pulling, which has been described already, is an important effect, and to minimize it the cavity must be tuned carefully. The collision-induced, spin-exchange process gives a frequency shift that varies with Q_L in the same way as the cavity pulling. Thus, the cavity-tuning procedure also eliminates this shift. Collisions with the cavity walls produce an effect called the "wall shift," which is difficult to predict and may be the ultimate limiting factor in the absolute precision of the maser frequency (Vessot and Levine 1970). This shift depends on the temperature and wall coating material. Its fractional value is about 10^{-11}. The first-order Doppler effect cancels, but the second-order Doppler effect does not because of its v^2/c^2 dependence (see Kleppner, Goldenberg, and Ramsey 1962). The fractional frequency shift is about equal to $-1.4 \times 10^{-13} T_g$. Finally, there is no first-order Zeeman effect in the $(F = 1, M_F = 0)$-to-$(F = 0, M_F = 0)$ transition. However, the second-order Zeeman fractional-frequency shift is $2.0 \times 10^2 B^2$, where B is the magnetic field in tesla.

Local Oscillator Stability

Local oscillator signals are generated by multiplying a signal from the locked oscillator of the frequency standard. The multipliers must have exceptional stability, as mentioned in Section 7.2, to avoid the introduction of additional noise and drift. Imperfect multipliers are sensitive to vibration and temperature and may have modulation at harmonics of the power line frequency. In an ideal multiplier, a signal of the form of Eq. (9.75) is converted to

$$V(t) = \cos\left[2\pi M\nu_0 t + M\phi(t)\right], \tag{9.124}$$

where M is the multiplication factor, ν_0 is the fundamental frequency, and ϕ is the random phase noise of the frequency standard. If the phase noise is small, $M\phi(t) \ll 1$, then the single-sided power spectrum of $V(t)$ is given by

$$\mathcal{S}_v(\nu) = \delta(\nu - M\nu_0) + M^2\mathcal{S}_\phi(\nu - M\nu_0), \tag{9.125}$$

where δ is a delta function representing the desired signal, and $\mathcal{S}_\phi$ is the power spectrum of the phase noise. Thus, the noise power increases as the square of the multiplication factor. In the general case, $\mathcal{S}_v$ can be written (Lindsey and Chie 1978)

$$\mathcal{S}_v(\nu) = \delta(\nu - M\nu_0) + \sum_{n=1}^{\infty} \frac{M^{2n}}{n!}\left[\mathcal{S}_\phi(\nu - M\nu_0) * \mathcal{S}_\phi(\nu - M\nu_0) * \cdots\right],$$

$$\tag{9.126}$$

where the term in brackets contains n replications of the same function convolved together. When only the leading term in the summation is retained, Eq. (9.126) reduces to Eq. (9.125). The higher-order terms in Eq. (9.126) represent a series of approximately Gaussian components because of the repeated convolutions.

Phase Calibration System

One way to check the integrity of the entire system is to inject an RF signal that is independently derived from the frequency standard into the front end of the receiver. The RF test signal can be derived by driving a step-recovery diode with, say, a 1-MHz signal from the frequency standard so as to generate a pulse train with 1-μsec period. Such a signal has harmonics at 1-MHz intervals throughout the microwave region, all of which have the same phase at the reference intervals. When the RF band is mixed down to baseband, one of the injected harmonics can be made to appear at a convenient frequency of order 10 kHz and compared with a reference signal from the frequency standard. This phase calibration signal can be continuously injected during

VLBI recording since a low enough level can be used that it can only be detected by very narrowband filtering in the processor ($\simeq$ 10-Hz bandwidth). The calibration allows one to compensate for variations such as those caused by thermal effects in cables (Whitney et al. 1976). Similar methods are used in some linked-element interferometers.

9.5 TIME SYNCHRONIZATION

The clocks at VLBI stations must be synchronized accurately enough to avoid time-consuming searches for interference fringes. What is needed is a continuously available time transfer system with an accuracy comparable to the reciprocal of the recorded bandwidth, which can be used to set the station clock and monitor the long-term drift of the frequency standard. High-frequency (HF) radio time services, such as WWV in the United States, which transmits on 2.5, 5, 10, 15, 20, and 25 MHz, are convenient for coarse synchronization because of the simplicity of the format, that is, 1-sec time ticks with verbal identification. Because of uncertainties in sky propagation, HF signals are only accurate to about ± 1 msec. Much higher accuracy can be achieved by monitoring signals from navigational systems such as Loran C, TRANSIT, OMEGA, or the NAVSTAR Global Positioning System (GPS) (Klepczynski 1983). There is a close connection between the problems of time transfer and navigation because of the finite propagation speed of radio signals. The GPS is an example of a general navigation system in which the user receives signals at 1.23 or 1.57 GHz from a number of satellites whose positions are known and whose clocks are synchronized to Coordinated Universal Time (UTC; see Chapter 12). If timing measurements from four satellites are made, and corrected for propagation effects in the atmosphere, users can determine their positions in three coordinates and their clock errors. The accuracies available to civilian users are better than 100 m in position and 100 nsec in time (Parkinson and Gilbert 1983). An analysis of the time transfer problem, including relativistic effects, has been performed by Ashby and Allan (1979).

Loran C has been widely used to monitor time at VLBI stations. Loran, an acronym for *Long Range Navigation*, is a system originally developed during World War II for ocean navigation (Pierce, McKenzie, and Woodward 1948). The transmission frequency is 100 kHz. The relative time of arrival of signals from three stations defines the observer's location on the earth's surface. Thus, Loran C stations have been deployed in chains, each consisting of a master station and three or four slave stations. These stations are synchronized to UTC, so they are equally useful for time transfer. Complete information is published by the U.S. Naval Observatory (1982). In order to set the local observatory clock, corrections must be made for (1) Loran transmission error, (2) propagation delay, and (3) receiver delay. The transmission errors, usually less than 1 μsec, are published by the U.S. Naval Observatory in monthly bulletins. To first order, the propagation delay is the distance divided by the propagation speed, for which an index of refraction of 1.0003 can be assumed. There is a second-order correction based on the diffraction of waves propagating over a smooth spherical surface. This effect gives an additional delay that is approximately proportional to distance and equals 5 μsec at 1000 km for land propagation and 2 μsec at 1000 km for sea propagation. The variation of arrival times of the ground wave is about 0.1 μsec, for all times of day and most weather conditions. Interference from the sky wave signal, especially at night, can lead to larger uncertainties. For this reason clock drift rates are based on measurements taken at the same time each day. Time transfer on paths up to 8000 km can be made from the sky wave signal to an accuracy of about ± 50 μsec. Detailed discussion of Loran C can be found in articles by Frank (1983), Potts and Wieder (1972), and Shapiro and Fisher (1970).

A discussion of other time transfer techniques including VLF, LF, HF broadcast systems, power lines, television, and moon bounce is given by Blair (1974). For time scales of a year, the accuracy of timing from pulsar observations approaches 1 part in 10^{14} (Davis et al. 1985). Ultimately, the best time transfer may be obtainable from the processed VLBI data (Clark et al. 1979).

9.6 RECORDING SYSTEMS

The basic consideration for any recording system is the representation of the signal and the method of incorporating the time information. Recording can be either digital or analog, and various data storage technologies are available. Here we discuss only digital recording on magnetic tape because the technologies involved are well suited to VLBI and are widely used.

A basic parameter of a recording system is its data rate, ν_b (bits sec^{-1}). This parameter limits the number of bits that can be recorded in a given time, and thus also the sensitivity of continuum observations in which the potential IF bandwidth is larger than $\nu_b/2N_b$, where N_b is the number of bits per sample. The signal is represented by samples having Q quantization levels taken at β times the Nyquist rate. For N samples there are Q^N possible data configurations, which require a minimum of $N \log_2 Q$ bits. Therefore, as noted in Chapter 8, the maximum RF bandwidth is

$$\Delta\nu = \frac{\nu_b}{2\beta N_b} = \frac{\nu_b}{2\beta \log_2 Q}. \tag{9.127}$$

The signal-to-noise ratio obtained in time τ is proportional to $\eta_Q\sqrt{\Delta\nu\,\tau}$, where η_Q is the quantization efficiency introduced in Chapter 8. From Eq. (9.127),

$$\eta_Q\sqrt{\Delta\nu\,\tau} = \eta_Q\sqrt{\frac{\nu_b\tau}{2\beta N_b}}. \tag{9.128}$$

If τ is the recording time for the tape, $\nu_b\tau$ is equal to the number of recorded bits on the tape. The quantity $\eta_Q/\sqrt{\beta N_b}$ thus provides an indication of the performance per bit, which should, of course, be maximized. For two- and four-level sampling, the obvious encoding schemes are one bit and two bits per sample, respectively. For three-level sampling, a problem arises since encoding one sample (one in three possible states) in two data bits (representing four possible states) is inefficient. Putting three samples into five bits or five samples into eight bits gives data rates of 1.67 and 1.60 bits per sample compared to the theoretical optimum value of $\log_2 3 = 1.585$. The values of $\eta_Q/\sqrt{\beta N_b}$ for various values of Q and β, and several encoding schemes, are listed in Table 9.3. The highest signal-to-noise ratio is achieved with three-level sampling at the Nyquist rate, although two- and four-level sampling give almost the same performance.

In addition to the encoding schemes discussed above, in which the number of bits required for a given number of samples is constant, one can also envisage a scheme in which the number of bits depends on the sample values, that is, a variable length code. For example, D'Addario (1984) has suggested encoding the $+1, 0$, and -1 values in three-level quantization as the binary numbers 11, 0, and 10, respectively. It is possible to decode such a data string uniquely, since all one-bit representations begin with 0 and all two-bit representations with 1. The average number of bits per sample depends on the amplitude probability distribution of the signal waveform and the threshold level settings. For a given number of bits, the threshold settings that maximize

TABLE 9.3   Performance of Various Signal Representations as a Function of Number of Quantization Levels, Sampling Rate, and Encoding Format[a]

Signal Representation		η_Q	N_b	$\dfrac{\eta_Q}{\sqrt{\beta N_b}}$
Sampling at Nyquist Rate ($\beta = 1$)				
Two-level		0.637	1.0	0.637
Three-level	"Ideal" encoding[b]	0.810	1.585	0.643
	5 samples/8 bit	0.810	1.60	0.640
	3 samples/5 bit	0.810	1.667	0.627
	1 sample/2 bit	0.810	2.0	0.573
Four-level	All products	0.881	2.0	0.623
	Low-level omitted	0.87	2.0	0.61
Sampling at 2 × Nyquist Rate ($\beta = 2$)				
Two-level		0.74	1.0	0.52
Three-level	"Ideal" encoding[b]	0.89	1.585	0.50
	5 samples/8 bit	0.89	1.60	0.50
	3 samples/5 bit	0.89	1.667	0.49
	1 sample/2 bit	0.89	2.0	0.45
Four-level	All products	0.94	2.0	0.47

[a] η_Q = quantization efficiency; N_b = number of bits per sample; β = oversampling factor.
[b] N samples encoded in $N \log_2 3$ bits.

TABLE 9.4   Characteristics of Some VLBI Systems

System		Period of Use	Basic Description	Tape Recorder	Sample Rate[a] (10^6 sec^{-1})	Tape Time (min)	Reference
NRAO Mark I[b]		1967–1978	Digital recording, IBM computer compatible format	Ampex TM-12	0.72	3.2	Bare et al. (1967)
NRAO Mark II	(A)	1971–1978	Digital recording on TV recorders	Ampex VR660C	4	190	Clark (1973)
	(B)	1976–1982		IVC 800	4	64	
	(C)	1979–		RCA VCT 500	4	246	
Canadian		1971–	Analog recording on TV recorder	IVC 800	8	64	Broten et al. (1967), Moran (1976)
MIT/NASA/NRAO Mark III		1977–	Digital recording on instrumentation recorder	Honeywell 96	112[c]	13.6	Rogers et al. (1983)
	(A)	1984–		Honeywell 96[d]	112[c]	164	Clark et al. (1985)

[a]All these systems use two-level quantization so that the sample rate equals the bit rate and also equals twice the recorded bandwidth.

[b]A similar system was developed in the Soviet Union (Kogan and Chesalin 1981).

[c]Lower sampling rates available for spectral-line VLBI.

[d]Trackwidth is 40 μm.

the signal-to-noise ratio, in general, are not the same as those derived in Chapter 8 that are optimum for a given number of samples. With D'Addario's encoding scheme, the best performance is achieved with the threshold set such that $\eta_Q = 0.769$ and $N_b = 1.370$ bits/sample so that the performance factor $\eta_Q/\sqrt{\beta N_b}$ is equal to 0.657. Thus, an increase in sensitivity of about 3% compared with the use of the scheme with 1.6 bits/sample could be achieved. However, the effects of bit errors, or interfering signals that change the amplitude distribution, could be more serious. Finally, the data could be encoded statistically in large blocks that would allow a theoretically optimal value of N_b of 1.317 bits/sample, which, with η_Q of 0.769, would give a performance factor of 0.670 (D'Addario 1984).

In practice, the desirability of a simple encoding scheme and other design considerations have usually resulted in the choice of two-level quantization. All three VLBI systems developed in the United States during the period 1968–1984 (Mark I, Mark II, and Mark III) use two-level sampling. For spectral-line observations where the bandwidth of the signal is small with respect to the bandwidth of the recording system, multilevel sampling is advantageous. Note (Table 9.3) that multilevel sampling is a more effective way of using recording capacity than sampling faster than the Nyquist rate.

Each data sample must have either an implicit or explicit time tag. Although an error rate of 10^{-3} in decoding the data bits is acceptable, a one-bit shift in the time axis can be a serious defect and is not acceptable. In virtually all recording systems the data are blocked into records. Each new record begins at a precise time so that the temporal registration of the data stream can be recovered if it is lost during the previous record. These record lengths are: Mark I—0.2 sec (144,000 bits); Mark II—16.7 msec (66,600 bits); and Mark III—5 msec (20,000 bits). In the Mark I system, which used standard computer tape format, the accuracy of recording was very high, and the time of any bit was obtained by counting bits from the beginning of the record, and records from the beginning of the tape. In the Mark II system, which uses video recorders, the data are recorded with a self-clocking code, while in the Mark III system, which uses instrumentation recorders, the data transitions themselves serve as the clock. The characteristics of several systems are given in Table 9.4.

9.7 PROCESSING SYSTEMS AND ALGORITHMS

A VLBI processor has two main functions: (1) reproduction of smooth data streams and (2) cross-correlation analysis of the data streams. The data stream from a tape recorder can be expected to have time-base irregularities of up to 100 μsec, caused by jitter in the mechanical playback system, and to be subject to dropouts because of tape imperfections. The processor must derive the true time base either from the encoded clock transitions in the case of a self-clocking code, or from the data transitions themselves when a bit synchronizer is used. There must be enough buffer storage to handle at least the mechanical jitter. The geometrical delay can be corrected with minimal buffer space by shifting the playback time, thereby retaining the data on the tape until they are needed by the correlator. If the data are read in synchronism from the tapes, a buffer memory of sample capacity about 5×10^4 times the clock rate in megahertz is needed for geometrical delay compensation.

The major differences between the design of the correlation part of the processor for VLBI and for a conventional interferometer are related to the fact that fringe rotation and delay compensation are usually performed in the VLBI processor on the quantized and sampled signal. This leads to special problems which we discuss here. Digitization of the signals introduces several signal-to-noise loss factors: η_Q, the loss factor associated with amplitude quantization of the recorded signals discussed in Chapter 8; η_R, the loss factor incurred by quantizing the phase of the fringe rotation waveform; η_S, the loss factor incurred by inadequate sideband rejection caused by having a finite number of delays in the correlator; and η_D, the loss caused by compensating the geometrical delay in discrete steps. We consider the case in which the signal has been quantized to two levels and sampled at the Nyquist rate.

Fringe rotation and delay compensation can be done on the analog signals at the telescope before recording. For example, the fringe rotation can be done at the telescopes by offsetting the local oscillators as described in Chapter 7 for a connected-element array. The advantage of this arrangement is that only a real correlation function (with both positive and negative delays) needs to be calculated (see Sections 8.9 and 9.1). Hence only half the correlator circuits are required. Also, the sensitivity loss from a digital fringe rotator is not incurred. A disadvantage is that the output of the correlator must be averaged over a short enough interval to accommodate the residual fringe frequency of a source anywhere in the primary beams of the antennas. The maximum residual fringe frequency of a source at the half power point of the primary beam is $\Delta \nu_f \simeq D\omega_e/d$ [see Eq. (12.20)], where D is the baseline length, d is the antenna diameter, and ω_e is the angular velocity of the earth in rad per sec. Hence, the averaging time of the correlator output must be less than $1/2\,\Delta\nu_f$ or 30 msec for a baseline equal to the earth's diameter and $d = 25$ m. The correlation functions can be averaged further after they have been passed through a fringe rotator, which removes the residual fringe frequency. Also, the unit at the telescope that continually changes the local oscillator frequency must be carefully designed so that full phase accountability is provided for astrometric work. Further information on VLBI systems and processing algorithms can be found in Thomas (1981) and Herring (1983).

Fringe Rotation Loss (η_R)

Fringe rotation is used to reduce to near zero the frequency of the fringe component of the correlated signals (see Chapter 6). Here we consider the fringe frequency to include the effect of offsets in the frequency standards.

Fringe rotation in the processor can be implemented in a number of ways as shown in Fig 9.17. If the fringe rotator is placed after the correlator (Fig. 9.17a), then the correlation function from the correlator must be averaged over an interval short with respect to the fringe period. Such a short averaging time necessitates a high data rate from the correlator, making this arrangement unattractive. Alternatively, before correlation one of the data streams can be

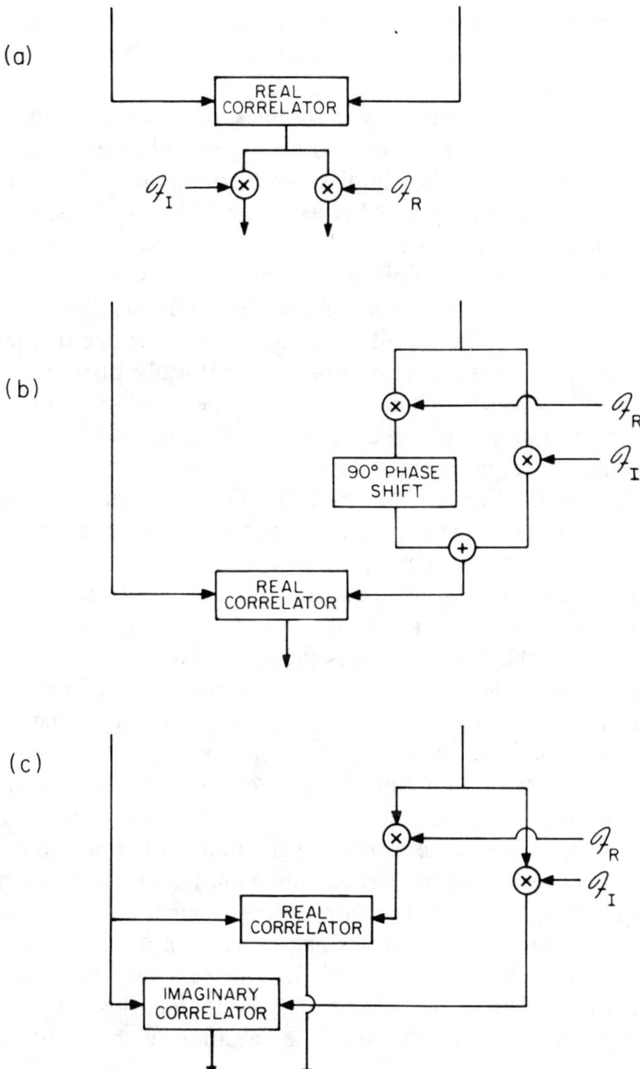

Figure 9.17   Various processor configurations showing possible locations of fringe rotator. If local oscillators at the antennas are offset to reduce the fringe frequency to near zero, scheme (a) is the most useful one. $\mathscr{F}_R$ and $\mathscr{F}_I$ are cosine and sine representations of the fringe function.

passed through a digital single-sideband mixer that shifts the Fourier components of the signal by the appropriate fringe frequency as shown in Fig. 9.17b. The 90° phase shift in this mixer is difficult to implement without introducing spectral distortion so this type of fringe rotator is rarely used (see also Section 8.7). We consider the fringe rotation scheme shown in Fig. 9.17c. Application of fringe rotation to the quantized signal by this method introduces two complications. First, the fringe function with which the signal is multipled must be coarsely quantized so as not to increase the number of bits per sample. Second, the multiplication introduces an unwanted noise sideband, which is described below under *Fringe Sideband Rejection Loss*. We now consider the first of these effects.

The data stream is multiplied by a complex function $\mathscr{F}$ whose real and imaginary parts, $\mathscr{F}_R$ and $\mathscr{F}_I$, approximate $\cos\phi$ and $\sin\phi$, where ϕ is the desired phase function. In the simplest approximation these functions are square waves with the appropriate frequency and phases. Thus, as shown in Fig. 9.18, the quantized signal is multiplied by a fringe rotation function whose amplitude is constant but whose phase steps by 90° every quarter cycle instead of smoothly progressing. The resulting visibility function then has a phase component that has a 90° sawtooth modulation at the fringe frequency. This resembles phase noise in which the phase is uniformly distributed between $\pm 45°$. Therefore, the average signal amplitude is degraded by $\sin(\pi/4)/(\pi/4)$ = 0.900. Another approach to calculating the loss in signal-to-noise ratio is to calculate the harmonics in the fringe rotation function. The first harmonic of $\mathscr{F}_R$ or $\mathscr{F}_I$ has an amplitude of $4/\pi = 1.273$. Only the signal mixed with the first harmonic appears in the processor output since the other harmonics are removed by time averaging. Thus part of the signal is scattered out of the fringe passband. The fraction retained is the square root of the ratio of the power in the first harmonic to the total power of the fringe rotation function, which is $\sqrt{8}/\pi = 0.900$. This represents the loss in signal-to-noise ratio. There is also a scale-factor change since the fringe amplitudes are increased by the action of the fringe rotator. Thus the fringe amplitudes must be divided by $4/\pi$, the relative amplitude of the first harmonic of $\mathscr{F}_R$.

A better fringe rotation function is the three-level approximation of a sine wave (Clark, Weimer, and Weinreb 1972) shown in Fig. 9.18. When the fringe rotation function is zero, the correlator is inhibited. Since the real and imaginary parts of $\mathscr{F}$ are never zero simultaneously, all data bits are used at least once. This fringe rotation function can be thought of as a phasor whose tip traces out a square such that it has phase jumps in 45° increments and its amplitude alternates between $\sqrt{2}$ and 1. The resulting jitter in phase is uniformly distributed between $\pm 22.5°$ and results in a loss of signal amplitude of $\sin(\pi/8)/(\pi/8) = 0.974$. Also, the variation in the amplitude of the phasor introduces a nonuniform weighting of the signal samples. This reduces the signal-to-noise ratio by a further factor equal to $(1 + \sqrt{2})/\sqrt{6} = 0.986$. The net loss in signal-to-noise ratio is 0.960. The reduction in signal-to-noise ratio is also equal to the square root of the ratio of the power in the first harmonic to the total power in $\mathscr{F}_R$. The first harmonic of $\mathscr{F}_R$ is

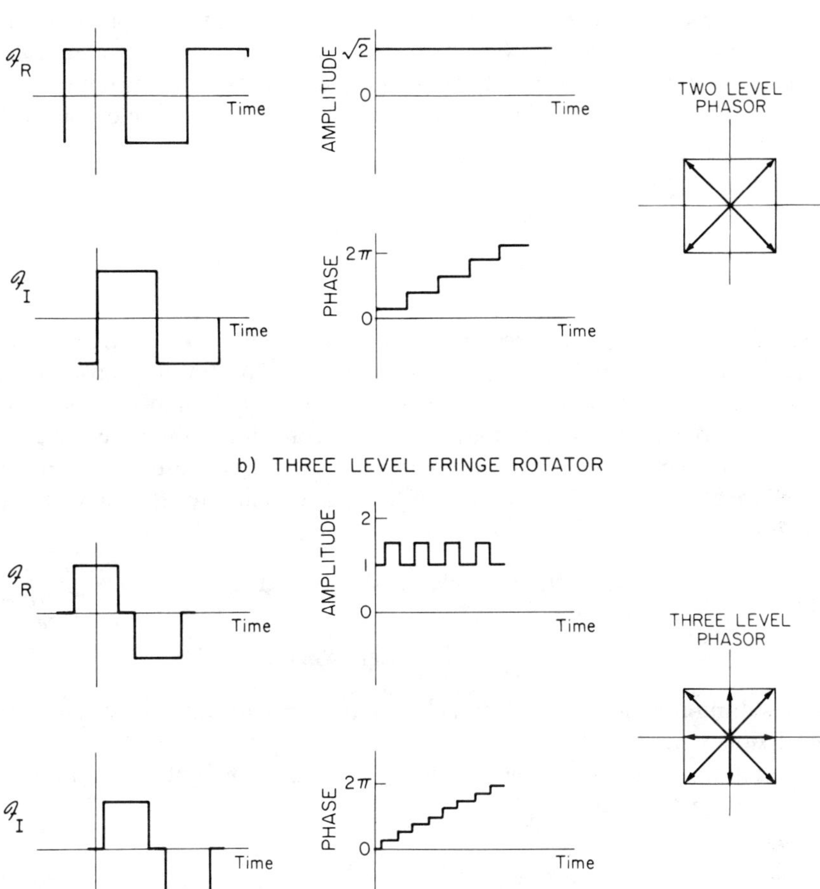

Figure 9.18 (a) Mathematical model of two-level fringe rotator showing $\mathscr{F}_R$ and $\mathscr{F}_I$, functions that approximate $\cos\phi$ and $\sin\phi$ (left), the amplitude and phase representation of $\mathscr{F}$ (middle), and the phasor plot of $\mathscr{F}$ (right). (b) Same plots for a three-level fringe rotator.

$(4/\pi)\cos(\pi/8) = 1.18$, which is the scale factor correction for the visibility. The three-level fringe function considered here is used in many VLBI processors. The fringe period is divided into 16 parts to generate $\mathscr{F}$. The transitions in $\mathscr{F}$, which then occur at integral multiples of 1/16 of the fringe period, are not optimally located but this approximation results in no more than 0.1% additional loss.

Fringe Sideband Rejection Loss (η_S)

The digital fringe rotator shown in Fig. 9.17c is not a complete single-sideband mixer. Thus as well as the wanted output, shifted in frequency by the fringe frequency, an unwanted component of noise corresponding to the image response of a mixer also appears. To understand the effect of this noise, consider the cross power spectrum of the correlator output. For upper-sideband operation, the cross power spectrum of the signal is given by Eq. (9.21), which is nonzero only for $v' > 0$. However, there will be noise at both positive and negative frequencies. Thus, the cross power spectrum of the output of the correlator is

$$\mathscr{S}'_{12}(v') = \mathscr{S}(v')e^{j\Phi(v')} + n_u(v'), \qquad v' > 0$$

$$= n_\ell(v'), \qquad\qquad\qquad v' < 0 \qquad (9.129)$$

where $\Phi(v')$ is the phase term in Eq. (9.21) and n_u and n_ℓ are the noise spectra for $v' > 0$ and $v' < 0$, respectively. For spectral-line observations, $\mathscr{S}'_{12}(v')$ must be computed, and the noise at $v' < 0$ is simply ignored. For continuum observations, the noise at $v' < 0$ contributes excess noise in the correlation function and must be removed. A straightforward way to remove the noise at $v' < 0$ is to compute $\mathscr{S}'_{12}(v')$ and multiply it by the filtering function

$$H_F(v') = 1, \qquad 0 < v' < \Delta v$$

$$(9.130)$$

$$= 0, \qquad \text{elsewhere.}$$

The resulting function, $\mathscr{S}'_{12}(v')H_F(v')$, can be Fourier transformed back into a correlation function. Alternatively, the filtering can be applied by convolving the correlation function at the output of the correlator by the Fourier transform of $H_F(v')$, which is

$$h_F(\tau) = \Delta v\, e^{j\pi\Delta v\,\tau}\left(\frac{\sin \pi\Delta v\,\tau}{\pi\Delta v\,\tau}\right), \qquad (9.131)$$

or

$$h_F(\tau) = F_1(\tau) + jF_2(\tau), \qquad (9.132)$$

where F_1 and F_2 are defined in Eq. (9.18). The convolution leaves the desired signal unchanged but removes the negative-sideband noise. Thus, the resulting correlation function still has the form of Eq. (9.20), plus the positive-sideband noise that cannot be removed.

The role of $h_F(\tau)$ can be understood in a different way. The correlation function at the output of the correlator is computed at discrete delays at intervals of $(2\Delta v)^{-1}$. Therefore the correlation function in Eq. (9.20) has a full width at half maximum of about three delay steps. In order to estimate the amplitude and phase of the correlation function, one would like to do more than just take these values from the peak of $\rho'_{12}(\tau)$. Rather, one would like to use all the information provided by the correlation function at various delays.

$h_F(\tau)$ is the appropriate interpolation function that properly weights the correlation function, gathering up the power at different delays to provide an optimal estimate of the fringe amplitude, phase, and delay. Note that $h_F(\tau)$ and $\rho'_{12}(\tau)$ are identical forms except for the unknown amplitude, phase, and delay. Thus these unknown quantities can be estimated by the usual procedure of matched filtering or, equivalently, least-mean-squares analysis in which the correlation function is convolved with $h_F(\tau)$. However, $\rho'_{12}(\tau)$ is measured only over a finite number of delay steps M and some information is lost and the signal-to-noise ratio reduced. Assume that the system lowpass response is rectangular and the delay errors $\Delta\tau_g$ and τ_e are zero, so that the correlation function is centered in the delay range of the correlator, which has M delay steps. The loss factor η_S is the ratio of the signal-to-noise ratio when M values of the correlation function are available, to the signal-to-noise ratio when the entire function is available:

$$\eta_S = \sqrt{\frac{\sum\limits_{k=-M'}^{M'} |h_F(\tau_k)|^2}{\sum\limits_{k=-\infty}^{\infty} |h_F(\tau_k)|^2}}, \tag{9.133}$$

where $\tau_k = k/2\,\Delta\nu$, $M' = (M-1)/2$, and M is an odd integer. The denominator in Eq. (9.133) equals $2\,\Delta\nu^2$ [e.g., see Eq. (A8.5)], so

$$\eta_S = \sqrt{\frac{1}{2} + \sum\limits_{k=1}^{M'} \left[\frac{\sin\left(\frac{\pi k}{2}\right)}{\frac{\pi k}{2}}\right]^2}. \tag{9.134}$$

For $M = 1$, $\eta_S = 1/\sqrt{2}$, which corresponds to the case of no image rejection. M must be at least 3 to ensure that the peak of the correlation function can be determined and $M \simeq 7$, for which $\eta_S = 0.975$, is adequate for most purposes. Note that because we assumed the correlation function was exactly centered, its value will be zero at delay steps $2, 4, 6, 8, \ldots$. This suggests that, for example, a 9-delay correlator is no better than a 7-delay correlator. In practice, the 9-delay correlator is better because the correlation function is rarely aligned perfectly in the correlator. In general, η_S is slightly smaller than given in Eq. (9.134) if the correlation function is not perfectly aligned (Herring 1983).

Discrete Delay Step Loss (η_D)

The delay introduced to align the bit streams is quantized at the sampling rate, which we assume to be the Nyquist rate. Thus there is a periodic sawtooth delay error with a peak-to-peak amplitude equal to the sampling period. This effect is also known as the fractional bit shift error. The delay error gives rise to a periodic phase shift that is a function of the baseband frequency, as shown in Fig. 9.19. The phase error has a peak-to-peak value of

$$\phi_{pp} = \frac{\pi \nu'}{\Delta \nu}, \tag{9.135}$$

and the sawtooth frequency is proportional to the fringe frequency and has a maximum value of

$$\nu_{ds(max)} = \frac{2\,\Delta\nu D\omega_e}{c} \quad \text{(delay steps per sec)}, \tag{9.136}$$

where D is the baseline length and ω_e is the angular velocity of the earth's rotation in rad sec^{-1}. If nothing is done to correct for this effect and the fringe amplitude is averaged over many times $1/\nu_{ds}$, then the phase at any frequency ν' is uniformly distributed over ϕ_{pp}. The amplitude loss as a function of baseband frequency is

$$L(\nu') = \frac{\sin(\phi_{pp}/2)}{\phi_{pp}/2}, \tag{9.137}$$

and the net signal-to-noise reduction over a baseband response of width $\Delta\nu$ is, using Eqs. (9.135) and (9.137),

$$\eta_D = \frac{1}{\Delta\nu} \int_0^{\Delta\nu} \frac{\sin(\pi\nu'/2\,\Delta\nu)}{\pi\nu'/2\,\Delta\nu}\, d\nu' = 0.873. \tag{9.138}$$

Unless the fringe amplitude averaging is done over an integral number of fringe periods, there is also a residual phase error whose magnitude decreases with the number of periods. At times when the fringe frequency is near zero, this phase error can be significant.

The effect of the discrete delay step can be compensated, and no sensitivity loss need occur. The delay error caused by delay quantization is a known quantity that introduces a phase slope in the cross power spectrum. Therefore, if the cross power spectra are calculated on a period short with respect to $1/\nu_{ds}$, which can be as small as 0.1 sec on a 5000-km baseline with $\Delta\nu = 2$ MHz [see Eq. (9.136)], then the effect of the discrete delay step can be removed by adjusting the slope of the phase of the cross power spectrum. This correction is easily done in spectral-line work where spectra are calculated anyway. Note that if this correction is not made, the sensitivity loss factor is 0.64 at the high-frequency edge of the band as given by Eq. (9.137). In this case, the amplitude response should be compensated by dividing the cross power spectra by $L(\nu')$. In continuum work, the correction is sometimes omitted because of the need to Fourier transform to the frequency domain and then back to cross correlation.

A way to compensate partially for the effect of discrete delay steps is to move the frequency at which the phase is unperturbed from zero frequency to

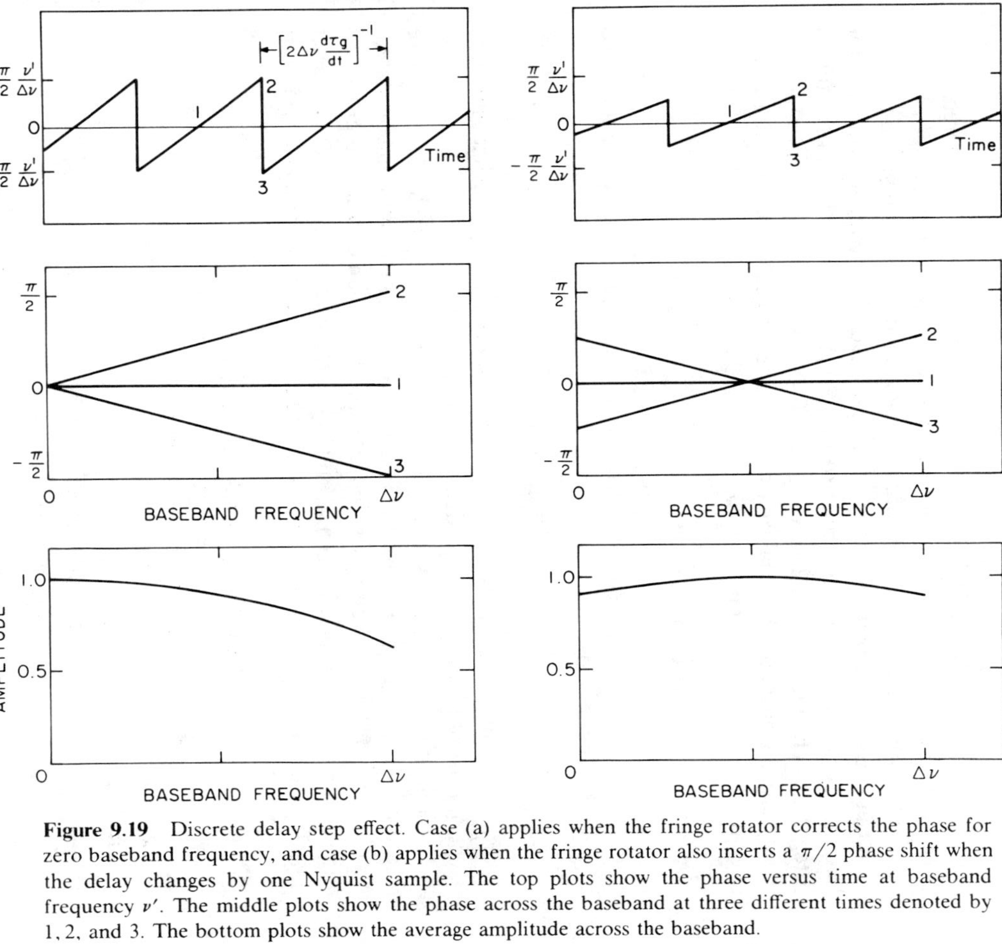

Figure 9.19 Discrete delay step effect. Case (a) applies when the fringe rotator corrects the phase for zero baseband frequency, and case (b) applies when the fringe rotator also inserts a $\pi/2$ phase shift when the delay changes by one Nyquist sample. The top plots show the phase versus time at baseband frequency ν'. The middle plots show the phase across the baseband at three different times denoted by 1, 2, and 3. The bottom plots show the average amplitude across the baseband.

$\Delta\nu/2$, the baseband center. The phase of the fringe rotator is increased by $\pi\,\Delta\nu\,\Delta\tau_s$, where $\Delta\tau_s$ is the delay error. Thus, when the delay changes by one sampling interval, a phase jump of $\pi/2$ is inserted in the fringe rotator. The resulting loss at the band edges is then only 0.90. The average loss over the band is given by an equation similar to Eq. (9.138), but with the upper limit of integration changed to $\Delta\nu/2$, and equals 0.966. Note that for a symmetrical bandpass response, the residual phase error is zero because the net phase shift over the band at any instant is zero.

Summary of Processing Losses

The loss factors that we have considered are all multiplicative, so the total loss is given by the equation

$$\eta = \eta_Q\,\eta_R\,\eta_S\,\eta_D, \tag{9.139}$$

where η_Q = quantization loss, η_R = fringe rotation loss, η_S = fringe sideband rejection loss, and η_D = discrete delay step loss.

If there are fringe rotators in each signal path to the correlator, the fringe rotation loss will be η_R^2 because the fringe rotator phases will be uncorrelated. A summary of the loss factors is given in Table 9.5. As an example, a processor might have two-level sampling ($\eta_Q = 0.637$), three-level fringe rotators in each signal path ($\eta_R = 0.922$), 11-channel correlation function ($\eta_S = 0.983$), and

TABLE 9.5   Signal-to-Noise Loss Factors

1.	*Quantization Loss* (η_Q)[a]	
	(a) Two-level	0.637
	(b) Three-level	0.810
	(c) Four-level, all products	0.881
2.	*Fringe Rotation Loss* (η_R)	
	(a) Two-level, one path	0.900
	(b) Three-level, one path	0.960
	(c) Two-level, both paths	0.810
	(d) Three-level, both paths	0.922
3.	*Fringe Sideband Rejection Loss* (η_S)	
	(a) 1 channel	0.707
	(b) 3 channels	0.952
	(c) 7 channels	0.975
	(d) 11 channels	0.983
4.	*Discrete Delay Step Loss* (η_D)	
	(a) Spectral correction	1.000
	(b) Video bandcenter correction	0.966
	(c) No correction	0.873

[a] Discussed in Chapter 8.

TABLE 9.6   Normalization Factors[a]

1.	*Quantization*[b]	
	(a) Two-level	1.57
	(b) Three-level	1.23
	(c) Four-level	1.13
2.	*Fringe Rotation*	
	(a) Two-level, one path	0.786
	(b) Three-level, one path	0.850
	(c) Two-level, both paths	0.617
	(d) Three-level, both paths	0.723

[a] Multiply correlator output by listed value
to obtain normalized correlation function.
[b] See Section 8.3.

bandcenter delay compensation ($\eta_D = 0.966$), giving a net loss of 0.558. Thus, the sensitivity is worse than that of an ideal analog system with the same bandwidth by about a factor of 2.

There are other loss factors that we have not discussed here. The passband will not in reality be perfectly flat, or the response zero for frequencies above half the Nyquist sampling frequency. These imperfections introduce loss, which for an ideal 9-pole Butterworth filter amounts to 2% (Rogers 1980). The frequency responses will not be perfectly matched for different antennas (see Section 7.3). The phase settings of the fringe rotator may be calculated exactly at convenient intervals and extrapolated by Taylor series: this approximation will introduce periodic phase jumps. The local oscillators may have power-line harmonic and noise sidebands that put some fringe power outside the usual fringe filter passband. Empirical values of η typical of the first decade of VLBI development are about 0.4 (Cohen 1973).

The η values refer to loss in signal-to-noise ratio. The fringe amplitudes must be corrected for scale changes due to signal quantization and fringe rotation. We summarize the multiplicative normalization factors to be applied to the fringe amplitudes in Table 9.6.

9.8 BANDWIDTH SYNTHESIS

For geodetic and astrometric purposes, it is useful to measure the geometrical group delay,

$$\tau_g = \frac{1}{2\pi} \frac{\partial \phi}{\partial \nu}, \tag{9.140}$$

as accurately as possible. With a single RF band, the delay can be found by fitting a straight line to the phase versus frequency of the cross power spectrum. The uncertainty in this delay, from the usual application of least-mean-squares analysis, is

$$\sigma_\tau = \frac{\sigma_\phi}{2\pi\,\Delta\nu_{\mathrm{rms}}}, \tag{9.141}$$

where σ_ϕ is the rms phase noise for a bandwidth $\Delta\nu$ and $\Delta\nu_{\mathrm{rms}}$ is the rms bandwidth, which for a single band of width $\Delta\nu$ is equal to $\Delta\nu/\sqrt{12}$ (see Appendix 12.1). σ_ϕ can be obtained from Eq. (6.57), and if processing losses are neglected Eq. (9.141) becomes

$$\sigma_\tau = \frac{T_S}{\beta T_A\sqrt{\Delta\nu_{\mathrm{rms}}^3\,\tau}}, \tag{9.142}$$

where β is a constant equal to $\pi(768)^{1/4} \simeq 16.5$ [see derivation of Eq. (A12.33)], and T_S and T_A are the geometric mean system and antenna temperatures. A much higher value of $\Delta\nu_{\mathrm{rms}}$ can be realized by observing at several different radio frequencies. This can be accomplished by switching the local oscillator of a single-band system sequentially in time among N frequencies, or by dividing up the recorded signal into N simultaneous RF bands (channels), which are spread over a wide frequency interval. The temporal switching method has the disadvantage that phase changes during the switching cycle degrade or bias the delay estimate. These methods are commonly referred to as bandwidth synthesis (Rogers 1970, 1976).

In a practical system, signals from a small number of RF bands ($\simeq 10$) are recorded. The problem of determining the optimum distribution of these bands in frequency is similar to the problem of finding a minimum-redundancy

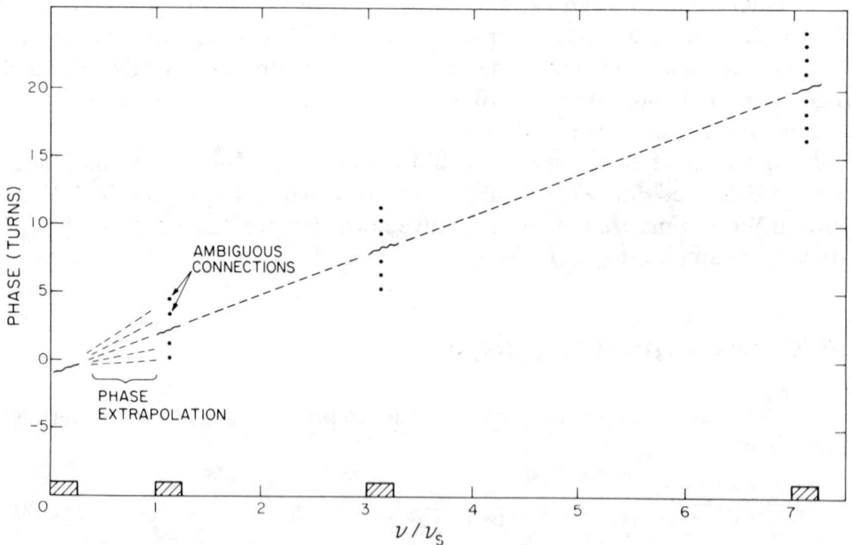

Figure 9.20 Fringe phase versus frequency for a bandwidth synthesis system. The phase is measured over discrete bands (crosshatched) spaced at multiples of the fundamental band separation frequency ν_s. The turn ambiguities give rise to sidelobes in the delay resolution function defined in Eq. (9.144) and shown in Fig. 9.21.

distribution of antenna spacing in a linear array, as discussed in Section 5.5. However, here we do not need to have all multiples of the unit (frequency) spacing up to the maximum value, and some gaps are not necessarily detrimental. From the spectral point of view, we wish to have the bands placed in some geometrical sequence of increasing separation so that phase can be extrapolated from one band to the next, as shown in Fig. 9.20, without having any 2π ambiguities in the phase connection process. The rms bandwidth depends critically on the unit spacing, which depends on the minimum signal-to-noise ratio. The delay accuracy for a multiband system is obtained from Eq. (9.141) in the same way as Eq. (9.142) but without the condition $\Delta\nu_{rms} = \Delta\nu/\sqrt{12}$. Thus we obtain

$$\sigma_\tau = \frac{T_S}{\sqrt{8\pi^2}\, T_A \sqrt{\Delta\nu\,\tau}\,\Delta\nu_{rms}}, \tag{9.143}$$

where $\Delta\nu_{rms}$ for a typical bandwidth synthesis system is approximately 40% of the total frequency interval spanned, $\Delta\nu$ is the total bandwidth, and τ is the integration time for each band. To avoid explicitly the problem of phase connection, we can form an equivalent delay function from the cross power spectra [see Eq. (9.21)] of the various bands observed:

$$D_R(\tau) = \sum_{i=1}^{N} \int_0^{\Delta\nu} \mathscr{S}_{12i}(\nu - \nu_i) e^{j2\pi\nu\tau}\, d\nu, \tag{9.144}$$

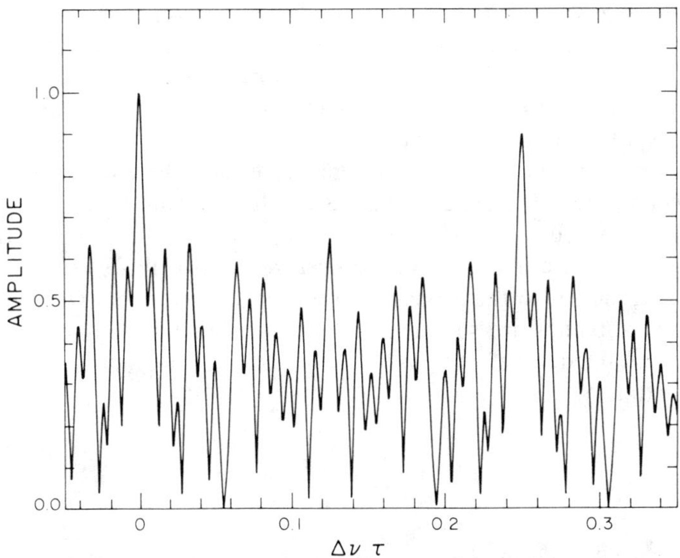

Figure 9.21   Delay resolution function for a five-channel system with a unit spacing $\nu_s = 4\,\Delta\nu$ and spacing of 0, 1, 3, 7, and 15 ν_s, as shown in part in Fig. 9.20. The "grating" lobe at $\tau\Delta\nu = 0.25$ need only be reduced sufficiently below unity to avoid delay ambiguity.

where the ν_i are the local oscillator frequencies relative to the lowest one, and $\nu - \nu_i$ is the baseband frequency. The maximum of $|D_R(\tau)|$ gives the maximum likelihood estimate of the interferometer delay (Rogers 1970). The a priori normalized delay resolution function, obtained by setting $\mathscr{S}_{12} = 1$ at frequencies where it is measured and $\mathscr{S}_{12} = 0$ otherwise in Eq. (9.144), is

$$|D_R(\tau)| = \Delta\nu \frac{\sin \pi \Delta\nu \tau}{\pi \Delta\nu \tau} \left| \sum_{i=1}^{N} e^{j2\pi\nu_i\tau} \right|. \qquad (9.145)$$

The sinc-function envelope is the delay resolution function for a single channel. The frequencies ν_i should be chosen to minimize the width of $D_R(\tau)$ while not allowing any subsidiary maximum to rise above a level such that it could be confused with the principal peak. In situations with low signal-to-noise ratio, the minimum unit spacing should be about four times the bandwidth of a single channel. The delay resolution function for a five-channel system is shown in Fig. 9.21.

9.9   PHASED ARRAYS AS VLBI ELEMENTS

It is important to consider the use of phased arrays as VLBI elements for two reasons. First, the elements of a connected-element array such as the VLA or the Westerbork array can be combined to form a phased array, thus improving the signal-to-noise ratio of a very-long-baseline interferometer in which they participate as a single station. Second, if elements of very large collecting areas are desired to achieve a high signal-to-noise ratio on each baseline, then it may be advantageous to build phased arrays rather than monolithic antennas because the cost of a parabolic reflector antenna increases approximately as the diameter to the power 2.7 (Meinel 1979).

Consider an array of N identical elements, each with system temperature T_S and collecting area A. Each antenna has a delay line and fringe rotator set for the source being observed. Thus, for a source that is unresolved by the array, the signals are those that would have been received if the antennas were in a plane perpendicular to the propagation vector to the source. The signals from the various antennas are summed, rather than combined in correlators. The resulting combined waveform can be treated as a normal radiometer signal and can be written

$$V_1 = \sum_{i=1}^{N} s_i(t) + n_i(t), \qquad (9.146)$$

where $\langle s_i^2 \rangle \propto T_A$, the antenna temperature for one element, and $\langle n_i^2 \rangle \propto T_S$. If this phased array were used with a total power radiometer, then the signal-to-noise ratio would be

$$\mathcal{R}_{\text{sn}} = \frac{NT_A\sqrt{\Delta\nu\,\tau}}{T_S + NT_A}. \tag{9.147}$$

This result is identical to that for a single element with antenna temperature NT_A. The beamwidth of the phased array is of order λ/D, where D is the width of the array, and the sidelobes depend on the element distribution. If V_1 is recorded and correlated with a signal from another VLBI element, then the signal-to-noise ratio will be $\sqrt{N}$ times that obtained with only one element of the phased array.

The elements in the phased array need not have the same collecting areas and system temperatures. In order to achieve maximum sensitivity in this case, the signal component from each element combined in the summing network should be set so that its voltage is proportional to its signal-to-noise ratio, that is, proportional to $\sqrt{T_{Ai}}/T_{Si}$. When the array is combined with another VLBI antenna, the resulting signal-to-noise ratio at the output of the VLBI correlator is

$$\mathcal{R}_{\text{sn}} = \sqrt{\frac{2\,\Delta\nu\,\tau T_A}{T_S} \sum_{i=1}^{N} \frac{T_{Ai}}{T_{Si}}}, \tag{9.148}$$

where T_S and T_A are the system and antenna temperatures of the other VLBI antenna.

If the array is instrumented for synthesis observations as a connected-element system, the phasing and calibration are conveniently accomplished using the hardware and software available for synthesis mapping. A calibration source, unresolved by the array, is observed in the usual correlator mode. From the correlator output phases, a set of N instrumental phases is calculated that is used to correct the signal phases before they are summed. If the source to be observed in the VLBI experiment is strong enough to be used as a calibrator, then the array can be continuously phased during the observation. If the array is not continuously phased, then it will slowly become dephased as a result of instrumental and atmospheric changes, so a calibrator must be observed periodically. In a correctly phased array the signals from a source combine coherently (the voltages add); in a randomly phased array they combine incoherently (the powers add), just as the noise components do. A randomly phased array is, on average, no more sensitive than a single antenna because the ratio of the signal to noise is not increased. Also, poor phasing makes calibration difficult.

In arrays that incorporate digital signal processing, the combined signal can be obtained by summing the quantized signals from the individual elements. With Q quantization levels and N elements, the combined signal can be represented by a number of quantization levels of order NQ. If N is, say, five or more, the combined waveform will have almost the appearance of an analog signal. For digital recording this waveform must be quantized again, and thus quantization losses are incurred twice.

BIBLIOGRAPHY

Biraud, F., Ed., *Very Long Baseline Interferometry Techniques*, Cepadues, Toulouse, France, 1983.

Brady, F. B., Ed., *Proc. IEEE*, Special Issue on Global Navigation Systems, Vol. 71, No. 10, 1983.

Chi, A. R., Ed., *Proc. IEEE*, Special Issue on Frequency Stability, Vol. 54, No. 2, 1966.

Fanti, R., K. Kellermann, and G. Setti, Eds., VLBI and Compact Radio Sources, *IAU Symposium 110*, D. Reidel, Dordrecht, 1984.

Kroupa, V. F., Ed., *Frequency Stability: Fundamentals and Measurement*, IEEE Press, New York, 1983.

Morris, D., Ed., *Radio Sci.*, Special Issue Devoted to the Open Symposium on Time and Frequency, Vol. 14, No. 4, 1979.

NASA, *Radio Interferometry Techniques for Geodesy*, NASA Conference Publication 2115, National Aeronautics and Space Administration, Washington, DC, 1980.

REFERENCES

Allan, D. W., Statistics of Atomic Frequency Standards, *Proc. IEEE*, **54**, 221–230, 1966.

Ashby, N. and D. W. Allan, Practical Implications of Relativity for a Global Coordinate Time Scale, *Radio Sci.*, **14**, 649–669, 1979.

Bare, C., B. G. Clark, K. I. Kellermann, M. H. Cohen, and D. L. Jauncey, Interferometer Experiments with Independent Local Oscillators, *Science*, **157**, 189–191, 1967.

Barnes, J. A., A. R. Chi, L. S. Cutler, D. J. Healey, D. B. Leeson, T. E. McGunigal, J. A. Mullen, W. L. Smith, R. L. Sydnor, R. F. C. Vessot, and G. M. R. Winkler, Characterization of Frequency Stability, *IEEE Trans. Instrum. Meas.*, **IM-20**, 105–120, 1971.

Blair, B. E., *Time and Frequency: Theory and Fundamentals*, NBS Monograph 140, U. S. Government Printing Office, Washington, DC, 1974, pp. 223–313.

Broten, N. W., T. H. Legg, J. L. Locke, C. W. McLeish, R. S. Richards, R. M. Chisholm, H. P. Gush, J. L. Yen, and J. A. Galt, Long Baseline Interferometry: A New Technique, *Science*, **156**, 1592–1593, 1967.

Cannon, W. H., The Classical Analysis of the Response of a Long Baseline Radio Interferometer, *Geophys. J. R. Astr. Soc.*, **53**, 503–530, 1978.

Clark, B. G., Radio Interferometers of Intermediate Type, *IEEE Trans. Antennas Propag.*, **AP-16**, 143–144, 1968.

Clark, B. G., The NRAO Tape-Recorder Interferometer System, *Proc. IEEE*, **61**, 1242–1248, 1973.

Clark, B. G., R. Weimer, and S. Weinreb, The Mark II VLB System, *NRAO Electronics Division Internal Report 118*, National Radio Astronomy Observatory, Green Bank, West Virginia, 1972.

Clark, T. A., B. E. Corey, J. L. Davis, G. Elgered, T. A. Herring, H. F. Hinteregger, C. A. Knight, J. I. Levine, G. Lundqvist, C. Ma, E. F. Nesman, R. B. Phillips, A. E. E. Rogers, B. O. Rönnäng, J. W. Ryan, B. R. Schupler, D. B. Shaffer, I. I. Shapiro, N. R. Vandenberg, J. C. Webber, and A. R. Whitney, Precision Geodesy Using the Mark III Very Long Baseline Interferometry System, *IEEE Trans. Geodesy Remote Sensing*, **GE-23**, 438–449, 1985.

Clark, T. A., C. C. Counselman, P. G. Ford, L. B. Hanson, H. F. Hinteregger, W. J. Klepczynski, C. A. Knight, D. S. Robertson, A. E. E. Rogers, J. W. Ryan, I. I. Shapiro, and A. R. Whitney, Synchronization of Clocks by Very Long Baseline Interferometry, *IEEE Trans. Instrum. Meas.*, **IM-28**, 184–187, 1979.

Cohen, M. H., Introduction to Very-Long-Baseline Interferometry, *Proc. IEEE*, **61**, 1192–1197, 1973.

Cohen, M. H. and D. B. Shaffer, Positions of Radio Sources from Long-Baseline Interferometry, *Astron. J.*, **76**, 91–100, 1971.

Cutler, L. S. and C. L. Searle, Some Aspects of the Theory and Measurement of Frequency Fluctuations in Frequency Standards, *Proc. IEEE*, **54**, 136–154, 1966.

D'Addario, L. R., *Minimizing Storage Requirements for Quantized Noise*, VLBA Memo. No. 332, National Radio Astronomy Observatory, Charlottesville, Virginia, 1984.

Davis, M. M., J. H. Taylor, J. M. Weisberg, and D. C. Backer, High Precision Timing of the Millisecond Pulsar PSR 1937 + 21, *Nature*, **315**, 547–550, 1985.

Dutta, P. and P. M. Horn, Low-Frequency Fluctuations in Solids: $1/f$ Noise, *Rev. Mod. Phys.*, **53**, 497–516, 1981.

Edson, W. A., Noise in Oscillators, *Proc. IRE*, **48**, 1454–1466, 1960.

Forman, P., Atomichron: The Atomic Clock from Concept to Commercial Product, *Proc. IEEE*, **73**, 1181–1204, 1985.

Frank, R. L., Current Developments in Loran-C, *Proc. IEEE*, **71**, 1127–1139, 1983.

Hellwig, H., Microwave Time and Frequency Standards, *Radio Sci.*, **14**, 561–572, 1979.

Herring, T. A., *Precision and Accuracy of Intercontinental Distance Determinations Using Radio Interferometry*, Air Force Geophysics Laboratory, Hanscom Field, Mass., AFGL-TR-84-0182, 1983.

Kartashoff, P. and J. A. Barnes, Standard Time and Frequency Generation, *Proc. IEEE*, **60**, 493–501, 1972.

Keshner, M. S., $1/f$ Noise, *Proc. IEEE*, **70**, 212–218, 1982.

Klemperer, W. K., Long Baseline Radio Interferometry with Independent Frequency Standards, *Proc. IEEE*, **60**, 602–609, 1972.

Klepczynski, W. J., Modern Navigation Systems and Their Relation to Timekeeping, *Proc. IEEE*, **71**, 1193–1198, 1983.

Kleppner, D., H. C. Berg, S. B. Crampton, N. F. Ramsey, R. F. C. Vessot, H. E. Peters, and J. Vanier, Hydrogen-Maser Principles and Techniques, *Phys. Rev. A.*, **138**, 972–983, 1965.

Kleppner, D., H. M. Goldenberg, and N. F. Ramsey, Theory of the Hydrogen Maser, *Phys. Rev.*, **126**, 603–615, 1962.

Kogan, L. R. and L. S. Chesalin, Software for VLBI Experiments for CS-Type Computers, *Sov. Astron.*, **25**, 510–513, 1982, translated from *Astron. Zh.*, **58**, 898–903, 1981.

Lesage, P. and C. Audoin, Characterization and Measurement of Time and Frequency Stability, *Radio Sci.*, **14**, 521–539, 1979.

Lindsey, W. C. and C. M. Chie, Frequency Multiplication Effects on Oscillator Instability, *IEEE Trans. Instrum. Meas.*, **IM-27**, 26–28, 1978.

Meinel, A. B., Multiple Mirror Telescopes of the Future, in *MMT and the Future of Ground-Based Astronomy*, T. C. Weeks, Ed., *SAO Special Report*, **385**, Harvard-Smithsonian Astrophysical Obs., Cambridge, Mass., 1979, pp. 9–22.

Moran, J. M., Spectral-Line Analysis of Very-Long-Baseline Interferometric Data, *Proc. IEEE*, **61**, 1236–1242, 1973.

Moran. J. M., Very Long Baseline Interferometric Observations and Data Reduction, in *Methods of Experimental Physics*, Vol. 12C, M. L. Meeks, Ed., Academic Press, New York, 1976, pp.

228–260.

Papoulis, A., *Probability, Random Variables and Stochastic Processes*, McGraw-Hill, New York, 1965.

Parkinson, B. W. and S. W. Gilbert, NAVSTAR: Global Positioning System—Ten Years Later, *Proc. IEEE*, **71**, 1177–1186, 1983.

Pierce, J. A., A. A. McKenzie, and R. H. Woodward, *Loran*, Radiation Laboratory Series, Vol. 4, McGraw-Hill, New York, 1948.

Potts, C. E. and B. Wieder, Precise Time and Frequency Dissemination via the Loran C System, *Proc. IEEE*, **60**, 530–539, 1972.

Press, W. H., Flicker Noises in Astronomy and Elsewhere, *Comments on Astrophys.*, **7**, 103–119, 1978.

Reid, M. J., A. D. Haschick, B. F. Burke, J. M. Moran, K. J. Johnston, and G. W. Swenson, Jr., The Structure of Interstellar Hydroxyl Masers: VLBI Synthesis Observations of W3(OH), *Astrophys. J.*, **239**, 89–111, 1980.

Rogers, A. E. E., Very Long Baseline Interferometry with Large Effective Bandwidth for Phase Delay Measurements, *Radio Sci.*, **5**, 1239–1247, 1970.

Rogers, A. E. E., Theory of Two-Element Interferometers, in *Methods of Experimental Physics*, Vol. 12C, M. L. Meeks, Ed., Academic Press, New York, 1976, pp. 139–157.

Rogers, A. E. E., The Sensitivity of a Very Long Baseline Interferometer, *Radio Interferometry Techniques for Geodesy*, NASA Conference Publication 2115, National Aeronautics and Space Administration, Washington, DC, 1980, pp. 275–281.

Rogers, A. E. E., R. J. Cappallo, H. F. Hinteregger, J. I. Levine, E. F. Nesman, J. C. Webber, A. R. Whitney, T. A. Clark, C. Ma, J. Ryan, B. E. Corey, C. C. Counselman, T. A. Herring, I. I. Shapiro, C. A. Knight, D. B. Shaffer, N. R. Vandenberg, R. Lacasse, R. Mauzy, B. Rayhrer, B. R. Schupler, and J. C. Pigg, Very-Long-Baseline Interferometry: The Mark III System for Geodesy, Astrometry, and Aperture Synthesis, *Science*, **219**, 51–54, 1983.

Rogers, A. E. E. and J. M. Moran, Coherence Limits for Very-Long-Baseline Interferometry, *IEEE Trans. Instrum. Meas.*, **IM-30**, 283–286, 1981.

Rutman, J., Characterization of Phase and Frequency Instability in Precision Frequency Sources: Fifteen Years of Progress, *Proc. IEEE*, **66**, 1048–1075, 1978.

Shapiro, I. I., Estimation of Astrometric and Geodetic Parameters, in *Methods of Experimental Physics*, Vol. 12C, M. L. Meeks, Ed., Academic Press, New York, 1976, pp. 261–276.

Shapiro, L.D. and D. O. Fisher, Using Loran-C Transmissions for Long Baseline Synchronization, *Radio Sci.*, **5**, 1233–1238, 1970.

Shimoda, K., T. C. Wang, and C. H. Townes, Further Aspects of the Theory of the Maser, *Phys. Rev.*, **102**, 1308–1321, 1956.

Siegman, A. E., *An Introduction to Lasers and Masers*, McGraw-Hill, New York, 1971, p. 404.

Thomas, J. B., *An Analysis of Radio Interferometry with the Block 0 System*, JPL Publication 81-49, Jet Propulsion Laboratory, Pasadena, California, 1981.

U.S. Naval Observatory, *Time Service Announcement*, Series 9, No. 203, 1982.

Vanier, J., M. Têtu, and L. G. Bernier, Transfer of Frequency Stability from an Atomic Frequency Reference to a Quartz-Crystal Oscillator, *IEEE Trans. Instrum. Meas.*, **IM-28**, 188–193, 1979.

Vessot, R. F. C., Frequency and Time Standards, in *Methods of Experimental Physics*, Vol. 12C, M. L. Meeks, Ed., Academic Press, New York, 1976, pp. 198–227.

Vessot, R. F. C., Relativity Experiments with Clocks, *Radio Sci.*, **14**, 629–647, 1979.

Vessot, R. F. C. and M. W. Levine, A Method for Eliminating the Wall Shift in the Atomic Hydrogen Maser, *Metrologia*, **6**, 116–117, 1970.

Vessot, R. F. C., M. W. Levine, E. M. Mattison, T. E. Hoffman, E. A. Imbier, M. Têtu, G.

Nystrom, J. J. Kelt, H. F. Trucks, and J. L. Vaniman, Space-Borne Hydrogen Maser Design, *Proceedings of the Eighth Annual Precise Time and Time Interval Meeting*, U. S. Naval Research Laboratory, X-814-77-149, 1976, pp. 227–333.

Whitney, A. R., A. E. E. Rogers, H. F. Hinteregger, C. A. Knight, J. I. Levine, S. Lippincott, T. A. Clark, I I. Shapiro, and D. S. Robertson, A Very Long Baseline Interferometer System for Geodetic Applications, *Radio Sci.*, **11**, 421–432, 1976.

Wineland, D. J., Trapped Ions, Laser Cooling, and Better Clocks, *Science*, **226**, 395–400, 1984.

Yen, J. L., K. I. Kellermann, B. Rayhrer, N. W. Broten, D. N. Fort, S. H. Knowles, W. B. Waltman, and G. W. Swenson, Jr., Real-Time, Very-Long-Baseline Interferometry Based on the Use of a Communications Satellite, *Science*, **198**, 289–291, 1977.

10

CALIBRATION AND FOURIER TRANSFORMATION OF VISIBILITY DATA

In this chapter we discuss in more detail the calibration and Fourier transformation of visibility data. The use of the fast algorithm for the discrete Fourier transform (FFT) and methods for evaluation of the visibility at rectangular grid points are included. Some practical hints on the recognition and avoidance of errors in maps and the planning of observations are also discussed. The terms map, image, and brightness distribution are used interchangeably to denote the output of the mapping process.

10.1 CALIBRATION OF THE VISIBILITY

The purpose of calibration is to remove, insofar as possible, the effect of instrumental factors in the measurements. Such factors largely depend on the individual antennas or antenna pairs, so correction must be applied to the visibility data before they are combined into a map. Editing the data to delete any that show evidence of interference or equipment malfunction is usually performed before the calibration proper. This mainly entails examining samples of the data for unexpected levels or variations. Data taken on calibration sources are particularly useful here, since the response to the calibrator is predictable and should vary only slowly and smoothly with time.

In the calibration procedure we first consider instrumental factors that are stable with time over periods of weeks or more. These include the following:

1. Antenna position coordinates that specify the baselines.
2. Antenna pointing corrections resulting from tolerances on axis alignments and so on.
3. Zero-point settings of the instrumental delays; that is, the settings for which the delays from the antennas to the correlator inputs are equal.

The parameters above vary only as a result of major changes such as the relocation of an antenna. They can be calibrated by observing sources with known positions. We assume here that they have been determined in advance of the mapping observation, and accurate values are used in controlling the array. We also assume that correction for the nonlinearity of signal quantization, if necessary, is applied automatically as the data are accumulated.

Corrections for Predictable or Directly Measurable Effects

Calibration of the visibility measurements for effects that vary during an observation principally involves correction of the complex gains of the antenna pairs. Such factors can be divided into those for which the behavior can be predicted or directly measured and those for which it cannot. The following effects are largely predictable:

1. The constant component of atmospheric attenuation as a function of zenith angle (see Chapter 13).
2. Variation of antenna gain as a function of elevation caused by elastic deformation of the structure.
3. Shadowing of one antenna by another at close spacings and low elevation angles.

Effects that can be directly estimated during an observation include:

1. Variation of system noise temperature and the resulting variation of gain from the ALC action (see Section 7.6).
2. Phase variations in the local oscillator system monitored by the round-trip phase measurement (see Section 7.2).
3. The variable component of atmospheric delay monitored by using water vapor radiometers (see Chapter 13) mounted at the antennas.

Corrections for the above effects are usually the first stage of the calibration procedure.

Use of Calibration Sources

Further steps in the calibration require the observation of one or more calibration sources. From Eq. (4.5) we can write the expression for the interferometer response as follows:

$$[\mathcal{V}(u, v)]_{\text{uncal}} = G_{mn}(t) \int_{-\infty}^{\infty} \int_{-\infty}^{\infty} \frac{A_N(\xi, \eta) B(\xi, \eta)}{\sqrt{1 - \xi^2 - \eta^2}} e^{-j2\pi(u\xi + v\eta)} d\xi\, d\eta.$$

(10.1)

Here the uncalibrated visibility includes a complex gain factor $G_{mn}(t)$ that is a function of the antenna pair (m, n) and of time. [From here on we generally omit the factor $A_N(\xi, \eta)/\sqrt{1 - \xi^2 - \eta^2}$ in the brightness–visibility relationship since it is close to unity.] To calibrate $G_{mn}(t)$ an unresolved source can be observed, with the phase reference position (field center) chosen to coincide with the source. Then the response is

$$\mathcal{V}_c(u, v) = G_{mn}(t) S_c,$$

(10.2)

where the subscript c indicates the calibrator, for which S_c is the flux density. To remove the system gain from the measurements of the source being mapped we can write

$$\mathcal{V}(u, v) = \frac{[\mathcal{V}(u, v)]_{\text{uncal}}}{G_{mn}(t)}$$

$$= [\mathcal{V}(u, v)]_{\text{uncal}} \left(\frac{S_c}{\mathcal{V}_c} \right).$$

(10.3)

Here $[\mathcal{V}(u, v)]_{\text{uncal}}$ has been corrected only for the predictable or directly measurable effects. These same corrections should also be applied to the calibrator data $\mathcal{V}_c$, for which we have omitted the (u, v) dependence in Eq. (10.3) on the assumption that the calibrator is unresolved. Since $G_{mn}(t)$ varies with time it is necessary to interrupt the mapping observations for periodic measurements of the calibrator. Values of $G_{mn}(t)$ can then be interpolated as required. The interval between calibration observations depends on the stability of the instrument and typically falls within the range of 15 min to several hours.

The desirable charcteristics of a calibration source are the following.

1.   The calibrator should be strong, so that a good signal-to-noise ratio is obtained in a short time, to minimize the (u, v) coverage lost from the source

being mapped. The gaps in the (u, v) coverage are more serious for a linear array, in which complete sectors are lost, than for a two-dimensional array in which the instantaneous coverage is more widely distributed (e.g., see Fig. 5.18).

2.   The calibrator should be unresolved, so that precise measurements of its visibility are not required. As indicated by Eq. (7.27), the measured gains for antenna pairs can be used to determine gain factors for the individual antennas. With this technique some of the spacings can be omitted from the calibration so long as each of the antennas is included.

3.   The position of the calibrator should be close to that of the source being mapped: an angular distance no greater than a few degrees is desirable. Effects in the atmosphere or antennas that cause the gain to vary with pointing angle are then more effectively removed, and time lost in driving the antennas between the map and calibrator positions is kept small.

It is not always possible to find a calibrator that satisfies all the above requirements. In particular, there are few unresolved calibrators suitable for VLBI observations with milliarcsecond resolution. Angular structure on this scale is sometimes variable over periods of months, and caution is necessary if a previously measured, partially resolved source is to be used as a calibrator. An alternative approach to amplitude calibration of VLBI data is to determine accurately the antenna collecting areas and system temperatures and, by using an independent measurement of the flux density of the source being mapped, to calculate the expected response if it were unresolved. The limited coherence times of independent frequency standards may make the calibration of the phase impractical. However, mapping techniques based on phase closure relationships as discussed in Chapter 11 enable information to be derived from the uncalibrated phase data.

Calibration of Spectral Line Data

In principle, the procedures for continuum observations described above could be applied individually to each channel of a spectral-line observation. However, the channel bandwidth is typically less than that of the full IF band by one or two orders of magnitude, and it would take an equivalent increase in the calibrator observing time to maintain the signal-to-noise ratio. Generally, therefore, one makes use of the fact that the channel-to-channel differences are relatively stable with time and need not be calibrated as frequently as the overall gain. Two steps are then required in the calibration. First, the responses of the spectral channels are summed to simulate the continuum response, and the continuum calibration procedures discussed above are applied equally to all channels. The second step is a bandpass calibration observation to determine the relative gains of the spectral channels. The bandpass calibration source should be unresolved, strong enough to provide good signal-to-noise ratio in the spectral channels, and have a flat spectrum. However, it need not be close in position to the source being mapped and can be observed before or after the mapping observations.

The quality of the bandpass calibration is the limiting factor in the accuracy with which spectral features can be detected. It is particularly critical in the study of spectral lines that are observed in the presence of a strong continuum background, such as hydrogen recombination lines that are emitted by ionized nebulae. In brightness these lines are only a few percent of the nebular thermal continuum, and without adequate bandpass calibration variations in the instrumental response to the continuum may be mistaken for true line features. See also the discussion of instrumental effects on spectra and the use of smoothing functions in Section 8.9.

Calibration of Polarization Data

In the calibration of polarization measurements two special requirements arise. These are the calibration of pairs of antennas with different polarizations, and the calibration and removal of instrumental polarization. As described in Section 4.9 it is necessary to use combinations of various antenna polarizations to measure the visibility components of the four Stokes parameters. Various combinations have been used in different instruments, and we consider two important examples; rotatable, linearly polarized feeds and circularly polarized feeds with both senses of rotation received at each antenna.

In either case some of the data are taken with nominally identical polarizations, and for these data the calibration follows the general procedures outlined earlier. Where appropriate, the same procedures are also applied to the data obtained with opposite polarizations. If the calibrator is polarized the amount and position angle should be known so that the appropriate flux density values can be used in Eq. (10.3). With rotatable linear polarizations the gain factors determined from the calibration observations with parallel feeds can be applied to the crossed polarization observations of the source being mapped, since the rotation of the feed should not involve any change in gain or electrical length of the signal paths. Observations using crossed linear polarizations must be corrected also for the instrumental polarization $P_{nmp}I$ introduced in Section 4.9. This is most easily done by observing an unpolarized calibration source, since Q, U, and V are then zero and the instrumental polarization is measured directly. The instrumental component may vary only slowly with time, in which case it is necessary to measure it only once or twice during the mapping period. A different method used by Hargrave and Ryle (1974) removes the first-order instrumental polarization by interchanging the polarization angles of the antennas. For example, the data for U are the averages of the data obtained with polarization angles $(0°, 90°)$ and $(90°, 0°)$.

Calibration of an array with circularly polarized antennas involves some different procedural details. We assume that for each pair of antennas the four combinations, RR, LL, RL, and LR, are recorded simultaneously. (The calibration for crossed linear feeds with all four combinations recorded can follow a similar approach.) Periodic observations of a calibrator are used to determine the gain variation with time, and the individual antenna voltage gains g_m, g_n are calculated from the pair gains G_{mn}. For each antenna there are two gain factors, one for the right-hand channel and one for the left. From the antenna gains the pair gains are obtained for both like and opposite hands by combining the appropriate gain factors. However, calibration of the gains for the opposite-hand combinations requires a further step. The calibration observations usually measure only the differences of the signal phases for like pairs of antennas and do not include any combination of opposite hands. Therefore, the pair gains for opposite hands contain an unknown phase term that is a constant for all RL pairs and has the opposite sign for the LR pairs. This phase offset is determined by an observation of a source that contains a known component of linear polarization. Finally, it is necessary to measure the instrumental component of the opposite-hand combinations. As described for the linearly polarized case this can be done by observing an unpolarized source. However, if altazimuth mounts are used the antennas rotate relative to the sky as a source is tracked, and the measurement can then be made on a source that contains an *unknown* linearly polarized component. The response in this case consists of two parts: the response of the instrumental polarization to the unpolarized component of the source, and the opposite-hand response to the linear component of the source given by the expressions in (4.49). The former of these is independent of the antenna rotation but the second varies as a sinusoid of twice the parallactic angle. If the calibration source is observed at intervals over several hours the two parts of the response can be separated.

Several further complications that apply to polarization measurements should be mentioned. First, the instrumental polarization usually varies over the antenna beam, so it is simplest to work only near the beam center. At high frequencies the pointing accuracy of the antenna beams may thus become critical. Also, caution is necessary in interpreting polarization maps of wide sources. Second, at frequencies up to a few gigahertz there is measurable rotation of the plane of linear polarization by the Faraday effect in the ionosphere, which is discussed further in Chapter 13. At long wavelengths it is useful to monitor this rotation by periodically observing a source with known polarization. Finally, care must be exercised in estimating derived parameters such as the percentage of polarization and the position angle of linear polarization, since for low signal-to-noise ratios these have non-Gaussian probability distributions. The quantity $m_\ell I = \sqrt{Q^2 + U^2}$ has a Rice distribution of the form of Eq. (6.56a), and the position angle has a distribution of the form of Eq. (6.56b). In particular, the percentage polarization can be overestimated, and a correction must be applied: see Wardle and Kronberg (1974). The discussion in Section 9.3 at the end of *Noise in VLBI Observations* is also relevant to this problem.

10.2   DERIVATION OF BRIGHTNESS FROM VISIBILITY

In this section we consider derivation of the brightness distribution from visibility data that have been calibrated as described above.

Model Fitting

The fitting of brightness models to visibility data was practiced extensively in early radio interferometry, especially when the visibility phase was poorly calibrated or the data were not sufficiently complete to allow Fourier transformation. Examples of models are shown in Figs. 1.5, 1.8, and 1.12. In the absence of phase information there is always an ambiguity of 180° in position angle of the model.

Gaussian functions are convenient model components for source brightness. They are always positive and vary smoothly with angle as do many of the structures in nebulae and radio galaxies. A circularly symmetrical Gaussian function of half-amplitude width $\sqrt{8\ln 2}\ \sigma$, centered at (ξ_1, η_1), is represented by

$$B_G(\xi, \eta) = B_0 \exp\left[\frac{-(\xi - \xi_1)^2 - (\eta - \eta_1)^2}{2\sigma^2}\right]. \tag{10.4}$$

The corresponding visibility function is

$$\mathscr{V}_G(u, v) = \sqrt{2\pi}\ \sigma B_0 \exp\left\{-\left[2\pi^2\sigma^2(u^2 + v^2) + j2\pi(u\xi_1 + v\eta_1)\right]\right\}. \tag{10.5}$$

The visibility has real and imaginary components that are sinusoidal corrugations, the ridges of which are normal to the radius vector to the point (ξ_1, η_1) in the brightness domain. These visibility components are modulated in amplitude by a Gaussian function centered on the (u, v) origin and of width inversely proportional to σ. Examination of the visibility distribution can thus indicate the form and position of the main brightness components. For discussions and examples of this type of model fitting see, for example, Maltby and Moffet (1962), Fomalont (1968), and Fomalont and Wright (1974).

A relationship between the visibility function and the moments of the brightness distribution provides some further insight into model fitting. In one dimension, for simplicity, the visibility function can be expressed as a Taylor series:

$$\mathscr{V}(u) = \mathscr{V}(0) + u\mathscr{V}'(0) + \frac{u^2}{2!}\mathscr{V}''(0) + \cdots + \frac{u^n}{n!}\mathscr{V}^{(n)}(0) + \cdots \tag{10.6}$$

The derivatives of the visibility are related to the moments of the brightness distribution as follows:

$$\mathscr{V}^{(n)}(0) = (-j2\pi)^n \int_{-\infty}^{\infty} \xi^n B_1(\xi)\, d\xi. \tag{10.7}$$

Equation (10.7) follows from a general relationship between the derivatives of a function at the origin and the moments of its Fourier transform (e.g., Bracewell 1965). Thus from Eqs. (10.6) and (10.7) we can write

$$\mathscr{V}(u) = S - j2\pi u \int_{-\infty}^{\infty} \xi B_1(\xi)\, d\xi - 2\pi^2 u^2 \int_{-\infty}^{\infty} \xi^2 B_1(\xi)\, d\xi$$

$$+ \cdots + \frac{(-j2\pi)^n}{n!} \int_{-\infty}^{\infty} \xi^n B_1(\xi)\, d\xi + \cdots \tag{10.8}$$

The zero-order moment is equal to the flux density S, the odd-order moments contribute to the imaginary components of the visibility, and the even-order moments contribute to the real part. If the source is symmetrical in ξ, the odd-order terms are zero. If, in addition, the source is only slightly resolved the decrease in $\mathscr{V}$ results mainly from the second-moment term. Then the source can be represented by any symmetrical model with an appropriate second moment (Moffet 1962).

For certain types of sources the radio emission can be specified with reasonable accuracy in terms of a physical model that involves only a small number of parameters. An example is thermal radiation from stellar envelopes. In cases where the source is only slightly resolved, a brightness map obtained by Fourier transformation differs from the point source response by only a very slight broadening. The parameters of the physical model can then best be determined by fitting the model visibility function directly to the observed values; see, for example, White and Becker (1982).

Mapping by Direct Fourier Transformation

The most straightforward method of obtaining a brightness distribution from measured visibility data is by *direct* Fourier transformation; that is, by performing the transformation without putting the visibility into any special form such as that for the fast algorithm described in Section 4.10. The measured visibility $\mathscr{V}_m(u, v)$ can be written

$$\mathscr{V}_m(u, v) = W(u, v)w(u, v)\mathscr{V}(u, v), \tag{10.9}$$

where $W(u, v)$ is the transfer function or spectral sensitivity function introduced in Chapter 5, and $w(u, v)$ represents any applied weighting. The Fourier transform of Eq. (10.9) is the measured brightness distribution which is

$$B_m(\xi, \eta) = B(\xi, \eta) ** b_0(\xi, \eta). \tag{10.10}$$

b_0 is the synthesized beam which is the Fourier transform of the weighted transfer function:

$$b_0(\xi, \eta) \rightleftharpoons W(u, v)w(u, v). \tag{10.11}$$

We assume that effects such as those of non-coplanar baselines, the signal bandwidth, and the visibility averaging can be neglected or considered separately.

The visibility is measured at an ensemble of n_d pairs of points symmetric about the (u, v) origin, and the direct Fourier transform of these data is represented by

$$\sum_{i=1}^{n_d} w_i \left[\mathcal{V}_m(u_i, v_i) e^{j2\pi(u_i\xi + v_i\eta)} + \mathcal{V}_m(-u_i, -v_i) e^{-j2\pi(u_i\xi + v_i\eta)} \right].$$

$$\tag{10.12}$$

Since $\mathcal{V}_m(u_i, v_i) = \mathcal{V}_m^*(-u_i, -v_i)$ the derived brightness is real (we assume that the antennas are identically polarized). The weighting factor w_i is introduced to control the form of the synthesized beam, and its value is related to the area density of the data at any point. The area density $\rho_\sigma(u, v)$ is defined such that the number of points in the range $u \pm \frac{1}{2}du$, $v \pm \frac{1}{2}dv$ is $\rho_\sigma(u, v)du\,dv$ (Thompson and Bracewell 1974). Although ρ_σ at any point depends on the size of the increments du and dv, it is usually simple to specify the variation of relative density with sufficient accuracy to obtain a satisfactory beam. As a simple example, in the observation of a high-declination source with an east–west array in which the antenna spacings are nonredundant integral multiples of a unit value, the visibility points lie on concentric circles as in Fig. 10.1. Then if the visibility is measured at uniform increments in hour angle, the area density at any ring is inversely proportional to the radius of the ring. With $w(u, v)$ proportional to $1/\rho_\sigma(u, v)$, the effective density of the data is uniform within a circle of radius $u_{\max}$ determined by the maximum spacing. The beam then closely approximates the Fourier transform of a circular disk function and is given by

$$\frac{J_1(2\pi\xi u_{\max})}{\pi\xi u_{\max}}, \tag{10.13}$$

where J_1 is the Bessel function of the first kind and first order. Such a response has a first sidelobe of 13.2% of the main beam. The full width of the beam at half-maximum is 0.705 $u_{\max}^{-1}$, or $0.705\lambda/D_{\max}$, where $D_{\max}$ is the baseline length corresponding to $u_{\max}$.[†] Similarly, if the effective density is uniform

---

[†] This result should not be confused with the calculation of the gain of a uniformly illuminated parabolic antenna of diameter D, which is proportional to $[2J_1(\pi D\xi/\lambda)/(\pi D\xi/\lambda)]^2$ and has a full width at half-maximum of $1.02\lambda/D$, first null at $1.22\lambda/D$, and first sidelobe of 1.7%.

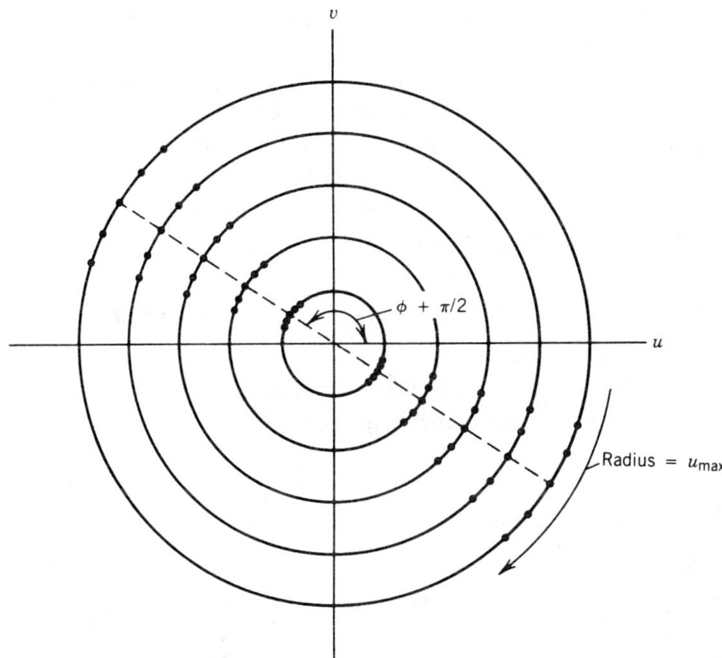

Figure 10.1 Transfer function (spacing loci) in the (u, v) plane for observations of a high-declination source using an east–west array with uniform increments in antenna spacing. The points indicate visibility measurements, and their (u, v) positions reflected through the origin, for uniform intervals of time. The angle ϕ indicates data for a specific hour angle. If the visibility values are weighted in proportion to the radii of the loci, the density of the visibility data is effectively uniform out to a radius u_{max}. This diagram also applies to observations at any declination if u and v are replaced by u' and v' which represent the projection of the spacings onto the equatorial plane.

within a rectangular area of dimensions $2u_{max}$ by $2v_{max}$ the beam is closely approximated by

$$\frac{\sin(2\pi u_{max}\xi)}{2\pi u_{max}\xi} \times \frac{\sin(2\pi v_{max}\eta)}{2\pi v_{max}\eta}. \tag{10.14}$$

The beam is not circularly symmetrical and the first sidelobe has a maximum value of 22% in the east–west and north–south directions through the beam center. Maps derived from visibility data that are effectively uniformly distributed within the measured area of the (u, v) plane are referred to as the *principal solution* or *principal response* for the visibility data (Bracewell and Roberts 1954, Thompson and Bracewell 1974).

The sidelobes of the beam functions in Expressions (10.13) and (10.14) can be reduced at the expense of some increase in the width of the main beam by introducing a Gaussian or similar taper into the weighting function. The effect of such tapering of the visibility is shown in Fig. 10.2. The taper can be specified in terms of the amplitude of the tapering function at a distance u_{max} from the (u, v) origin; a taper to a few tenths of the central value is commonly used. With such a taper the weighting $w(u, v)$ is the product of two functions: $w_u(u, v)$, the weighting required to obtain uniform effective density, and $w_t(u, v)$, the tapering function. Thus, the synthesized beam is the Fourier transform of $W(u, v)w_u(u, v)w_t(u, v)$:

$$b_0(\xi, \eta) = \overline{W}(\xi, \eta) ** \overline{w}_u(\xi, \eta) ** \overline{w}_t(\xi, \eta), \qquad (10.15)$$

where the bar denotes a Fourier transform. The Fourier transform of $W(u, v)w_u(u, v)$ is simply the beam obtained from the uniform effective

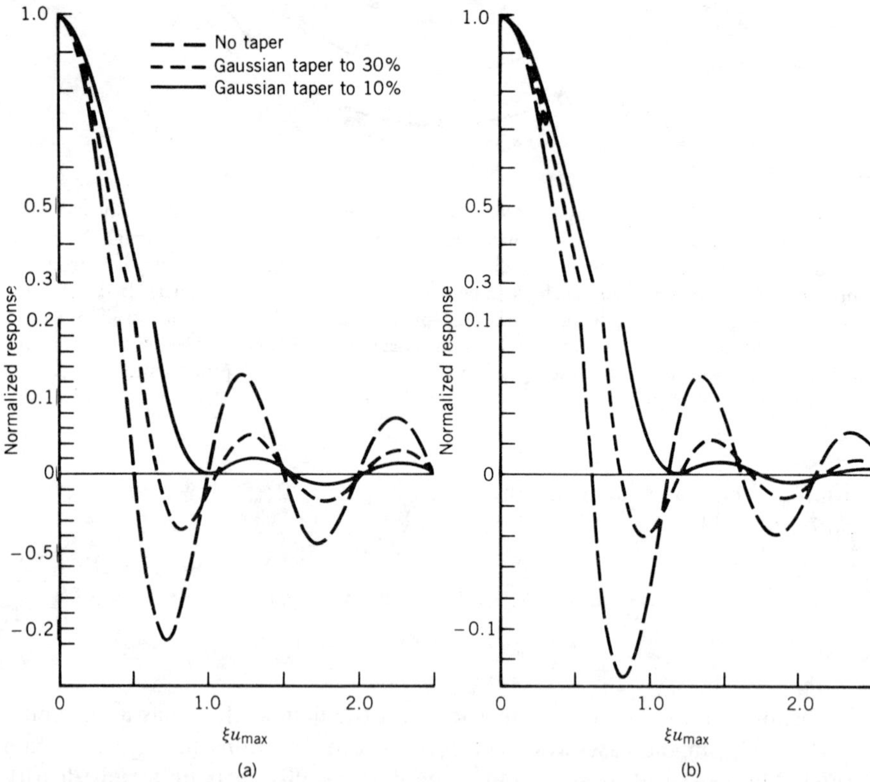

Figure 10.2 Examples of synthesized-beam profiles. Curves for no taper correspond to a visibility distribution that is uniform within (a) a rectangular area of width $2u_{max}$ and (b) a circular area of diameter $2u_{max}$. These curves show the variation with ξ in Expressions (10.14) for (a) and (10.13) for (b) and correspond to the principal response. The effects of Gaussian tapers which reduce the visibility at the edge of the distribution to 30% and to 10% are also shown. Note the differences in the ordinate scales.

density, for example, Expression (10.13) or (10.14). If $w_t(u, v)$ is a two-dimensional Gaussian function its Fourier transform is also a Gaussian. Thus the sidelobe reduction results from convolution with a Gaussian. The variances of functions are additive under convolution (e.g., Bracewell 1965) so the beam obtained by convolution with $\bar{w}_t$ is broader than that with no tapering, as is evident in Fig. 10.2. The Gaussian taper is often used when it is necessary to degrade the resolution, for example, to match that of another map, or when searching for an extended, low-brightness source.

With a tracking array the area density of the visibility data is usually greatest near the (u, v) origin because the spacing vectors move more slowly with hour angle in that region; this is particularly obvious for an east–west array as in Fig. 10.1. Thus the inverse density weighting $w_u(u, v)$ and the tapering $w_t(u, v)$ to some extent cancel one another. However, if both are omitted the synthesized beam is simply the Fourier transform of the transfer function. This situation is sometimes referred to as natural weighting. As shown in Section 6.2, it maximizes the signal-to-noise ratio for a point source, but the resolution is degraded and the beam often has undesirably broad skirts. The loss in sensitivity in improving the beamshape is usually small, and attention to the correct weighting is important.

Implementation of the Direct Fourier Transform

In Fourier transformation of the visibility the brightness is usually computed at points in a rectangular grid with uniform increments in ξ and η, since this is a very convenient form for subsequent processing. The direct implementation of the transformation in Expression (10.12) is straightforward, but it is interesting to review some particular cases. In arrays where the data gathering rate is not too high it is possible to perform the mapping transformation as the data are gathered, since Fourier transformation is a linear (additive) process. If the averaged visibilities are read out from the correlators at intervals of a few seconds, the calibration and transformation can be applied and the results added into an array in which the brightness distribution is accumulated before the next set of visibility measurements appears. Only a small computer, such as the one used to control the array, may be needed. In the case of an east–west linear array the visibility data at any instant are located on a straight line through the origin in the (u, v) plane; see Fig. 10.1. One such set of points contains the information in a fan-beam scan of the source, and the Fourier transform of the points takes the form of a corrugated surface as shown in Fig. 10.3. This function must be sampled on a rectangular grid in ξ and η before addition into the brightness array. An application of this method to the Cambridge Five-Kilometer telescope is described by Kenderdine (1974).

If no weighting were applied to the visibility in the above method, the map would consist of a summation of functions of the form of that in Fig. 10.3, the profiles of which are identical to fan-beam scans of the field with varying position angle. These would combine to form a map with the undesirable

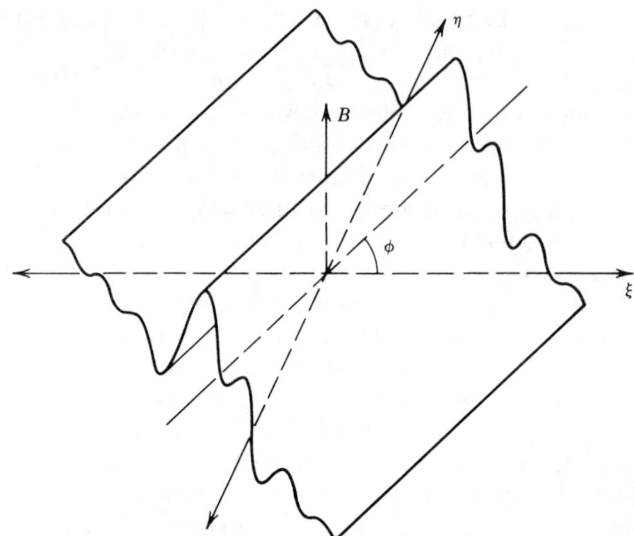

Figure 10.3   A surface in the (ξ, η) plane which is the Fourier transform of visibility data in the (u, v) plane measured along a line making an angle $\phi + \pi/2$ with the u axis, as shown by the broken line in Fig. 10.1.

natural weighting. However, the weighting can be adjusted easily for data in the (u, v) plane. It is interesting to recall that during the 1950s a number of phased arrays were in use that produced fan-beam scans as a direct output, and the combination of such data into two-dimensional maps with a desirable weighting was a laborious process. Christiansen and Warburton's (1955) solar map, the first example of the use of earth rotation in two-dimensional mapping, involved Fourier transformation, weighting, and retransformation of the data by manual calculation. A method of combining fan-beam scans without Fourier transformation was later devised by Bracewell and Riddle (1967) using a convolution to adjust the visibility weighting. For discussion of some basic relationships between one- and two-dimensional responses see Bracewell (1956a). The same concepts are applicable to image processing in other fields, for example, tomography (Bracewell and Wernecke 1975).

Frater and Skellern (1978) and Frater (1978) have discussed the design of special hardware to perform the transformation of data distributed in (u, v, w) space in the manner of interferometer observations. The aim is to produce a special-purpose computer for a lower cost than a general-purpose computer that would perform the same operation. This goal is most simply achieved for linear arrays of antennas in which the data at any instant lie on a straight line in (u, v, w) space. Such special-purpose computers may be rather inflexible and difficult to adapt to evolving methods of image processing.

Analog devices for data processing have also been considered for synthesis mapping but are not widely used. These are based on optical, acoustic, or electron-beam processes for Fourier transformation. Like special-purpose digital hardware, the analog elements provide increased speed at the expense of flexibility. In the analog case, a further problem is limitation of the *dynamic range*, which is the ratio of the highest to the lowest brightness levels of reliable detail in the image. Maintaining image quality in any iterative process that requires successive Fourier transformation and retransformation (see Chapter 11) requires high precision. For a review of analog processing possibilities see Cole (1979).

Mapping by Discrete Fourier Transformation

The speed of the fast algorithm for the discrete Fourier transform (FFT) described in Section 4.10 is a major advantage in computing large maps. However, the use of the FFT introduces two complications in addition to those discussed for the direct transform: (1) the necessity to evaluate the visibility at points on a rectangular grid and (2) the resulting possibility of aliasing of parts of the image from outside the synthesized field. The evaluation at the grid points is sometimes referred to as *gridding*. The output of such a process can usually be represented by the following expression:

$$\frac{w(u,v)}{\Delta u\,\Delta v}\,{}^2\text{III}\!\left(\frac{u}{\Delta u},\frac{v}{\Delta v}\right)\{C(u,v)**[W(u,v)\mathscr{V}(u,v)]\}. \qquad (10.16)$$

Here the visibility $\mathscr{V}(u,v)$, measured at the points denoted by the transfer function $W(u,v)$, is convolved with a function $C(u,v)$ to produce a continuous visibility distribution. This is then resampled at points in a rectangular grid with incremental spacings Δu and Δv. The resampling is here represented by the two-dimensional shah function ${}^2\text{III}$ (Bracewell 1956b) defined by

$$ {}^2\text{III}\!\left(\frac{u}{\Delta u},\frac{v}{\Delta v}\right) = \Delta u\,\Delta v \sum_{m=-\infty}^{\infty}\sum_{n=-\infty}^{\infty} {}^2\delta(u - m\,\Delta u, v - n\,\Delta v), $$

$$(10.17)$$

where ${}^2\delta$ is the two-dimensional delta function. The weighting to optimize the beam is applied to the resampled data. Although the above process is described mathematically in terms of convolution and resampling, in practice, the convolution is evaluated only at the grid points.

The Fourier transform of Expression (10.16) represents the measured brightness:

$$B_m(\xi,\eta) = {}^2\text{III}(\xi\,\Delta u, \eta\,\Delta v)**\overline{w}(\xi,\eta)**\{\overline{C}(\xi,\eta)[\overline{W}(\xi,\eta)**B(\xi,\eta)]\}.$$

$$(10.18)$$

The brightness function is convolved with the Fourier transform of the transfer function, multiplied by a function that is the Fourier transform of the convolving function, and then convolved with the Fourier transforms of the weighting and resampling functions. The final convolution causes the whole map to be replicated at intervals Δu^{-1} in ξ and Δv^{-1} in η, as in Fig. 4.14. These intervals are equal to the dimensions of the map in the (ξ, η) plane, that is, $\Delta u^{-1} = M\Delta\xi$, $\Delta v^{-1} = N\Delta\eta$, for an $M \times N$ point array. The function $\overline{C}(\xi, \eta)$ takes the form of a taper applied to the map, and if this function does not vary greatly on the scale of width of $\overline{w}(\xi, \eta)$, which is usually the case for large maps, then $\overline{w}(\xi, \eta)$ in Eq. (10.18) can be convolved directly with $\overline{W}(\xi, \eta) ** B(\xi, \eta)$ and Eq. (10.18) becomes

$$B_m(\xi, \eta) = {}^2\mathrm{III}(\xi\,\Delta u, \eta\,\Delta v) ** \left\{ \overline{C}(\xi, \eta)\left[B(\xi, \eta) ** b_0(\xi, \eta)\right]\right\},$$

$$(10.19)$$

where $b_0(\xi, \eta)$ enters through the relationship in (10.11). Comparison with Eq. (10.10) shows that the effect of the gridding and resampling is to multiply the map by $\overline{C}(\xi, \eta)$ and replicate it. This replication introduces the aliasing.

Returning to the estimation of the visibility at the grid points, we may perhaps expect the best technique to be some form of exact interpolation so that the resulting values are equal to those that would be obtained by measurement at the grid points. A method of this type has been described by Thompson and Bracewell (1974). However, the problem of aliasing remains, and the most effective way to deal with this is to convolve the data in the (u, v) plane with the Fourier transform of a function that in the (ξ, η) plane varies very little over the map, and then falls off rapidly at the map edges. We therefore look for a convolving function $C(u, v)$ for which the Fourier transform $\overline{C}(\xi, \eta)$ has these properties. An ideal function with infinitely sharp cutoff at the field edges would completely eliminate the aliasing since there would be no overlap of the replicated maps. Unfortunately, this ideal is not practical because the required convolving function is not bounded in the (u, v) plane. Nevertheless, a very worthwhile degree of suppression of the aliasing is possible with more compact functions. The combination into a single operation of the gridding and the convolution to minimize aliasing is a convenient technique that is widely used. Note, however, that the grid point estimation cannot be described as interpolation since the function $C(u, v) ** [W(u, v)\mathscr{V}(u, v)]$, in general, does not pass through the measured visibility points. Note also that although convolution is effective in suppressing artifacts that result from gridding of the data, it does not reduce sidelobe or ringlobe responses to sources located outside the area of the map.

Convolving Functions and Aliasing

From the foregoing discussion we can conclude that the point of principal concern in the use of the FFT is the choice of convolving function. A detailed discussion of convolving functions is given by Schwab (1984). It is convenient to consider those that are separable into one-dimensional functions of the same form for u and v, that is,

$$C(u, v) = C_1(u)C_1(v). \qquad (10.20)$$

We therefore discuss some examples of the function C_1.

Rectangular Function.   This function is the one used in cell averaging discussed in Section 4.10. It can be written

$$C_1(u) = (\Delta u)^{-1}\Pi\left(\frac{u}{\Delta u}\right), \qquad (10.21)$$

where Π is the unit rectangle function defined by

$$\Pi(x) = \begin{cases} 1, & |x| \le \frac{1}{2} \\ 0, & |x| > \frac{1}{2}. \end{cases} \qquad (10.22)$$

The Fourier transform of $C_1(u)$ is

$$\overline{C}_1(\xi) = \frac{\sin(\pi\,\Delta u\,\xi)}{\pi\,\Delta u\,\xi}. \qquad (10.23)$$

At the edge of the synthesized field $\xi = (2\Delta u)^{-1}$ and $\overline{C}_1(1/2\,\Delta u) = 2/\pi$. The map is tapered by a sinc-function profile in the ξ and η directions and a sinc-squared profile along the diagonals. Equation (10.23) is plotted in Fig. 10.4, and the value at the first maximum outside the edge of the map is 0.22 of the value at the map center. The effect of aliasing is shown more directly in Fig. 10.5a which is a plot of $\overline{C}_1(\xi)/\overline{C}_1[f(\xi)]$, where $f(\xi)$ is the value of ξ within the map (i.e., $|f(\xi)| < (2\,\Delta u)^{-1}$) at which the alias of a feature at ξ would appear. This quantity gives the relative response to an aliased feature in a map that has been corrected for the taper imposed by $\overline{C}_1(\xi)$. It is clear that cell averaging performs poorly in suppressing aliasing, but the computation required is minimal.

Gaussian Function.   Here we have

$$C_1(u) = \frac{1}{\alpha\,\Delta u\sqrt{\pi}}\,e^{-(u/\alpha\,\Delta u)^2} \qquad (10.24)$$

and

$$\overline{C}_1(\xi) = e^{-(\pi\alpha\,\Delta u\,\xi)^2}. \qquad (10.25)$$

The value of the constant α can be chosen to vary the widths of the functions as desired. If α is too small $C_1(u)$ will be too narrow, and only visibility

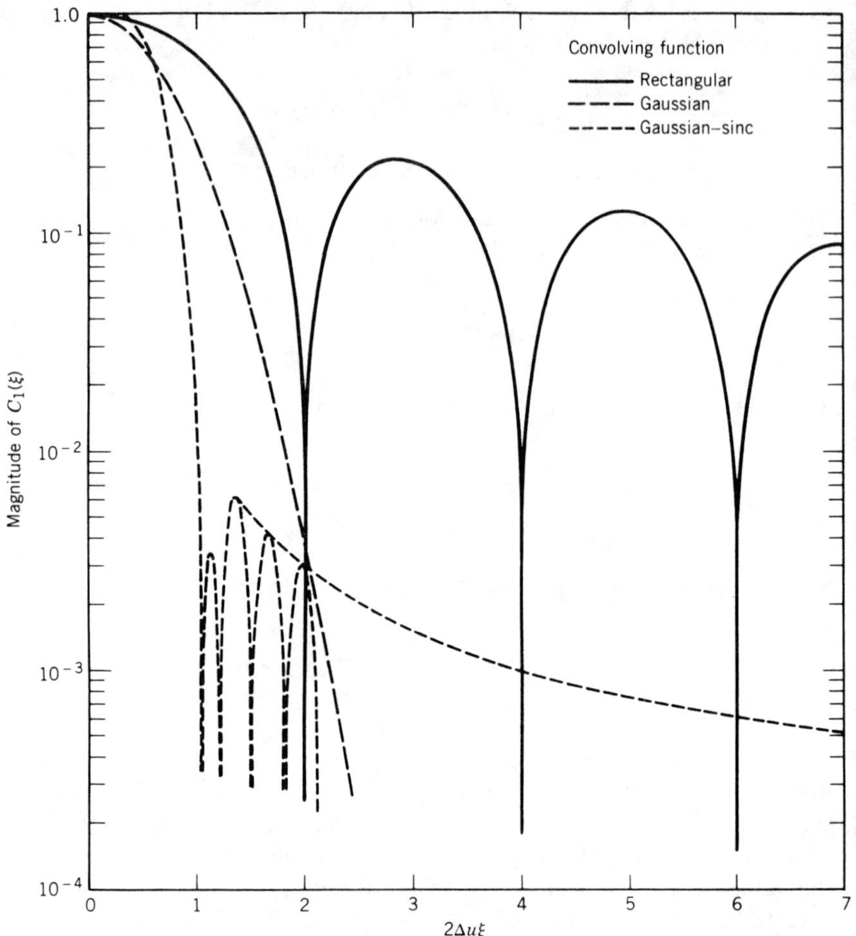

Figure 10.4   Three examples of the tapering function $\overline{C}_1(\xi)$ which is the Fourier transform of the convolving function $C_1(u)$. For the Gaussian convolving function, $\alpha = 0.75$. For the Gaussian–sinc convolving function, $\alpha_1 = 1.55$, $\alpha_2 = 2.52$, and beyond the fourth subsidiary maximum only the envelope of the maxima is shown. On the abscissa scale the center of the map is at zero and the edge at 1.0. The data for the Gaussian–sinc function were computed by F. R. Schwab.

measurements that are close to grid points will be used effectively in the mapping. If α is too large the function $C_1(u)$ will encompass a large number of measured points and the computation required will be excessive.

The Gaussian convolving function was widely used in the early years of the Westerbork array with $\alpha = 2\sqrt{\ln 4}\,/\pi = 0.750$ (Brouw 1971). The value of the factor $e^{-(u/\alpha\,\Delta u)^2}$ in $C_1(u)$ is then equal to 0.41 for a point on a diagonal in the (u, v) plane midway between two grid points. Thus, all measured points enter into the map with significant weights, and at the edge of the map the tapering factor $\overline{C}_1 = \frac{1}{4}$. A curve for this Gaussian function is shown in Fig. 10.4.

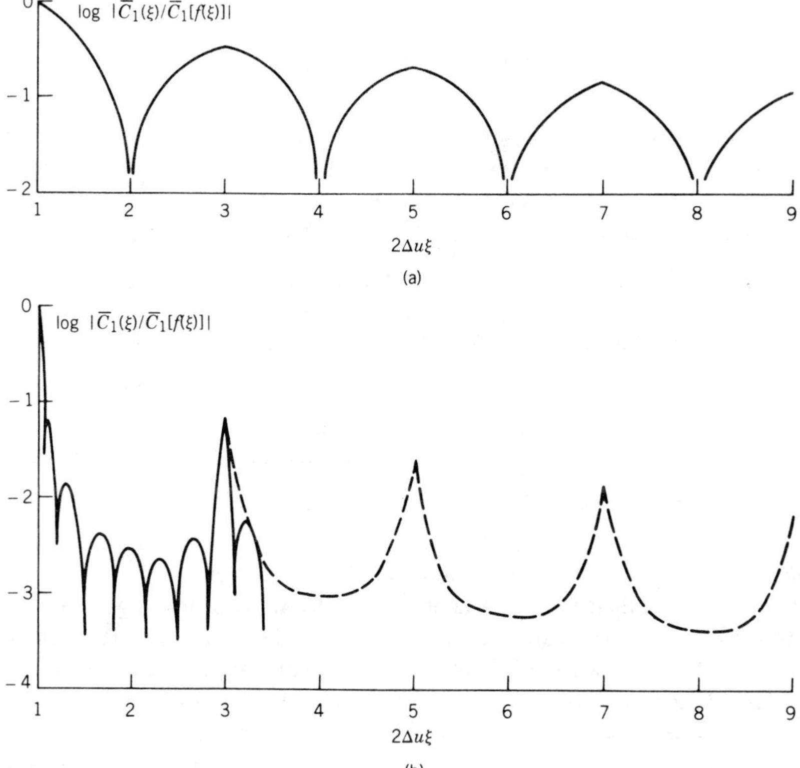

Figure 10.5   Logarithmic plot of the factor by which the amplitudes of structures outside the map are multiplied when aliased into the map. On the abscissa scale 1.0 is the edge of the map and $2, 4, 6, \ldots$ are the centers of the adjacent replications. (a) Aliasing factor for a rectangular convolving function of width equal to Δu (cell averaging). (b) Aliasing factor for a Gaussian–sinc convolving function with the optimized parameters given in the text. The broken line indicates the envelope of the maxima. Data computed by F. R. Schwab.

Gaussian-Sinc Function.   The ideal form for the map tapering function $\overline{C}_1(\xi)$ would be a rectangle which corresponds to convolution with a sinc function in the (u, v) plane. However, the envelope of a sinc function falls to zero too slowly as its argument increases and the computation required for the convolution is much too large. Truncation of the sinc function is undesirable because in the ξ domain the desired rectangular function is convolved with the Fourier transform of the truncation function, and this destroys the sharp cutoff at the map edges. A better procedure is to multiply the sinc function with a Gaussian, which gives

$$C_1(u) = \frac{\sin(\pi u / \alpha_1 \Delta u)}{\pi u} e^{-(u/\alpha_2 \Delta u)^2} \tag{10.26}$$

and

$$\overline{C}_1(\xi) = \Pi(\alpha_1 \Delta u\, \xi) * \left[\sqrt{\pi}\, \alpha_2 \Delta u\, e^{-(\pi \alpha_2 \Delta u\, \xi)^2}\right]. \tag{10.27}$$

Good performance is obtained with $\alpha_1 = 1.55$ and $\alpha_2 = 2.52$ if the convolving function extends over an area about $6\,\Delta u$ in width. Corresponding curves for $\overline{C}_1(\xi)$ and the resulting aliasing are given in Figs. 10.4 and 10.5b. This convolving function is much better than either of the two previous examples.

Spheriodal Functions.   Various other functions can be found that have the features desirable for convolution. As a measure of the effectiveness of the suppression of aliasing, Brouw (1975) has suggested the following quantity:

$$\frac{\displaystyle\iint_{\text{map}} [\overline{C}(\xi, \eta)]^2 d\xi\, d\eta}{\displaystyle\int_{-\infty}^{\infty} \int_{-\infty}^{\infty} [\overline{C}(\xi, \eta)]^2 d\xi\, d\eta}, \tag{10.28}$$

which shows the fraction of the integrated squared amplitude of the tapering function that falls within the map. A criterion for choosing a convolving function is the maximization of Expression (10.28). This approach led to the prolate spheroidal wave functions (e.g., Slepian and Pollack 1961) and the spheroidal functions (Rhodes 1970). Schwab (1984) found that among functions he investigated the latter provide the best approach to an optimum convolving function. The spheroidal functions are solutions to certain differential equations and are not expressible in simple analytic form. In applying such functions for convolution of visibility data, they are computed in advance to provide a look-up table. Comparison of some functions of this type with the Gaussian-sinc function shows that the aliasing factor $\overline{C}_1(\xi)/\overline{C}_1[f(\xi)]$ falls off about as rapidly from the center to the edge of the map, but as ξ increases it reaches values roughly an order of magnitude lower than those for the Gaussian-sinc function. Computational capacity complicates the choice of the optimal function, since it limits the area of the (u, v) plane over which the convolution can be performed. Also, round-off errors in the Fourier transform are amplified in the removal of the tapering function and may limit the allowable taper at the map edges.

Aliasing and the Signal-to-Noise Ratio

Features aliased into a map from outside the boundary include not only the images of features on the sky but also the random variations resulting from the system noise. If we consider a direct Fourier transform of the noise component of the visibility, it is clear from Expression (10.12) that for any point (ξ, η) the visibility data are weighted by complex exponential factors all of which have the same modulus. Since the noise is independent at each data point in the

(u, v) plane the variance of the noise in the (ξ, η) plane is statistically constant in all parts of the map. If the FFT is used, however, the rms noise level across the map is multiplied by the function $\overline{C}(\xi, \eta)$, and details beyond the map edge are aliased into the map. Note that the noise contributions combine additively in the variance. Thus, in one dimension the noise variance as a function of ξ is proportional to

$$\text{III}(\xi\,\Delta u) * |C_1(\xi)|^2. \tag{10.29}$$

The replication resulting from the FFT can also be written in terms of a summation, and the variance of the noise at a point ξ within the map is then proportional to

$$\sum_{i=-\infty}^{\infty} |\overline{C}_1(\xi + i\,\Delta u^{-1})|^2. \tag{10.30}$$

Usually $\overline{C}_1(\xi)$ decreases sufficiently with ξ that only the noise from the adjacent replication of the map makes a serious contribution through aliasing. This contribution is greatest near the edge of the map as shown in Fig. 10.6.

If the convolving function is the Gaussian–sinc type we see from Fig. 10.5b that, except for values of $2\xi\,\Delta u$ between 1.0 and 1.1, aliased features are reduced in amplitude by a factor $< 10^{-1}$, and in the square of the amplitude by $< 10^{-2}$. Thus, there is no significant increase in the noise level as a result of aliasing except in a narrow zone at the edge of the map.

At the other extreme, the aliasing is most serious in the case of cell averaging, $C_1(u)$ being the sinc function given by Eq. (10.23). Expression

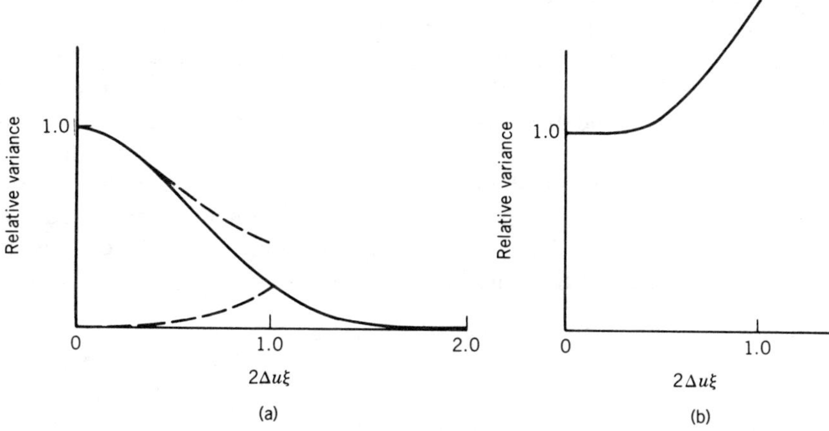

Figure 10.6   Effect of aliasing on the variance of the noise across a map. The abscissa in each case is ξ in units of half the map width: the map center is at 0, the edge at 1.0, and the center of the adjacent replication at 2.0. (a) Solid curve shows the taper for a Gaussian convolving function C_1, and broken curves show the effect of aliasing. (b) Variance of the noise including aliased component after correction for taper C_1. After Napier and Crane (1982).

(10.30) then becomes

$$\sum_{i=-\infty}^{\infty} \frac{\sin^2[\pi(\Delta u\,\xi + i)]}{[\pi(\Delta u\,\xi + i)]^2} = 1, \qquad (10.31)$$

which indicates that the aliasing exactly cancels the taper, and the variance of the noise is constant with ξ, that is, before any correction for tapering of the image is applied. This result could also be deduced from the fact that in cell averaging each visibility measurement contributes to one grid point only, and the noise components of the visibility at the grid points are therefore independent. However, the brightness distribution of the sky within the field being mapped is tapered by the function $\overline{C}_1(\xi)$, and correction for this taper then causes the noise to increase toward the map edges. For the sinc-function taper the noise is increased by a factor of $\pi/2$ at the edge of the map on the ξ and η axes and by $(\pi/2)^2$ at the corners. At the center of the map the aliased contribution originates at points for which $2\xi\,\Delta u$ is an even integer in the plots in Fig. 10.5, and in both cases shown the aliasing factor $\overline{C}_1(\xi)/\overline{C}_1[f(\xi)]$ drops to a very low value. With any of the convolving functions that we have considered there is no significant increase in the noise at the center of the map, and the signal-to-noise ratio for a source at that point is determined by the factors discussed in Section 6.2.

Interpretation of the Brightness Variable in the Map

The quantity measured in a synthesized map is, of course, the radio brightness, but $\mathscr{V}$ is usually calibrated in terms of the equivalent flux density of a point source, and the brightness unit in the resulting map is the peak response to a point source of unit flux density. The response to an extended source is the convolution of the sky brightness $B(\xi, \eta)$ with the normalized synthesized beam $b_N(\xi, \eta)$, for which $b_N(0,0) = 1$. Note that since there is usually no measured visibility value at the (u, v) origin, the integral of $b_N(\xi, \eta)$ over all angles is zero, that is to say, there is no response to a uniform level of brightness. At any point on the extended source where the brightness varies slowly compared with the width of the synthesized beam, the convolution with $b_N(\xi, \eta)$ results in a flux density that is approximately $B\Omega_0$, where Ω_0 is the equivalent solid angle of the main lobe of the synthesized beam:

$$\Omega_0 = \iint_{\substack{\text{main}\\\text{lobe}}} \frac{b_N(\xi, \eta)\,d\xi\,d\eta}{\sqrt{1 - \xi^2 - \eta^2}}. \qquad (10.32)$$

Thus the scale of the map can also be interpreted as brightness measured in flux density units per beam area Ω_0. An approximate brightness scale can thereby be assigned. However, this is appropriate for structure of size comparable with the main beam. For broader structures the negative sidelobes that inevitably result from the lack of data at the (u, v) origin reduce the response until it is zero for a constant brightness level. As a result, the response in the map is not accurately proportional to the sky brightness, and the assignment of a brightness scale is approximate at best. The map can be improved by adding data at and near the (u, v) origin, by using single-antenna observations as discussed in the following section. Also, several procedures discussed in Chapter 11 can reduce or eliminate the negative sidelobes of the synthesized beam. To assign a brightness scale one can then Fourier transform the map to obtain visibility values and choose a brightness scale on the basis of agreement with the measured visibility data. The agreement can be maximized for spatial frequencies corresponding to the scale size of structure of particular interest.

The signal-to-noise ratio for an extended source can be obtained by determining the response in units of flux density and dividing by the rms noise level given by Eq. (6.55). The result is approximately equal to $B\Omega_0/S_{\rm rms}$. This holds true when the source is wider than the beam, and the signal-to-noise ratio can then be improved by increasing Ω_0, that is, by degrading the angular resolution.

10.3 PRACTICAL CONSIDERATIONS

This section is concerned with the causes of errors in maps and some remarks on planning and reduction of observations. Information of this type can also be found in the observers' handbooks of instruments such as the VLA and the Westerbork array: see the bibliography at the end of this chapter.

Possible Causes of Errors in Maps

Even after careful editing and calibration of the data, the possibility of errors in a map should not be discounted. Synthesis mapping involves complicated equipment in which occasional malfunctions must be expected, and the effects of the troposphere and ionosphere can introduce phase variations that are difficult or impossible to eliminate by direct calibration. Hamaker (1979) and Ekers (1982) have discussed mapping errors for specific instruments.

A very useful aid in identifying erroneous features is a familiarity with the behavior of functions under Fourier transformation. For example, the (u, v) distribution of the visibility corresponding to a suspicious feature in a map may indicate a problem related to a particular group of antennas or a specific period in the observing. A persistent error in one antenna pair will, for an east–west spacing, be distributed along an elliptical ring centered on the (u, v) origin, and in the (ξ, η) plane will give rise to an elliptical feature with a radial profile in the form of the zero-order Bessel function. An error of short duration, on the other hand, introduces two delta functions representing the measurement and its conjugate. In the map these produce a sinusoidal corrugation over the (ξ, η) plane. The amplitude in the map plane is small since in an

$M \times N$ point visibility matrix the effect of the two erroneous points is diluted by a factor $2(MN)^{-1}$, which is usually of order 10^{-3}–10^{-6}. Thus a single short-duration error may be entirely negligible compared with the system noise. Conversely, a persistent, low-level error that is widely distributed in the visibility plane could compound into a significant spurious feature in a localized region in the map. This would require some systematic distribution in the phase of the erroneous visibility. A random distribution in phase results in a largely uniform distribution over the map as in the case of the system noise.

The errors just mentioned are of an additive nature, that is to say, they combine by addition with the true visibility values. In the map the Fourier transform of the error distribution $\varepsilon_a(u, v)$ is added to the brightness distribution, and we have

$$\mathscr{V}(u, v) + \varepsilon_a(u, v) \rightleftharpoons B(\xi, \eta) + \bar{\varepsilon}_a(\xi, \eta). \qquad (10.33)$$

Other types of additive errors result from interference, cross coupling of system noise between antennas, and correlator offset errors. The sun is many orders of magnitude stronger than most radio sources and can produce interference of a different character from that of terrestrial sources because of its diurnal motion. The response to the sun is governed mainly by the sidelobes of the primary beam, the difference in fringe frequencies for the sun and the source, and the bandwidth and visibility-averaging effects. Solar interference is most severe for low-resolution arrays with narrow bandwidths. Cross coupling of noise (cross talk) occurs only between closely spaced antennas and is most severe for low-elevation angles when shadowing of antennas occurs. Under such conditions one antenna points toward the back surface of another and coupling can occur by radiation leakage through the gaps between reflector panels. Steady offsets in the correlator outputs can be caused by misadjustment of analog correlators or misadjustment of samplers in digital systems. These offsets are greatly reduced by phase switching, but residual effects can occur. Since they remain largely constant during an observation they result in errors near the (ξ, η) origin in the map.

A second class of errors comprises those that combine with the visibility in a multiplicative manner, and for these we can write

$$\mathscr{V}(u, v) \varepsilon_m(u, v) \rightleftharpoons B(\xi, \eta) * * \bar{\varepsilon}_m(\xi, \eta). \qquad (10.34)$$

The Fourier transform of the error distribution is convolved with the brightness distribution, and the resulting distortion produces erroneous structure connected with the main features in the map. In contrast, the distribution of errors of the additive type is unrelated to the true brightness pattern. Multiplicative errors mainly involve the gain constants of the antennas and result from calibration errors or from phase errors introduced by the atmosphere.

Distortions that increase with distance from the center of the map constitute a third category of errors. These include the effects of non-coplanar baselines (see Section 4.1), bandwidth (see Section 6.3), and visibility averaging (see Section 6.4), which are predictable and therefore somewhat different in nature from the other distortions mentioned above. Errors in the pointing of the antennas introduce errors in the map that increase toward the half-power points on the beam, where the rate of change of collecting area with angle is generally greatest.

Visibility at Low Spatial Frequencies

A problem common to all synthesis arrays in which the antennas are of uniform size is that the minimum antenna spacing cannot be less than the diameter of an antenna, and for practical reasons is usually somewhat greater. This limit does not apply to the increments between the longer spacings, but there is a central hole in the (u, v) coverage which exacerbates the lack of a visibility value at the origin. The missing data result in broad negative sidelobes on the synthesized beam, such that the beam appears to be situated in the center of a shallow bowl-shaped depression. This effect is most noticeable when the field to be mapped is wide enough that there are several empty (u, v) cells within the central area. One way of improving the situation is to insert the visibility at the origin from an independent measurement of the flux density within the field. Sufficient weight should be applied to this value so that the effective data density is comparable to that in other parts of the (u, v) plane. Although the datum at the origin has only the effect of adding a constant to the Fourier transform of the visibility, it facilitates the removal of the bowl-shaped structure by the CLEAN procedure which is described in Chapter 11. Short spacing visibility values from independent measurements can also be added. The most straightforward method of obtaining such data is from an interferometer with smaller antennas. An appropriate convolving function can be used to suppress responses from any sources that are outside the synthesized field, but within the wider beams of the smaller antennas. It is also possible to obtain short spacing visibility values from scans made with a single large antenna of diameter greater than the spacing required. Fourier transformation of a map made with a single antenna gives the visibility function weighted by the spectral sensitivity of the antenna. The latter can be obtained by Fourier transformation of the response to a point source. The spectral sensitivity function of an aperture antenna falls away as the spatial frequency increases and is zero for spatial frequencies greater than the antenna diameter measured in wavelengths. In practice, it is difficult to determine the lower levels of the spectral sensitivity with adequate accuracy to correct the visibility, and it is commonly recommended that the antenna diameter should be at least twice as great as the maximum spacing for which the visibility is required. For further discussion see, for example, Welch and Thornton (1985) or Bajaja and van Albada (1979).

Miscellaneous Hints on Planning and Reduction of Observations

In choosing the observing bandwidth for continuum observations the radial smearing effect should be considered since the signal-to-noise ratio for a point source near the edge of the field is not necessarily maximized by maximizing the bandwidth. Then in choosing the data averaging time the resulting circumferential smearing can be about equal to the radial effect. The required condition is obtained from Eqs. (6.64) and (6.70) and for high declinations is

$$\frac{\Delta\nu}{\nu_0} \simeq \omega_e \tau_a. \tag{10.35}$$

In some arrays such as the VLA the resolution at a given frequency can be varied by changing the spacings of the antennas. When attempting to detect a weak source of measurable angular diameter, or an extended background emission, it is important not to choose an angular resolution that is too high. The signal-to-noise ratio for an extended source is approximately proportional to $B\Omega_0$, as discussed in the previous section. The observing time required to obtain a given signal-to-noise ratio is proportional to Ω_0^{-2} or to θ_b^{-4}, where θ_b is the synthesized beamwidth.

The total observing time to obtain the necessary signal-to-noise ratio for a particular study may be much less than 12 hr. If the observation is made with an east–west linear array it will be necessary to observe over a large range of hour angles to obtain the required (u, v) coverage, and it may be possible to interleave short-duration observations of several different areas of sky within a single 12-hr period. This procedure is most likely to be possible if the fields to be synthesized are small: the spacings Δu and Δv between the visibility data in the discrete Fourier transformation are then large, so the time intervals between visibility measurements can be correspondingly long. If the observation is made with a two-dimensional array such as the VLA, in which the instantaneous (u, v) coverage is sufficient to provide a fairly satisfactory synthesized beam, there is a choice between making a single observation of the required duration or dividing the observation into several shorter ones spread out in time. The latter course will improve the (u, v) coverage. However, if the total observing time is less than the time scale of variation of large atmospheric structures, it may be best to confine the observation to the shortest possible period. This will reduce the scatter in the resulting phase errors, which are then likely to produce a position offset rather than degradation of the resolution.

If the antenna beam contains a source that is much stronger than the features to be studied, the response to the strong source can be subtracted, provided it is a point source or one that can be accurately modeled. This is best done by subtracting the computed visibility before gridding the measurements for the FFT. The subtracted response will then accurately include the effect of the sidelobes of the synthesized beam. Nevertheless, the precision of the operation will be reduced if the source response is significantly affected by bandwidth, visibility averaging, and similar effects, unless these are included in source-subtraction programs. It may be simpler to place the strong source at the center of the field, or reduce the bandwidth and integration time, so that it can be accurately removed.

Removal of strong features from a map is sometimes a helpful procedure in locating low-level errors in the visibility. This can be done by subtracting a source as described above or by retransforming a critical low-level area of a map back to the visibility domain.

When observing a very weak source it may be advisable to place the source a few beamwidths away from the (ξ, η) origin to avoid confusion with residual errors from correlator offsets. Experience with any particular instrument will show whether this is necessary.

As part of the procedure in making any map it is useful also to make a low-resolution map covering the entire area of the primary antenna beam. For this map, the data can be heavily tapered in the (u, v) plane to reduce the resolution and thus also the computation. Such a map will reveal any sources outside the field of the final map which may introduce aliased responses in the FFT. Aliasing of these sources can be suppressed by subtraction of the visibility or use of a suitable convolving function. The sidelobe or ringlobe responses to such a source are also eliminated by subtraction of the source but not by convolution in the (u, v) plane. The low-resolution map will also emphasize any extended low-brightness features that might otherwise be over-looked.

BIBLIOGRAPHY

Westerbork Synthesis Radio Telescope Users Manual, A. G. Willis and H. C. Kahlman, Eds., Netherlands Foundation for Radio Astronomy, Dwingeloo, Netherlands, 1980.

An Introduction to the NRAO Very Large Array, R. M. Hjellming, Ed., National Radio Astronomy Observatory, Socorro, New Mexico, 1982.

Synthesis Mapping, Proceedings of NRAO Workshop No. 5, Socorro, New Mexico, June 21–25, 1982, A. R. Thompson and L. R. D'Addario, Eds., National Radio Astronomy Observatory, Green Bank, West Virginia, 1982.

REFERENCES

Bajaja, E., and G. D. van Albada, Complementing Aperture Synthesis Radio Data by Short Spacing Components from Single Dish Observations, *Astron. Astrophys.*, **75**, 251–254, 1979.

Bracewell, R. N., Strip Integration in Radio Astronomy, *Aust. J. Phys.*, **9**, 198–217, 1956a.

Bracewell, R. N., Two-dimensional Aerial Smoothing in Radio Astronomy, *Aust. J. Phys.*, **9**, 297–314, 1956b.

Bracewell, R. N., *The Fourier Transform and Its Applications*, McGraw-Hill, New York, 1965 (2nd ed., 1978).

Bracewell, R. N. and A. C. Riddle, Inversion of Fan-Beam Scans in Radio Astronomy, *Astrophys. J.*, **150**, 427–434, 1967.

Bracewell, R. N. and J. A. Roberts, Aerial Smoothing in Radio Astronomy, *Aust. J. Phys.*, **7**, 615–640, 1954.

Bracewell, R. N. and S. J. Wernecke, Image Reconstruction over a Finite Field of View, *J. Opt. Soc. Am.* **65**, 1342–1346, 1975.

Brouw, W. N., Data Processing for the Westerbork Synthesis Radio Telescope, Doctoral Thesis, University of Leiden, 1971.

Brouw, W. N., Aperture Synthesis, in *Methods of Computational Physics*, Vol. 14, B. Alder, S. Fernbach, and M. Rotenberg, Eds., Academic Press, New York, 1975, pp. 131–175.

Christiansen, W. N., and J. A. Warburton, The Distribution of Radio Brightness over the Solar Disk at a Wavelength of 21 cm. III. The Quiet Sun—Two-Dimensional Observations, *Aust. J. Phys.*, **8**, 474–486, 1955.

Cole, T. W., Analog Processing Methods for Synthesis Observations, in *Image Formation From Coherence Functions in Astronomy*, C. van Schooneveld, Ed., D. Reidel, Dordrecht, Holland, 1979, pp. 123–141.

Ekers, R. D., Common Image Defects, in *Synthesis Mapping, Proceedings of NRAO Workshop No. 5, Socorro, New Mexico, June 21–25, 1982*, A. R. Thompson and L. R. D'Addario, Eds., National Radio Astronomy Observatory, Green Bank, West Virginia, 1982.

Fomalont, E. B., The East–West Structure of Radio Sources at 1425 MHz, *Astrophys. J. Suppl.*, **15**, 203–274, 1968.

Fomalont, E. B. and M. C. H. Wright, Interferometry and Aperture Synthesis, in *Galactic and Extragalactic Radio Astronomy*, G. L. Verschuur and K. I. Kellermann, Eds., Springer-Verlag, New York, 1974, pp. 256–290.

Frater, R. H., Direct Transform Hardware Processing of Rotational Synthesis Data—II, *Astron. Astrophys.*, **68**, 397–403, 1978.

Frater, R. H. and D. J. Skellern, Direct Transform Hardware Processing of Rotational Synthesis Data—I, *Astron. Astrophys.*, **68**, 391–396, 1978.

Hamaker, J. P., Measurement Errors in Rotational Synthesis and Their Effects in the Map Plane, in *Image Formation from Coherence Functions in Astronomy*, C. van Schooneveld, Ed., D. Reidel, Dordrecht, Holland, 1979, pp. 27–45.

Hargrave, P. J. and M. Ryle, Observations of Cygnus A with the 5-km Radio Telescope, *Mon. Not. R. Astron. Soc.*, **166**, 305–327, 1974.

Kenderdine, S., Fourier Transform Supersynthesis, *Astron. Astrophys. Suppl.*, **15**, 413–415, 1974.

Maltby, P. and A. T. Moffet, Brightness Distribution in Discrete Radio Sources. III. The Structure of the Sources, *Astrophys. J. Suppl.*, **7**, 141–163, 1962.

Moffet, A. T., Brightness Distribution in Discrete Radio Sources, I. Observations with an East–West Interferometer, *Astrophys. J. Suppl.*, **7**, 93–123, 1962.

Napier, P. J. and P. C. Crane, Signal-to-Noise Ratios, in *Synthesis Mapping, Proceedings of NRAO Workshop No. 5, Socorro, New Mexico, June 21–25, 1982*, A. R. Thompson and L. R. D'Addario, Eds., National Radio Astronomy Observatory, Green Bank, West Virgina, 1982.

Rhodes, D. R., On the Spheriodal Functions, *J. Res. Natl. Bur. Stand. (U.S.) B*, **74**, 187–209, 1970.

Schwab, F. R., Optimal Gridding of Visibility Data in Radio Interferometry, in *Indirect Imaging*, J. A. Roberts Ed., Cambridge University Press, Cambridge, England 1984, pp. 333–346.

Slepian, D. and H. O. Pollak, Prolate Spheroidal Wave Functions, Fourier Analysis and Uncertainty. I, *Bell Syst. Tech. J.*, **40**, 43–63, 1961.

Thompson, A. R. and R. N. Bracewell, Interpolation and Fourier Transformation of Fringe Visibilities, *Astron. J.*, **79**, 11–24, 1974.

Wardle, J. F. C. and P. P. Kronberg, Linear Polarization of Quasi-Stellar Radio Sources at 3.71 and 11.1 Centimeters, *Astrophys. J.*, **194**, 249–255, 1974.

Welch, W. J., and D. D. Thornton, An Introduction to Millimeter and Submillimeter Interferometry and a Summary of the Hat Creek System, *International Symposium on Millimeter and Submillimeter Wave Radio Astronomy*, International Scientific Radio Union, Institut de Radio Astronomie Millimetrique, Granada, Spain, 1985, pp. 53–64.

White, R. L. and R. H. Becker, The Resolution of P Cygni's Stellar Wind, *Astrophys. J.*, **262**, 657–662, 1982.

11

IMAGE PROCESSING AND ENHANCEMENT

In the preceding chapter we discussed the construction of the radio image from the measured visibility data. As a means of optimizing the result we considered only the weighting of the visibility and its effects on the sidelobes of the synthesized beam. This chapter is concerned with techniques of processing that further improve the image, and image construction from limited visibility data. There are two principal deficiencies in the visibility data that limit the accuracy of synthesis maps. These are (1) the limited distribution of spatial frequencies u and v and (2) errors in the measurements themselves. Because of atmospheric effects such errors are usually more serious in the phase, and in VLBI the phase is usually uncalibrated. Since these deficiencies introduce certain obvious errors such as negative brightness into the map, it is possible to improve the result by injecting some a priori information about the brightness characteristics. This idea together with closure relationships for amplitude and phase provide the basis of the techniques to be discussed.

11.1 LIMITATION OF (u, v) COVERAGE AND METHODS OF COMPENSATION

The principal response, as discussed in Chapter 10, is that form of the brightness distribution obtained by transformation of visibility data for which the product of weight and area density are uniform over the sampled areas of the (u, v) plane. An interesting property of this weighting is that it minimizes the mean squared deviation of the resulting brightness from the true brightness, within the constraint that unmeasured visibility values remain zero. This can be understood as follows. Since the true brightness distribution $B(\xi, \eta)$ and the true visibility function $\mathscr{V}(u, v)$ are a Fourier pair, and the weighted measured visibility and the derived brightness $B_0(\xi, \eta)$ are a Fourier pair, it follows that the differences between these quantities in the two domains are also a Fourier pair, to which we can apply Parseval's theorem. Recall that $W(u, v)$ is the transfer function, $w_u(u, v)$ is the weighting required to obtain effective uniform density of data in the (u, v) plane, and $w_t(u, v)$ is an applied taper. Thus, we can write

$$\int_{-\infty}^{\infty} \int_{-\infty}^{\infty} \left| \mathscr{V}(u, v) - \mathscr{V}(u, v) W(u, v) w_u(u, v) w_t(u, v) \right|^2 du\, dv$$

$$= \int_{-\infty}^{\infty} \int_{-\infty}^{\infty} \left| B(\xi, \eta) - B_0(\xi, \eta) \right|^2 d\xi\, d\eta. \tag{11.1}$$

The right-hand side of Eq. (11.1) is proportional to the mean squared difference between the true and observed brightness distributions. The left-hand side of Eq. (11.1) can be envisaged as two components that correspond to the sampled and unsampled areas of the (u, v) plane, that is, the areas within which $W(u, v)$ is nonzero or zero, respectively. If the tapering function w_t, which was introduced in Chapter 10 to control the sidelobes, is equal to unity, then the values that are transformed to obtain B_0 represent the untapered visibility within the sampled areas of the (u, v) plane. B_0 then corresponds to the principal response, and the left-hand side of Eq. (11.1) represents only the contribution from the unsampled areas. If, however, a taper is introduced, w_t decreases monotonically from unity at the (u, v) origin, and there is a further contribution to the left-hand side of Eq. (11.1) from the data within the sampled areas.

Although tapering increases the mean squared deviation of the measured brightness from the true brightness, the principal response is not necessarily the most useful form of the image. As pointed out in Chapter 10, the strong sidelobes of the corresponding beam (see Fig. 10.2) obscure low-level detail and thereby reduce the dynamic range. Thus tapering may improve the image with respect to criteria such as sidelobe minimization, but only to the extent that it can be obtained by adjusting the measured data. In what follows we discuss the methods in which the constraint that the unmeasured data must be zero no longer applies.

CLEAN Algorithm

Although the sidelobes in the synthesized beam limit the dynamic range of a map, the low-level detail is not totally lost, since the profile of the beam is calculable from the weighted transfer function. If we ignore the effects introduced by finite receiving bandwidth, visibility averaging, atmospheric phase errors, and so on, the observed brightness can be regarded as the true brightness convolved with the synthesized beam $b_0(\xi, \eta)$:

$$B_0(\xi, \eta) = B(\xi, \eta) * * b_0(\xi, \eta). \tag{11.2}$$

Knowing $B_0(\xi, \eta)$ and $b_0(\xi, \eta)$ can we solve for $B(\xi, \eta)$? An analytic procedure for deconvolving two functions is to take the Fourier transform of the convolution, which is equal to the product of the Fourier transforms of the components, divide out the Fourier transform of the known function, and transform back. From Eq. (11.2) we have

$$B(\xi, \eta) * * b_0(\xi, \eta) \rightleftharpoons \mathscr{V}(u, v)\left[W(u, v)w_u(u, v)w_t(u, v)\right]. \tag{11.3}$$

However, the weighted transfer function contains large areas of zeros, so we cannot divide it out to obtain $\mathscr{V}(u, v)$. The unmeasured visibilities present a fundamental problem, and any procedure that improves the derived brightness other than visibility tapering must involve placing nonzero visibility values in the unmeasured (u, v) areas.

Bracewell and Roberts (1954) pointed out that there are an infinite number of solutions to the convolution equation (11.2) since one can add any arbitrary visibility values in the unsampled areas of the (u, v) plane. The Fourier transform of these added values constitutes an invisible distribution that cannot be detected by any instrument with corresponding zero areas in the transfer function. It may be argued that in interpreting observations from any radio telescope one should maintain only zeros in the unmeasured regions of spectral sensitivity, to avoid arbitrarily generating information. On the other hand, the zeros are themselves arbitrary values, some of which are certainly wrong. What is wanted is a procedure that allows the visibility at the unmeasured points to take values consistent with the most reasonable or likely brightness distribution, without adding arbitrary and possibly erroneous detail. Since judgments can be made about what is "reasonable" in a brightness distribution, such a priori information can be incorporated into the map. Extensive sinusoidal structure and, of course, negative brightness values are examples of instrumental artifacts.

One of the more successful of such procedures is the algorithm CLEAN devised by Högbom (1974). This is basically a numerical deconvolving process applied in the (ξ, η) domain. The procedure is to break down the brightness distribution into point-source responses, and then replace each one with the corresponding response to a "clean" beam, that is, a beam free of sidelobes. The principal steps are as follows.

1. Compute the map and the response to a point source by Fourier transformation of the visibility and the weighted transfer function. These functions, the synthesized brightness and the synthesized beam,

are sometimes referred to as the "dirty map" and the "dirty beam", respectively. The spacing of the sample points in the (ξ, η) plane should not exceed about one-third of the synthesized beamwidth.

2.   Find the highest brightness point on the map and subtract the response to a point source, including the full sidelobe pattern, centered on that position. The peak amplitude of the subtracted point source is equal to γ times the corresponding map amplitude. γ is called the loop gain by analogy with negative feedback in electrical systems, and commonly has a value of 0.25 or less.

3.   Return to step 2 and repeat the procedure iteratively until all significant source structure has been removed from the map. As an indicator of this condition, one can compare the highest peak with the rms level of the remaining brightness, look for the first time that the rms level fails to decrease when a subtraction is made, or note when significant numbers of negative components start to be removed.

4.   Add the removed components, in the form of clean-beam responses, to the residual brightness distribution to obtain the new map. The clean beam is most often chosen to be a Gaussian with a half-amplitude width equal to that of the original synthesized beam.

It is assumed that each beam-function profile subtracted represents the response to a point source. As discussed in Section 4.5, the visibility function of a point source is a pair of real and imaginary sinusoidal corrugations that extend to infinity in the (u, v) plane. Any brightness feature for which the visibility function is the same within the (u, v) area sampled by the transfer function would produce a response in the map identical to the point-source response. For example, a feature resembling the principal-response beamshape would behave in this manner. However, it is certainly unlikely that real features would resemble a synthesized beam, especially if the latter contains negative sidelobes. Högbom (1974) has pointed out that much of the sky is a random distribution of point sources on an empty background, and CLEAN was initially developed for this situation. Nevertheless, experience shows that CLEAN also works well on extended and complicated sources.

The result of the first three steps in the CLEAN procedure outlined above can be represented by a model brightness distribution that consists of a series of delta functions with magnitudes and positions representing the subtracted components. Note that the visibility values derived from these clean components agree with the measured values to within the noise. Since the modulus of the visibility of each delta function extends uniformly to infinity in the (u, v) plane, the visibility is extrapolated as required beyond the cutoff of the transfer function.

The delta function components do not constitute a satisfactory model for astronomical purposes. The observations are sufficient only to show that the finest structure is smaller than the synthesized beam. Groups of delta functions with separations no greater than the beamwidth may actually represent extended structure. Convolution of the delta-function model by the clean beam, which occurs in step 4 of the procedure above, removes the danger of overinterpretation, that is, it deemphasizes the high spatial frequencies that may be spuriously extrapolated from the measured data. Thus CLEAN performs a function resembling interpolation in the (u, v) plane. Desirable characteristics of a clean beam are that it should be free from sidelobes, particularly negative ones, and that its Fourier transform should be constant inside the sampled region of the (u, v) plane and rapidly fall to a low level outside it. These characteristics are essentially incompatible since a sharp cutoff in the (u, v) plane results in oscillations in the (ξ, η) plane, and the usual compromise is a Gaussian function. Convolution with such a beam introduces a Gaussian taper in the (u, v) plane. This function tapers the unmeasured data generated by CLEAN. It also tapers the measured data, and the resulting brightness distribution no longer agrees with the measured visibility data. However, the absence of sidelobes is deemed of greater importance for most astronomical purposes.

It is instructive to compare the CLEAN algorithm with the process of removing the effect of the transfer function as discussed in relation to Eq. (11.3). We cannot directly divide out the weighted transfer function on the right-hand side of Eq. (11.3) because it is truncated to zero outside the areas of measurement. The way in which this problem is solved in CLEAN is by analyzing the measured visibility into sinusoidal visibility components and then removing the truncation so that they extend over the full (u, v) plane. Selecting the highest peak in the (ξ, η) plane is equivalent to selecting the largest complex sinusoid in the (u, v) plane. Thus, as Schwarz (1979) points

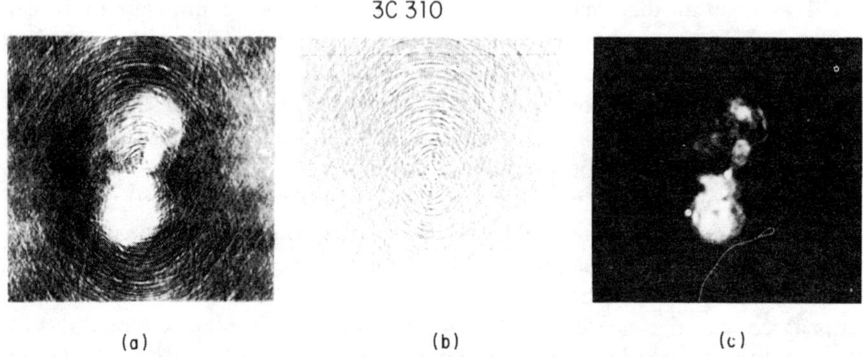

Figure 11.1   Performance of the CLEAN algorithm illustrated by images of the radio galaxy 3C310 from observations with the VLA. (a) Fourier transform of the calibrated visibility data. Sidelobe structure is at the 2% level. (b) The corresponding synthesized beam. (c) The image produced by the Cornwell (1983) version of the CLEAN algorithm. From Napier, Thompson, and Ekers (1983); © 1983 IEEE.

out, CLEAN can be regarded as a harmonic analysis in the (u, v) plane as well as a deconvolution in the (ξ, η) plane.

At the point that the component subtraction is stopped it is generally assumed that the residual brightness distribution consists mainly of the noise. Retaining the residual distribution within the map is, like the convolution with the clean beam, a nonideal procedure that is necessary to prevent misinterpretation of the final result. Without the residual background there would be an amplitude cutoff in the structure corresponding to the lowest subtracted component. Also, the presence of the background fluctuations provides an indication of the level of uncertainty in the brightness values. An example of the effect of processing with the CLEAN algorithm is shown in Fig. 11.1.

Implementation and Performance of the CLEAN Algorithm

The CLEAN algorithm was conceived as an empirical means of reducing the effect of limited (u, v) coverage in mapping, and as such its logical basis is easy to understand. Being highly nonlinear, however, CLEAN does not yield readily to a complete mathematical analysis. Some important conclusions have been derived by Schwarz (1978, 1979) who has shown that the condition for convergence of CLEAN is that the Fourier transform of the synthesized beam, that is, the weighted transfer function, must be non-negative. This condition is fulfilled in the usual synthesis procedure. Schwartz's analysis also indicates that if the number of delta-function components in the CLEAN model does not exceed the number of independent visibility data in the deviation of the uncleaned map, CLEAN converges to a solution that is the least-squares fit of the Fourier transforms of the delta-function components to the measured visibility. In enumerating the visibility data, either the real and imaginary parts or the conjugate values (but not both) are counted independently. In maps made using the FFT algorithm there are equal numbers of grid points in the (u, v) and (ξ, η) planes, but not all (u, v) grid points contain visibility measurements. To maintain the condition for convergence it is a common procedure to apply CLEAN only within a limited area or "window" of the original map.

In order to clean a map of a given dimension it is necessary to have a beam pattern of twice the map dimensions so that a point source can be subtracted from any location in the map. However, in practical implementations of the CLEAN algorithm it is often convenient for the map and beam to be the same size. In this case only the central quarter of the map can be properly processed. Thus, it is commonly recommended that the map obtained from the initial Fourier transform should have twice the dimensions required for the final map. As mentioned above, the use of such a window also helps to ensure that the number of components removed does not exceed the number of visibility data and, in the absence of noise, allows the residuals within the window area to approach zero (Schwarz 1979).

Several arbitrary choices influence the result of the CLEAN process. These include the parameter γ, the window area, and the criterion for termination. γ

is usually assigned a value between 0.1 and 0.5, and it is a matter of general experience that CLEAN responds better to extended structure if the loop gain is in the lower part of this range. The computation time for CLEAN increases rapidly as γ is decreased, because of the increasing number of subtraction cycles required. If the signal-to-noise ratio is $\mathcal{R}_{sn}$, then the number of cycles required for one point source is $-\log \mathcal{R}_{sn}/\log(1 - \gamma)$. Thus, for example, with $\mathcal{R}_{sn} = 100$ and $\gamma = 0.2$, a point source requires 21 cycles. Complex maps typically require thousands of subtraction cycles.

A well-known problem of CLEAN is the generation of spurious structure in the form of spots or ridges as a modulation upon broad features. A heuristic explanation of this effect is given by Clark (1982). The algorithm locates the maximum in the broad feature and removes a point-source component as shown in Fig. 11.2. The negative sidelobes of the beam add new maxima which are selected in subsequent cycles, and, thus, there is a tendency for the component-subtraction points to be located at intervals equal to the spacing of the first sidelobe of the original synthesized beam. The resulting map contains a lumpy artifact introduced by CLEAN, but the map is consistent with the measured visibility data. Cornwell (1983) has introduced a modification of the CLEAN algorithm called smoothness-stabilized clean which reduces this unwanted modulation. The original CLEAN algorithm minimizes

$$\sum_i w_i|\mathcal{V}_i - \mathcal{V}_i'|^2, \tag{11.4}$$

where $\mathcal{V}_i$ is the measured visibility at (u_i, v_i), w_i is the applied weighting, $\mathcal{V}_i'$ is the corresponding visibility of the CLEAN-derived model, and the summation is taken over the points with nonzero data in the input transformation for the uncleaned map. Cornwell's algorithm minimizes

$$\sum_i w_i|\mathcal{V}_i - \mathcal{V}_i'|^2 - \kappa s, \tag{11.5}$$

where s is a measure of smoothness and κ is an adjustable parameter. Cornwell finds that the mean squared brightness of the model, taken with a negative sign, is an effective implementation of s, and he shows that the inclusion of the smoothness term requires only a modification of the original synthesized beam before performing the normal CLEAN procedure.

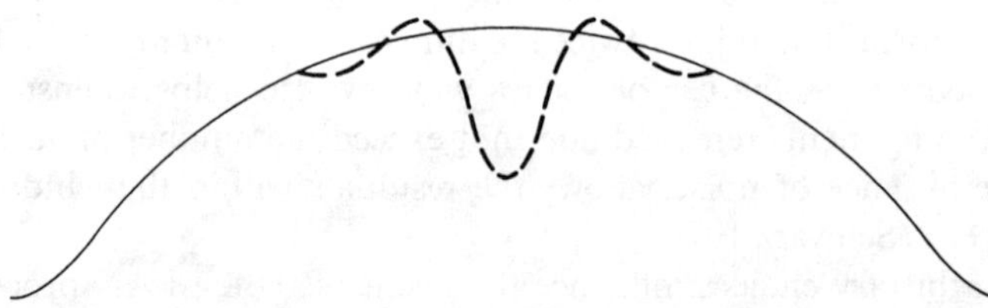

Figure 11.2 Subtraction of the point-source response (broken line) at the maximum of a broad feature, as in the process CLEAN. After Clark (1982).

The effects of visibility tapering appear in both the original map and beam, and thus the magnitudes and positions of the components subtracted in the CLEAN process should be largely independent of any taper. Since tapering reduces the resolution, however, it is a common practice to use no taper (the principal solution) for maps that are to be processed using CLEAN. Alternatively, in difficult cases such as those involving extended, smooth structure, reduction of sidelobes by tapering may improve the performance of CLEAN.

An important reduction in the computation required for CLEAN has been introduced by Clark (1980). This is based on subtraction of the point-source responses in the (u, v) plane and using the FFT for moving data between the (u, v) and (ξ, η) domains. The procedure involves two cycles. A series of minor cycles is used to locate the components to be removed by performing approximate subtractions in the (ξ, η) domain using only the main beam and first sidelobes of the synthesized beam. Then in a major cycle the series of identified point-source responses are subtracted, without approximation, in the (u, v) plane. Clark devised this technique for use with data from the VLA and found that it reduced the computation by a factor of 2–10 compared with the original CLEAN algorithm.

To summarize the characteristics of CLEAN we note that it is simple to understand from a qualitative viewpoint, straightforward to implement, and its usefulness has been proven by wide usage for over a decade. On the other hand, a full analysis of its response, as distinct from that of the least-squares fit to which it converges, is difficult. The response of CLEAN is not unique, and it can produce spurious artifacts. It is sometimes used in conjunction with model fitting techniques: for example, a disk model can be removed from the image of a planet and the residual brightness processed by CLEAN. It is also used as part of more complex image construction techniques, which are described in Section 11.2.

Constrained Optimization Techniques

An important class of image restoration algorithms operate to produce a map that agrees with the measured visibility to within the noise level, while constraining the result to maximize some measure of image quality. Of these the maximum entropy method has received particular attention in radio astronomy, and we now discuss it as an example of this general class. If $B'(\xi, \eta)$ is the brightness distribution derived by the maximum entropy method, a function $F(B')$ is defined which is referred to as the entropy of the distribution. $F(B')$ is determined entirely by the distribution of B' as a function of solid angle and takes no account of structural forms within the map. In constructing the map, $F(B')$ is maximized within the constraint that the Fourier transform of B' should fit the observed visibility values.

In astronomical image formation an early application of the maximum entropy method is that of Frieden (1972) to optical images. In radio astronomy the earliest discussions are by Ables (1974) and Ponsonby (1973). The aim of the technique, as described by Ables, is to obtain a brightness distribution consistent with all relevant data but minimally committal with regard to missing data. Thus, $F(B')$ must be chosen so that maximization introduces legitimate a priori information but allows the visibility in the unmeasured areas to assume values that minimize the detail introduced.

Two forms of $F(B')$ have been discussed extensively. They are

$$F_1 = -\sum_i \frac{B_i'}{B_s'} \log\left(\frac{B_i'}{B_s'} \right) \tag{11.6}$$

and

$$F_2 = \sum_i \log B_i', \tag{11.7}$$

where $B_i' = B'(\xi_i, \eta_i)$, $B_s' = \sum_i B_i'$, and the sums are taken over all discrete values in the map. The literature contains a substantial number of papers that discuss the derivation of the two expressions from theoretical considerations, which we touch upon only briefly. The subject invokes Bayesian statistics which are used to relate the probabilities of the brightness distribution prior to, and after, the introduction of measurements: for a general discussion see Jaynes (1968, 1982). Maximization of entropy can be interpreted as a means of introducing prior information on the sky brightness. The entropy expressions are derived from statistical considerations of the a priori probability distribution on the sky. Maximization of entropy is thus an attempt to choose the most likely brightness distribution consistent with the measurements. Gull and Daniell (1979) consider the distributions of brightness quanta scattered randomly on the sky, and they derive the form F_1 which is also used by Frieden (1972). Note that such randomization results in a high probability of largely uniform brightness, which is quite different to the point-source sky model which was the initial basis for CLEAN. The entropy form F_2 is obtained by Ables (1974) and Wernecke and D'Addario (1977). An approach based on Bose–Einstein statistics is used by Kikuchi and Soffer (1977) who obtain both forms F_1 and F_2 as special cases of a more general expression. Examples of radio maps made using the F_1 form are given by Gull and Daniell (1978) and using F_2 by Wernecke and D'Addario (1977). The maximum entropy method is clearly shown to be an effective mapping algorithm, and both forms of the entropy measure appear to work about equally well.

The disagreement over the entropy forms has persisted because the methods of derivation introduce concepts that are better understood in other applications, so that the physical basis of the arguments appears tenuous. This situation has led a number of investigators to what may be described as a pragmatic approach to the maximum entropy method (Högbom 1979; Subrahmanya 1979; Nityananda and Narayan 1982). They view the method as an effective algorithm, but question whether there is any fundamental relationship between the entropy expressions and the brightness distributions of actual astronomical objects. As Högbom (1979) points out, both F_1 and F_2 contain the required mathematical characteristics: the first derivatives tend to infinity as B' approaches zero, so maximizing F_1 or F_2 produces positivity in the image. The second derivatives are everywhere negative, which favors uniformity in the brightness. In the extreme case where the visibility is known only at the (u, v) origin, maximizing F_1 or F_2 results in a uniform brightness level. The tendency to uniformity is the characteristic required to suppress the oscillations in the principal response and is consistent with Ables' goal of minimizing the information represented by the extrapolated visibility data. Although this minimization provides a good general principle in image construction, it cannot, of course, be expected to provide the correct solution in all cases. Narayan and Nityananda (1984) studied a general class of functions F which have the properties $d^2F/dB'^2 < 0$ and $d^3F/dB'^3 > 0$. F_1 and F_2 discussed above are members of this class. Many functions of this class are effective for 'maximum entropy' image processing as a result of their mathematical properties, even though there may be no underlying physical or information-theoretic basis for their choice.

In the implementation of the maximum entropy method the usual procedure is to maximize an expression of the form

$$F(B') - \Lambda \sum_k \frac{|\mathscr{V}_k - \mathscr{V}_k'|^2}{\sigma_k^2}, \tag{11.8}$$

where $\mathscr{V}_k = \mathscr{V}(u_k, v_k)$, $\mathscr{V}_k'$ is the corresponding visibility for the model B', σ_k^2 is the variance of the noise in $\mathscr{V}_k$, and the summation is taken over the visibility data. The parameter Λ is chosen so that the agreement between the model and the measurements is sufficiently close. It governs the relative importance of maximizing the entropy and of minimizing the mean squared residuals. The maximization involves an iterative procedure in which Λ is used as a Lagrange multiplier: for descriptions of the procedure see Wernecke and D'Addario (1977), Wernecke (1977), and Gull and Daniell (1978). Nityananda and Narayan (1982) discuss a wider range of methods.

A feature of maps derived by the maximum entropy method is that, in general, the point-source response varies with position, so the angular resolution is not constant over the map. Comparison of maximum entropy maps with those obtained using direct Fourier transformation often shows higher angular resolution in the former. The extrapolation of the visibility values can provide some increase in resolution over more conventional mapping techniques, but this effect is limited by the noise in the data.

During the first decade after the introduction of the CLEAN and other optimization techniques, CLEAN has been the most widely used, partly because it generally requires less computing. However, maximum entropy and similar optimization techniques become more competitive for maps with broad, smooth features for which CLEAN requires many subtractions, and they do not share CLEAN's problem of introducing spurious fine structure in broad features. The problem of the response to noise is also more tractable in the maximum entropy methods; see, for example Bryan and Skilling (1980).

11.2 MAPPING WITH INCOMPLETE PHASE DATA

The CLEAN and other optimization procedures are aimed mainly at compensation for limited (u, v) coverage. We now discuss different methods that have been developed to obtain the best interpretation of visibility data for which only the modulus exists, or for which the phase is inaccurate or uncalibrated. Pearson and Readhead (1984) have also reviewed this subject. For an interesting discussion of image formation using phase data only, see Oppenheim and Lim (1981).

Mapping With the Visibility Modulus Only

Circumstances in which phase data are entirely missing include measurements with uncalibrated phase using only two antennas, in which case the closure relationships to be discussed later do not apply, and measurements with the intensity interferometer to be discussed in Chapter 15. These conditions do not usually occur with large synthesis arrays, but the problem of mapping without phase data is worth discussing for the general insight that it provides. Also, it is likely to be of importance in optical interferometry, which is discussed in Chapter 15.

The Fourier transform of the squared modulus of the visibility is equal to the autocorrelation of the brightness distribution $B\star\star B$:

$$|\mathscr{V}(u, v)|^2 = \mathscr{V}(u, v)\mathscr{V}^*(u, v) \rightleftharpoons B(\xi, \eta) * * B(-\xi, -\eta)$$

$$= B(\xi, \eta)\star\star B(\xi, \eta). \qquad (11.9)$$

The problem of mapping with $|\mathscr{V}|$ only is therefore one of interpreting a map of the autocorrelation of B. The question of whether it is possible to obtain a unique solution for B from its autocorrelation function has been a longstanding one in image processing. Without phase data the position of the center of the field cannot be determined, and there is a 180° rotational ambiguity in the position angle of the map. Accepting these limitations, we consider the solution to be unique if the form of the image is uniquely defined.

The basic question to be investigated is whether it is possible to construct the brightness distribution $B(\xi)$ unambiguously from the modulus of the visibility $\mathscr{V}(u)$. For simplicity we first consider the one-dimensional case. An answer can be found by examining some of the general properties of $\mathscr{V}$ and B as functions of the complex variable, $z = x + jy$. B is real, positive (unless it represents Stokes parameters $Q, U,$ or V), and in most cases localized. The product of B and the antenna beam pattern can certainly be considered localized, so the integral of the measured brightness function over an infinite range of ξ remains finite. The Fourier transform, $\mathscr{V}$, is an entire function, that is, it is analytic for all finite z. In general, $\mathscr{V}$ is also analytic over the entire upper-half complex plane, which is the condition that its real and imaginary parts should be a Hilbert transform pair (e.g., see Morse and Feshbach 1953). A similar connection between the modulus and the phase can be obtained by writing

$$\mathscr{V} = |\mathscr{V}| e^{j\phi_v}, \tag{11.10}$$

$$\ln \mathscr{V} = \ln|\mathscr{V}| + j(\phi_v + 2n\pi), \tag{11.11}$$

where n is an integer. In this case, the Hilbert transform relationship between $\ln|\mathscr{V}|$ and ϕ_v is complicated by the fact that $\ln \mathscr{V}$ is generally not analytic in the upper-half plane, since it is singular at the zeros in $\mathscr{V}$ which occur at infinite z and at some finite values also. Various ways of circumventing this problem are discussed by Burge et al. (1976).

Among the important investigations of the phase problem are those of Bates (1969) and his collaborators. Bates noted that an entire function of the type of $\mathscr{V}(z)$ can be specified by the positions of its zeros in the complex plane. $B(\xi)$ is represented by a Fourier series in which the number of terms M is finite because both the extent of B in the ξ domain and the extent of $\mathscr{V}$ in the u domain (which determines the resolution) are limited. The visibility derived from the series representation of B is expressible as a product of two functions. One of these is a polynomial of order $2M$ that has complex zeros, and the other has zeros on the real axis only. The zeros of the first function occur in pairs symmetrically disposed about the imaginary axis: zeros in these pairs occur at values of z that have real parts of opposite sign and identical imaginary parts. Furthermore, if any of these zeros is replaced by a zero at the complex conjugate value of z, that is, the zero is shifted to its mirror-image position in the real axis, the value of $|\mathscr{V}|$ is unchanged. (The corresponding function for $|\mathscr{V}|^2$ contains zeros at both sets of positions.) If L is the number of complex zeros of $|\mathscr{V}|$, then $L \leq M$, and by moving zeros between conjugate values of z, 2^L solutions for $\mathscr{V}$, and correspondingly for B, may be obtained. This is reduced to 2^{L-1} if we omit those solutions that differ only by a 180° rotation from another one. Bates (1969) showed how a solution for B can be obtained, and given one solution the others can also be determined. Napier (1972) used this technique to compute all possible solutions for a number of one-dimensional cases, and he found that for some of them more than half of the solutions could be rejected because of significant negative features. However, no unique solution was obtained unless the distribution was a very simple one.

For two-dimensional data the situation is different. Bates' complex-zero theory is developed for the one-dimensional case only, but Napier and Bates

(1974) were able to apply it to a two-dimensional image by taking a series of one-dimensional profiles at different position angles. Only one solution without negative features was found. Bruck and Sodin (1979) also deduced that in two dimensions the solution is much more likely to be unique, basing this conclusion on an analytical argument. They pointed out that a polynomial that represents B or $\mathcal{V}$ in one dimension can always be factored into a serial product form. The complex zeros that control the multiple solutions in Bates' theory arise from such factors. In two dimensions, however, the equivalent polynomial is much less likely to yield to factoring except in cases in which there is a form of geometrical repeatability in the image.

Further work by Bates (1984) indicates that in almost all practical two-dimensional cases, positivity is sufficient to provide a solution that is unique except for the 180° rotational uncertainty and the loss of absolute position information. This result applies only to images of limited angular extent, and the spacing of the sample points in u and v must be no more than half of that specified by the sampling theorem (see Section 4.11) for the source in order to sample adequately the Fourier transform of the autocorrelation function of the brightness, which extends over twice the dimensions of the brightness distribution in (ξ, η). Only the moduli of the visibility values are measured, but it is possible to deduce the phases by a procedure which starts by assuming that the phase at the origin is zero. An arbitrary sign is then chosen for the phase at one of the points neighboring the (u, v) origin. This choice is allowed because of the 180° rotational ambiguity in the resulting brightness distribution. The phases at two adjacent points are then deduced unambiguously using a combinatorial algorithm which is recursively continued across the (u, v) plane. Bates concluded that the positivity constraint on the brightness almost always ensures that the most compact image is the appropriate one. Other algorithms for reconstruction without phase data have been developed and are used mainly in fields other than radio astronomy; see, for example, Fienup (1978). Algorithms for phaseless reconstruction are likely to be of more practical importance in optical than in radio mapping, since in the latter the closure relationships to be discussed later in this chapter can usually be invoked, and these additional data also improve the sensitivity. However, the fact that positivity leads to a unique solution in two dimensions is important since it helps to explain why iterative mapping techniques that embody this constraint converge satisfactorily.

One other study of mapping without phase information should be described since it illustrates the ambiguities from a different viewpoint, namely, point sources in the sky rather than zeros in the visibility. Baldwin and Warner (1976) have investigated the problem for survey observations which mainly contain large numbers of sources that appear as unresolved points. The autocorrelation function then takes the form of a distribution of point-source responses: the response is most easily visualized in terms of the autocorrelation. The case where the field contains one source much stronger than the others is of special interest. Let the strong source have unit flux density and position (ξ_0, η_0), and let the other sources have flux densities S_k and positions (ξ_k, η_k). Consider the strong source and two others. The combined visibility is

$$\mathcal{V}(u, v) = e^{j\phi_0} + S_1 e^{j\phi_1} + S_2 e^{j\phi_2}, \tag{11.12}$$

where $S_k \ll 1$ and

$$\phi_k = -2\pi(u\xi_k + v\eta_k). \tag{11.13}$$

The modulus of the visibility is

$$|\mathcal{V}(u, v)| = [\mathcal{V}(u, v)\mathcal{V}^*(u, v)]^{1/2}$$

$$= \left[1 + S_1^2 + S_2^2 + 2S_1\cos(\phi_0 - \phi_1)\right.$$

$$\left. + 2S_2\cos(\phi_0 - \phi_2) + 2S_1 S_2\cos(\phi_1 - \phi_2)\right]^{1/2}. \tag{11.14}$$

Then, neglecting second-order terms in S_1 and S_2, one obtains

$$|\mathcal{V}(u, v)| = 1 + S_1\cos(\phi_0 - \phi_1) + S_2\cos(\phi_0 - \phi_2)$$

$$= 1 + \left(\frac{S_1}{2}\right)e^{j(\phi_0 - \phi_1)} + \left(\frac{S_1}{2}\right)e^{-j(\phi_0 - \phi_1)}$$

$$+ \left(\frac{S_2}{2}\right)e^{j(\phi_0 - \phi_2)} + \left(\frac{S_2}{2}\right)e^{-j(\phi_0 - \phi_2)}. \tag{11.15}$$

Baldwin and Warner consider the map which results from the visibility modulus with no phase data, that is, the Fourier transform of Eq. (11.15). By inspection of Eq. (11.15) one can see that the map will contain the following components: (1) the response to the strong source which appears at the origin, (2) a half-amplitude response to each of the two weaker sources located at its correct position relative to the strong source, and (3) a half-amplitude response to each of the weaker sources symmetrically displaced with respect to the strong source (i.e., a mirror image reflected across the origin). Note that a similar set of responses would occur for the autocorrelation which is the Fourier transform of $|\mathcal{V}|^2$. The strong source response can be accepted as a true map feature, and since we tolerate the 180° rotational ambiguity, any one of the other four responses, with its amplitude doubled, can also be accepted. It is then possible to identify the appropriate position for the third source if second-order features that we have neglected can be detected. The term containing $S_1 S_2$ in Eq. (11.14) produces responses symmetrically displaced about the origin at positions corresponding to the position difference of the two weaker sources. Alternatively, it may be possible to suppress the spurious images by placing the strong source at the edge of the antenna beams. With larger numbers of sources the situation becomes complicated, but, in principle, the true responses and their images can be sorted out down to a level determined by the noise, or by confusion if source pairs with similar vector spacings occur. In a further analysis Baldwin and Warner (1978) considered cases in which there is no dominant source. They demonstrated a procedure in which a few of the stronger sources are identified in the autocorrelation map, and these form a trial map. Low-accuracy phase measurements, if available, may be useful in identifying strong sources at this stage. A hybrid map is then computed using the measured visibility amplitudes, and visibility phases computed from the trial map. The hybrid map contains (1) the features of the trial map with correct positions and amplitudes, (2) other sources in their correct positions at half the correct amplitudes, and (3) the images of the sources in (2) but at fractional amplitudes because they are formed by combination with numerous other sources, not just the dominant strong one as in the earlier example. The stronger sources that are in the hybrid map but not in the trial map are likely to be real and therefore are added to the trial map with doubled amplitudes. A new hybrid map using the measured amplitudes and phases from the new trial map is derived and the process repeated until the identification of true features is limited by the noise. Baldwin and Warner (1978, 1979) show examples of maps that demonstrate the practicality of the above technique and discuss the implementation in detail.

We can conclude from the above discussions that if the lack of absolute position measurement and the 180° ambiguity can be accepted, it is indeed feasible in two dimensions to obtain maps from visibility modulus data alone. Constraints on the positivity and confinement of the image are required. Of course, the lack of phase data reduces the sensitivity, and the nonlinearity of the procedures may be expected to place additional lower limits on the brightness of features that can be mapped reliably.

Mapping with Uncalibrated Phase Data

We now consider the case of great practical importance in radio mapping, namely, that where both phase and amplitude measurements of the visibility exist but errors in calibration limit the accuracy of the resulting map. With connected-element arrays the calibration of the visibility amplitude is often accurate to a few percent, but phase errors expressed as a fraction of a radian may be much larger as a result of variations in the ionosphere or troposphere. In the usual process of calibration the reference source is rarely close enough to the source being mapped to share the same atmospheric irregularities, and it is often impractical to alternate between the map position and the reference source on a sufficiently short time scale. In VLBI observations, instrumental limits discussed in Chapter 9 inhibit calibration of the phase and may complicate the amplitude calibration. Nevertheless, the relative values of the uncalibrated visibility measured simultaneously on a number of baselines contain information about the brightness distribution that can be extracted through the closure relationships.

We have shown in Section 7.3 that the correlator output for antenna pair (m, n) can be written

$$r_{mn} = G_{mn}\mathcal{V}_{mn} = g_m g_n^* \mathcal{V}_{mn},$$
(11.16)

where G_{mn} is the complex gain for the antenna pair and the g terms are gain factors for the individual antennas. Considering first the phase relationships we represent the arguments of the exponential terms of r_{mn}, g_m, g_n, and $\mathcal{V}_{mn}$ by ϕ_{rmn}, ϕ_{gm}, ϕ_{gn}, and ϕ_{vmn} respectively. Thus, we can write

$$\phi_{rmn} = \phi_{gm} - \phi_{gn} + \phi_{vmn}.$$
(11.17)

Now for three antennas m, n, and p the phase closure relationship is

$$\phi_{rmn} + \phi_{rnp} + \phi_{rpm} = \phi_{vmn} + \phi_{vnp} + \phi_{vpm}$$
(11.18)

The antenna gain terms, g_m and so on, here contain the effect of the atmospheric paths to the antennas, and since these terms do not appear in Eq. (11.18) it is evident that the combination of the three correlator-output phases constitutes an observable quantity that depends only on the phase function of the visibility. This property of the phase closure relationships was first recognized and made use of by Jennison (1958) in the experiments mentioned in Chapter 1.

If we have n_a antennas with no redundant spacings, and we measure the correlation of all pairs, the number of independent phase closure relationships is equal to the number of correlator-output phases less the number of unknown instrumental phases, one of which can be arbitrarily chosen. Thus, the number of closure phase relationships is

$$\tfrac{1}{2}n_a(n_a - 1) - (n_a - 1) = \tfrac{1}{2}(n_a - 1)(n_a - 2).$$
(11.19)

This number can also be obtained by taking one antenna and considering the number of different groups of three that can be formed that include it. Each of these groups must contain one baseline that is not in any other group, so the relationships are independent.

An amplitude closure relationship involves four antenna pairs, for which four antennas m, n, p, and q are required:

$$\frac{|r_{mn}||r_{pq}|}{|r_{mp}||r_{nq}|} = \frac{|\mathcal{V}_{mn}||\mathcal{V}_{pq}|}{|\mathcal{V}_{mp}||\mathcal{V}_{nq}|}.$$
(11.20)

Here the moduli of the g_m terms cancel out. There are two independent closure relationships for the four antennas: a second one may be obtained by replacing the subscripts in the numerator (or denominator) of Eq. (11.20) by mq and np. The number of independent amplitude closure relationships for n_a antennas with no redundant baselines is equal to the number of measured amplitudes, $\frac{1}{2}n_a(n_a - 1)$ less the number of unknown antenna gain factors n_a, that is, $\frac{1}{2}n_a(n_a - 1) - n_a = \frac{1}{2}n_a(n_a - 3)$. For earlier usage of the principle of taking ratios of observed visibility amplitudes to eliminate instrumental gains, see Smith (1952) and Twiss, Carter, and Little (1960).

The rekindling of interest in closure techniques in the 1970s began with their rediscovery by Rogers et al. (1974) who used closure phases to derive model parameters for VLBI data. Fort and Yee (1976) and several later groups incorporated closure data into iterative mapping techniques, of which we shall describe that by Readhead et al. (1980). Figure 11.3 shows a flow chart of the procedure which is as follows:

1. Obtain an initial trial map based on inspection of visibility amplitudes and any a priori data such as a map at a different wavelength or epoch. If the trial map is inaccurate the convergence will be slow, but if

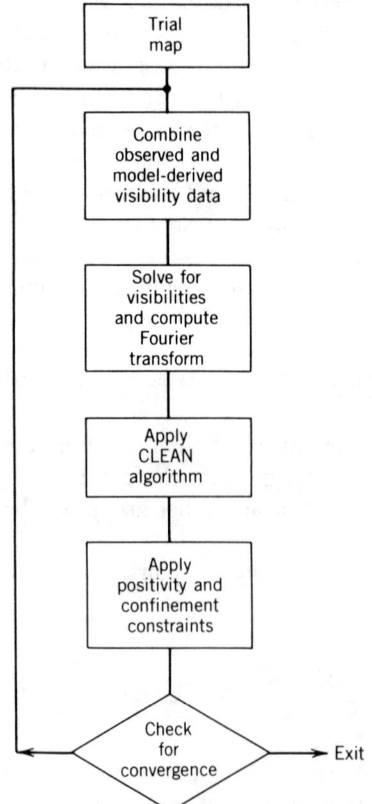

Figure 11.3   Flow chart of the hybrid mapping procedure.

necessary an arbitrary trial map such as that of a single point source will generally suffice.

2. For each visibility integration period determine a complete set of independent amplitude and/or phase closure equations. For each such set compute a sufficient number of visibility values from the model such that when added to the closure relationships the total number of independent equations is equal to the number of antenna spacings.

3. Solve for the complex visibility corresponding to each antenna spacing and make a map from the visibility data by Fourier transformation.

4. Apply CLEAN and select the positive components within an area judged to contain the source as the new trial map. The residuals are not used.

5. Test for convergence and return to step 2 as necessary.

Note that the solution improves with iteration because of the constraints of confinement and positivity introduced in step 4. These nonlinear processes can be envisioned as spreading the errors in the model-derived visibility values throughout the visibility data, so that they are diluted when combined with the observed values in the next iterative cycle.

In the process described, and most variants of it, the map is formed by using some data from the model and some from direct measurements, and following Baldwin and Warner (1978) the name hybrid mapping is widely used as a generic description. The use of closure phases eliminates the $180°$ rotational ambiguity that occurs when there is no phase data, but there is no absolute position measurement. With the use of closure amplitude, only relative levels of brightness are determined, but it is generally not difficult to calibrate enough of the data to establish a brightness scale. In many cases the amplitude data are sufficiently accurate as observed, and only the phase closure relationships need be used: Readhead and Wilkinson (1978) have described a version of the above program using phase closure only. Other versions of this technique, which differ mainly in detail of implementation from that described, have been developed by Cotton (1979) and Rogers (1980). If there is some redundancy in the baselines the number of free parameters is reduced, which can be advantageous, as discussed by Rogers.

The number of antennas is obviously an important factor in mapping by the closure relationships since it affects the efficiency with which the data are used. We can quantify this efficiency by considering the number of closure data as a fraction of the number of data that would be available if full calibration were possible, as a function of n_a. This fraction is also equal to the ratio of observed data to observed plus model-derived data in each iteration of the hybrid mapping procedure. For the phase data the fraction is

$$\frac{\frac{1}{2}(n_a - 1)(n_a - 2)}{\frac{1}{2}n_a(n_a - 1)} = \frac{n_a - 2}{n_a}. \tag{11.21}$$

For the amplitude data the fraction is

$$\frac{\frac{1}{2}n_a(n_a - 3)}{\frac{1}{2}n_a(n_a - 1)} = \frac{n_a - 3}{n_a - 1}.$$ (11.22)

The above expressions are plotted in Fig. 11.4. For $n_a = 4$ the closure relationships yield only 50% of the possible phase data and only 33% of the amplitude data. For $n_a = 10$, however, the corresponding figures are 80% and 78%. Thus, in any array in which the atmosphere or instrumental effects may limit the accuracy of calibration by a reference source, it is desirable that the number of antennas should be at least five and preferably 10 or more. The number of iterations required to obtain a solution with the hybrid technique depends on the complexity of the source, the number of antennas, the accuracy of the initial model, and other factors including details of the algorithm used. However, these solutions generally converge rapidly.

Another group of image construction programs that basically performs the same function as hybrid mapping, but with a different approach, is usually described as self-calibration. Here the complex antenna gains are regarded as free parameters to be calibrated using information from the mapping observations, and explicitly derived together with the brightness. In certain cases the process is easily explained. For example, in mapping an extended source containing a compact component (as in many radio galaxies) the broad structure is resolved with the longer antenna spacings, leaving only the compact source. This can be used as a calibrator to provide the relative phases of the long-spacing antenna pairs, but not the absolute phase since the position is not known. Then, if there is a sufficient number of long spacings in the array, the gain factors of the antennas can be obtained using long spacings only. Such

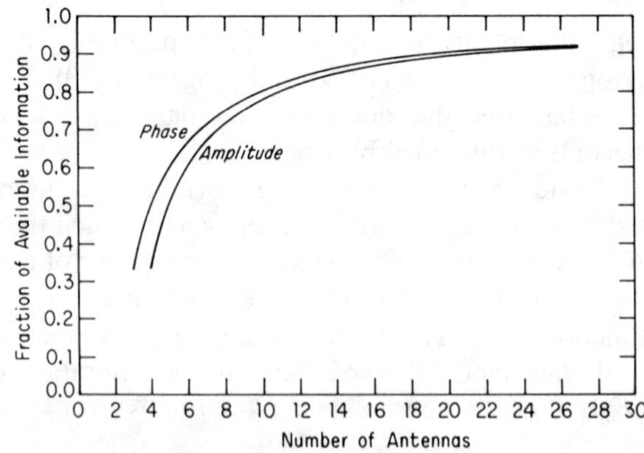

Figure 11.4   Visibility data that can be obtained through adaptive calibration techniques expressed as a fraction of those available from a fully calibrated array. The curves correspond to Eqs. (11.21) and (11.22).

a special brightness distribution, however, is not essential to the method, and with an iterative technique it is possible to use almost any source as its own calibrator. Programs of this type were developed by Schwab (1980) and by Cornwell and Wilkinson (1981).

The procedure in self-calibration is to use a least-squares method to minimize the square of the modulus of the difference between the observed visibilities $\mathscr{V}_{mn}$ and the corresponding values $\mathscr{V}'_{mn}$ for the derived model. The expression that is minimized is

$$\sum_{\text{time}} \sum_{m<n} w_{mn} |\mathscr{V}_{mn} - g_m g_n^* \mathscr{V}'_{mn}|^2, \tag{11.23}$$

where the weighting coefficient w_{mn} is usually chosen to be inversely proportional to the variance of $\mathscr{V}_{mn}$, and the quantities shown are all functions of time within the observing period. Expression (11.23) can be written

$$\sum_{\text{time}} \sum_{m<n} w_{mn} |\mathscr{V}'_{mn}|^2 |X_{mn} - g_m g_n^*|^2, \tag{11.24}$$

where

$$X_{mn} = \frac{\mathscr{V}_{mn}}{\mathscr{V}'_{mn}}. \tag{11.25}$$

If the model is accurate, the ratio X_{mn} of the uncalibrated observed visibility to the visibility predicted by the model is independent of u and v but proportional to the antenna gains. Thus, in effect, the values of X_{mn} simulate the response to a point-source calibrator and enable the gains to be determined. However, since the initial model is only approximate, the desired result must be approached by iteration.

The self-calibration procedure is as follows:

1. Make an initial map as for hybrid mapping.
2. Compute the X_{mn} factors for each visibility integration period within the observation.
3. Determine the antenna gain factors for each integration period.
4. Use the gains to calibrate the observed visibility values and make a map.
5. Use CLEAN and select components to provide positivity and confinement of the image: Cornwell (1982) recommends omitting all features for which $|B(\xi, \eta)|$ is less than that for the most negative feature.
6. Test for convergence and return to step 2 as necessary.

The numbers of independent measurements used in the above procedure, $\frac{1}{2} n_a (n_a - 1)$ less n_a or $(n_a - 1)$ for amplitudes or phases, respectively, are the same as the numbers of independent closure relationships in the hybrid

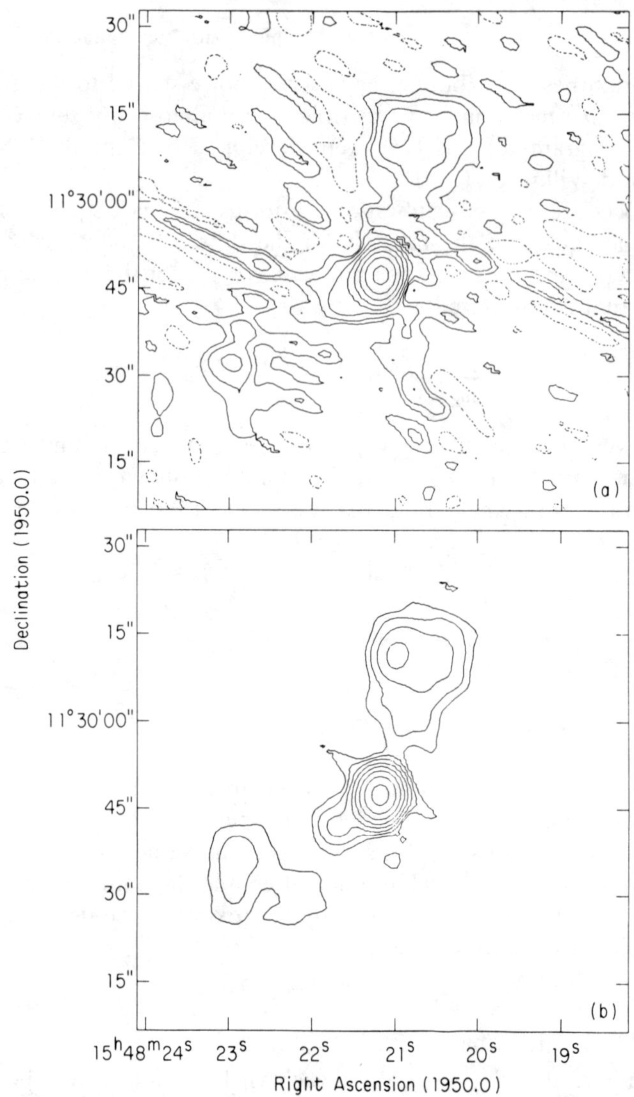

Figure 11.5   Effect of self-calibration on a VLA radio image of the quasar 1548 + 115. (a) Image obtained by normal calibration techniques which has spurious detail at the level of 1% of the peak brightness. (b) Image obtained by the self-calibration technique, in which the level of spurious detail is reduced below the 0.2% level. In both (a) and (b) the lowest contour level is 0.6%. From Napier, Thompson, and Ekers (1983); © 1983 IEEE.

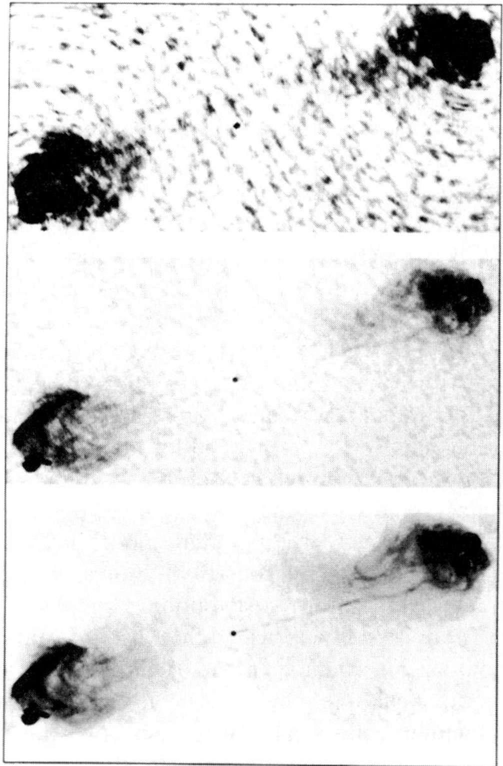

Figure 11.6   Three stages in the reduction of the observation of Cygnus A shown in Fig. 1.15 (Perley, Dreher, and Cowan, 1984). The top image is the result of transformation of the calibrated visibility data using the FFT algorithm. The calibration source was approximately 3° from Cygnus A. The center image shows the effect of reduction using the maximum entropy algorithm. This compensates principally for the undersampling in the spatial frequencies and thereby removes sidelobes from the synthesized beam. The result is similar to that obtainable using the CLEAN algorithm. The bottom image shows the effect of the self-calibration technique, the maximum-entropy image being used as the initial model. This final step improves the dynamic range by a factor of 3. In observations where the initial calibration is not as good as in this case, self-calibration usually provides a greater improvement. The long dimension of the field is 2.1 arcmin and contains approximately 1000 pixels. Reproduced by permission of NRAO/AUI.

mapping schemes. Thus the two types of procedures are basically equivalent and differ only in details of approach and implementation. The efficiency as a function of the number of antennas (Fig. 11.4) applies to both. Examples of the performance of the self-calibration technique are shown in Figs. 11.5 and 11.6. Following Schwab (1980) we use the term *adaptive calibration* for both the hybrid mapping and self-calibration techniques.

Treating the gain factors, which are the fundamental unknown quantities, as free parameters as in self-calibration is a rather more direct approach than that of hybrid mapping. A global estimate of the instrumental factors is obtained using the entire data set. Cornwell (1982) points out that it is easier to deal correctly with the noise when considering complex visibility as a vector quantity, as in self-calibration, than when considering amplitude and phase separately, as in hybrid mapping. The noise combines additively in the vector components resulting in a Gaussian distribution, whereas in the modulus and phase the more complicated probability distributions of Eqs. (6.56) result. Cornwell and Wilkinson (1981) have developed a form of adaptive calibration that takes account of the different probability distributions of the amplitude and phase fluctuations, including system noise, for the different antennas. It has been used with the MERLIN array that incorporates antennas of different sizes and designs (Davies, Anderson, and Morison 1980). The probability distributions of the antenna-associated errors are legitimate a priori information, which can be empirically determined for an array.

Experience shows that adaptive calibration techniques in many cases converge to a satisfactory result using only a single point source as a starting model, although inaccuracy in the initial model increases the number of iterative cycles required. A point source, of course, is a good model for the phase of a symmetrical brightness distribution. However, a symmetrical model can sometimes produce artifacts when the visibility data correspond to an asymmetrical source. For example, a true brightness distribution comprising a strong source and a weak one may appear with a second weak source added, the two being symmetrically located with respect to the strong source. Such configurations arise in a similar manner to the symmetrical pairs in the Baldwin and Warner method discussed earlier in this section.

Although the adaptive calibration techniques require some care and experience to use correctly, they produce results of high quality. For one of the most important early examples see the VLBI maps of 3C273 in Fig 1.16. The ability of the methods to remove phase errors of atmospheric origin enables maps of dynamic range exceeding 40 dB to be obtained under favorable circumstances. It should be noticed, however, that these techniques can best mitigate atmospheric phase variations that, at any instant, are uniform over the whole field under observation, because only in such circumstances can the phase effects be represented by a single gain term for each antenna. This requirement is essentially the same as the isoplanatic condition of optical observations which is discussed in Chapter 15. At centimeter and shorter wavelengths the requirement is usually met. At meter wavelengths the situation can be more complicated, since antenna beams are wider and the density of strong sources on the sky is greater. However, it is possible to relax the isoplanatism requirement at the expense of more complicated procedures (Schwab 1984). It must also be remembered that the accuracy of the closure relationships depends on the accuracy of the matching of the frequency responses and polarization parameters discussed in Sections 7.3 and 7.4. In general, any effect that cannot be represented by a single gain factor for each antenna, for example, anomalous behavior of a correlator, degrades the closure accuracy.

11.3 MISCELLANEOUS APPLICATION NOTES

In implementing CLEAN the Fourier transform of the weighted transfer function is used as the point-source response for deconvolution. Errors in the measured visibility, particularly in the phase, cause the actual synthesized beam to deviate from the theoretical response, and therefore it may appear preferable to deconvolve with the response measured by mapping a point source. However, because of the inevitable component of noise in the measured response, the theoretical beam is almost always preferred. In cases where examination of the calibrator data indicates periods of poor phase stability, usually attributable to varying atmospheric conditions, it may be advantageous to omit the poor quality data to obtain more accurate subtraction of the source components by CLEAN.

The CLEAN procedure is helpful in removing or reducing the effect of missing short-spacing visibility values discussed in Section 10.3. Since the theoretical beam contains negative sidelobes resulting from the missing spacings, the application of CLEAN is very effective in reducing the undesirable negative areas surrounding the sources in a map. Inserting short-spacing or zero-spacing visibility values obtained independently is always a good procedure when possible, but it should not be forgotten that in any subsequent application of CLEAN the additional data must be included in deriving the point-source response to be subtracted.

In spectral-line data it is often necessary to subtract maps at different frequencies, or subtract a continum map from a line map, to examine spectral features. In such cases CLEAN should not be applied independently to the maps before subtraction since it generates low-level structure that depends on the number of iterations and other parameters. These erroneous details differ from one map to another and can become serious in the difference map. To subtract a continuum map from a line-plus-continuum map, a good procedure is to apply CLEAN to the continuum map and then remove the same components from the line-plus-continuum map, using the appropriate point-source response. If the subtraction is performed in the (u, v) plane, the difference between the point-source responses of the two maps is compensated for automatically. The continuum residuals should also be subtracted from the line-plus-continuum map.

A final point on the use of CLEAN concerns the methods of handling the w component discussed in Section 4.1. These procedures can result in a map over which the beam varies, and which therefore cannot be processed by CLEAN in the usual manner. For example, in the method in which the observing period is subdivided into intervals over each of which the spacing vectors can be considered coplanar, the maps for the different intervals will require different corrections to the angular scale. When combined they result in a beam of varying profile.

In using adaptive calibration techniques, the integration period of the data must not be longer than the coherence time of the phase variations, otherwise the visibility amplitude may be reduced. The coherence time may be governed by the atmosphere, or in the case of VLBI observations by the frequency standards. In order for the mapping procedure to work, the field under observation must contain structure fine enough to provide a phase reference and bright enough to be detected with satisfactory signal-to-noise ratio within the coherence time. Thus, adaptive calibration does not solve all problems, and cannot be used for the detection of a very weak source in an otherwise empty field.

In the application of adaptive calibration to spectral-line data the signal-to-noise ratio in a single channel may be too low to provide a satisfactory result. It is better to make a continuum map using an average over those channels that do not include prominent line features. By this procedure complex gains of the antennas averaged over the corresponding channels are obtained. The antenna gains for the individual channels can then be determined from the bandpass calibration.

The techniques discussed in this chapter can be largely characterized as quasi-empirical since the procedural details have not all been fully analyzed, and they include user-selected parameters, the most effective choice of which rests on experience. Their development has been limited by available computing capability, and as this increases new techniques will continue to arise. Thus, the principles illustrated in this chapter, rather than the procedures themselves, are likely to be of lasting importance.

BIBLIOGRAPHY

Roberts, J. A., Ed., *Indirect Imaging*, Cambridge University Press, Cambridge, England, 1984.

van Schooneveld, C., Ed., *Image Formation from Coherence Functions in Astronomy*, D. Reidel, Dordrecht, Holland, 1979.

REFERENCES

Ables, J. G., Maximum Entropy Spectral Analysis, *Astron. Astrophys. Suppl.*, **15**, 383–393, 1974.

Baldwin, J. E. and P. J. Warner, Aperture Synthesis Without Phase Measurements, *Mon. Not. R. Astron. Soc.*, **175**, 345–353, 1976.

Baldwin, J. E. and P. J. Warner, Phaseless Aperture Synthesis, *Mon. Not. R. Astron. Soc.*, **182**, 411–422, 1978.

Baldwin, J. E. and P. J. Warner, Fundamental Aspects of Aperture Synthesis with Limited or No Phase Information, in *Image Formation from Coherence Functions in Astronomy*, C. van Schooneveld, Ed., D. Reidel, Dordrecht, Holland, 1979, pp. 67–82.

Bates, R. H. T., Contributions to the Theory of Intensity Interferometry, *Mon. Not. R. Astron. Soc.*, **142**, 413–428, 1969.

Bates, R. H. T., Uniqueness of Solutions to Two-Dimensional Fourier Phase Problems for Localized and Positive Images, *Comp. Vision, Graphics Image Process.*, **25**, 205–217, 1984.

Bracewell, R. N. and J. A. Roberts, Aerial Smoothing in Radio Astronomy, *Aust. J. Phys.*, **7**, 615–640, 1954.

Bruck, Y. M. and L. G. Sodin, On the Ambiguity of the Image Reconstruction Problem, *Opt. Commun.*, **30**, 304–308, 1979.

Bryan, R. K. and J. Skilling, Deconvolution by Maximum Entropy, as Illustrated by Application to the Jet of M87, *Mon. Not. R. Astron. Soc.*, **191**, 69–79, 1980.

Burge, R. E., M. A. Fiddy, A. H. Greenaway, and G. Ross, The Phase Problem, *Proc. R. Soc. Lond. A.*, **350**, 191–212, 1976.

Clark, B. G., An Efficient Implementation of the Algorithm "CLEAN," *Astron. Astrophys.*, **89**, 377–378, 1980.

Clark, B. G., Large Field Mapping, in *Synthesis Mapping, Proceedings of NRAO Workshop No. 5, Socorro, New Mexico, June 21–25, 1982*, A. R. Thompson and L. R. D'Addario, Eds., National Radio Astronomy Observatory, Green Bank, West Virginia, 1982.

Cornwell, T. J., Self Calibration, in *Synthesis Mapping, Proceedings of NRAO Workshop No. 5, Socorro, New Mexico, June 21–25, 1982*, A. R. Thompson and L. R. D'Addario, Eds., National Radio Astronomy Observatory, Green Bank, West Virginia, 1982.

Cornwell, T. J., A Method of Stabilizing the Clean Algorithm, *Astron. Astrophys.*, **121**, 281–285, 1983.

Cornwell, T. J. and P. N. Wilkinson, A New Method for Making Maps with Unstable Radio Interferometers, *Mon. Not. R. Astron. Soc.*, **196**, 1067–1086, 1981.

Cotton, W. D., A Method Of Mapping Compact Structure in Radio Sources Using VLBI Observations, *Astron. J.*, **84**, 1122–1128, 1979.

Davies, J. G., B. Anderson, and I. Morison, The Jodrell Bank Radio-Linked Interferometer Network, *Nature*, **288**, 64–66, 1980.

Fienup, J. R., Reconstruction of an Object from the Modulus of its Fourier Transform, *Opt. Lett.*, **3**, 27–29, 1978.

Fort, D. N. and H. K. C. Yee, A Method of Obtaining Brightness Distributions from Long Baseline Interferometry, *Astron. Astrophys.*, **50**, 19–22, 1976.

Frieden, B. R., Restoring with Maximum Likelihood and Maximum Entropy, *J. Opt. Soc. Am.*, **62**, 511–518, 1972.

Gull, S. F. and G. J. Daniell, Image Reconstruction from Incomplete and Noisy Data, *Nature*, **272**, 686–690, 1978.

Gull, S. F. and G. J. Daniell, The Maximum Entropy Method, in *Image Formation from Coherence Functions in Astronomy*, C. van Schooneveld, Ed., D. Reidel, Dordrecht, Holland, 1979, pp. 219–225.

Högbom, J. A., Aperture Synthesis with a Non-Regular Distribution of Interferometer Baselines, *Astron. Astrophys. Suppl.*, **15**, 417–426, 1974.

Högbom, J. A., The Introduction of A Priori Knowledge in Certain Processing Algorithms, in *Image Formation from Coherence Functions in Astronomy*, C. van Schooneveld, Ed., D. Reidel, Dordrecht, Holland, 1979, pp. 237–239.

Jaynes, E. T., Prior Probabilities, *IEEE Trans. Syst. Sci. Cybern.* **SSC-4**, 227–241, 1968.

Jaynes, E. T., On the Rationale of Maximum-Entropy Methods, *Proc. IEEE*, **70**, 939–952, 1982.

Jennison, R. C., A Phase Sensitive Interferometer Technique for the Measurement of the Fourier Transforms of Spatial Brightness Distributions of Small Angular Extent, *Mon. Not. R. Astron. Soc.*, **118**, 276–284, 1958.

Kikuchi, R. and B. H. Soffer, Maximum Entropy Image Restoration. I. The Entropy Expression, *J. Opt. Soc. Am.*, **67**, 1656–1665, 1977.

Morse, P. M. and H. Feshbach, *Methods of Theoretical Physics*, McGraw-Hill, New York, 1953, pp. 370–372.

Napier, P. J., The Brightness Temperature Distributions Defined by a Measured Intensity Interferogram, *New Zealand J. Sci.*, **15**, 342–355, 1972.

Napier, P. J. and R. H. T. Bates, Inferring Phase Information from Modulus Information in Two-Dimensional Aperture Synthesis, *Astron. Astrophys. Suppl.*, **15**, 427–430, 1974.

Napier, P. J., A. R. Thompson, and R. D. Ekers, The Very Large Array: Design and Performance of a Modern Synthesis Radio Telescope, *Proc. IEEE*, **71**, 1295–1320, 1983.

Narayan, R. and R. Nityananda, Maximum Entropy—Flexibility Versus Fundamentalism, in *Indirect Imaging*, J. A. Roberts, Ed., Cambridge University Press, Cambridge, England, 1984, pp. 281–290.

Nityananda, R. and R. Narayan, Maximum Entropy Image Reconstruction—A Practical Non-Information-Theoretic Approach, *J. Astrophys. Astron.*, **3**, 419–450, 1982.

Oppenheim, A. V. and J. S. Lim, The Importance of Phase in Signals, *Proc. IEEE*, **69**, 529–541, 1981.

Pearson, T. J. and A. C. S. Readhead, Image Formation by Self Calibration in Radio Astronomy, *Ann. Rev. Astron. Astrophys.*, **22**, 97–130, 1984.

Perley, R. A., J. W. Dreher, and J. J. Cowan, The Jet and Filaments in Cygnus A, *Astrophys, J.*, **285**, L35–L38, 1984.

Ponsonby, J. E. B., An Entropy Measure for Partially Polarized Radiation and its Application to Estimating Radio Sky Polarization Distributions from Incomplete "Aperture Synthesis" Data by the Maximum Entropy Method, *Mon. Not. R. Astron. Soc.*, **163**, 369–380, 1973.

Readhead, A. C. S., R. C. Walker, T. J. Pearson, and M. H. Cohen, Mapping Radio Sources with Uncalibrated Visibility Data, *Nature*, **285**, 137–140, 1980.

Readhead, A. C. S. and P. N. Wilkinson, The Mapping of Compact Radio Sources from VLBI Data, *Astrophys. J.*, **223**, 25–36, 1978.

Rogers, A. E. E., Methods of Using Closure Phases in Radio Aperture Synthesis, *Soc. Photo-Opt. Inst. Eng.*, **231**, 10–17, 1980.

Rogers, A. E. E., H. F. Hinteregger, A. R. Whitney, C. C. Counselman, I. I. Shapiro, J. J. Wittels, W. K. Klemperer, W. W. Warnock, T. A. Clark, L. K. Hutton, G. E. Marandino, B. O. Rönnäng, O. E. H. Rydbeck, and A. E. Niell, The Structure of Radio Sources 3C273B and 3C84 Deduced from the "Closure" Phases and Visibility Amplitudes Observed with Three-Element Interferometers, *Astrophys. J.*, **193**, 293–301, 1974.

Schwab, F. R., Adaptive Calibration of Radio Interferometer Data, *Soc. Photo-Opt. Inst. Eng.*, **231**, 18–24, 1980.

Schwab, F. R., Relaxing the Isoplanatism Assumption in Self-Calibration: Application to Low Frequency Radio Interferometry, *Astron. J.*, **89**, 1076–1081, 1984.

Schwarz, U. J., Mathematical-Statistical Description of the Iterative Beam Removing Technique (Method CLEAN), *Astron. Astrophys.*, **65**, 345–356, 1978.

Schwarz, U. J., The Method "CLEAN"—Use, Misuse, and Variations, in *Image Formation from Coherence Functions in Astronomy*, C. van Schooneveld, Ed., D. Reidel, Dordrecht, Holland, 1979, pp. 261–275.

Smith, F. G., The Measurement of the Angular Diameter of Radio Stars, *Proc. Phys. Soc. B*, **65**, 971–980, 1952.

Subrahmanya, C. R., An Optimum Deconvolution Method, in *Image Formation from Coherence Functions in Astronomy*, C. van Schooneveld, Ed., D. Reidel, Dordrecht, Holland, 1979, pp. 287–290.

Twiss, R. Q., A. W. L. Carter, and A. G. Little, Brightness Distribution over some Strong Radio Sources at 1427 Mc/s, *Observatory*, **80**, 153–159, 1960.

Wernecke, S. J., Two-Dimensional Maximum Entropy Reconstruction of Radio Brightness, *Radio Sci.*, **12**, 831–844, 1977.

Wernecke, S. J. and L. R. D'Addario, Maximum Entropy Image Reconstruction, *IEEE Trans. Comput.*, **C-26**, 351–364, 1977.

12

INTERFEROMETER TECHNIQUES FOR ASTROMETRY AND GEODESY

The output fringe pattern of an interferometer provides a measure of the scalar product of the baseline and source-position vectors $\mathbf{D} \cdot \mathbf{s}$, and of the rotational motion of the earth. Up to this point we have assumed that all these factors are describable by constants that can be specified with high accuracy. However, the measurement of source positions to an accuracy of milliarcseconds requires that variation in the earth's rotation vector be taken into account. The required baseline accuracy approaches that at which variation in the antenna positions resulting from crustal motions of the earth can be detected. The calibration of the baseline and the measurement of source positions can be accomplished in a single observing period of one or more days. Geodetic[†] data are obtained from repetition of this procedure over intervals of months or years, which reveals the variation in the baseline and earth-rotation parameters.

In this chapter we are concerned with the techniques by which angular positions can be measured with the greatest possible accuracy, and with the design of interferometers for optimum determination of source-position, baseline, and geodetic parameters. We do not consider the interpretation of these results.

The redefinition of the meter has an interesting implication for the units of baseline lengths derived from interferometric data. An interferometer measures the relative time of arrival of the signal wavefront at the two antennas, that is,

[†] For simplicity we use the term *geodetic* to include geodynamic and static phenomena regarding the shape and orientation of the earth.

the geometrical delay. Baselines determined from interferometric data are therefore in units of lightseconds. Conversion to meters formerly depended on the value chosen for c. However, in 1983 the Conférence Générale des Poids et Mesures adopted a new definition of the meter: "the meter is the length of the path travelled by light in vacuum during a time interval of 1/299,792,458 of a second." The second and the speed of light are now primary quantities, and the meter is a derived quantity. Thus baseline lengths can be given properly in meters. Issues related to fundamental units are discussed by Petley (1983).

12.1   REQUIREMENTS FOR ASTROMETRY

Position measurements of radio sources accurate to no better than a few tens of arcseconds are mainly of historical interest and have been mentioned in Chapter 1. In early studies of this kind, the baseline vector was often established by surveying the antenna locations, the instrumental phase was estimated by calibration of the transmission lines, and the positions of the fringe maxima on the sky were thereby deduced. An informative review of these techniques, including various procedures for minimizing instrumental errors, is given by Smith (1952). In this chapter, we are concerned with more recent procedures for which the precision is a few tenths to better than a few thousandths of an arcsecond. We begin with a heuristic discussion of how baseline and source position parameters may be determined. A formal discussion is given in Section 12.2.

In the measurement of declination with a tracking interferometer, it is possible to solve for both the source-position and baseline parameters independently. This can be illustrated simply by the following consideration. The phase of the fringe pattern for an interferometer is $2\pi w$, where w is the spacing component given by Eq. (4.17). The phase can be written

$$\phi = 2\pi D_\lambda \left[\sin d \sin \delta + \cos d \cos \delta \cos(H - h)\right] + \phi_{in}, \qquad (12.1)$$

where D_λ is the length of the baseline in wavelengths, H and δ are the hour angle and declination of the source, h and d are the hour angle and declination of the baseline, and ϕ_{in} is an instrumental phase term. For an east–west baseline, $h = -\pi/2$ when measured from the local meridian, $d = 0$, and the phase reduces to

$$\phi = -2\pi D_\lambda \cos \delta \sin H + \phi_{in}. \qquad (12.2)$$

Thus, the dependence of ϕ on δ is proportional to $\cos \delta$, and by observing sources close to the celestial equator, where the dependence is small, D_λ can be established with high accuracy (e.g., see Ryle and Elsmore 1973). Positions of sources at higher declinations can then be determined, and these can be used to calibrate a north–south baseline for more accurate measurement of low-declination sources.

In the determination of right ascension, interferometer observations provide relative measurements, that is, the differences in right ascension between different sources. The zero circle of right ascension is defined as the great circle through the pole and through the intersection of the celestial equator and the ecliptic at the vernal equinox, where the apparent position of the sun moves from the southern to the northern celestial hemisphere. This direction can be located in terms of the motions of the planets, which are well-defined objects for optical observations. It has been related to the positions of bright stars that provide a reference system for optical measurements of celestial position. Relating the radio measurements to the zero of right ascension is less easy, since solar system objects are generally too weak or too wide when viewed as radio sources. Results have been obtained from the lunar occultation of the source 3C273B (Hazard et al. 1971) and from measurements of the weak radio emission from nearby stars such as Algol (Ryle and Elsmore 1973, Elsmore and Ryle 1976).

There are two promising techniques for determining the direction of the vernal equinox directly from passive radio techniques. The first method involves radio interferometric observations of the minor planets (Johnston, Seidelmann, and Wade 1982). The second method involves a comparison of the positions of pulsars obtained from pulse timing measurements with those obtained from radio interferometry (Fomalont et al. 1984, Bartel et al. 1985). Since the timing measurements yield positions relative to the coordinate frame defined by the earth's orbit, while the interferometer measurements refer to the frame of the rotating earth, the direction of the vernal equinox and other dynamical parameters of the earth's orbit can be determined from this comparison.

In many radio astrometric catalogues, the right ascension of 3C273B is fixed at the value given by Hazard et al. (1971). 3C273B is not an ideal reference source because it has time-variable structure at the level of about a milliarcsecond (see Fig. 1.16). Another approach is to compare the radio and optical positions of many identified radio galaxies and quasars. The right ascension scale can be chosen to minimize the rms differences between the radio and optical positions for a number of sources, thereby minimizing the effect of any real differences between the radio and optical centers. As of the mid-1980s, the relation of the right ascension scales had been determined within about 0.1 arcsec and should be further improved as instrumental development continues. Note that the most distant radio objects offer a better ultimate reference for celestial positions than stars within our galaxy, for which the angular motions are greater.

In the reduction of interferometer measurements in astrometry, the visibility data are interpreted basically in terms of the positions of point sources. The data processing is equivalent, in effect, to model fitting using delta-function brightness components, the visibility function for which has been discussed in Section 4.5. The essential position data are determined from the calibrated visibility phase or, in the case of VLBI observations, from the geometrical delay as measured by maximization of the cross correlation of the signals (i.e., the use of the bandwidth pattern) and from the fringe frequency. Because the position information is contained in the visibility phase, the phaseless mapping techniques and the principle of closure phase discussed in Chapter 11 are of use in astrometry and geodesy only insofar as they can provide a means of correcting for the effects of source structure. Uniformity of (u, v) coverage is less important than in mapping because high dynamic range maps are not needed. Determination of the position of an unresolved source depends on interferometry with precise phase calibration and sufficient spacings to avoid ambiguity in the position.

12.2 SOLUTION FOR BASELINE AND SOURCE-POSITION VECTORS

In determining an interferometer baseline, it is convenient to use calibrators whose angular positions are known with accuracy comparable to that required for the baseline. However, this is not essential, and as will now be shown it is possible, and often necessary, to solve for source and baseline parameters simultaneously.

Connected-Element Systems

Consider an observation with a tracking interferometer of arbitrary baseline in which the source is unresolved. Let $\mathbf{D}_\lambda$ be the assumed baseline vector and $(\mathbf{D}_\lambda - \Delta\mathbf{D}_\lambda)$ be the true vector. Similarly, let $\mathbf{s}$ be a unit vector indicating the assumed position of the source, and let $(\mathbf{s} - \Delta\mathbf{s})$ indicate the true position. The expected fringe phase, using the assumed positions, is $2\pi\mathbf{D}_\lambda \cdot \mathbf{s}$. The observed phase, measured relative to the expected phase, is a function of the hour angle H of the source given by

$$\Delta\phi(H) = 2\pi\left[(\mathbf{D}_\lambda - \Delta\mathbf{D}_\lambda) \cdot (\mathbf{s} - \Delta\mathbf{s}) - \mathbf{D}_\lambda \cdot \mathbf{s}\right] + \phi_{in}$$

$$= \phi_{in} - 2\pi(\Delta\mathbf{D}_\lambda \cdot \mathbf{s} + \mathbf{D}_\lambda \cdot \Delta\mathbf{s}). \tag{12.3}$$

A second-order term involving $\Delta\mathbf{D}_\lambda \cdot \Delta\mathbf{s}$ has been neglected since we assume that fractional errors in $\mathbf{D}_\lambda$ and $\mathbf{s}$ are small.

The baseline vector can be written in terms of the coordinate system introduced in Section 4.2:

$$\mathbf{D}_\lambda = \begin{bmatrix} X_\lambda \\ Y_\lambda \\ Z_\lambda \end{bmatrix} \quad \text{and} \quad \Delta\mathbf{D}_\lambda = \begin{bmatrix} \Delta X_\lambda \\ \Delta Y_\lambda \\ \Delta Z_\lambda \end{bmatrix}, \tag{12.4}$$

where X, Y, and Z are a right-handed coordinate system with Z parallel to the earth's spin axis and X in the meridian plane of the interferometer. The source-position vector can be specified in the (X, Y, Z) system in terms of the hour angle H and declination δ of the source by using Eq. (4.16):

$$\mathbf{s} = \begin{bmatrix} s_X \\ s_Y \\ s_Z \end{bmatrix} = \begin{bmatrix} \cos \delta \cos H \\ -\cos \delta \sin H \\ \sin \delta \end{bmatrix}. \tag{12.5}$$

Taking the differential of Eq. (12.5), we can write

$$\Delta \mathbf{s} \simeq \begin{bmatrix} -\sin \delta \cos H \Delta \delta + \cos \delta \sin H \Delta \alpha \\ \sin \delta \sin H \Delta \delta + \cos \delta \cos H \Delta \alpha \\ \cos \delta \Delta \delta \end{bmatrix}, \tag{12.6}$$

where $\Delta \alpha$ and $\Delta \delta$ are the angular errors in right ascension and declination, and $\Delta \alpha$ has the opposite sign to the corresponding error in hour angle. By substituting Eqs. (12.4)–(12.6) into (12.3), it is found that the expression for the measured phase can be written

$$\Delta \phi (H) = \phi_0 + \phi_1 s_X - \phi_2 s_Y$$

$$= \phi_0 + \phi_1 \cos \delta \cos H + \phi_2 \cos \delta \sin H, \tag{12.7}$$

where

$$\phi_0 = \phi_{\text{in}} - 2\pi (\Delta Z_\lambda \sin \delta + Z_\lambda \Delta \delta \cos \delta), \tag{12.8}$$

$$\phi_1 = 2\pi (-\Delta X_\lambda + X_\lambda \Delta \delta \tan \delta - Y_\lambda \Delta \alpha), \tag{12.9}$$

and

$$\phi_2 = 2\pi (\Delta Y_\lambda - X_\lambda \Delta \alpha - Y_\lambda \Delta \delta \tan \delta). \tag{12.10}$$

From Eq. (12.7) it is seen that $\Delta \phi (H)$ is a sinusoid in H with an offset ϕ_0. Thus, three parameters—amplitude, phase, and offset—can be measured for any source by observing periodically or continuously for approximately 12 hr. If m sources are observed, $3m$ quantities are obtained. The number of unknown parameters required to specify the m positions, the baseline, and the instrumental phase (assumed to be constant) is $2m + 3$, the right ascension of one source being arbitrarily chosen. Thus, if $m \geq 3$, it is possible to solve for all the unknown quantities. Note that the sources should have as wide a range in declination as possible in order to distinguish ΔZ_λ from ϕ_{in} in Eq. (12.8). In an early study by Wade (1970), the instrumental parameters in Eqs. (12.7)–(12.10) were separated from the unknown source positions and determined independently. The source positions were then obtained. An error analysis for this procedure is given by Brosche, Wade, and Hjellming (1973). A more common procedure is to solve for the instrumental and source-position parameters simultaneously using a least-mean-squares analysis. Usually, many more than three sources are observed so there is redundant information, and variation of the instrumental phase with time as well as other parameters can be included in the solution. A discussion of the method of least-mean-squares analysis can be found in the appendix to this chapter.

Measurements with VLBI Systems

The use of independent local oscillators at the antennas in VLBI systems does not permit calibration of the fringe phase, except in special circumstances. The earliest method used for obtaining positional information in VLBI was the analysis of the fringe frequency (fringe rate). The fringe frequency is the time rate of change of the interferometer phase. Thus, from Eq. (12.1) the fringe frequency is

$$\nu_f = \frac{1}{2\pi}\frac{d\phi}{dt} = -\omega_e D_\lambda \cos d \cos \delta \sin(H - h) + \nu_{in}, \qquad (12.11)$$

where ω_e is the angular velocity of the earth (dH/dt), and ν_{in} is an instrumental term equal to $d\phi_{in}/dt$. ν_{in} largely results from the differences in frequency of the standards at the antennas.

The quantity $D_\lambda \cos d$ is the projection of the baseline in the equatorial plane, denoted D_E. Thus, Eq. (12.11) can be rewritten

$$\nu_f = -\omega_e D_E \cos \delta \sin(H - h) + \nu_{in}. \qquad (12.12)$$

The polar component of the baseline (the projection of the baseline along the polar axis) does not appear in the equation for fringe frequency. An interferometer with a baseline parallel to the spin axis of the earth has lines of constant phase parallel to the celestial equator and the interferometer phase does not change with hour angle. Therefore, the polar component of the baseline cannot be determined from the analysis of fringe frequency.

The usual practice in VLBI is to refer hour angles to the Greenwich meridian. We follow this convention and use a right-handed coordinate system with X through the Greenwich meridian and with Z towards the north celestial pole. Thus, in terms of the Cartesian coordinates for the baseline, Eq. (12.12) becomes

$$\nu_f = -\omega_e \cos \delta (X_\lambda \sin H + Y_\lambda \cos H) + \nu_{in}. \qquad (12.13)$$

The residual fringe frequency $\Delta\nu_f$, that is, the difference between the observed and expected fringe frequency, can be calculated by taking the derivatives of Eq. (12.13) with respect to δ, H, X_λ, and Y_λ, and also including the unknown quantity ν_{in}. We thereby obtain

$$\Delta\nu_f = \nu_{in} + a_1 \cos H + a_2 \sin H, \qquad (12.14)$$

where

$$a_1 = \omega_e(Y_\lambda \sin\delta\,\Delta\delta + X_\lambda\cos\delta\,\Delta\alpha - \cos\delta\,\Delta Y_\lambda) \tag{12.15}$$

and

$$a_2 = \omega_e(X_\lambda \sin\delta\,\Delta\delta - Y_\lambda\cos\delta\,\Delta\alpha - \cos\delta\,\Delta X_\lambda). \tag{12.16}$$

Note that $\Delta\nu_f$ is a diurnal sinusoid and that the average value of $\Delta\nu_f$ is the instrumental term ν_{in}. Information about source positions and baselines must come from the two parameters a_1 and a_2. Therefore, unlike the case of fringe phase [Eq. (12.7)] where three parameters per source are available, it is not possible to solve for both source and baseline parameters with fringe-frequency data. For example, from observations of m sources, $2m + 1$ quantities are obtained. The total number of unknowns (two baseline parameters, $2m$ source parameters, and ν_{in}) is $2m + 3$. If the position of one source is known, the rest of the source positions and X_λ, Y_λ, and ν_{in} can be determined. Note that the accuracy of the measurements of source declinations is reduced for sources close to the celestial equator because of the $\sin\delta$ terms in Eqs. (12.15) and (12.16).

As an illustration of the order of magnitude of the parameters involved in fringe-frequency observations, consider two antennas with an equatorial component of spacing equal to 1000 km and an observing wavelength of 3 cm. Then $D_E \simeq 3 \times 10^7$ wavelengths, and the fringe frequency for a low-declination source is about 2 kHz. Assume that the coherence time of the independent frequency standards is about 10 min. In this period 10^6 fringe cycles can be counted. If we suppose that the phase can be measured to 0.1 turn, ν_f will be obtained to a precision of 1 part in 10^7. The corresponding errors in D_E and angular position are 10 cm and 0.02 arcsec, respectively.

To overcome the limitations of fringe-frequency analysis, techniques for the precise measurement of the relative group delay of the signals at the antennas were developed. The use of bandwidth synthesis to improve the accuracy of delay measurements has been discussed in Section 9.8. The group delay is equal to the geometrical delay τ_g except that, as measured, it also includes unwanted components resulting from clock offsets at the antennas and atmospheric differences in the signal paths. The fringe phase measured with a connected-element interferometer observing at frequency ν is $2\pi\nu\tau_g$, modulo 2π. Except for the dispersive ionosphere, the group delay therefore contains the same type of information as the fringe phase, without the ambiguity resulting from the modulo 2π restriction. Thus, group delay measurements permit a solution for baselines and source positions similar to that discussed above for connected-element systems, except that clock offset terms also must be included.

In most astrometric experiments, measurements of the group delay and the fringe frequency (or, equivalently, the rate of change of phase delay) are analyzed together. The intrinsic precisions with which these quantities can be measured are derived in Appendix 12.1 [Eqs. (A12.27) and (A12.34)] and can be written

$$\sigma_f = \sqrt{\frac{3}{2\pi^2}} \left(\frac{T_S}{T_A}\right) \frac{1}{\sqrt{\Delta\nu\,\tau^3}} \tag{12.17}$$

and

$$\sigma_\tau = \frac{1}{\sqrt{8\pi^2}} \left(\frac{T_S}{T_A}\right) \frac{1}{\sqrt{\Delta\nu\,\tau}\,\Delta\nu_{\mathrm{rms}}}, \tag{12.18}$$

where σ_f and σ_τ are the rms errors in fringe frequency and delay, T_S and T_A are the system and antenna temperatures, $\Delta\nu$ is the IF bandwidth, $\Delta\nu_{\mathrm{rms}}$ is the rms bandwidth (which is typically 40% of the spanned bandwidth) introduced in Section 9.8 (see also Appendix 12.1), and τ is the integration time. For a single rectangular RF band, $\Delta\nu_{\mathrm{rms}} = \Delta\nu/\sqrt{12}$. The geometrical delay is $\tau_g = (D/c)\sin\theta$, where θ is the angle between the source vector and the plane perpendicular to the baseline. Thus, the sensitivity of the delay to angular changes is

$$\frac{\Delta\tau_g}{\Delta\theta_\tau} = \frac{D}{c}\cos\theta, \tag{12.19}$$

where $\Delta\theta_\tau$ is the increment in θ corresponding to an increment $\Delta\tau_g$ in τ_g. Similarly, the sensitivity of the fringe frequency to angular changes [since $\nu_f = \nu(d\tau_g/dt)$] is (for an east–west baseline)

$$\frac{\Delta\nu_f}{\Delta\theta_f} = D_\lambda\omega_e\cos\theta, \tag{12.20}$$

where $\Delta\theta_f$ is the increment in θ corresponding to an increment $\Delta\nu_f$ in ν_f. Thus by setting $\Delta\nu_f = \sigma_f$ and $\Delta\tau_g = \sigma_\tau$ and ignoring geometric factors, we obtain the equation

$$\frac{\Delta\theta_\tau}{\Delta\theta_f} \simeq 2\pi\frac{\tau/T_e}{\Delta\nu/\nu}, \tag{12.21}$$

where $T_e = 2\pi/\omega_e$ is the period of the earth's rotation. Equation (12.21) describes the relative precision of delay and fringe-frequency measurements. In practice, measurements of delay are generally more accurate because of the noise imposed by the atmosphere. Measurements of fringe frequency are sensitive to the time derivative of atmospheric path length, and in a turbulent atmosphere this derivative can be large, while the average path length is relatively constant. Note that fringe-frequency and delay measurements are complementary. For example, with a VLBI system of known baseline and instrumental parameters, the position of a source can be found from a single observation using the delay and fringe frequency because these quantities constrain the source position in approximately orthogonal directions. The earliest analyses of fringe-frequency and delay measurements to determine source positions and baselines were made by Cohen and Shaffer (1971) and Hinteregger et al. (1972).

The accuracy with which the group delay can be measured is a fraction of the reciprocal bandwidth. The accuracy with which the fringe phase can be measured is a similar fraction of the observing frequency. The ratio of the observing frequency to the bandwidth, including effects of bandwidth synthesis, is commonly one to two orders of magnitude. On the other hand, the antenna spacings used in VLBI are one to two orders of magnitude greater than those used in connected-element systems. Thus the accuracy of source positions estimated from group delay measurements with VLBI systems is comparable with the accuracy of those estimated from fringe phase measurements on connected-element systems having much shorter baselines.

The ultimate limitations on ground-based interferometry are imposed by the atmosphere. Dual-frequency-band measurements effectively remove ionospheric phase noise (see Chapter 13). The rms phase noise of the troposphere increases about as $d^{5/6}$, where d is the projected baseline length, for baselines shorter than a few kilometers (see Chapter 13). In this regime, measurement accuracies of angles improve only slowly with increasing baseline length. For baselines greater than $\sim$ 100 km, the effects of the troposphere above the interferometer elements are uncorrelated and the measurement accuracy might be expected to improve more rapidly with baseline length. However, for widely spaced elements, the zenith angle can be very different, and the atmospheric model becomes very important. As of the mid-1980s, the angular accuracy achievable with connected-element systems approaches 10^{-2} arcsec (Kaplan et al. 1982, Perley 1982), and with VLBI it approaches 10^{-3} arcsec (Fanselow et al. 1984, Herring et al. 1985).

For measurements of relative positions of closely spaced sources, it is possible to measure the relative fringe phases in a VLBI system and thus obtain the full accuracy, of microarcsecond order, inherent in the long baseline. The most accurate measurements can be made when the sources are sufficiently close that both fall within the antenna beams (e.g., Marcaide and Shapiro 1983) or when they are close enough that measurements of each can be closely interleaved (Shapiro et al. 1979, Bartel et al. 1984). Pulsar astrometry has been done using quasars that are nearby in angle as calibrators (Taylor et al. 1984, Bartel et al. 1985). For widely spaced sources, a technique using two antennas working with a common frequency standard at each station, one tracking each source, has been successfully demonstrated by Counselman et al. (1974) in an experiment to measure gravitational deflection of radio waves.

12.3   TIME AND THE MOTION OF THE EARTH

We now consider the effect of changes in the magnitude and direction of the earth's rotation vector on interferometric measurements. These changes cause variations in the apparent celestial coordinates of sources, the baseline vectors of the antennas, and universal time. The variations of the earth's rotation can be divided into three categories. First, there are variations in the direction of the rotation axis, resulting mainly from precession and nutation of the spinning body. Since the direction of the axis defines the location of the pole of the celestial coordinate system, the result is a variation in the right ascension and declination of celestial objects. Second, the axis of rotation varies slightly with respect to the earth; that is, the positions on the earth at which this axis intersects the earth's surface vary. This effect is known as polar motion. Since the (X, Y, Z) coordinate system of baseline specification introduced in Section 4.2 takes the direction of the earth's axis as the Z axis, polar motion results in a variation of the measured baseline vectors (but not, of course, in the baseline length). It also results in a variation in universal time. Third, the rate of rotation varies as a result of atmospheric and crustal effects, and this again results in variation in universal time. We briefly discuss these effects. Detailed discussions from a geophysical viewpoint can be found in Munk and MacDonald (1960) and Lambeck (1980).

Precession and Nutation

The gravitational effects of the sun, moon, and planets on the nonspherical earth produce a variety of perturbations in its orbital and rotational motions. To take account of these effects it is necessary to know the resulting variation of the ecliptic, which is defined by the plane of the earth's orbit, as well as the variation of the celestial equator, which is defined by the rotational motion of the earth. The gravitational effects of the sun and moon on the equatorial bulge (quadrupole moment) of the earth result in a precessional motion of the earth's axis around the pole of the ecliptic.

The earth's rotation vector is inclined at about 23.5° to the pole of its orbital plane, the ecliptic. The period of the resulting precession is approximately 26,000 years, corresponding to a motion of the rotation vector of 20 arcsec per year [$2\pi \sin(23.5°)/26{,}000$ yr]. The 23.5° obliquity is not constant, but is currently decreasing at a rate of 47 arcsec per century due to the effect of the planets, which also cause a further component of precession. The luni-solar and planetary precessional effects, together with a smaller relativistic precession, are known as the general precession. Precession results in the motion of the line of intersection of the ecliptic and the celestial equator. This line, called the line of nodes, defines the equinoxes and the zero of right ascension, which precess at a rate of 50 arcsec per year. In addition, the time-varying luni-solar gravitation effects cause nutation of the earth's axis with periods of up to 18.6 years and a total amplitude of about 9 arcsec. The principal variations of the ecliptic and equator are those just described, but other smaller effects also occur. The general accuracy within which positional variations can be calculated is better than 1 milliarcsec (Herring, Gwinn, and Shapiro 1985). Expressions for precession can be found in Lieske et al. (1977) and for nutation in Wahr (1981). The required procedures are discussed in texts on spherical astronomy, such as Woolard and Clemence (1966), Taff (1981), and the *Astronomical Almanac*.

Since precession and nutation result in variations in celestial coordinates that can be as large as 50 arcsec per year for objects at low declinations, these effects must be taken into account in almost all observational work, whether astrometric or not. Positions of objects in astronomical catalogs are therefore reduced to the coordinates of standard epochs, B1900.0, B1950.0, or J2000.0. These dates denote the beginning of a Besselian year or Julian year, as indicated by the B or J. The positions correspond to the mean equator and equinox for the specified epoch, where "mean" indicates the positions of the equator and equinox resulting from the general precession, but not including nutation. In most large instruments, the B1950.0 or J2000.0 coordinates of sources are used when an observing program is entered into the control computer. For further explanation and a discussion of a method of conversion between standard epochs, see the Supplement section in the *Astronomical Almanac* for 1984. Correction is also required for aberration, that is, for the apparent shift in position resulting from the finite velocity of light and the motion of the observer. Two components are involved: annual aberration resulting from the earth's orbital motions, which has a maximum value of about 20 arcsec, and diurnal aberration resulting from the rotational motion, which has a maximum value of 0.3 arcsec. The retarded baseline concept (Section 9.3) used in VLBI data reduction accounts for the diurnal aberration. For the nearer stars, corrections for proper motion (i.e., actual motion of the star through space), and in some cases also for the parallax resulting from the changing position of the earth in its orbit, are required. The impact of radio techniques, particularly VLBI, is resulting in refinement of the classical expressions and parameters. Effects such as the deflection of electromagnetic waves in the sun's gravitational field must also be included in positional work of the highest accuracy (see Section 13.4).

Polar Motion

The term *polar motion* denotes the variation of the *geographic* poles of the earth, as distinct from the precessional and other motions of the celestial pole. Polar motion results from motion of the axis of rotation relative to the axis of the figure of the earth and is largely, though not totally, of geophysical origin. The motion of the geographic pole around the pole of the earth's figure is irregular, but over the last century the distance between these two poles has wandered by up to 0.5 arcsec, or 15 m on the earth's surface. In a year's time, the excursion of the figure axis is typically 6 m or less. The motion can be

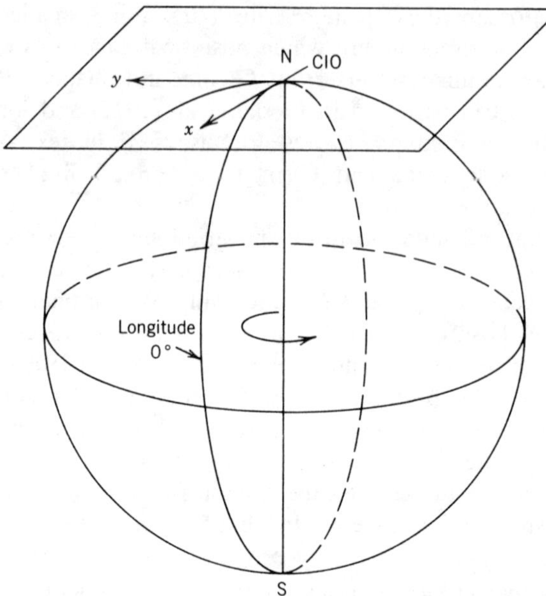

Figure 12.1   Coordinate system for the measurement of polar motion. The x coordinate is in the Greenwich meridian and y axis is 90° to the west. CIO is the conventional international origin.

analyzed into several components, some regular and some highly irregular, and not all are understood. The two major components have periods of 12 and 14 months. The 12-month component is a forced motion due to the annual redistribution of water and of atmospheric angular momentum, and is far from any resonance. The 14-month component, known as the Chandler wobble (Chandler 1891), is the motion at a resonance frequency whose driving force is unknown. For a more detailed description, see Woolard and Clemence (1966).

The motion of the geographic pole is measured in angle or distance in the x and y directions as shown in Fig. 12.1. The (x, y) origin is the mean pole of 1900–1905, which is referred to as the *conventional international origin* (CIO), and the x axis is in the plane of the Greenwich meridian (Markowitz and Guinot 1968). Since polar motion is a small angular effect, it can often be ignored in mapping observations, especially if the visibility is measured with respect to a calibrator that is only a few degrees from the center of the field being mapped.

Universal Time

Like the motion of the earth, the system of timekeeping based on earth rotation is a complicated subject, and for a detailed discussion one can refer to Smith (1972) or to the texts mentioned in the discussion of precession and nutation above. We shall briefly review some essentials.

Solar time is defined in terms of the rotation of the earth with respect to the sun. In practice, the stars present more convenient objects for measurement, so solar time is derived from measurement of the sidereal rotation. The positions of stars or radio sources used for such measurements are adjusted for precession, nutation, and so on, and the resulting time measurements thus depend only on the angular velocity of the earth and on polar motion. When converted to the solar time scale, these measurements provide a form of universal time (UT) known as UT0: this is not truly "universal" since the effects of polar motion, which can amount to about 35 msec, depend on the location of the observatory. When UT0 is corrected for polar motion, the result is known as UT1. Since it is a measure of the rotation of the earth relative to fixed celestial objects, UT1 is the form of time required in astronomical observing, including the analysis of interferometric observations, navigation, and surveying. However, UT1 contains the effects of small variations in the earth's rotation rate attributable largely to geophysical effects such as the seasonal variations in the distribution of water between the surface and atmosphere. Fluctuations in the length of day over the period of a year are typically about 1 msec. To provide a more uniform measure of time, UT2 is derived from UT1 by attempting to remove seasonal variations. UT2 is rarely used. UT1 and UT2 include the effect of the gradual decrease of the rotation rate of the earth. This causes the length of the UT1/UT2 day to increase slightly when compared with International Atomic Time (IAT), which is based on the frequency of the cesium line (see Section 9.4). The IAT second is the basis for another form of UT, Coordinated Universal Time (UTC), which is offset from IAT so that $|UT1 - UTC| < 1$ sec. This relationship is maintained by inserting 1-sec discontinuities (leap seconds) in UTC when required on specified days of the year. Most time services such as Loran C, WWV, and GPS (see Section 9.5) transmit UTC.

The practice at many observatories is to maintain UTC or IAT using an atomic standard and then obtain UT1 from the published values of $\Delta UT1 = UT1 - UTC$. Since $\Delta UT1$ is measured rather than computed, in principle it can be determined only after the fact. However, it is possible to predict it by extrapolation with satisfactory accuracy for periods of 1 or 2 weeks, and thus to implement UT1 in real time. Values of $\Delta UT1$ are available from the Bureau International de l'Heure (BIH), which was established in 1912 at the Paris Observatory to coordinate international timekeeping, and from the U.S. Naval Observatory. Rapid service data are available from these institutions with a timeliness suitable for extrapolation.

Measurement of Polar Motion and UT1

The classical optical methods of measuring polar motion and UT1 are by timing the meridian transits of stars of known positions. Observations at different longitudes using stars at more than one declination are required to determine all three parameters $(x, y, \Delta UT1)$. During the 1970s it became evident that such astrometric tasks can also be performed by radio interferometry (McCarthy and Pilkington 1979). Programs using radio interferometry devoted full-time to monitoring UT1 and polar motion were started in the late 1970s (Johnston et al. 1978, Carter and Strange 1979).

To specify the baseline components of an interferometer, we use the (X, Y, Z) system of Section 4.2, rotated so that the X axis lies in the Greenwich meridian instead of the local meridian. Let ΔX, ΔY, and ΔZ be the changes in the baseline components resulting from polar motion (x, y) (in radians) and a time variation (UT1 $-$ UTC) corresponding to Θ radians. Then we may write

$$\begin{bmatrix} \Delta X \\ \Delta Y \\ \Delta Z \end{bmatrix} = \begin{bmatrix} 0 & -\Theta & -x \\ \Theta & 0 & y \\ x & -y & 0 \end{bmatrix} \begin{bmatrix} X \\ Y \\ Z \end{bmatrix}, \tag{12.22}$$

where the square matrix is a three-dimensional rotational matrix valid for small angles of rotation. Θ, x and y are the rotation angles about the Z, $-Y$, and X axes, respectively. From Eq. (12.22) we obtain

$$\Delta X = -\Theta Y - xZ,$$

$$\Delta Y = \Theta X + yZ,$$

$$\Delta Z = xX - yY. \tag{12.23}$$

Thus, if one observes a series of sources at periodic intervals and determines the variation in baseline parameters, Eqs. (12.23) can be used to determine UT1 and polar motion. For an interferometer with an east–west baseline ($Z = 0$), one can determine Θ but cannot separate the effects of x and y. An east–west interferometer located on the Greenwich meridian ($X = Z = 0$) would yield measures of Θ and y but not of x. If it had a north–south component of baseline $Z \neq 0$, one could still measure y but would not be able to separate the effects of x and Θ. In general, one cannot measure all three quantities with a single baseline, since a single direction is specified by two parameters only. Systems suitable for a complete solution might be, for example, two east–west interferometers separated by about 90° in longitude or a three-element non-colinear interferometer. An example of interferometric measurements of the pole position is shown in Fig. 12.2.

The methods just described are applicable, of course, to observations using connected-element interferometers in which the phase can be calibrated, and also to VLBI observations in which the bandwidth is sufficient to obtain accurate group delay measurements. If only the fringe frequency is measurable, as in narrow bandwidth VLBI systems, the result is insensitive to the Z component of the baseline. Thus, in Eqs. (12.23), one has measurements of ΔX and ΔY only, and in general it is not possible to separate the effects of polar motion and variation of UT1. However, if $Z = 0$ (east–west baseline), UT1

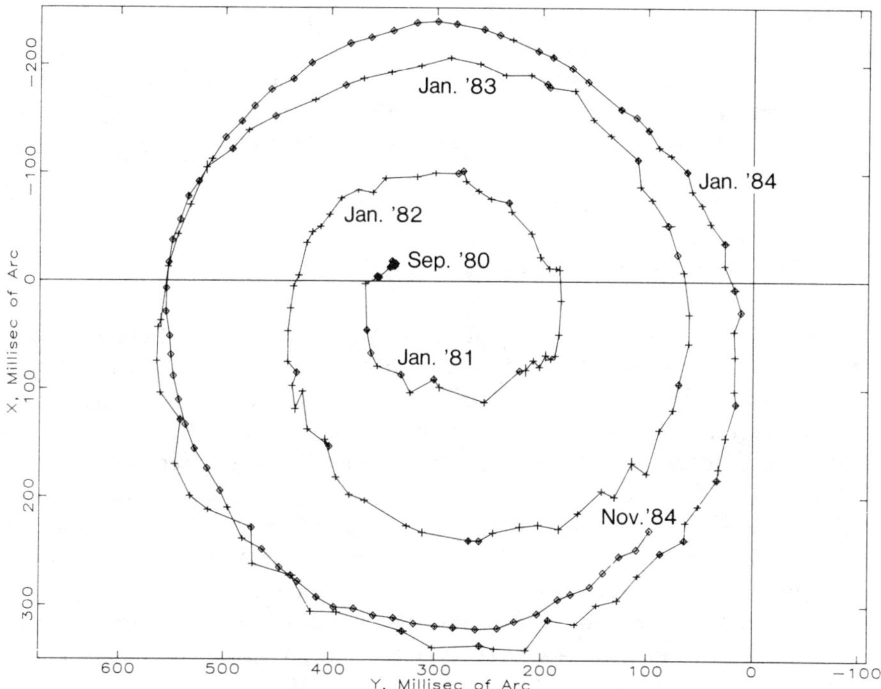

Figure 12.2 VLBI determination of the position of the pole. The diamonds indicate measurements in which three or more stations participated in the observations so that both pole coordinates could be determined from the observations. The crosses mark observations that employed only one baseline for which only the x component of the pole was determined. The corresponding y components were obtained from the BIH. Note that 100 milliarcsec corresponds to 3.2 m. Figure from Carter, Robertson, and MacKay (1985).

can be derived. A comparison of determinations of UT1 and polar motion by VLBI, satellite laser ranging, and BIH analyses of standard astrometric data is given in Robertson et al. (1983) and Carter et al. (1984).

12.4 GEODETIC MEASUREMENTS

Certain geophysical phenomena, for example, earth tides (Melchior 1978) and movements of tectonic plates, can result in variations in the baseline vector of a VLBI system. Variations in the length of the baseline are clearly attributable to such phenomena, whereas variations in the direction can also result from polar motion and rotational variations. Magnitudes of the effects are of order 1–10 cm yr^{-1} for plate motions and 30 cm (diurnal) for earth tides, and are thus measurable using the most accurate techniques of VLBI. Earth tides were first detected by Shapiro et al. (1974), and refined measurements were reported by Herring et al. (1983). In addition to solid-earth tides, displacement of land masses resulting from tidal shifts of water masses, called ocean loading, is believed to be a measurable effect. Evidence for the motion of tectonic plates in the present era was reported by Herring et al. (1986). For reviews of geodetic applications of VLBI, see Shapiro (1976), Counselman (1976), and Clark et al. (1985).

12.5 MAPPING ASTRONOMICAL MASERS

In the envelopes of many newly formed stars, and those of highly evolved stars, the radio emission from molecules such as H_2O and OH is caused by a maser process. The frequency spectrum of the emission is often complicated, containing many spectral features or components caused by clouds of gas moving at different line-of-sight velocities. Maps of strong maser sources reveal hundreds of compact components with brightness temperatures approaching 10^{15} K, angular sizes as small as 10^{-4} arcsec, and flux densities as high as 10^6 Jy. The components are typically distributed over an area of several arcseconds diameter and a Doppler velocity range of 10–300 km sec^{-1} (0.7–20 MHz for the H_2O maser transition at 22 GHz). Individual features have linewidths of about 1 km sec^{-1} or less (74 kHz at 22 GHz). The physics and phenomenology of masers are discussed by Reid and Moran (1981) and Moran (1982). The processing and analysis of maser data require large correlator systems because the ratio of required bandwidth to spectral resolution is large (10^2–10^4). They also require prodigious amounts of image processing because the ratio of the field of view to the spatial resolution is large (10^2–10^4). As an extreme example, the H_2O maser in W49 has at least 400 features (Walker, Matsakis, and Garcia-Barreto 1982) distributed over 3 arcsec. The complete mapping of this source at a resolution of 10^{-3} arcsec with 3 pixels per resolution interval would require the production of 600 maps, each with at least 10^8 pixels. However, most of the map cells would contain no signal. Thus, the usual procedure is to measure the positions of the features crudely by fringe-frequency analysis, and then map small fields around these locations by Fourier synthesis techniques. Examples of maps made by fringe-frequency analysis can be found in Walker, Matsakis, and Garcia-Barreto (1982); by phase analysis in Genzel et al. (1981) and Norris and Booth (1981); and by Fourier synthesis in Reid et al. (1980) and Norris, Booth, and Diamond (1982). We shall briefly discuss some of the techniques used in mapping masers and their accuracies. Note that geometrical (group) delays cannot be measured accurately because of the narrow bandwidths of the maser lines.

In mapping masers, we must explicitly consider the frequency dependence of the fringe visibility. We assume that a maser source consists of a number of point sources. Furthermore, we assume that the measurements are made with a VLBI system and that the desired RF band is converted to a single baseband channel. Adapting Eq. (9.23), we can write the residual fringe phase of one maser component at frequency ν as

$$\Delta\phi(\nu) = 2\pi\left[\nu\Delta\tau_g(\nu) + (\nu - \nu_{LO})\tau_e + \nu\tau_a\right] + \phi_{in} + 2\pi n, \quad (12.24)$$

where τ_e is the relative delay error due to clock offsets, τ_a is the differential atmospheric delay, $\Delta\tau_g(\nu)$ is the difference between the true geometrical delay of the source $\tau_g(\nu)$ and the expected (reference) delay, ν_{LO} is the local oscillator frequency, ϕ_{in} is the instrumental phase, which includes the local oscillator frequency difference and can be a rapidly varying function of time, and $2\pi n$ represents the phase ambiguity. A frequency can usually be found that has only one unresolved maser component, and this component can then be used as a phase reference. The use of a phase reference feature is fundamental to all maser analysis procedures, and it allows maps of the relative positions of maser components to be made with high accuracy. The difference in residual fringe phase between a maser feature at frequency ν and the reference feature at frequency ν_R is

$$\Delta^2\phi(\nu) = \Delta\phi(\nu) - \Delta\phi(\nu_R), \quad (12.25)$$

which, with the use of Eq. (12.24), becomes

$$\Delta^2\phi(\nu) = 2\pi\left\{\nu\left[\tau_g(\nu) - \tau_g(\nu_R)\right]\right.$$
$$\left. + (\nu - \nu_R)\left[\tau_g(\nu_R) - \tau_g'(\nu_R)\right] + (\nu - \nu_R)\left[\tau_e + \tau_a\right]\right\},$$

$$(12.26)$$

where $\tau_g'(\nu_R)$ is the expected delay of the reference feature. The frequency-independent terms ϕ_{in} and $2\pi n$ cancel in Eq. (12.26). However, there are residual terms in Eq. (12.26) that are proportional to the difference in frequency between the feature of interest and the reference feature. These terms arise because phases at different frequencies are differenced in Eq. (12.25). Following the notation of Eq. (12.3), we can write Eq. (12.26) as

$$\Delta^2\phi(\nu) = \frac{2\pi\nu}{c}\mathbf{D}\cdot\Delta\mathbf{s}_{\nu R} - \frac{2\pi\nu}{c}\Delta\mathbf{D}\cdot\Delta\mathbf{s}_{\nu R}$$

$$- \frac{2\pi}{c}\left[(\nu - \nu_R)(\Delta\mathbf{D}\cdot\mathbf{s}_R + \mathbf{D}\cdot\Delta\mathbf{s}_R)\right] + 2\pi(\nu - \nu_R)(\tau_e + \tau_a),$$

$$(12.27)$$

where $\mathbf{D}$ is the assumed baseline, $\Delta\mathbf{D}$ is the baseline error, $\Delta\mathbf{s}_{\nu R}$ is the separation vector from the reference feature to the feature at frequency ν, $\mathbf{s}_R$ is the assumed direction of the reference feature, and $\Delta\mathbf{s}_R$ is the error in the reference feature direction. The first term on the right-hand side of Eq. (12.27) is the desired quantity from which the position of the feature relative to the reference feature can be determined, and the rest of the terms describe the phase errors introduced by uncertainty in baseline, source position, clock offset, and atmospheric delay. These phase-error terms can be converted approximately to angular errors by dividing them by $c/2\pi\nu D$. Thus, for example, a 0.3-m error in a baseline component would cause a delay error of about 1 nsec in the term $\Delta\mathbf{D}\cdot\mathbf{s}_R$ in Eq. (12.27) and a phase error of 10^{-3} turns for features separated by 1 MHz. This phase error corresponds to a nominal error of 10^{-6} arcsec on a baseline of 2500 km at 22 GHz, which provides a fringe spacing of 10^{-3} arcsec. Similarly, a clock or atmospheric error of 1 nsec would cause the same positional error. The same baseline error also causes additional positional errors through the $\Delta\mathbf{D}\cdot\Delta\mathbf{s}_{\nu R}$ term of 10^{-7} arcsec per arcsecond separation of the features. A detailed discussion of mapping errors caused by this calibration method can be found in Genzel et al. (1981).

Another method of calibrating the fringe phase is to scale the phase of the reference feature to the frequency of the feature to be calibrated. That is, let

$$\Delta^2\phi(\nu) = \Delta\phi(\nu) - \Delta\phi(\nu_R)\frac{\nu}{\nu_R}. \tag{12.28}$$

This method of calibration is more accurate than the method of Eq. (12.25) because error terms proportional to $\nu - \nu_R$ do not appear. However, there are additional terms involving the phase ambiguity and the instrumental phase. Thus, this calibration method is applicable only if the fringe phase can be followed carefully enough to avoid the introduction of phase ambiguities.

Maps of lower accuracy and sensitivity than those obtainable from phase data can be made with fringe-frequency data. Suppose that the interferometer is well calibrated. The differential fringe frequency, that is, the difference in fringe frequency between the feature at frequency ν and the reference feature, can then be written [using Eq. (12.14)]

$$\Delta^2\nu_f(\nu) \simeq \dot{u}\,\Delta\alpha'(\nu) + \dot{v}\,\Delta\delta(\nu), \tag{12.29}$$

where $\dot{u}$ and $\dot{v}$ are the time derivatives of the projected baseline components, $\Delta\alpha'(\nu)$ and $\Delta\delta(\nu)$ are the coordinate offsets from the reference feature, and $\Delta\alpha'(\nu) = \Delta\alpha(\nu)\cos\delta$. The relative positions of the maser feature can then be found by fitting Eq. (12.29) to a series of fringe-frequency measurements at various hour angles. This technique was first employed by Moran et al. (1968) for the mapping of an OH maser. The errors in fringe-frequency measurements decrease as $\tau^{3/2}$, where τ is the length of an observation, but, for large values of τ, $\Delta^2\nu_f$ is not constant because $\ddot{u}$ and $\ddot{v}$ are not zero. Thus, there is a limited field of view available for accurate mapping with fringe-frequency measurements. This field of view can be estimated by equating the rms fringe-frequency error in Eq. (A12.27) with τ times the derivative of the differential fringe frequency with respect to time. Therefore, for an east–west baseline,

$$D_\lambda \omega_e^2 \, \Delta\theta\tau \cos\theta \simeq \sqrt{\frac{3}{2\pi^2}} \left(\frac{T_S}{T_A}\right) \frac{1}{\sqrt{\Delta\nu\tau^3}}. \qquad (12.30)$$

For $\sqrt{2\pi^2/3} \, \cos\theta \simeq 1$, the field of view is

$$\Delta\theta \simeq \frac{T_S}{D_\lambda T_A \omega_e^2 \tau^2 \sqrt{\Delta\nu\tau}}, \qquad (12.31)$$

or

$$\Delta\theta \simeq \frac{1}{\mathscr{R}_{sn} D_\lambda \omega_e^2 \tau^2}, \qquad (12.32)$$

where $\mathscr{R}_{sn}$ is the signal-to-noise ratio. Let $\mathscr{R}_{sn} = 10$ and $\tau = 100$ sec. The field of view is then about equal to 2000 times the fringe spacing. This restriction is often important. Usually when a feature is found the phase center of the field is moved to the estimated position of the feature, and the position is then redetermined. Only components that are detected on individual observations on each baseline can be mapped with the fringe-frequency mapping technique. Thus, fringe-frequency mapping is less sensitive than synthesis mapping, in which fully coherent sensitivity is achieved.

The fringe-frequency analysis procedure can be extended to handle the case in which there are many point components in one frequency channel. From each observation (i.e., a measurement on one baseline lasting for a few minutes), the fringe-frequency spectrum is calculated. Multiple components will appear as distinct fringe-frequency features, as shown in Fig. 12.3. The fringe frequency of each feature defines a line in $\Delta\alpha'$, $\Delta\delta$ space on which a maser component lies. The slope of the line is $\tan^{-1}(\dot{v}/\dot{u})$. As the projected baseline changes, the slopes of the lines change. The intersections of the lines define the source positions (see Fig. 12.3). For this method to work, the components must be sufficiently separated to produce separate peaks in the fringe-frequency spectrum. The fringe-frequency resolution is about τ^{-1}, which defines an effective beam of width

$$\Delta\theta_f = \frac{1}{D_\lambda \omega_e \tau \cos\theta}. \qquad (12.33)$$

Fringe-frequency mapping is discussed in detail by Giuffrida et al. (1981) and Walker (1981).

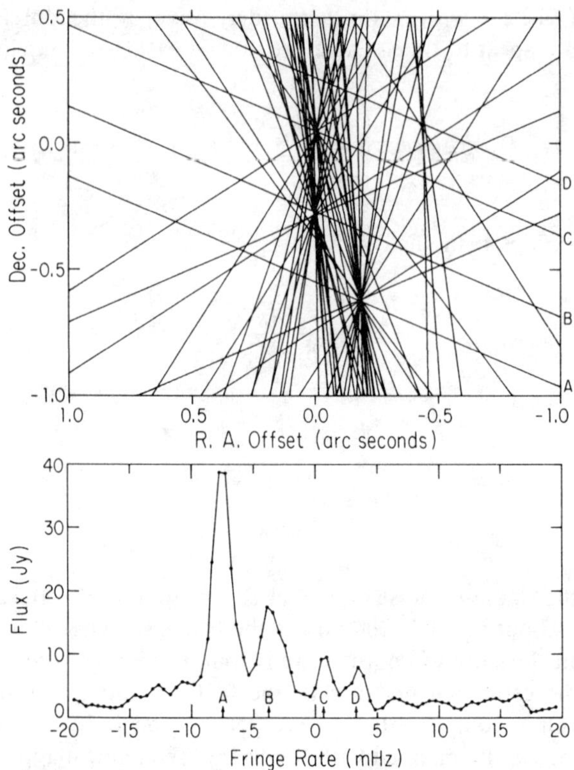

Figure 12.3   The lower plot is the fringe-frequency spectrum at one particular hour angle and one frequency in the spectrum of the water vapor maser in W49N. The ordinate is flux density. There are four peaks, each corresponding to a separate feature on the sky. The upper plot shows such lines from many scans. The peaks in the lower plot and their corresponding lines in the upper plot are labeled A–D. There are at least four separate features at the frequency of these data. Their positions are marked by the locations where many lines intersect. The feature corresponding to line D is sufficiently far from the phase center that its fringe frequency changes enough during the 20-min integration to degrade significantly the estimate of the feature position. The window in which accurate positions can be determined is 0."5 in right ascension and 2" in declination. The window can be moved by shifting the phase center of the data. Figure from Walker (1981), in the *Astronomical Journal*.

APPENDIX 12.1   LEAST-MEAN-SQUARES ANALYSIS

The principles of least-mean-squares analysis play a fundamental role in astrometry where the goal is to extract a number of parameters from a set of noisy measurements. We briefly discuss these principles in an elementary way, ignoring mathematical subtleties, and apply them to the problems encountered in interferometry. Detailed discussions of the statistical analysis of data can be found in books such as Bevington (1969) and Hamilton (1964).

Suppose we wish to measure a quantity m. We make a set of measurements y_i that are the sum of the desired quantity m and a noise contribution n_i, that is,

$$y_i = m + n_i, \tag{A12.1}$$

where n_i is a Gaussian random variable with zero mean and variance σ_i^2. The probability that the ith measurement will be y_i is therefore

$$p(y_i) = \frac{1}{\sqrt{2\pi}\,\sigma_i} e^{-(y_i - m)^2/2\sigma_i^2}. \tag{A12.2}$$

If all the measurements are independent, then the probability that an experiment will yield a set of N measurements $y_1, y_2, \ldots, y_N$ is

$$L = \prod_{i=1}^{N} p(y_i), \tag{A12.3}$$

where the $\prod$ denotes the product of the $p(y_i)$ terms. L, viewed as a function of m, is called the likelihood function. The method of maximum likelihood is based on the assumption that the best estimate of m is the one that maximizes L. Maximizing L is the same as maximizing $\ln L$, where

$$\ln L = \sum_{i=1}^{N} \ln \frac{1}{\sqrt{2\pi}\,\sigma_i} - \frac{1}{2} \sum_{i=1}^{N} \frac{(y_i - m)^2}{\sigma_i^2}. \tag{A12.4}$$

Since the first summation term on the right-hand side of Eq. (A12.4) is a constant and the second summation term is multiplied by $-\frac{1}{2}$, the maximization of L is equivalent to the minimization of the second summation term in Eq. (A12.4). Thus, we wish to minimize the quantity χ^2 given by

$$\chi^2 = \sum_{i=1}^{N} \frac{(y_i - m)^2}{\sigma_i^2}. \tag{A12.5}$$

In the more general problem discussed later in this appendix, m is replaced by a function with one or more parameters describing the system model. With this generalization, Eq. (A12.5) becomes the fundamental equation of the method of weighted least squares. In this method the parameters of the model are determined by minimizing the sum of the differences between the measurements and the model, weighted by the variances of the measurements. The quantity χ^2, which is a random variable whose mean value equals the number of data points less the number of parameters when the model adequately describes the measurements, indicates the goodness of the fit. The method of least squares, appropriate when the noise is a Gaussian random process, is a special case of the more general method of maximum likelihood. Gauss invented the method of least squares perhaps as early as 1795, using arguments similar to those given here, for the purpose of estimating the orbital parameters of planets and comets (Gauss 1809). The method was independently developed by Legendre in 1806 (Hall 1970).

Returning to Eq. (A12.5) we can estimate m by setting the derivative of χ^2 with respect to m equal to zero. The resulting estimate of m is

$$m_e = \frac{\sum \dfrac{y_i}{\sigma_i^2}}{\sum \dfrac{1}{\sigma_i^2}},\tag{A12.6}$$

where the summation goes from $i = 1$ to N. Using Eq. (A12.2) we note that $\langle y_i \rangle = m$ and $\langle y_i^2 \rangle = m^2 + \sigma_i^2$. Therefore, by calculating the expectation of Eq. (A12.6), it is clear that $\langle m_e \rangle = \langle y_i \rangle = m$, and it is easy to show that

$$\langle m_e^2 \rangle = m^2 + \left(\sum \frac{1}{\sigma_i^2} \right)^{-1}.\tag{A12.7}$$

Hence the variance of the estimate of m_e is

$$\sigma_m^2 = \langle m_e^2 \rangle - \langle m_e \rangle^2 = \left(\sum \frac{1}{\sigma_i^2} \right)^{-1}.\tag{A12.8}$$

Equation (A12.8) shows that when poor quality or noisy data are added to better data, the value of σ_m may be reduced only slightly. If all the measurement errors are equal to σ, then Eq. (A12.8) reduces to the well-known result

$$\sigma_m = \frac{\sigma}{\sqrt{N}},\tag{A12.9}$$

and m_e is the average of the measurements. In many instances σ is not known. An estimate of σ is

$$\sigma_e^2 = \frac{1}{N} \sum (y_i - m)^2.\tag{A12.10}$$

However, m is not known, only its estimate, m_e. If m_e were used in place of m in Eq. (A12.10), the value of σ_e^2 would be an underestimate of σ^2 because of the manner in which m_e was determined in minimizing χ^2. The unbiased estimate of σ^2 is

$$\sigma_e^2 = \frac{1}{N-1} \sum (y_i - m_e)^2.\tag{A12.11}$$

It is easy to show by substitution of Eq. (A12.6) into Eq. (A12.11) that $\langle \sigma_e^2 \rangle = \sigma^2$. The term $N - 1$ appears in Eq. (A12.11) because there are N data points and one free parameter.

Consider a model described by the function $f(x; p_1, \ldots, p_n)$, where x is the independent variable, which takes values x_i, where $i = 1$ to N, at the sample points, and $p_1, \ldots, p_n$ are a set of parameters. We assume that the values of the independent variable are exactly known. If the function f correctly models the measurement system, the measurement set is given by

$$y_i = f(x_i;\, p_1, \ldots, p_n) + n_i, \tag{A12.12}$$

where n_i represents the measurement error. The general problem is to find the values of the parameters for which χ^2, given by the generalization of Eq. (A12.5),

$$\chi^2 = \sum \frac{[y_i - f(x_i)]^2}{\sigma_i^2}, \tag{A12.13}$$

is a minimum.

A simple example of this problem is the fitting of a straight line to a data set. Let

$$f(x;\, a, b) = a + bx, \tag{A12.14}$$

where a and b are the parameters to be found. Minimizing χ^2 is accomplished by solving the equations

$$\frac{\partial \chi^2}{\partial a} = -\sum \frac{2(y_i - a - bx_i)}{\sigma_i^2} = 0 \tag{A12.15a}$$

and

$$\frac{\partial \chi^2}{\partial b} = -\sum \frac{2(y_i - a - bx_i)x_i}{\sigma_i^2} = 0 \tag{A12.15b}$$

In matrix notation we have

$$\begin{bmatrix} \sum \dfrac{y_i}{\sigma_i^2} \\[2ex] \sum \dfrac{x_i y_i}{\sigma_i^2} \end{bmatrix} = \begin{bmatrix} \sum \dfrac{1}{\sigma_i^2} & \sum \dfrac{x_i}{\sigma_i^2} \\[2ex] \sum \dfrac{x_i}{\sigma_i^2} & \sum \dfrac{x_i^2}{\sigma_i^2} \end{bmatrix} \begin{bmatrix} a_e \\[2ex] b_e \end{bmatrix}, \tag{A12.16}$$

where we distinguish between the true values of the parameters and their estimates by the subscript e. The solution is

$$a_e = \frac{1}{\Delta}\left[\left(\sum \frac{x_i^2}{\sigma_i^2}\right)\left(\sum \frac{y_i}{\sigma_i^2}\right) - \left(\sum \frac{x_i}{\sigma_i^2}\right)\left(\sum \frac{x_i y_i}{\sigma_i^2}\right)\right] \tag{A12.17}$$

and

$$b_e = \frac{1}{\Delta}\left[\left(\sum \frac{1}{\sigma_i^2}\right)\left(\sum \frac{x_i y_i}{\sigma_i^2}\right) - \left(\sum \frac{x_i}{\sigma_i^2}\right)\left(\sum \frac{y_i}{\sigma_i^2}\right)\right], \tag{A12.18}$$

where Δ is the determinant of the square matrix in Eq. (A12.16) given by

$$\Delta = \left(\sum \frac{1}{\sigma_i^2} \right) \left(\sum \frac{x_i^2}{\sigma_i^2} \right) - \left(\sum \frac{x_i}{\sigma_i^2} \right)^2. \tag{A12.19}$$

Estimates of the errors in the parameters a_e and b_e can be calculated from Eqs. (A12.17) and (A12.18) and are given by

$$\sigma_a^2 = \langle a_e^2 \rangle - \langle a_e \rangle^2 = \frac{1}{\Delta} \sum \frac{x_i^2}{\sigma_i^2} \tag{A12.20}$$

and

$$\sigma_b^2 = \langle b_e^2 \rangle - \langle b_e \rangle^2 = \frac{1}{\Delta} \sum \frac{1}{\sigma_i^2}. \tag{A12.21}$$

Note that a_e and b_e are random variables, and in general $\langle a_e b_e \rangle$ is not zero so that the parameter estimates are correlated. The error estimates in Eqs. (A12.20) and (A12.21) include the deleterious effects of the correlation between parameters. In this particular example, the correlation can be made equal to zero by adjusting the origin of the x axis so that $\Sigma(x_i/\sigma_i^2) = 0$.

The above analysis can be used to estimate the accuracy of measurements of fringe frequency and delay made with an interferometer. Fringe frequency, the rate of change of fringe phase with time,

$$\nu_f = \frac{1}{2\pi} \frac{\partial \phi}{\partial t}, \tag{A12.22}$$

can be estimated by fitting a straight line to a sequence of uniformly spaced measurements of phase with respect to time. The fringe frequency is proportional to the slope of this line. Assume that N measurements of phase, ϕ_i, each having the same rms error σ_ϕ, are made at times t_i, spaced by interval T, running from time $-NT/2$ to $NT/2$, such that the total time of the observation is $\tau = NT$. From Eq. (A12.21) and the above definitions, including Eq. (A12.22), the error in the fringe-frequency estimate is

$$\sigma_f^2 = \frac{\sigma_\phi^2}{(2\pi)^2 \sum t_i^2}, \tag{A12.23}$$

since $\Sigma t_i = 0$. The term Σt_i^2 is approximately given by

$$\sum t_i^2 \simeq \frac{1}{T} \int_{-\tau/2}^{\tau/2} t^2 dt = \frac{1}{T} \frac{\tau^3}{12} = \frac{N\tau^2}{12}. \tag{A12.24}$$

$\tau/\sqrt{12}$ can be thought of as the rms time span of the data. Thus, Eq. (A12.23) becomes

$$\sigma_f^2 = \frac{12\sigma_\phi^2}{(2\pi)^2 N\tau^2}. \tag{A12.25}$$

The expression for σ_ϕ, given in Eq. (6.57) for the case when the source is unresolved and there are no processing losses, is

$$\sigma_\phi = \frac{T_S}{T_A\sqrt{2\Delta\nu T}}, \tag{A12.26}$$

where T_S is the system temperature, T_A is the antenna temperature due to the source, and $\Delta\nu$ is the bandwidth. Substitution of Eq. (A12.26) into Eq. (A12.25) yields

$$\sigma_f = \sqrt{\frac{3}{2\pi^2}} \left(\frac{T_S}{T_A}\right) \frac{1}{\sqrt{\Delta\nu\,\tau^3}} \quad \text{(Hz)}. \tag{A12.27}$$

Note that this result does not depend on the details of the analysis procedure such as the choice of N. Equivalently, one can estimate the fringe frequency by finding the peak of the fringe-frequency spectrum, that is, the peak of the Fourier transform of $e^{j\phi_i}$.

The delay is the rate of change of phase with frequency,

$$\tau = \frac{1}{2\pi}\frac{\partial\phi}{\partial\nu}. \tag{A12.28}$$

Thus, the delay can be estimated by finding the slope of a straight line fitted to a sequence of phase measurements as a function of frequency. For a single band, such data can be obtained from the cross power spectrum, the Fourier transform of the cross-correlation function. Assume that N measurements of phase are made at frequencies ν_i, each with a bandwidth $\Delta\nu/N$ and with an error σ_ϕ. In this calculation, only the relative frequencies are important. It is convenient for the purpose of analysis to set the zero of the frequency axis such that $\Sigma\nu_i = 0$. The error in delay [from Eqs. (A12.19), (A12.21), and (A12.28)] is

$$\sigma_\tau^2 = \frac{\sigma_\phi^2}{(2\pi)^2\sum_i\nu_i^2}. \tag{A12.29}$$

Using a calculation for $\Sigma\nu_i^2$ analogous to the one in Eq. (A12.24), we can write Eq. (A12.29) as

$$\sigma_\tau^2 = \frac{12\sigma_\phi^2}{(2\pi)^2 N\Delta\nu^2}. \tag{A12.30}$$

Thus, substitution of Eq. (A12.26) (with an integration time of τ and bandwidth $\Delta\nu/N$) into Eq. (A12.30) yields

$$\sigma_\tau = \sqrt{\frac{3}{2\pi^2}}\left(\frac{T_S}{T_A}\right)\frac{1}{\sqrt{\Delta\nu^3\tau}}\,. \tag{A12.31}$$

We can define the rms bandwidth as

$$\Delta\nu_{\mathrm{rms}} = \sqrt{\frac{1}{N}\sum\nu_i^2} \tag{A12.32}$$

and obtain from Eqs. (A12.26) and (A12.29) the result quoted in Chapter 9 [Eq. (9.142)],

$$\sigma_\tau = \frac{1}{\beta}\left(\frac{T_S}{T_A}\right)\frac{1}{\sqrt{\Delta\nu_{\mathrm{rms}}^3\tau}}, \tag{A12.33}$$

where $\beta = \pi(768)^{1/4}$. (Note that in Section 9.8, σ_ϕ applies to the full bandwidth $\Delta\nu$.) The expressions for σ_τ in Eqs. (A12.30), (A12.31), and (A12.33) incorporate the condition $\Delta\nu_{\mathrm{rms}} = \Delta\nu/\sqrt{12}$ and apply to a continuous passband of width $\Delta\nu$.

In bandwidth synthesis, which is described in Section 9.8, the measurement system consists of N channels of width $\Delta\nu/N$, which are not in general contiguous. The rms delay error is obtained by substituting Eqs. (A12.26) and (A12.32) into Eq. (A12.29), yielding

$$\sigma_\tau = \frac{1}{\sqrt{8\pi^2}}\left(\frac{T_S}{T_A}\right)\frac{1}{\sqrt{\Delta\nu\tau}\,\Delta\nu_{\mathrm{rms}}}, \tag{A12.34}$$

where $\Delta\nu_{\mathrm{rms}}$ is given by Eq. (A12.32) and $\Delta\nu$ is the total bandwidth. $\Delta\nu_{\mathrm{rms}}$ is generally equal to about 40% of the total frequency range spanned.

A general formulation of the linear least-squares solution can be found when the model function f is a linear function of the parameters p_k, that is, when

$$f(x; p_1,\ldots,p_n) = \sum_{k=1}^{n}\frac{\partial f}{\partial p_k}p_k, \tag{A12.35}$$

where n is the number of parameters. For example, the model could be a cubic polynomial,

$$f(x; p_0, p_1, p_2, p_3) = p_0 + p_1 x + p_2 x^2 + p_3 x^3, \tag{A12.36}$$

in which case $\partial f/\partial p_k = x^k$ for $k = 0, 1, 2$, and 3. If the parameters appear as linear multiplicative factors, then the minimization of Eq. (A12.13) leads to a set of n equations of the form

$$\frac{\partial\chi^2}{\partial p_k} = 0, \qquad k = 1, 2,\ldots, n. \tag{A12.37}$$

Substitution of Eq. (A12.13) into Eq. (A12.37) and use of Eq. (A12.35) yield the set of n equations

$$D_k = \sum_{j=1}^{n} T_{kj} p_j, \qquad k = 1, 2, \ldots, n \tag{A12.38}$$

where

$$D_k = \sum_{i=1}^{N} \frac{y_i}{\sigma_i^2} \frac{\partial f(x_i)}{\partial p_k} \tag{A12.39}$$

and

$$T_{jk} = \sum_{i=1}^{N} \frac{1}{\sigma_i^2} \frac{\partial f(x_i)}{\partial p_j} \frac{\partial f(x_i)}{\partial p_k}, \tag{A12.40}$$

and the summations are carried out over the set of N independent measurements. In matrix notation, the equation set (A12.38) is

$$[D] = [T][P_e], \tag{A12.41}$$

where $[D]$ is a column matrix with elements D_k, $[P_e]$ is a column matrix containing the estimates of the parameters p_{ek}, and $[T]$ is a symmetric square matrix with elements T_{jk}. For obvious reasons, $[T]$ is sometimes called the matrix of the normal equations. Note that Eq. (A12.41) is a generalization of Eq. (A12.16). The matrices $[T]$ and $[D]$ are sometimes written as the product of other matrices (Hamilton 1964, Chapter 4). Let $[M]$ be the variance matrix (size $N \times N$) whose diagonal elements are σ_i^2 and whose off-diagonal elements are zero; let $[F]$ be a column matrix containing the data y_i; and let $[A]$ be the partial derivative matrix (size $n \times N$) whose elements are $\partial f(x_i)/\partial p_k$. Then one can write $[T] = [A]^T [M]^{-1} [A]$ and $[D] = [A]^T [M]^{-1} [F]$ where $[A]^T$ is the transpose of $[A]$ and $[M]^{-1}$ is the inverse of $[M]$. The analysis can be generalized to include the situation where the errors between measurements are correlated. In this case, $[M]$ is modified to include off-diagonal elements $\sigma_i \sigma_j \rho'_{ij}$, where ρ'_{ij} is the correlation coefficient for the ith and j^{th} measurement.

The solution to Eq. (A12.41) is

$$[P_e] = [T]^{-1} [D], \tag{A12.42}$$

where $[T]^{-1}$ is the inverse matrix of $[T]$, and $[P_e]$ is the column matrix containing the parameter estimates. The elements of $[T]^{-1}$ are denoted T'_{jk}. It can be shown by direct calculation that the estimates of the errors of the parameters σ_{ek}^2 are the diagonal elements of $[T]^{-1}$. Thus,

$$\sigma_{ek}^2 = T'_{kk}. \tag{A12.43}$$

The probability that parameter p_k will be within $\pm\sigma_k$ of its true value is 0.68, which is the integral under the one-dimensional Gaussian probability distribution between $\pm\sigma_k$. The probability that all the n parameters will be within $\pm\sigma$ of their true values (that is, within the error "box" in the n-dimensional space) is approximately 0.68^n when the correlations are moderate.

The normalized correlation coefficients between parameters are proportional to the off-diagonal elements of $[T]^{-1}$, that is,

$$\rho_{jk} = \frac{\langle(p_{ej} - p_j)(p_{ek} - p_k)\rangle}{\sigma_{ek}\sigma_{ej}} = \frac{T'_{jk}}{\sqrt{T'_{jj}T'_{kk}}}. \tag{A12.44}$$

For any two parameters, there is a Gaussian probability distribution that describes the distribution of errors,

$$p(\epsilon_j, \epsilon_k) = \frac{1}{2\pi\sigma_j\sigma_k\sqrt{1 - \rho_{jk}^2}} \exp\left\{ -\frac{1}{2(1 - \rho_{jk}^2)}\left[\frac{\epsilon_j^2}{\sigma_j^2} + \frac{\epsilon_k^2}{\sigma_k^2} - \frac{2\rho_{jk}\epsilon_j\epsilon_k}{\sigma_j\sigma_k}\right]\right\},$$

$$\tag{A12.45}$$

where $\epsilon_k = p_{ek} - p_k$, $\epsilon_j = p_{ej} - p_j$. The contour of $p(\epsilon_k, \epsilon_j) = p(0,0)e^{-1/2}$

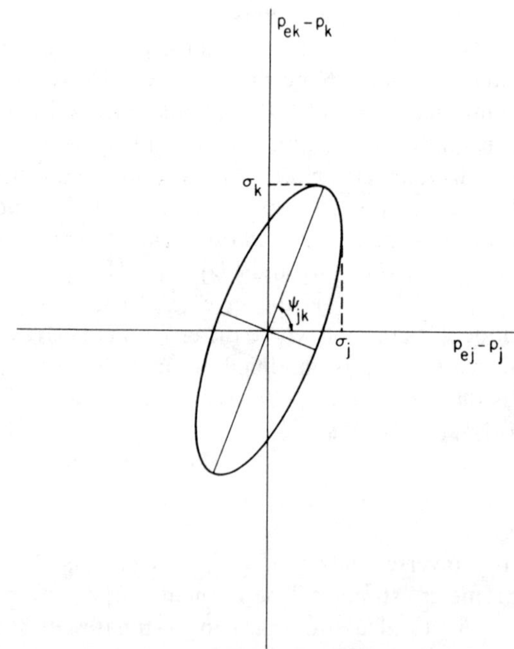

Figure A12.1   The error ellipse, or contour, defining the e^{-1} level of the joint probability function [Eq. (A12.45)] for the estimates of parameters p_k and p_j. $p_{ek} - p_k$ and $p_{ej} - p_j$ are the parameter estimates minus their true values. The angle ψ_{jk} is defined by Eq. (A12.46).

defines an ellipse, shown in Fig. A12.1, which is known as the error ellipse. The probability that both parameters will lie within the error ellipse is the integral of Eq. (A12.45) over the area of the error ellipse, which equals 0.46. The orientation of the error ellipse is given by

$$\psi_{jk} = \tfrac{1}{2}\tan^{-1}\left(\frac{2\rho_{jk}\sigma_j\sigma_k}{\sigma_j^2 - \sigma_k^2}\right). \tag{A12.46}$$

The errors in the parameters p_k are completely determined by the matrix $[T]^{-1}$ through Eqs. (A12.43)–(A12.45). The elements of $[T]^{-1}$ depend only on the partial derivatives of the model function and the values of the measurement errors, which can usually be predicted in advance based on the characteristics of the measurement apparatus. Therefore, once an experiment is planned, the errors in the parameters can be predicted from $[T]^{-1}$ without reference to the data. For this reason, $[T]^{-1}$ is sometimes called the design matrix. Studies of the design matrix for a specific experiment might reveal a very high correlation between two parameters leading to large errors in their estimated values. It is often possible to modify the experiment to obtain more data that will reduce the correlation. After the data are analyzed, the value of χ^2 can be computed. If the model is a good fit to the data, χ^2 should be approximately equal to $N - n$, the number of measurements minus the number of parameters. If it is not, the difficulty is often that the values of σ_i are estimated incorrectly or that the model does not describe adequately the measurement system—that is, the model has too few parameters or is not correct. Even if $\chi^2 \simeq N - n$, the derived errors in Eq. (A12.43) may not be realistic, and they are referred to as "formal errors." The formal errors describe the *precision* of the parameter estimates. The *accuracy* of the parameter measurements is the deviation between the estimates of the parameters and the true values of the parameters. The accuracy of the measurements is often difficult to determine. For example, an unknown effect that closely mimics the functional dependence of one of the model parameters may be present in an experiment. The model may appear to be a good one, but the accuracy of the particular model parameter in question will be much poorer than expected because of the systematic error introduced by the unmodeled effect.

The discussion of linear least-mean-squares analysis can be generalized to include nonlinear functions in a straightfoward manner. Assume that $f(x;p)$ has one nonlinear parameter p_n. For the purpose of discussion we can separate f into linear and nonlinear parts, $f_L(x;p_1,\ldots,p_{n-1})$ and $f_{NL}(x;p_n)$, and approximate the nonlinear function by the first two terms in a Taylor expansion,

$$f_{NL}(x;\,p_n) \simeq f_{NL}(x,p_{0n}) + \frac{\partial f_{NL}}{\partial p_n}\Delta p_n, \tag{A12.47}$$

where p_{0n} is the initial guess of parameter p_n and $\Delta p_n = p_n - p_{0n}$. We assume that the initial parameter guesses are accurate enough for Eq. (A12.47) to be valid. We replace the data with $y_i - f_{\mathrm{NL}}(x_i; p_{0n})$ and then compute the elements of the matrices $[D]$ and $[T]$ from the partial derivatives including $\partial f_{\mathrm{NL}}/\partial p_n$. The nth parameter in the matrix $[P_e]$ in Eq. (A12.42) will be the differential parameter Δp_n defined in Eq. (A12.47). The solution must be iterated with a new Taylor expansion centered on the parameter $p_{0n} + \Delta p_n$. Thus, nonlinear functions can be accommodated in the analysis through linearization, but initial guesses of the nonlinear parameters and solution iteration are required. In some cases nonlinear estimation problems can cause difficulties (e.g., Lampton, Margon, and Bowyer 1976).

We can envision how the principles of least-mean-squares analysis are applied to a large astrometric experiment. Consider a hypothetical VLBI experiment made on a three-station array. Suppose 10 recordings are made of each of 20 sources during observations made over 1 day (an epoch). The observations are repeated six times a year for 5 years. The data set would consist of 18,000 measurements (20 sources $\times 10$ observations $\times 3$ baselines $\times 30$ epochs) of delay and fringe frequency or 36,000 total measurements. The measurements of delay and fringe frequency can be combined in the analysis since, in the least-squares method, the relevant quantities are the squares of the measurements divided by their variances, which are dimensionless [Eq. (A12.13)]. Now we can count the number of parameters in the analysis model: 39 source coordinates (one right ascension fixed); 9 station coordinates; 90 atmospheric parameters (a zenith excess path length at each station at each epoch); 120 clock parameters (a clock error and clock rate error at two of the stations per epoch); 90 polar motion and UT1 $-$ UTC parameters, as well as several other parameters to model precession, nutation, solid-earth tides, gravitational deflection by the sun, movement of stations and other effects such as antenna axis offsets (see Section 4.7). The total number of parameters is about 360. The parameters within each observation epoch are linked because of the common clock and atmosphere parameters. Parameters among epochs are linked because of baseline, precession, and nutation parameters. Naturally, partial solutions from subsets of the data should be obtained before a grand global solution is attempted. Procedures are available for obtaining global solutions that do not require the inversion of matrices as large as the total number of parameters (e.g., Morrison 1969). Experiments of the scale described here have been carried out (e.g., Fanselow et al. 1984, Herring et al. 1985).

One final topic concerns the estimation of the coordinates of a radio source with a well-calibrated interferometer, which has accurately known baselines and instrumental phases. In this case, the differential interferometer phase is, from Eq. (12.1),

$$\Delta\phi = 2\pi D_\lambda\{[\sin d \cos\delta - \cos d \sin\delta \cos(H - h)]\Delta\delta$$

$$+ \cos d \cos\delta \sin(H - h)\Delta\alpha\}. \quad \text{(A12.48)}$$

Expressing the geometric quantities in terms of projected baseline components, we can write Eq. (A12.48) as

$$\Delta\phi = 2\pi(u\,\Delta\alpha' + v\,\Delta\delta), \tag{A12.49}$$

where $\Delta\alpha' = \Delta\alpha\cos\delta$. A set of phase measurements from one or more baselines can be analyzed by the method of least squares to determine $\Delta\alpha'$ and $\Delta\delta$. The partial derivatives are $\partial f/\partial p_1 = 2\pi u$ and $\partial f/\partial p_2 = 2\pi v$, where $p_1 = \Delta\alpha'$ and $p_2 = \Delta\delta$. From Eqs. (A12.40) and (A12.49), the normal-equation matrix is

$$[T] = \frac{4\pi^2}{\sigma_\phi^2}\begin{bmatrix} \sum u_i^2 & \sum u_i v_i \\ \sum u_i v_i & \sum v_i^2 \end{bmatrix}, \tag{A12.50}$$

where all the measurements are assumed to have the same uncertainty σ_ϕ given by Eq. (A12.26). The inverse of $[T]$ is

$$[T]^{-1} = \frac{1}{\Delta}\begin{bmatrix} \sum v_i^2 & -\sum u_i v_i \\ -\sum u_i v_i & \sum u_i^2 \end{bmatrix}, \tag{A12.51}$$

where

$$\Delta = \frac{4\pi^2}{\sigma_\phi^2}\left[\sum u_i^2 \sum v_i^2 - \left(\sum u_i v_i\right)^2\right]. \tag{A12.52}$$

The correlation coefficient, defined by Eq. (A12.44), is

$$\rho_{12} = \frac{-\sum u_i v_i}{\sqrt{\sum u_i^2 \sum v_i^2}}. \tag{A12.53}$$

The variances of the estimates of the parameters are given by the diagonal elements of Eq. (A12.51),

$$\sigma_{\alpha'}^2 = \frac{\sigma_\phi^2 \sum v_i^2}{4\pi^2\left[\sum v_i^2 \sum u_i^2 - \left(\sum u_i v_i\right)^2\right]}, \tag{A12.54}$$

and

$$\sigma_\delta^2 = \frac{\sigma_\phi^2 \sum u_i^2}{4\pi^2\left[\sum v_i^2 \sum u_i^2 - \left(\sum u_i v_i\right)^2\right]}. \tag{A12.55}$$

If the (u, v) loci are long (i.e., the observations extend over a large fraction of the day), then $\Sigma u_i v_i$ will be small compared to Σu_i^2 and Σv_i^2 so that

$$\sigma_{\alpha'} \simeq \frac{\sigma_\phi}{2\pi\sqrt{\Sigma u_i^2}} \tag{A12.56}$$

and

$$\sigma_\delta \simeq \frac{\sigma_\phi}{2\pi\sqrt{\Sigma v_i^2}}. \tag{A12.57}$$

Furthermore, if only one baseline is used on a high-declination source, then $u_i \simeq v_i \simeq D_\lambda$ and both errors reduce to the intuitive result

$$\sigma_{\alpha',\delta} \simeq \frac{\sigma_\phi}{2\pi\sqrt{N}\,D_\lambda}. \tag{A12.58}$$

Alternatively, the source position can be found by Fourier transformation of the visibility data. This procedure can be thought of as mapping or as multiplying the visibility data by the exponential factors $\exp[2\pi(u_i\Delta\alpha' + v_i\Delta\delta)]$ and summing over the data. The resulting "function" is maximized with respect to $\Delta\alpha'$ and $\Delta\delta$. In this latter view, it is easy to understand that (basic) mapping (i.e., no tapering or gridding of the data) is a maximum likelihood procedure for finding the position of a point source and therefore formally equivalent to the method of least squares. The synthesized beam b_0 for N measurements is

$$b_0(\Delta\alpha', \Delta\delta) = \frac{1}{N}\sum\cos\left[2\pi(u_i\Delta\alpha' + v_i\Delta\delta)\right]. \tag{A12.59}$$

The shape of b_0 near its peak can be found by expanding Eq. (A12.59) to second order,

$$b_0(\Delta\alpha', \Delta\delta) \simeq 1 - \frac{2\pi^2}{N}\left(\Delta\alpha'^2\sum u_i^2 + \Delta\delta^2\sum v_i^2 - 2\,\Delta\alpha'\Delta\delta\sum u_i v_i\right).$$

$$\tag{A12.60}$$

From Eq. (A12.60) it is easy to see that the contours of the synthesized beam are proportional to the error ellipse defined by Eqs. (A12.45), (A12.46), (A12.53), (A12.54), and (A12.55). Note that the method of least squares can be applied only in the regime of high signal-to-noise ratio, where phase ambiguities can be resolved. However, the Fourier synthesis method can be applied in any case.

BIBLIOGRAPHY

Fanselow, J. L. and O. J. Sovers, *Observation Model and Parameter Partials for the JPL VLBI Parameter Estimation Software MASTERFIT-V2.0*, JPL Publ. No. 83-39, Jet Propulsion Laboratory, Pasadena, California, 1985.

Herring, T. A., *Precision and Accuracy of Intercontinental Distance Determinations Using Radio Interferometry*, Air Force Geophysical Laboratory, Hanscom Field, Mass., AFGL-TR-84-0182, 1983.

High-Precision Earth Rotation and Earth–Moon Dynamics, O. Calame, Ed., D. Reidel, Dordrecht, Holland, 1982.

Multidisciplinary Use of the Very Long Baseline Array, National Academy of Sciences, PB84-163690, National Technical Information Services, Springfield, Virginia, 1983.

Radio Interferometry Techniques for Geodesy, NASA Conference Publication 2115, National Aeronautics and Space Administration, Washington, D.C., 1980.

Time and the Earth's Rotation, IAU Symposium 82, D. D. McCarthy and J. D. H. Pilkington, Eds., D. Reidel, Dordrecht, Holland, 1979.

REFERENCES

Astronomical Almanac, U. S. Government Printing Office, Washington, D.C., published annually.

Bartel, N., M. I. Ratner, I. I. Shapiro, R. J. Cappallo, A. E. E. Rogers, and A. R. Whitney, Pulsar Astrometry via VLBI, *Astron. J.*, **90**, 318–325, 1985.

Bartel, N., M. I. Ratner, I. I. Shapiro, T. A. Herring, and B. E. Corey, Proper Motion of Components of the Quasar 3C345, in IAU Symposium 110, *VLBI and Compact Radio Sources*, R. Fanti, K. Kellermann, and G. Setti, Eds., D. Reidel, Dordrecht, Holland, 1984, pp. 113–116.

Bevington, P. R., *Data Reduction and Error Analysis for the Physical Sciences*, McGraw-Hill, New York, 1969.

Brosche, P., C. M. Wade, and R. M. Hjellming, Precise Positions of Radio Sources. IV. Improved Solutions and Error Analysis for 59 Sources, *Astrophys. J.*, **183**, 805–818, 1973.

Carter, W. E., D. S. Robertson, and J. R. MacKay, Geodetic Radio Interferometric Surveying: Applications and Results, *J. Geophys. Res.*, **90**, 4577–4587, 1985.

Carter, W. E., D. S. Robertson, J. E. Pettey, B. D. Tapley, B. E. Schutz, R. J. Eanes, and M. Lufeng, Variations in the Rotation of the Earth, *Science*, **224**, 957–961, 1984.

Carter, W. E. and W. E. Strange, The National Geodetic Survey Project "Polaris," *Tectonophysics*, **52**, 39–46, 1979.

Chandler, S. C., On the Variation of Latitude, *Astron. J.*, **11**, 65–70, 1891.

Clark, T. A., B. E. Corey, J. L. Davis, G. Elgered, T. A. Herring, H. F. Hinteregger, C. A. Knight, J. I. Levine, G. Lundqvist, C. Ma, E. F. Nesman, R. B. Phillips, A. E. E. Rogers, B. O. Rönnäng, J. W. Ryan, B. R. Schupler, D. B. Shaffer, I. I. Shapiro, N. R. Vandenberg, J. C. Webber, and A. R. Whitney, Precise Geodesy Using the Mark-III Very-Long-Baseline Interferometer System, *IEEE Trans. Geosci. Remote Sensing*, **GE-23**, 438–449, 1985.

Cohen, M. H. and D. B. Shaffer, Positions of Radio Sources from Long-Baseline Interferometry, *Astron. J.*, **76**, 91–100, 1971.

Counselman, C. C., III, Radio Astrometry, *Ann. Rev. Astron. Astrophys.*, **14**, 197–214, 1976.

Counselman, C. C., III, S. M. Kent, C. A. Knight, I. I. Shapiro, T. A. Clark, H. F. Hinteregger, A. E. E. Rogers, and A. R. Whitney, Solar Gravitational Deflection of Radio Waves Measured by Very Long Baseline Interferometry, *Phys. Rev. Lett.*, **33**, 1621–1623, 1974.

Elsmore, B. and M. Ryle, Further Astrometric Observations with the 5-km Radio Telescope, *Mon. Not. R. Astron. Soc.*, **174**, 411–423, 1976.

Fanselow, J. L., O. J. Sovers, J. B. Thomas, G. H. Purcell, Jr., E. J. Cohen, D. H. Rogstad, L. J. Skjerve, and D. J. Spitzmesser, Radio Interferometric Determination of Source Positions Utilizing Deep Space Network Antennas — 1971 to 1980, *Astron. J.*, **89**, 987–998, 1984.

Fomalont, E. B., W. M. Goss, A. G. Lyne, and R. N. Manchester, Astrometry of 59 Pulsars: A Comparison of Interferometric and Timing Positions, *Mon. Not. R. Astron. Soc.*, **210**, 113–130, 1984.

Gauss, K. F., *Theoria Motus*, 1809. Reprinted in translation as *Theory of the Motion of the Heavenly Bodies Moving about the Sun in Conic Sections*, Dover, New York, 1963, p. 249.

Genzel, R., M. J. Reid, J. M. Moran, and D. Downes, Proper Motions and Distances of H_2O Maser Sources. I. The Outflow in Orion-KL, *Astrophys. J.*, **244**, 884–902, 1981.

Giuffrida, T. S., P. E. Greenfield, B. F. Burke, A. D. Haschick, J. M. Moran, O. E. H. Rydbeck, B. O. Rönnäng, L. B. Bååth, K. S. Yngvesson, L. I. Matveenko, V. I. Kostenko, L. R. Kogan, and I. G. Moiseev, The Structure of the H_2O Maser Source in W3(OH), *Sov. Astron. Lett.*, **7(3)**, 198–203, 1981.

Hall, T., *Karl Friedrich Gauss*, M.I.T. Press, Cambridge, Mass., 1970, p. 74.

Hamilton, W. C., *Statistics in Physical Science*, Ronald, New York, 1964.

Hazard, C., J. Sutton, A. N. Argue, C. M. Kenworthy, L. V. Morrison, and C. A. Murray, Accurate Radio and Optical Positions of 3C273B, *Nature Phys. Sci.*, **233**, 89–91, 1971.

Herring, T. A., B. E. Corey, C. C. Counselman III, I. I. Shapiro, A. E. E. Rogers, A. R. Whitney, T. A. Clark, C. A. Knight, C. Ma, J. W. Ryan, B. R. Schupler, N. R. Vandenberg, G. Elgered, G. Lundqvist, B. O. Rönnäng, J. Campbell, and P. Richards, Determination of Tidal Parameters from VLBI Observations, in *Proceedings of the Ninth International Symposium on Earth Tides*, J. Kuo, Ed., E. Schweizerbart'sche Verlagsbuchhandlung, Stuttgart, 1983, pp. 205–211.

Herring, T. A., C. R. Gwinn, and I. I. Shapiro, Geodesy by Radio Interferometry: Corrections to the IAU 1980 Nutation Series, *Proc. MERIT/COTES Symposium*, I. I. Mueller, Ed., Ohio State Univ. Press, Columbus, Ohio, 1985.

Herring, T. A., I. I. Shapiro, T. A. Clark, C. Ma, J. W. Ryan, B. R. Schupler, C. A. Knight, G. Lundqvist, D. B. Shaffer, N. R. Vandenberg, H. F. Hinteregger, R. B. Phillips, A. E. E. Rogers, J. C. Webber, A. R. Whitney, G. Elgered, B. O. Rönnäng, B. E. Corey, and J. L. Davis, Geodesy by Radio Interferometry: Evidence for Contemporary Plate Motion, *J. Geophys. Res.*, 1986.

Hinteregger, H. F., I. I. Shapiro, D. S. Robertson, C. A. Knight, R. A. Ergas, A. R. Whitney, A. E. E. Rogers, J. M. Moran, T. A. Clark, and B. F. Burke, Precision Geodesy Via Radio Interferometry, *Science*, **178**, 396–398, 1972.

Johnston, K. J., P. K. Seidelmann, and C. M. Wade, Observations of 1 Ceres and 2 Pallas at Centimeter Wavelengths, *Astron. J.*, **87**, 1593–1599, 1982.

Johnston, K., J. Spencer, C. Mayer, W. Klepczynski, G. Kaplan, D. McCarthy, and G. Westerhout, Requirements for a Radio Coordinate Reference Frame for the Determination of Earth Rotation Parameters, in *Modern Astrometry*, IAU Colloqium No. 48, F. V. Prochazka and R. H. Tucker, Eds., Institute of Astronomy, University Observatory, Vienna, 1978, pp. 171–174.

Kaplan, G. H., F. J. Josties, P. E. Angerhofer, K. J. Johnston, and J. H. Spencer, Precise Radio Source Positions from Interferometric Observations, *Astron. J.*, **87**, 570–576, 1982.

Lambeck, K., *The Earth's Variable Rotation: Geophysical Causes and Consequences*, Cambridge University Press, Cambridge, 1980.

Lampton, M., B. Margon, and S. Bowyer, Parameter Estimation in X-Ray Astronomy, *Astrophys. J.*, **208**, 177–190, 1976.

Lieske, J. H., T. Lederle, W. Fricke, and B. Morando, Expressions for the Precession Quantities Based Upon the IAU (1976) System of Astronomical Constants, *Astron. Astrophys.*, **58**, 1–16, 1977.

McCarthy, D. D. and J. D. H. Pilkington, Eds., *Time and the Earth's Rotation*, IAU Symposium No. 82, D. Reidel, Dordrecht, Holland, 1979. See papers on Radio Interferometry.

Marcaide, J. M. and I. I. Shapiro, High Precision Astrometry via Very-Long-Baseline Radio Interferometry: Estimate of the Angular Separation between the Quasars 1038 + 528A and B, *Astron. J.*, **88**, 1133–1137, 1983.

Markowitz, W. and B. Guinot, Eds., *Continental Drift, Secular Motion of the Pole, and Rotation of the Earth*, IAU Symposium No. 32, D. Reidel, Dordrecht, Holland, 1968, pp. 13–14.

Melchior, P., *The Tides of the Planet Earth*, Pergamon Press, Oxford, 1978.

Moran, J. M., Maser Action in Nature, in *CRC Handbook of Laser Science and Technology*, Vol. 1, *Lasers and Masers*, M. J. Weber, Ed., CRC Press, Boca Raton, Florida, 1982, pp. 483–506.

Moran, J. M., B. F. Burke, A. H. Barrett, A. E. E. Rogers, J. A. Ball, J. C. Carter, and D. D. Cudaback, The Structure of the OH Source in W3. *Astrophys. J. (Lett.)*, **152**, L97–L101, 1968.

Morrison, N., *Introduction to Sequential Smoothing and Prediction*, McGraw-Hill, New York, 1969, p. 645.

Munk, W. H. and G. J. F. MacDonald, *The Rotation of the Earth*, Cambridge University Press, Cambridge, 1960.

Norris, R. P., and R. S. Booth, Observations of OH Masers in W3OH, *Mon. Not. R. Astron. Soc.*, **195**, 213–226, 1981.

Norris, R. P., R. S. Booth, and P. J. Diamond, MERLIN Spectral Line Observations of W3OH, *Mon. Not. R. Astron. Soc.*, **201**, 209–222, 1982.

Perley, R. A., The Positions, Structures, and Polarizations of 404 Compact Radio Sources, *Astron. J.*, **87**, 859–880, 1982.

Petley, B. W., New Definition of the Metre, *Nature*, **303**, 373–376, 1983.

Reid, M. J., A. D. Haschick, B. F. Burke, J. M. Moran, K. J. Johnston, and G. W. Swenson, Jr., The Structure of Interstellar Hydroxyl Masers: VLBI Synthesis Observations of W3(OH), *Astrophys. J.*, **239**, 89–111, 1980.

Reid, M. J. and J. M. Moran, Masers, *Ann. Rev. Astron. Astrophys.*, **19**, 231–276, 1981.

Robertson, D. S., W. E. Carter, R. J. Eanes, B. E. Schutz, B. D. Tapley, R. W. King, R. B. Langley, P. J. Morgan, and I. I. Shapiro, Comparison of Earth Rotation as Inferred from Radio Interferometric, Laser Ranging, and Astrometric Observations, *Nature*, **302**, 509–511, 1983.

Ryle, M. and B. Elsmore, Astrometry with the 5-km Telescope, *Mon. Not. R. Astron. Soc.*, **164**, 223–242, 1973.

Shapiro, I. I., Estimation of Astrometric and Geodetic Parameters, in *Methods of Experimental Physics*, Vol. 12C, M. L. Meeks, Ed., Academic Press, New York, 1976.

Shapiro, I. I., D. S. Robertson, C. A. Knight, C. C. Counselman III, A. E. E. Rogers, H. F. Hinteregger, S. Lippincott, A. R. Whitney, T. A. Clark, A. E. Niell, and D. J. Spitzmesser, Transcontinental Baselines and the Rotation of the Earth Measured by Radio Interferometry, *Science*, **186**, 920–922, 1974.

Shapiro, I. I., J. J. Wittels, C. C. Counselman III, D. S. Robertson, A. R. Whitney, H. F. Hinteregger, C. A. Knight, A. E. E. Rogers, T. A. Clark, L. K. Hutton, and A. E. Niell, Submilliarcsecond Astrometry via VLBI. I. Relative Position of the Radio Sources 3C345 and NRAO512, *Astron. J.*, **84**, 1459–1469, 1979.

Smith, F. G., The Determination of the Position of a Radio Star, *Mon. Not. R. Astron. Soc.*, **112**, 497–513, 1952.

Smith, H. M., International Time and Frequency Coordination, *Proc. IEEE*, **60**, 479–487, 1972.

Taff, L. G., *Computational Spherical Astronomy*, Wiley, New York, 1981.

Taylor, J. H., C. R. Gwinn, J. M. Weisberg and L. A. Rawley, Pulsar Astrometry, in *VLBI and Compact Radio Sources*, IAU Symposium 110, R. Fanti, K. Kellermann, and G. Setti, Eds., D. Reidel, Dordrecht, Holland, 1984, pp. 347–353.

Wade, C. M., Precise Positions of Radio Sources. I. Radio Measurements, *Astrophys. J.*, **162**, 381–390, 1970.

Wahr, J. M., The Forced Nutations of an Elliptical, Rotating, Elastic and Oceanless Earth, *Geophys. J. R. Astron. Soc.*, **64**, 705–727, 1981.

Walker, R. C., The Multiple-Point Fringe-Rate Method of Mapping Spectral-Line VLBI Sources with Application to H_2O Masers in W3-IRS5 and W3(OH), *Astron. J.*, **86**, 1323–1331, 1981.

Walker, R. C., D. N. Matsakis, and J. A. Garcia-Barreto, H_2O Masers in W49N. I. Maps. *Astrophys. J.*, **255**, 128–142, 1982.

Woolard, E. W. and G. M. Clemence, *Spherical Astronomy*, Academic Press, New York, 1966.

13

PROPAGATION EFFECTS

The neutral and ionized media lying between a radio source and the surface of the earth often have profound effects on the radiation fields traversing them. The most important of these media are the neutral lower atmosphere or troposphere, the ionosphere, the ionized interplanetary medium, and the ionized interstellar medium. We are concerned with three types of effects of these media. First, the large-scale structures in the media give rise to refractive effects. These effects, which can be analyzed in terms of geometrical optics and Fermat's principle, are the deflection of the radio waves, the change of the propagation velocity, and the rotation of the plane of polarization. Second, radiation can be absorbed. Finally, radiation can be scattered by the turbulent structure in the media. The phenomenon of scattering results in scintillation, or seeing.

In the troposphere, water vapor plays a particularly important role in radio propagation. The refractivity of water vapor is about 20 times greater in the radio range than in the near-infrared or optical regimes. The phase fluctuations in radio interferometers at centimeter and millimeter wavelengths are predominantly caused by fluctuations in the distribution of water vapor. Water vapor is poorly mixed in the atmosphere, and the total column density of water vapor cannot be accurately sensed from surface meteorological measurements. Uncertainties in the water vapor content are a fundamental limitation to the accuracy of VLBI measurements. Small-scale (< 1 km) fluctuations in water vapor distribution limit the angular resolution of connected-element interferometers. Furthermore, spectral lines of water vapor cause substantial absorption at frequencies above 100 GHz and make the troposphere completely opaque at frequencies between 1 and 10 THz (300 and 30 μm). Thus, any discussion of the neutral atmosphere must be primarily concerned with the effects of water vapor. Propagation in the neutral atmosphere from the point of view of radio communications is discussed by Crane (1981) and Bohlander, McMillan, and Gallagher (1985).

Above the neutral atmosphere, radiation encounters three morphologically distinct plasmas: the ionosphere, the interplanetary medium, and the interstellar medium. Most plasma effects that concern us scale as ν^{-2}. Therefore, detrimental effects can be mitigated by carrying out investigations at the highest frequency possible. However, the effects of the ionosphere can easily be detected in VLBI measurements at frequencies up to at least 10 GHz. Furthermore, because of astrophysical requirements, many observations must be made at frequencies where plasma effects cause problems.

Our interest in the propagation media arises because the media degrade interferometric measurements of radio sources. Alternatively, observations of radio sources can be used to probe the characteristics of the propagation media. Radio interferometric measurements have been used widely for this purpose.

13.1 NEUTRAL ATMOSPHERE

In the lowest part of the atmosphere the temperature decreases monotonically from the surface at a rate of about 6.5 K km^{-1}, except for an occasional low-level inversion, until it reaches about 218 K at an altitude of approximately 11 km. This lowermost layer is called the troposphere. Above 11 km the temperature is constant for a distance of about 10 km in the region called the tropopause. Above the tropopause the temperature begins to rise with altitude in the stratosphere. Within the neutral atmosphere, the propagation of radio waves is most affected by the troposphere. Before discussing the refraction, absorption, and scattering of radio waves in the troposphere in detail, we introduce some basic physical concepts.

Basic Physics

Consider a plane wave propagating along the z direction in a uniform dissipative dielectric medium, as represented by the equation

$$\mathbf{E}(z, t) = \mathbf{E}_0 e^{j(k\hat{n}z - 2\pi\nu t)}, \tag{13.1}$$

where k is the propagation constant in free space and is equal to $2\pi\nu/c$, c is the speed of light, and $\mathbf{E}_0$ is the electric field amplitude. $\hat{n}$ is the complex index of refraction, equal to $n_R + jn_I$. If the imaginary part of the index of refraction is positive, the wave will decay exponentially. The power absorption coefficient is defined as

$$\alpha = \frac{4\pi\nu}{c} n_I. \tag{13.2}$$

The propagation constant in the atmosphere is k multiplied by the real part of the index of refraction, which can be written

$$kn_R = \frac{2\pi n \nu}{c} = \frac{2\pi \nu}{v_p}, \qquad (13.3)$$

where $n = n_R$ is the index of refraction when absorption is neglected, and v_p is the phase velocity. The phase velocity of the wave, c/n, is less than c by about 0.03% in the lower atmosphere. The extra time required to traverse a medium with index of refraction $n(z)$ compared with the time necessary to traverse the same distance in free space is

$$\Delta t = \frac{1}{c} \int (n - 1)\, dz, \qquad (13.4)$$

where we assume that the effect of the difference in physical length between the actual ray path and the straight line path is negligible. The *excess* path length is defined as $c\,\Delta t$ or

$$\mathscr{L} = 10^{-6} \int N(z)\, dz, \qquad (13.5)$$

where we have introduced the refractivity N, defined by $N = 10^6(n - 1)$. Note that the concept of *excess* path length, which is used extensively in this chapter, does not represent an actual physical path.

The refractivity of moist air in the radio range is given by the empirical formula (see discussion in this section under *Smith–Weintraub Equation*)

$$N = 77.6\frac{p_D}{T} + 64.8\frac{p_V}{T} + 3.776 \times 10^5 \frac{p_V}{T^2}, \qquad (13.6)$$

where T is the temperature in kelvins, p_D is the partial pressure of the dry air, and p_V is the partial pressure of water vapor in millibars (1 mb = 100 newtons per square meter = 100 pascals; 1 atmosphere = 1013 mb). The first two terms on the right-hand side of Eq. (13.6) arise from the displacement polarizations of the gaseous constituents of the air. The third term is due to the permanent dipole moment of water vapor. Equation (13.6) is accurate to better than 1% for frequencies below 100 GHz. The contributions of dispersive components of refractivity associated with resonances are very small (see discussion in this section under *Origin of Refraction*).

The refractivity can be expressed in terms of gas density, using the ideal gas law,

$$p = \frac{\rho R T}{M}, \qquad (13.7)$$

where p and ρ are the partial pressure and density of any constituent gas, R is the universal gas constant equal to 8.314 J mol^{-1} K^{-1}, and $\mathcal{M}$ is the molecular weight, which for dry air in the troposphere is $\mathcal{M}_D = 28.96$ g mol^{-1} and for water vapor is $\mathcal{M}_V = 18.02$ g mol^{-1}. Thus, $p_D = \rho_D RT/\mathcal{M}_D$ and $p_V = \rho_V RT/\mathcal{M}_V$, where ρ_D and ρ_V are the densities of dry air and water vapor, respectively. Since the total pressure P is the sum of the partial pressures, and the total density ρ_T is the sum of the constituent densities, Eq. (13.7) can be written $P = \rho_T RT/\mathcal{M}_T$, where

$$\mathcal{M}_T = \left(\frac{1}{\mathcal{M}_D} \frac{\rho_D}{\rho_T} + \frac{1}{\mathcal{M}_V} \frac{\rho_V}{\rho_T} \right)^{-1}. \tag{13.8}$$

Substitution of the appropriate forms of Eq. (13.7) and the equation $\rho_D = \rho_T - \rho_V$ into Eq. (13.6) yields

$$N = 0.2228\rho_T + 0.076\rho_V + 1742\frac{\rho_V}{T}, \tag{13.9}$$

where ρ_T and ρ_V are in g m^{-3}. Since the second term on the right-hand side of Eq. (13.9) is small with respect to the third term, it can be combined with the third term, to give, for $T = 280$ K,

$$N \simeq 0.2228\rho_T + 1763\frac{\rho_V}{T} = N_D + N_V. \tag{13.10}$$

Equation (13.10) defines the dry and wet refractivities, N_D and N_V, respectively. These definitions are not universally followed in the literature. Note that N_D is proportional to the total density and therefore has a contribution due to the induced dipole moment of water vapor. Graphs of mean values of N_V around the world are shown in Fig. 13.1.

The atmosphere obeys the equation of hydrostatic equilibrium to a high degree of accuracy (Humphreys 1940). A parcel of gas in static equilibrium between pressure and gravity obeys the equation

$$\frac{dP}{dh} = -\rho_T g, \tag{13.11}$$

where g is the acceleration due to gravity, approximately equal to 980 cm sec^{-2}, and h is the height above the earth's surface. Using the ideal gas law, we can integrate Eq. (13.11) assuming a specific form for the temperature profile and mixing ratio. If an isothermal atmosphere with constant mixing ratio is assumed, then ρ_T is an exponential function with a scale height of about 8.5 km, which is close to the observed scale height. Other models are described by Hess (1959). The excess path length caused by the dry component of refractivity does not depend on the height distribution of total density or temperature,

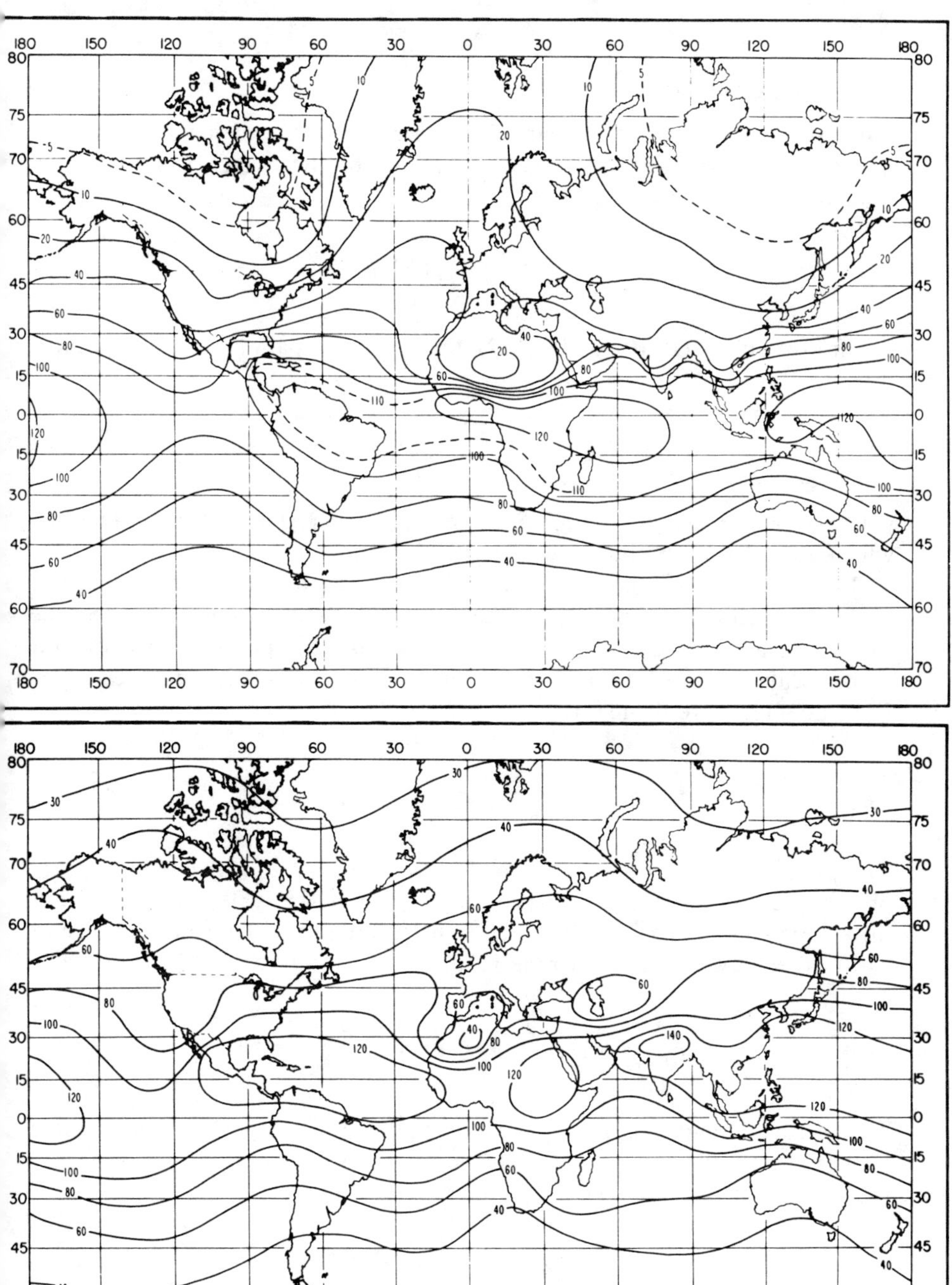

Figure 13.1 (Top) Worldwide distribution of mean sea level value of the wet refractivity N_V for February. (Bottom) N_V for August. From Bean et al. (1966). Note the seasonal variation in mean water vapor content.

but only on the surface pressure P_0, under conditions of hydrostatic equilibrium. If g is assumed to be constant with height, the surface pressure can be obtained by integrating Eq. (13.11),

$$P_0 = g \int_0^\infty \rho_T(h) \, dh. \tag{13.12}$$

From Eqs. (13.5), (13.10), and (13.12), the dry excess path length in the zenith direction is

$$\mathscr{L}_D = 10^{-6} \int_0^\infty N_D \, dh = AP_0, \tag{13.13}$$

where $A = 77.6 \, R/g\mathscr{M}_D = 0.228$ cm mb^{-1}. Under standard conditions for which $P_0 = 1013$ mb, the value of $\mathscr{L}_D$ is 231 cm.

The water vapor is not well mixed in the atmosphere and therefore is not well correlated with ground-based meteorological parameters (Reber and Swope 1972). On average, the water vapor density has an exponential distribution with a scale height of 2 km. The partial pressure and density of water vapor from Eq. (13.7) are related by

$$\rho_V = \frac{217 p_V}{T} \quad (\text{g m}^{-3}). \tag{13.14}$$

The partial pressure of water vapor for saturated air p_{VS} at temperature T, obtained from the Clausius–Clapeyron equation (Hess 1959), can be approximated to an accuracy of better than 1% within the temperature range of 240–310 K by the formula (Crane 1976)

$$p_{VS} = 6.11 \left(\frac{T}{273} \right)^{-5.3} e^{25.2(T-273)/T} \quad (\text{mb}). \tag{13.15}$$

The relative humidity is p_V/p_{VS}. The component of the path length resulting primarily from the permanent dipole moment of water vapor is proportional to the integral of N_V in Eq. (13.10), which is

$$\mathscr{L}_V = 1763 \times 10^{-6} \int_0^\infty \frac{\rho_V(h)}{T(h)} \, dh, \tag{13.16}$$

where the units of $\mathscr{L}_V$ are the same as those of h. If we assume that the atmosphere is isothermal and that ρ_V decreases exponentially with a scale height of 2 km, then, from Eqs. (13.14) and (13.16),

$$\mathscr{L}_V = 7.6 \times 10^4 \frac{p_{V0}}{T^2} \quad (\text{cm}), \tag{13.17}$$

where p_{V0} is the partial pressure of water vapor at the surface of the earth. Hence if $T = 273$ K, $\mathscr{L}_V$ in centimeters is approximately equal to p_{V0} in millibars.

The integrated water vapor density or the height of the column of water condensed from the atmosphere is given by

$$w = \frac{1}{\rho_w} \int_0^\infty \rho_V(h)\, dh, \tag{13.18}$$

where ρ_w is the density of water, 10^6 g m^{-3}. In an isothermal atmosphere at 280 K,

$$\mathscr{L}_V \simeq 6.3w. \tag{13.19}$$

The values of $\mathscr{L}_V$ under extreme conditions for a temperate, sea level site can be calculated. With $T = 303$ K (30°C), and relative humidity = 0.8, we have $p_{V0} = 34$ mb, $\rho_{V0} = 24$ g m^{-3}, $w = 4.9$ cm, and $\mathscr{L}_V = 28$ cm. With $T = 258$ K (-15°C), and relative humidity = 0.5, we have $p_{V0} = 1.0$ mb, $\rho_{V0} = 0.8$ g m^{-3}, $w = 0.15$ cm, and $\mathscr{L}_V = 1.1$ cm. The total zenith excess path length through the atmosphere is $\mathscr{L} \simeq \mathscr{L}_D + \mathscr{L}_V$, which from Eqs. (13.13) and (13.19) is

$$\mathscr{L} \simeq 0.228 P_0 + 6.3w \quad \text{(cm)}, \tag{13.20}$$

where P_0 is in millibars, and w is in centimeters. Equation (13.20) is reasonably accurate for estimation purposes because the fractional variation in the temperature of the lower atmosphere, and in the scale height of water vapor, is usually less than 10%. However, it is usually not accurate enough to predict the path length to a small fraction of a wavelength.

Refraction and Propagation Delay

If the vertical distributions of temperature and water vapor pressure are known, then precise estimates of the angle of arrival and excess propagation time for a ray impinging on the atmosphere at an arbitrary angle can be computed by ray tracing. Here we consider a few elementary cases in order to derive some simple analytic expressions. The simplest case is that of an interferometer in a uniform or plane-parallel atmosphere, as shown in Fig. 13.2. The refraction of the ray is governed by Snell's law, which is

$$n_0 \sin z_0 = \sin z, \tag{13.21}$$

where z is the zenith angle at the top of the atmosphere (where $n = 1$), and z_0 is the zenith angle at the surface (where $n = n_0$). The geometrical delay for an

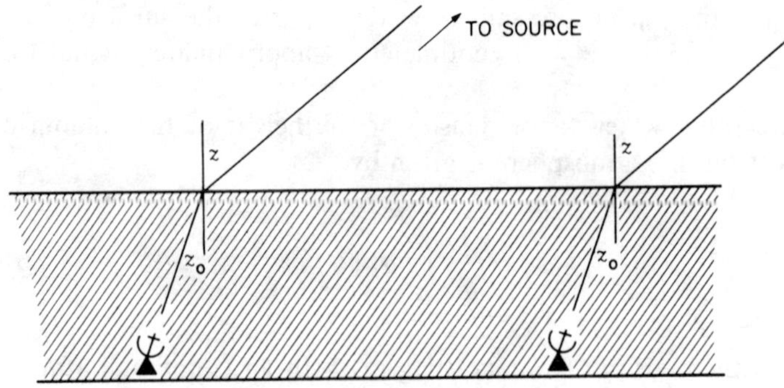

Figure 13.2   Two-element interferometer with the atmosphere modeled as a uniform flat slab. The geometrical delay is the same as it would be if the interferometer were in free space.

interferometer, as defined in Chapter 2, is

$$\tau_g = \frac{n_0 D}{c} \sin z_0 = \frac{D}{c} \sin z. \tag{13.22}$$

τ_g can be calculated from the angle of arrival z_0 and the speed of light at the earth's surface c/n_0, or from z and the speed of light in free space. Thus, if earth curvature is neglected and the atmosphere is uniform, then the resulting geometrical delay is the same as the free-space value. The angle of refraction need only be calculated to ensure that the antennas track the source properly. The angle of refraction, $\Delta z = z - z_0$, can be written, using Eq. (13.21),

$$\Delta z = z - \sin^{-1}\left(\frac{1}{n_0} \sin z\right). \tag{13.23}$$

This equation can be expanded in a Taylor series in $n_0 - 1$, which to first order gives

$$\Delta z \simeq (n_0 - 1)\tan z. \tag{13.24}$$

Since $n_0 - 1 \simeq 3 \times 10^{-4}$ at the surface of the earth, Eq. (13.24) can be written

$$\Delta z \text{ (arcmin)} \simeq \tan z. \tag{13.25}$$

The angle of refraction can also be calculated for more realistic cases. Ignore the curvature of the earth and consider the atmosphere to consist of a large number of plane-parallel layers numbered 0 through m, as shown in Fig. 13.3. Let the index of refraction at the surface be n_0 and at the top layer be $n_m = 1$. Snell's law applied to the various layers gives the following set of

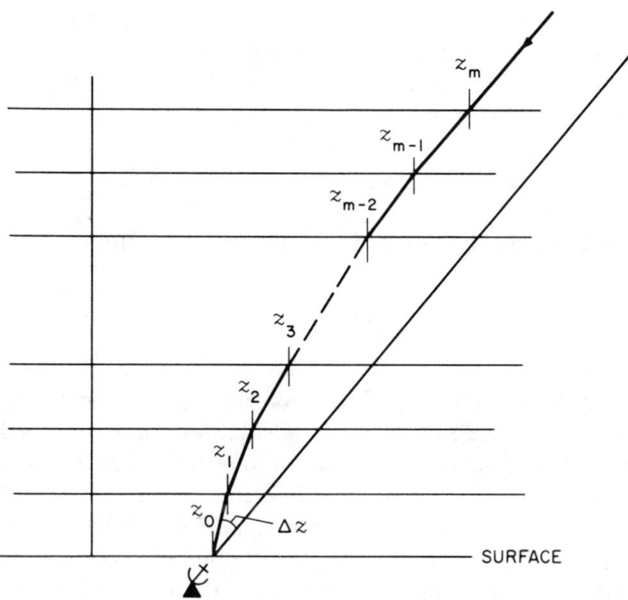

Figure 13.3   The atmosphere modeled as a set of thin uniform slabs. The angle of incidence on the topmost slab is z_m, which is equal to the free-space zenith angle z, and the angle of incidence at the surface is z_0. The total bending is $\Delta z = z - z_0$.

equations:

$$n_0 \sin z_0 = n_1 \sin z_1$$

$$n_1 \sin z_1 = n_2 \sin z_2$$

$$\vdots \qquad\qquad \vdots$$

$$n_{m-1} \sin z_{m-1} = \sin z, \tag{13.26}$$

where $z = z_m$. From the above equations, we see that $n_0 \sin z_0 = \sin z$. This result is identical to that for the homogeneous case. Thus, regardless of the vertical distribution of the index of refraction, the angle of refraction is given by Eq. (13.21), where n_0 is the surface value of the index of refraction. This result can also be obtained by an elementary application of Fermat's principle. An interesting application of this result is that if $n_0 = 1$, as would be the case if the measuring device were in a vacuum chamber at the surface of the earth, then there would be no net refraction, that is, $z_0 = z$.

For an atmosphere consisting of spherical layers, the angle of refraction is given by the formula (Smart 1962)

$$\Delta z = r_0 n_0 \sin z_0 \int_1^{n_0} \frac{dn}{n\sqrt{r^2 n^2 - r_0^2 n_0^2 \sin^2 z_0}}, \tag{13.27}$$

where r is the distance from the center of the earth to the layer where the index of refraction is n and r_0 is the radius of the earth. This result is derivable from Snell's law in spherical coordinates: $nr \sin z = $ constant (Smart 1962). For small zenith angles, expansion of Eq. (13.27) gives

$$\Delta z \simeq (n_0 - 1)\tan z_0 - a_2 \tan z_0 \sec^2 z_0, \tag{13.28}$$

where a_2 is a constant. Equation (13.28) can also be written

$$\Delta z \simeq a_1 \tan z_0 - a_2 \tan^3 z_0, \tag{13.29}$$

where $a_1 \simeq 56$ arcsec and $a_2 \simeq 0.07$ arcsec for a dry atmosphere under standard conditions (COESA 1976). The refraction at the horizon is about $0.46°$ (see Fig. 13.5). See Saastamoinen (1972a) for a more detailed treatment.

The differential delay induced in an interferometer by a horizontally stratified troposphere results from the difference in zenith angle of the source at the antennas. Consider two closely spaced antennas. If the excess path in the zenith direction is $\mathscr{L}_0$, then the excess path in other directions is approximately $\mathscr{L}_0 \sec z$. This approximation becomes inaccurate at large zenith angles. The difference in excess paths $\Delta\mathscr{L}$ is

$$\Delta\mathscr{L} = \mathscr{L}_0 \Delta z \frac{\sin z}{\cos^2 z}, \tag{13.30}$$

where Δz is the difference in zenith angles at the two antennas.

If the antennas are on the equator and the source has a declination of zero, then Δz is equal to the difference in longitudes or D/r_0, where D is the separation between antennas. For this case,

$$\Delta\mathscr{L} = \frac{\mathscr{L}_0 D}{r_0} \frac{\sin z}{\cos^2 z}. \tag{13.31}$$

If $D = 10$ km, $\mathscr{L}_0 = 230$ cm, $r_0 = 6370$ km, and $z = 80°$, then $\Delta\mathscr{L}$ is 12 cm. The calculation of the difference in excess paths can be easily generalized as follows. Let $\mathbf{r}_1$ and $\mathbf{r}_2$ be vectors from the center of the earth to each antenna. The geometrical delay is $(\mathbf{r}_1 \cdot \mathbf{s} - \mathbf{r}_2 \cdot \mathbf{s})/c$, where $\mathbf{s}$ is the unit vector in the direction of the source. Since $\cos z_1 = (\mathbf{r}_1 \cdot \mathbf{s})/r_0$ and $\cos z_2 = (\mathbf{r}_2 \cdot \mathbf{s})/r_0$, where z_1 and z_2 are the zenith angles at the two antennas, the geometrical delay can be written

$$\tau_g = \frac{r_0}{c}(\cos z_1 - \cos z_2) \simeq \frac{r_0}{c}\Delta z \sin z. \tag{13.32}$$

Substitution of Δz from Eq. (13.32) into Eq. (13.30) yields an expression for the difference in excess path lengths, valid for short-baseline interferometers

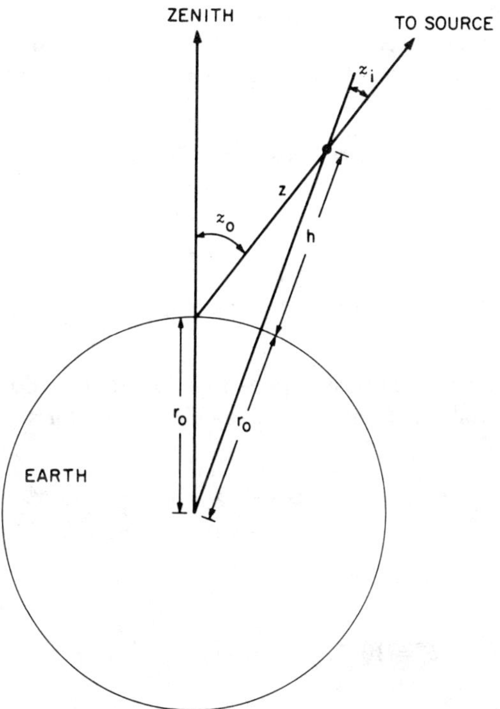

Figure 13.4   Geometry for calculating the propagation delay, taking into account the sphericity of the earth. The ray path along the z coordinate is assumed to be straight. The angle z_i is the zenith angle of the ray at height h. This angle is needed in the calculation of the excess path length through the ionosphere [Eqs. (13.129) and (13.130)].

and moderate values of zenith angle:

$$\Delta \mathcal{L} = \frac{c T_g \mathcal{L}_0}{r_0} \sec^2 z. \tag{13.33}$$

For very-long-baseline interferometers, the expression in Eq. (13.30) is not appropriate. The difference in excess path lengths is approximately $\Delta \mathcal{L} = \mathcal{L}_1 \sec z_1 - \mathcal{L}_2 \sec z_2$, where $\mathcal{L}_1$, $\mathcal{L}_2$, z_1, and z_2 are the excess zenith path lengths and the zenith angles at the two antennas. We now derive a more accurate expression for the excess path length to each antenna. Assume the index of refraction is exponentially distributed with a scale height h_0. The geometry is shown in Fig. 13.4. The excess path length is

$$\mathcal{L} = 10^{-6} N_0 \int_0^\infty \exp\left(-\frac{h}{h_0}\right) dz, \tag{13.34}$$

where N_0 is the refractivity at the earth's surface, h is the height above the surface, and dz is the differential length along the ray path. Bending of the ray is neglected. From the geometry of Fig. 13.4, one can show that

$$h \simeq z \cos z + \frac{z^2}{2r_0} \sin^2 z. \tag{13.35}$$

Therefore,

$$\mathscr{L} \simeq 10^{-6} N_0 \int_0^\infty \exp\left(-\frac{z}{h_0} \cos z \right) \exp\left(-\frac{z^2}{2r_0 h_0} \sin^2 z \right) dz. \tag{13.36}$$

The argument of the rightmost exponential function in Eq. (13.36) is small, and this exponential function can be expanded in Taylor series so that

$$\mathscr{L} \simeq 10^{-6} N_0 \int_0^\infty \exp\left(-\frac{z}{h_0} \cos z \right) \times \left(1 - \frac{z^2}{2r_0 h_0} \sin^2 z \cdots \right) dz. \tag{13.37}$$

Integration of Eq. (13.37) yields

$$\mathscr{L} \simeq 10^{-6} N_0 h_0 \sec z \left(1 - \frac{h_0}{r_0} \tan^2 z \right). \tag{13.38}$$

Equation (13.38) can also be written

$$\mathscr{L} \simeq 10^{-6} N_0 h_0 \left[\left(1 + \frac{h_0}{r_0} \right) \sec z - \frac{h_0}{r_0} \sec^3 z \right]. \tag{13.39}$$

Thus $\mathscr{L}$ is a function of odd powers of $\sec z$, whereas the bending angle, given in Eq. (13.29), is a function of odd powers of $\tan z$. Equations (13.38) and (13.39) both diverge as z approaches $90°$. For $z = 90°$, Eq. (13.35) shows that $h \simeq z^2/2r_0$. Hence, from Eq. (13.34), the excess path at the horizon is

$$\mathscr{L} \simeq 10^{-6} N_0 \sqrt{\frac{\pi r_0 h_0}{2}} \simeq 70 \mathscr{L}_0 \simeq 14 N_0 \quad \text{(cm)}, \tag{13.40}$$

for $r_0 = 6370$ km and $h_0 = 2$ km. A model incorporating both the dry atmosphere with a scale height $h_D = 8$ km and the wet atmosphere with a scale height of $h_V = 2$ km can be obtained by applying Eq. (13.38) to both the dry and wet component using Eqs. (13.13) and (13.17). This result is

$$\mathscr{L} = 0.228 P_0 \sec z \, (1 - 0.0013 \tan^2 z) + \frac{7.5 \times 10^4 p_{V0} \sec z}{T^2} (1 - 0.0003 \tan^2 z).$$

$$\tag{13.41}$$

More sophisticated models have been derived by Marini (1972), Saastamoinen

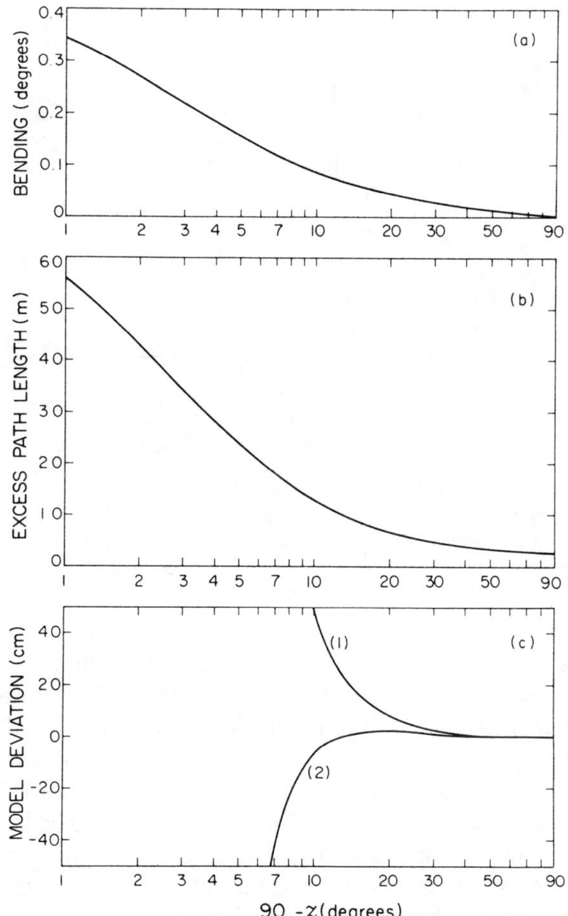

Figure 13.5 (a) The bending angle versus $90° - z$, where z is the zenith angle that the ray would have in the absence of refraction, calculated by a ray-tracing algorithm for a standard dry atmosphere (COESA 1976). (b) The excess path length versus $90° - z$ calculated by a ray-tracing algorithm. The zenith excess path is 2.31 m. (c) Deviation between the excess path length and (1) the $\mathcal{L}_0 \sec z$ model and (2) the model of Eq. (13.41), both for $\rho_{V0} = 0$ and the same zenith excess path as in (b).

(1972b), Davis et al. (1985), and others. A comparison of the approximate formula of Eq. (13.41) and a ray-tracing solution is given in Fig. 13.5.

Absorption

When the sky is clear the principal sources of atmospheric attenuation are the molecular resonances of water vapor, oxygen, and ozone. The resonances of water vapor and oxygen are pressure broadened and cause attenuation far from the resonance frequencies. A plot of the absorption versus frequency is

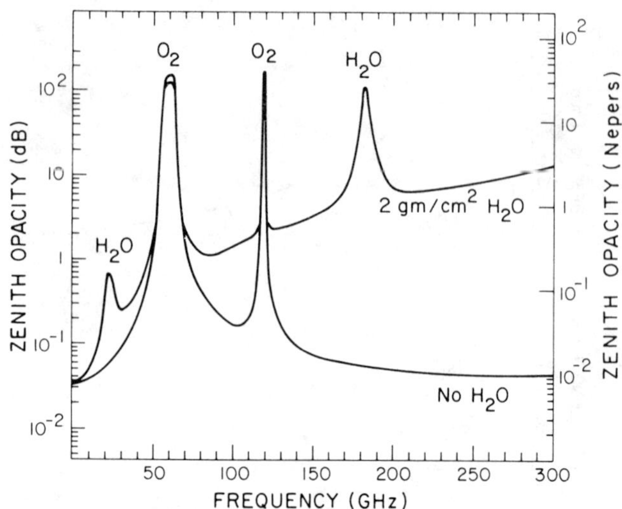

Figure 13.6   Atmospheric zenith opacity. The absorption from narrow ozone lines has been omitted. Adapted from Waters (1976). For zenith opacity at frequencies above 300 GHz, see Liebe (1981). Note that 2 gm/cm^2 of H$_2$O corresponds to $w = 2$ cm.

shown in Fig. 13.6. Below 30 GHz absorption is dominated by the weak 6_{16}–5_{23} transition of H$_2$O at 22.2 GHz (Liebe 1969). Absorption by this line rarely exceeds 20% in the zenith direction. The oxygen lines in the band 50–70 GHz are considerably stronger, and no astronomical observations can be made from the ground in this band. An isolated oxygen line at 118 GHz makes observations impossible in the band 116–120 GHz. At higher frequencies there is a series of strong water vapor lines. Observations in the windows between these lines can be made at dry locations usually found at high altitudes. The physics of atmospheric absorption is discussed in detail by Waters (1976), and a model of absorption at frequencies below 1000 GHz is given by Liebe (1981, 1985). We are concerned here only with the phenomenology of absorption and its calibration.

The absorption coefficient depends on the temperature, gas density, and total pressure. For example, the absorption coefficient for the 22-GHz H$_2$O line can be written (Staelin 1966)

$$\alpha = (3.24 \times 10^{-4} e^{-644/T}) \frac{\nu^2 P \rho_V}{T^{3.125}} \left(1 + 0.0147 \frac{\rho_V T}{P}\right)$$

$$\times \left[\frac{1}{(\nu - 22.235)^2 + \Delta\nu^2} + \frac{1}{(\nu + 22.235)^2 + \Delta\nu^2}\right]$$

$$+ 2.55 \times 10^{-8} \rho_V \nu^2 \frac{\Delta\nu}{T^{3/2}} \quad (\text{cm}^{-1}), \tag{13.42}$$

where $\Delta\nu$ is approximately the half width at half-maximum of the line in gigahertz, given by the equation

$$\Delta\nu = 2.58 \times 10^{-3}\left(1 + 0.0147\frac{\rho_V T}{P}\right)\frac{P}{(T/318)^{0.625}}, \qquad (13.43)$$

ν is the frequency in gigahertz, T is the temperature in kelvins, P is the total pressure in millibars, and ρ_V is the water vapor density in grams per cubic meter. The lineshape specified by Eq. (13.42) is called the Van Vleck–Weisskopf profile. A complete discussion of the absorption coefficient due to water vapor can be found in Waters (1976).

The intensity of a ray passing through an absorbing medium obeys the radiative transfer equation. We assume that the medium is in local thermodynamic equilibrium at temperature T and that scattering is negligible. In the domain where the Rayleigh–Jeans approximation to the Planck function is valid, so that the intensity is proportional to the brightness temperature, the equation of radiative transfer can be written (Rybicki and Lightman 1979)

$$\frac{dT_B}{dz} = -\alpha(T_B - T), \qquad (13.44)$$

where T_B is the brightness temperature and α is the absorption coefficient defined in Eqs. (13.2) and (13.42). The solution to Eq. (13.44) for radiation propagating along the z axis is

$$T_B(\nu) = T_{B0}(\nu)e^{-\tau_\nu} + \int_0^\infty \alpha(\nu, z)T(z)e^{-\tau_\nu'}\,dz, \qquad (13.45)$$

where T_{B0} is the brightness temperature in the absence of absorption, including the cosmic background component,

$$\tau_\nu' = \int_0^z \alpha(\nu, z')\,dz', \qquad (13.46)$$

and

$$\tau_\nu = \int_0^\infty \alpha(\nu, z')\,dz'. \qquad (13.47)$$

Here z is the distance measured from the observer. τ_ν is called the *optical depth* or *opacity*. The first term on the right-hand side of Eq. (13.45) describes the absorption of the signal, and the second describes the emission contribution of the atmosphere. Equation (13.45) illustrates the fundamental law that an absorbing medium must also radiate. If $T(z)$ is constant throughout the medium, then Eq. (13.45) can be written

$$T_B(\nu) = T_{B0}(\nu)e^{-\tau_\nu} + T(1 - e^{-\tau_\nu}). \tag{13.48}$$

The presence of absorption can have a very significant effect on system performance. If the receiver temperature is T_R, then the system temperature, which is the sum of T_R and the atmospheric brightness temperature (the effects of ground radiation being neglected) is

$$T_S = T_R + T(1 - e^{-\tau_\nu}). \tag{13.49}$$

Furthermore, if the brightness temperature scale is referenced to a point outside the atmosphere by multiplying the measurements of brightness temperature [see Eq. (13.48)] by e^{τ_ν}, then the effective system temperature is $T_S e^{\tau_\nu}$ or

$$T_S' = T_R e^{\tau_\nu} + T(e^{\tau_\nu} - 1). \tag{13.50}$$

In effect, the atmospheric loss is modeled by an equivalent attenuator at the receiver input. Suppose that $T_R = 30$ K, $T = 290$ K, and $\tau_\nu = 0.2$; then the effective system temperature is 100 K. In such a situation the atmosphere would degrade the system sensitivity by more than a factor of 3. Note that the loss in sensitivity results primarily from the increase in system temperature rather than the attenuation of the signal, which is only 20%. The emission from the atmosphere induces signals in spaced antennas that are uncorrelated and thus contributes only to the noise in the output of an interferometer.

The absorption can be estimated directly from measurements made with a radio telescope. One technique is called the tipping-scan method, in which the opacity is determined from the atmospheric emission. If the antenna is scanned from the zenith to the horizon, the observed brightness temperature, in the absence of background sources, will depend on the zenith angle since the opacity is proportional to the path length through the atmosphere, which varies approximately as sec z. Thus, the atmospheric brightness temperature is

$$T_B = T(1 - e^{-\tau_0 \sec z}), \tag{13.51}$$

where τ_0 is the zenith opacity. τ_0 is the negative of the slope of the curve of $\ln(T - T_B)$ plotted versus sec z since

$$\ln\left(1 - \frac{T_B}{T}\right) = -\tau_0 \sec z. \tag{13.52}$$

The opacity can also be estimated from measurements of the absorption suffered by a radio source over a range of zenith angles. The observed brightness temperature on-source minus the brightness temperature off-source at the same zenith angle is

$$\Delta T_B = T_{B0}e^{-\tau_0 \sec z}. \tag{13.53}$$

Since

$$\ln \Delta T_B - \ln T_{B0} = -\tau_0 \sec z, \tag{13.54}$$

τ_0 can be found without knowledge of T_{B0} if a sufficient range in $\sec z$ is covered.

Another technique, called the chopper-wheel method, is commonly used at millimeter wavelengths. A wheel consisting of alternate open and absorbing sections is placed in front of the feed horn. As the wheel rotates the radiometer alternately views the sky and the absorbing sections, and synchronously measures the difference in temperature between the sky at temperature T and the chopper wheel at temperature T_0. Thus, the on-source and off-source temperatures are

$$\Delta T_{\text{on}} = T_{B0}e^{-\tau_\nu} + T(1 - e^{-\tau_\nu}) - T_0 \tag{13.55}$$

and

$$\Delta T_{\text{off}} = T(1 - e^{-\tau_\nu}) - T_0. \tag{13.56}$$

These measurements can be combined to obtain T_{B0} and thereby eliminate the effect of atmospheric absorption. In the case where $T_0 = T$,

$$T_{B0} = \frac{\Delta T_{\text{off}} - \Delta T_{\text{on}}}{\Delta T_{\text{off}}} T_0. \tag{13.57}$$

When sensitivity is critical, the chopper wheel is used only to calibrate the output in the off-source position. $\Delta T_{\text{off}} - \Delta T_{\text{on}}$ in the numerator of Eq. (13.57) is then replaced by $T_{\text{off}} - T_{\text{on}}$. Measurement of T_{B0} provides the flux density of the source, which may be required to estimate the visibility at the origin of the (u, v) plane.

The opacity can be estimated also from surface meteorological measurements. This method is not as accurate as the direct radiometric measurement techniques described above, but has the advantage of not expending observing time. Waters (1976) has analyzed data of absorption versus surface water vapor density for various frequencies by fitting them to an equation of the form $\tau_0 = \alpha_0 + \alpha_1 \rho_V$. The coefficients α_0 and α_1 are listed in Table 13.1.

Origin of Refraction

For practical reasons, we have discussed separately the effects of the propagation delay and the absorption in the neutral atmosphere. However, the delay and the absorption are intimately related because they are derived from the

TABLE 13.1   Empirical Coefficients for Estimating Opacity from Surface Absolute Humidity[a]

ν (GHz)	α_0 (nepers)	α_1 (nepers m^3 g^{-1})
15	0.013	0.0009
22.2	0.026	0.011
35	0.039	0.0030
90	0.039	0.0090

Source: From Waters (1976).

[a] From equation $\tau_0 = \alpha_0 + \alpha_1 \rho_V$ fitted to opacity data derived from radiosonde measurements and measurements of surface absolute humidity ρ_V (g m^{-3}).

real and imaginary parts of the dielectric constant of the gas in the atmosphere. The real and imaginary parts of the dielectric constant are not independent but are related by the Kramers–Kronig relation, which is similar to the Hilbert transform (Van Vleck, Purcell, and Goldstein 1951). We now discuss this relationship from the physical viewpoint of the classical theory of dispersion. From this analysis it will become clear why the atmospherically induced delay is essentially independent of frequency, even in the vicinity of spectral lines that cause significant absorption.

A dilute gas of molecules can be modeled as bound oscillators. In each molecule an electron with mass m and charge $-e$ is harmonically bound to the nucleus, and the electron's motion is characterized by a resonance frequency ν_0 and damping constant $2\pi\Gamma$. The equation of motion with a harmonic driving force $-eE_0 e^{-j2\pi\nu t}$ caused by the electric field of an electromagnetic wave is

$$m\ddot{x} + 2\pi m\Gamma\dot{x} + 4\pi^2 m\nu_0^2 x = -eE_0 e^{-j2\pi\nu t}, \tag{13.58}$$

where x is the displacement of the bound electron, E_0 and ν are the amplitude and frequency of the applied electric field, and the dots denote time derivatives. The steady-state solution has the form $x = x_0 e^{-j2\pi\nu t}$, where

$$x_0 = \frac{eE_0/4\pi^2 m}{\nu^2 - \nu_0^2 + j\nu\Gamma}. \tag{13.59}$$

The magnitude of the dipole moment per unit volume, **P**, is equal to $-\mathcal{N}ex_0$, where $\mathcal{N}$ is the density of gas molecules. The dielectric constant[†] ε is

[†] In this section and in Section 13.2 we use rationalized MKSA units, also known as SI (Système International) units. In this system the constitutive relation between the displacement vector **D**, the electric field vector **E**, and the polarization vector **P** is $\mathbf{D} = \epsilon_0 \mathbf{E} + \mathbf{P} = \epsilon\mathbf{E}$, where ϵ_0 is the permittivity of free space and ϵ is the permittivity of the medium. The dielectric constant ε is ϵ/ϵ_0. A comparison of various systems of units and equations in electricity and magnetism can be found in Jackson (1975).

$1 + \mathbf{P}/(\epsilon_0 \mathbf{E})$, so that

$$\varepsilon = 1 - \frac{\mathcal{N}e^2/4\pi^2 m\epsilon_0}{\nu^2 - \nu_0^2 + j\nu\Gamma}.$$

(13.60)

This classical model predicts neither the resonance frequency nor the absolute amplitude of the oscillation. A full treatment of the problem requires the application of quantum mechanics. The proper quantum mechanical calculation for a system with many resonances yields a result that closely resembles Eq. (13.60) (e.g., Loudon 1973):

$$\varepsilon = 1 - \frac{\mathcal{N}e^2}{4\pi^2 m\epsilon_0} \sum_i \frac{f_i}{\nu^2 - \nu_{0i}^2 + j\nu\Gamma_i},$$

(13.61)

where f_i is the so-called oscillator strength of the ith resonance. The f_i's obey the sum rule, $\Sigma f_i = 1$.

The dielectric constant ($\varepsilon = \varepsilon_R + j\varepsilon_I$) and index of refraction ($n = n_R + jn_I$) are connected by Maxwell's relation,

$$n^2 = \varepsilon.$$

(13.62)

Thus, $\varepsilon_R = n_R^2 - n_I^2$ and $\varepsilon_I = 2n_In_R$. Since for a dilute gas $n_R \simeq 1$ and $n_I \ll 1$, we have $n_R \simeq \sqrt{\varepsilon_R}$ and $n_I \simeq \varepsilon_I/2$. Therefore, for a gas with a single resonance,

$$n_R \simeq 1 - \frac{\mathcal{N}e^2(\nu^2 - \nu_0^2)/8\pi^2 m\epsilon_0}{(\nu^2 - \nu_0^2)^2 + \nu^2\Gamma^2}$$

(13.63)

and

$$n_I \simeq \frac{\mathcal{N}e^2\nu\Gamma/8\pi^2 m\epsilon_0}{(\nu^2 - \nu_0^2)^2 + \nu^2\Gamma^2}.$$

(13.64)

The resonance is usually sharp, that is, $\Gamma \ll \nu_0$, and the expressions for n_R and n_I can be simplified by considering their behavior in the vicinity of the resonance frequency ν_0, in which case

$$\nu^2 - \nu_0^2 = (\nu + \nu_0)(\nu - \nu_0) \simeq 2\nu_0(\nu - \nu_0).$$

(13.65)

Thus,

$$n_R \simeq 1 - \frac{2b(\nu - \nu_0)}{(\nu - \nu_0)^2 + \Gamma^2/4}$$

(13.66)

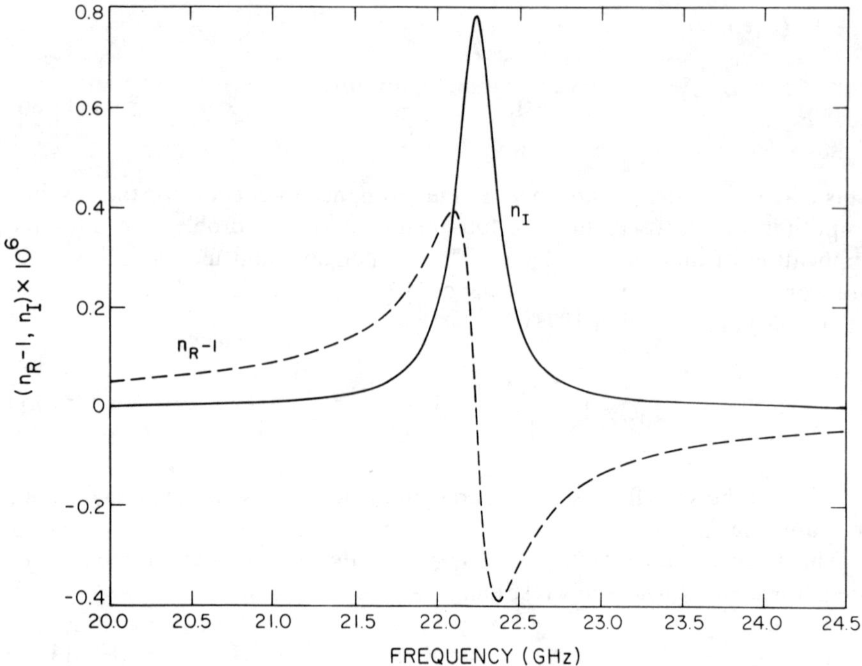

Figure 13.7   Real and imaginary parts of the index of refraction versus frequency for a single resonance given by Eqs. (13.63) and (13.64). The case shown is for the 6_{16}–5_{23} transition in *pure* water vapor with $\rho_V = 7.5$ g m^{-3}. In the atmosphere the line is much broader (Liebe 1969).

and

$$n_{\mathrm{I}} \simeq \frac{b\Gamma}{\left(\nu - \nu_0\right)^2 + \Gamma^2/4},$$

(13.67)

where $b = \mathscr{N}e^2/32\pi^2 m\epsilon_0\nu_0$.

Equation (13.67) defines an unnormalized Lorentzian profile for n_{I} that is symmetric about frequency ν_0 and has a full width at half-maximum of Γ and a peak amplitude of $4b/\Gamma$. The function $n_{\mathrm{R}} - 1$ is antisymmetric about frequency ν_0 and has extreme values of $\pm 2b/\Gamma$ at frequencies $\nu_0 \pm \Gamma/2$, respectively. The functions n_{R} and n_{I} are plotted in Fig. 13.7. Note that the peak deviation from unity in the real part of the index of refraction Δn is equal to one-half the peak value of n_{I}, denoted $n_{\mathrm{I\,max}}$. Thus, from Eq. (13.2), we see that the peak absorption coefficient, $\alpha_m = 4\pi n_{\mathrm{I\,max}}\nu_0/c$, is related to Δn by the formula

$$\Delta n = \frac{\alpha_m \lambda_0}{8\pi},$$

(13.68)

where λ_0 is the wavelength of the resonance, c/ν_0. The magnitude of the dispersive part of the index of refraction is equal to the peak absorption over a distance of $\lambda_0/8\pi$. Therefore, unless the resonance is extremely strong, Δn is negligible. Consider the 22-GHz water vapor line. The attenuation in the atmosphere when $\rho_V = 7.5$ g m^{-3} is 0.15 dB/km so $\alpha_m = 3.5 \times 10^{-7}$ cm^{-1}. Equation (13.68) then predicts that $\Delta n = 1.9 \times 10^{-8}$, which is the value measured in the laboratory (Liebe 1969). For the same value of ρ_V, the contribution of all transitions of water vapor to the value of the index of refraction at low frequencies ($10^{-6} N_V$), from Eq. (13.10), is equal to 4.4×10^{-5}. Thus the fractional change in refractivity near the 22-GHz line is only 1 part in 2000. The water vapor line at 500 GHz has an absorption coefficient of 29,000 dB/km and a change in N_V of 144. In the atmospheric "window" above 550 GHz, where radio astronomical observations are only possible from very dry sites, the refractive index can be noticeably different from the value at lower frequencies.

Equation (13.68) is an important result of very general validity. We derived it from a specific model [Eq. (13.58)] that led to an approximately Lorentzian profile for the absorption spectrum. However, Eq. (13.68) could be derived from the Kramers–Kronig relation. In practice, line profiles are found to differ slightly from the Lorentzian form, and more sophisticated models are needed to fit them exactly.

The low-frequency value of the index of refraction as given by the Smith–Weintraub equation [Eq. (13.9)] results from the contributions of many transitions. From Eq. (13.63) we see that the profile of n_R is not exactly symmetric, and while n_R tends to unity as ν tends to ∞, n_R tends to $1 + 4b/\nu_0 = 1 + 2\Delta n\Gamma/\nu_0 = 1 + (\lambda_0\alpha_m/4\pi)(\Gamma/\nu_0)$ as ν tends to zero. From Eq. (13.61) it is clear that the low-frequency value of the index of refraction, which is due to many resonances, each characterized by parameters Δn_i, Γ_i, and ν_{0i}, is

$$n_S = 1 + 2\sum_i \frac{\Delta n_i\Gamma_i}{\nu_{0i}}. \tag{13.69}$$

Thus, in moving down in frequency through the spectrum, we find that the index of refraction increases a small amount after passing through each resonance frequency, as shown schematically in Fig. 13.8. The water vapor molecule has a large number of strong rotational transitions in the band 30 μm to 1 mm (10 THz to 300 GHz). The atmosphere is opaque through most of this region because of these lines. The refractivity due to water vapor is small in the optical region and is greater by a factor of 22 in the radio region. Therefore, while the effects of water vapor are small in the optical region, they are very important in the radio region. The dry air refractivity, due primarily to resonances of oxygen and nitrogen in the ultraviolet, is nearly the same in the optical and radio regions.

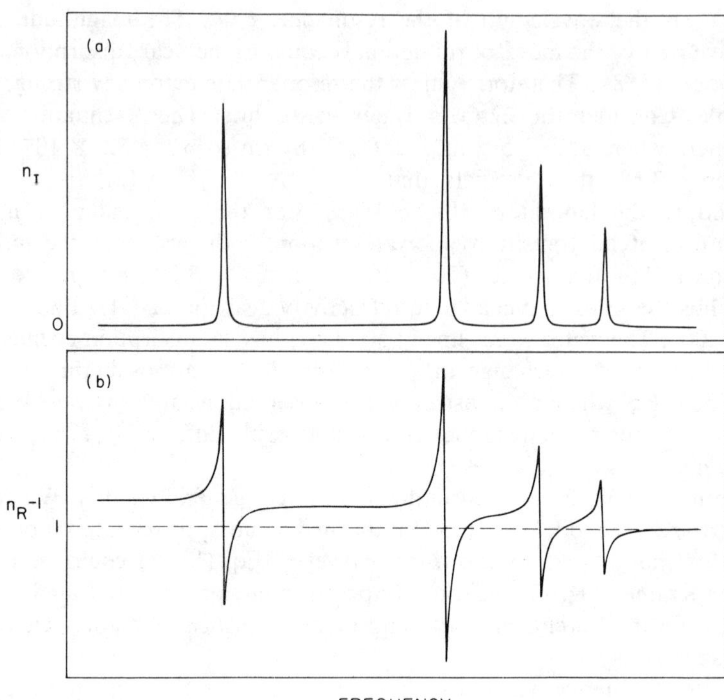

Figure 13.8 (a) Imaginary, and (b) real, parts of the index of refraction versus frequency for a dilute gas with many resonances. The buildup of the index of refraction at low frequencies is greatly exaggerated.

Smith–Weintraub Equation

Detailed discussions of the radio refractivity equation can be found in Bean and Dutton (1966), Thayer (1974), and Hill, Lawrence, and Priestley (1982). From the classic work of Debye (1929), it can be shown that the refractivity of molecules with induced dipole transitions varies as pressure and T^{-1}, and the refractivity of molecules with permanent dipole moments varies as pressure and T^{-2}. The principal constituents of the atmosphere, oxygen molecules O_2 and nitrogen molecules N_2, being homonuclear, have no permanent electric dipole moments. However, molecules such as H_2O and other minor trace constituents have permanent dipole moments. Thus, the general form of the refractivity equation is

$$N = \frac{K_1 p_D}{TZ_D} + \frac{K_2 p_V}{TZ_V} + \frac{K_3 p_V}{T^2 Z_V}, \tag{13.70}$$

where p_D and p_V are the partial pressures of the dry air and water vapor,

K_1, K_2, and K_3 are constants, and Z_D and Z_V are compressibility factors for dry air gases and water vapor which correct for nonideal gas behavior and deviate from unity in atmospheric conditions by less than 1 part in 10^3. These compressibility factors are given by the equations (Owens 1967)

$$Z_D^{-1} = 1 + p_D\left[57.90 \times 10^{-8}\left(1 + \frac{0.52}{T}\right) - 9.4611 \times 10^{-4}\frac{(T - 273)}{T^2}\right]$$

$$(13.71)$$

and

$$Z_V^{-1} = 1 + 1650\frac{p_V}{T^3}\left[1 - 0.01317(T - 273) + 1.75 \times 10^{-4}(T - 273)^2\right.$$

$$\left. + 1.44 \times 10^{-6}(T - 273)^3\right], \qquad (13.72)$$

where p_D and p_V are in millibars. The first and second terms in Eq. (13.70) are due to ultraviolet electronic transitions of the induced dipole type for dry air molecules and water vapor, respectively, and the third term is due to the permanent dipole infrared rotational transitions of water vapor. If we neglect terms other than unity in the Z factors, Eq. (13.70) becomes

$$N = 77.6\frac{p_D}{T} + 64.8\frac{p_V}{T} + 3.776 \times 10^5\frac{p_V}{T^2}. \qquad (13.73)$$

We can rewrite Eq. (13.73) in terms of the total pressure as

$$N = 77.6\frac{P}{T} - 12.8\frac{p_V}{T} + 3.776 \times 10^5\frac{p_V}{T^2}. \qquad (13.74)$$

For temperatures around 280 K, the last two terms on the right-hand side of Eq. (13.74) can be combined to give

$$N \simeq \frac{77.6}{T}\left(P + 4810\frac{p_V}{T}\right). \qquad (13.75)$$

Equation (13.75) is the well-known *Smith–Weintraub equation* (Smith and Weintraub 1953). This equation is accurate to about 1%, or about $\pm 1N$-unit, at frequencies below 100 GHz. The accuracy of Eqs. (13.74) and (13.75) can be improved by adding a small term that increases monotonically with frequency and that accounts for the effect of the wings of the infrared transitions. Hill and Clifford (1981) show that because of this effect the wet refractivity increases by about 0.5% at 100 GHz, and 2% at 200 GHz, over its value at low frequencies.

To obtain the optical refractivity, we omit the infrared transition term from Eq. (13.73) and obtain

$$N_{\text{opt}} \simeq 77.6 \frac{p_D}{T} + 64.8 \frac{p_V}{T}.$$
(13.76)

For precise work more accurate values for N_{opt} that include small terms having wavelength dependence to account for the effects of the wings of ultraviolet transitions can be found in Allen (1964). The ratio of the wet refractivity in the radio and optical regions is obtained by omitting the dry air terms from Eqs. (13.73) and (13.76): $N_{V\,\text{rad}}/N_{V\,\text{opt}} \simeq 1 + 5830/T$. For $T \simeq 280$ K, this ratio is about equal to 22, as mentioned in connection with the discussion following Eq. (13.69).

Phase Fluctuations

In the radio region, the most important nonuniformly distributed quantity in the troposphere is the water vapor density. Variations in the water vapor distribution over an interferometer cause phase fluctuations that degrade the measurements. In the optical region, variations in temperature rather than in water vapor content are the principal cause of phase fluctuations. The fluctuations along an initially plane wavefront that has traversed the atmosphere can be characterized by a so-called structure function of the phase. This function is defined as

$$\mathcal{D}_\phi(d) = \langle [\Phi(x) - \Phi(x - d)]^2 \rangle,$$
(13.77)

where $\Phi(x)$ is the phase at point x and $\Phi(x - d)$ is the phase at point $x - d$. We assume that $\mathcal{D}_\phi$ depends only on the magnitude of the separation between the measurement points, that is, the projected baseline length of the interferometer d. The rms deviation in the interferometer phase is

$$\sigma_\phi = \sqrt{\mathcal{D}_\phi(d)}.$$
(13.78)

For the sake of illustration, we assume a simple functional form for σ_ϕ given by

$$\sigma_\phi = \frac{2\pi a d^\beta}{\lambda}, \qquad d \le d_m,$$
(13.79a)

and

$$\sigma_\phi = \sigma_m, \qquad d > d_m,$$
(13.79b)

where a is constant, and $\sigma_m = 2\pi a d_m^\beta/\lambda$. The form of Eq. (13.79) is shown in

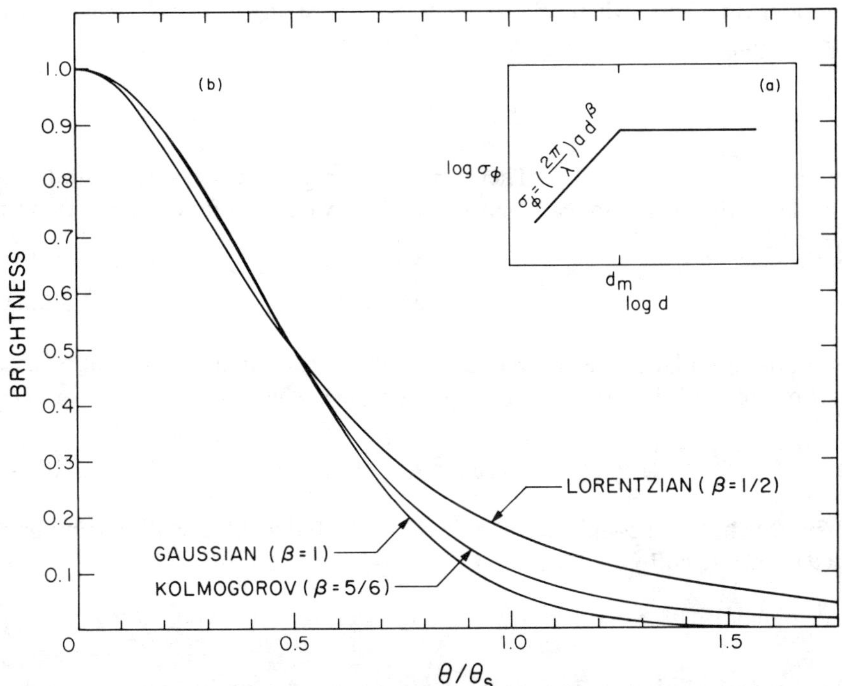

Figure 13.9 (a) Simple model for the rms phase fluctuation induced by the troposphere in an interferometer of baseline length d given by Eq. (13.79). (b) The point-source response function $\overline{w}_a(\theta)$ for various power-law models obtained by taking the Fourier transform of the visibility in the regime $d < d_m$. The values of θ_s, the full width at half-maximum of $\overline{w}_a(\theta)$, for each model are: Gaussian ($\beta = 1$), $(\sqrt{8 \ln 2})a$; modified Lorentzian ($\beta = \frac{1}{2}$), $1.53\pi\lambda^{-1}a^2$; Kolmogorov ($\beta = \frac{5}{6}$), $2.75\lambda^{-1/5}a^{6/5}$, where λ is the wavelength and a is the constant defined in Eq. (13.79).

Fig. 13.9a. This form can be derived by assuming a multiple-scale power-law model for the spectrum of the phase fluctuations. There must be a limiting distance d_m beyond which fluctuations do not increase noticeably; otherwise VLBI would not work. This limit, called the outer scale length of the fluctuations, is probably of the order of a few kilometers. Beyond this dimension the fluctuations in the path lengths become uncorrelated.

First, consider an interferometer that operates in the domain of baselines shorter than d_m. The measured visibilities $\mathcal{V}_m$ are related to the true visibilities by the equation

$$\mathcal{V}_m = \mathcal{V}e^{j\phi}, \tag{13.80}$$

where $\phi = \Phi(x) - \Phi(x - d)$ is a random variable describing the phase fluctuations introduced by the atmosphere. If we assume ϕ is a Gaussian random variable with zero mean, then the expectation of the visibility is

$$\langle \mathcal{V}_m \rangle = \mathcal{V}\langle e^{j\phi} \rangle = \mathcal{V}e^{-\sigma_\phi^2/2}. \tag{13.81}$$

Consider the case where $\beta = 1$, that is, where σ_ϕ is proportional to d. Substituting Eq. (13.79a) into Eq. (13.81) yields

$$\langle \mathscr{V}_m \rangle = \mathscr{V} e^{-2\pi^2 a^2 q^2}, \tag{13.82}$$

where $q = \sqrt{u^2 + v^2} = d/\lambda$. Therefore, on average, the measured visibility is the true visibility multiplied by an atmospheric weighting function $w_a(q)$ given by

$$w_a(q) = e^{-2\pi^2 a^2 q^2}. \tag{13.83}$$

In the image plane, the derived map is the convolution of the true source distribution and the Fourier transform of $w_a(q)$, which is

$$\overline{w}_a(\theta) \propto e^{-\theta^2/2a^2}, \tag{13.84}$$

where θ is here the conjugate variable of q. The full width at half-maximum of $\overline{w}_a(\theta)$ is θ_s, given by

$$\theta_s = \sqrt{8(\ln 2)}\, a. \tag{13.85}$$

Thus, the resolution is degraded since the derived map is convolved with a Gaussian beam of width θ_s (in addition to the effects of any other weighting functions as described in Chapter 10). θ_s is the seeing angle. Images with finer resolution than θ_s can sometimes be obtained by use of adaptive calibration procedures described in Chapter 11. Now from Eq. (13.79a),

$$a = \frac{\sigma_\phi \lambda}{2\pi d} = \frac{\sigma_d}{d}, \tag{13.86}$$

where $\sigma_d = \sigma_\phi \lambda / 2\pi$ is the rms uncertainty in path length. Thus we obtain

$$\theta_s = 2.35 \frac{\sigma_d}{d} \quad \text{(radians)}. \tag{13.87a}$$

Since σ_d/d is constant, θ_s is independent of wavelength. This independence results from the condition $\beta = 1$ in Eq. (13.79a). Let d_0 be the baseline length for which $\sigma_\phi = 1$ rad. From Eq. (13.86) we see that Eq. (13.85) can be written in the form

$$\theta_s = \frac{\sqrt{2\ln 2}}{\pi} \frac{\lambda}{d_0} \simeq 0.37 \frac{\lambda}{d_0}. \tag{13.87b}$$

For the case where β is arbitrary, we find $\overline{w}_a(\theta)$ by substituting Eq. (13.79a) into Eq. (13.81) and writing the two-dimensional Fourier transform as a

Hankel transform (Bracewell 1978). Thus

$$\overline{w}_a(\theta) \propto \int_0^\infty \exp\left[-2\pi^2 a^2 \lambda^{2(\beta-1)} q^{2\beta}\right] J_0(2\pi q\theta)\, q\, dq, \qquad (13.88)$$

where J_0 is the Bessel function of order zero. In general, $\overline{w}_a(\theta)$ cannot be evaluated analytically. However, it is easy to show by making appropriate substitutions in Eq. (13.88) that $\theta_s \propto a^{1/\beta}\lambda^{(\beta-1)/\beta}$. A case that can be treated analytically is the one for which $\beta = 0.5$. In this case we obtain (Bracewell 1978, p. 249)

$$\overline{w}_a(\theta) \propto \frac{1}{\left[\theta^2 + \left(\pi a^2/\lambda\right)^2\right]^{3/2}}, \qquad (13.89)$$

which represents a Lorentzian profile raised to the 3/2-power and has very broad skirts. The full width at half-maximum of $\overline{w}(\theta)$ is

$$\theta_s = \frac{1.53\pi a^2}{\lambda}, \qquad (13.90a)$$

or

$$\theta_s = \frac{0.77}{2\pi}\frac{\lambda}{d_0} \simeq 0.12\frac{\lambda}{d_0}. \qquad (13.90b)$$

In the case of Kolmogorov turbulence, which will be discussed later in this section, $\beta = 5/6$. Numerical integration of Eq. (13.88) yields

$$\theta_s \simeq 2.75 a^{6/5}\lambda^{-1/5}. \qquad (13.91)$$

Plots of $\overline{w}_a(\theta)$ for various power-law models of phase fluctuations are shown in Fig. 13.9b.

Now consider the case of an interferometer operating in the domain of baselines greater than d_m, where σ_ϕ is a constant equal to σ_m. This case is most applicable to VLBI arrays or to large connected-element arrays. If the time scale of the fluctuation is short with respect to the measurement time, then, on average, all the visibility measurements are reduced by a constant factor, $e^{-\sigma_m^2/2}$. Thus, this type of atmospheric fluctuation does not reduce the resolution. However, on average the measured flux density is reduced from the true value by the factor $e^{-\sigma_m^2/2}$. If the time scale of the fluctuations is long with respect to the measurement time, then each visibility measurement suffers a phase error $e^{j\phi}$. Assume that K visibility measurements are made of a point source of flux density S. The map of the source, considering only one dimension for simplicity, is

$$\overline{w}_a(\theta) = \frac{S}{K}\sum_{i=1}^{K} e^{j\phi_i} e^{j2\pi u_i\theta}. \qquad (13.92)$$

The expectation of $\bar{w}_a(\theta)$ at $\theta = 0$ is

$$\langle \bar{w}_a(0) \rangle = Se^{-\sigma_m^2/2}. \tag{13.93}$$

The measured flux density is less than S. The missing flux density is scattered around the map. This is immediately evident from Parseval's theorem,

$$\sum_i |\bar{w}_a(\theta_i)|^2 = \frac{1}{K} \sum_i |\mathscr{V}(u_i)|^2 = S^2. \tag{13.94}$$

Thus, the total flux density could be obtained by integrating the square of the image-plane response. The rms deviation in the flux density, measured at the peak response for a source at $\theta = 0$, is $\sqrt{\langle \bar{w}_a^2(\theta) \rangle - \langle \bar{w}_a(\theta) \rangle^2}$, which we call σ_S. This quantity can be calculated from Eq. (13.92) and is given by

$$\sigma_S = \frac{S}{\sqrt{K}} \sqrt{1 - e^{-\sigma_m^2}}, \tag{13.95}$$

which reduces to $\sigma_S \simeq S\sigma_m/\sqrt{K}$ when $\sigma_m \ll 1$.

The theory of propagation through a turbulent neutral atmosphere has been treated in detail in the seminal publications of Tatarski (1961, 1971). This theory has been developed and applied extensively to problems of optical seeing (e.g., Roddier 1981, Woolf 1982, Coulman 1985) and to infrared interferometry (Sutton, Subramanian, and Townes 1982). We confine the discussion here to a few central ideas concerning the structure function of phase, and indicate how it is related to other functions that are used to characterize atmospheric turbulence.

When the Reynolds number, a dimensionless parameter involving the viscosity and a characteristic scale size and velocity of a flow, exceeds a critical value, the flow becomes turbulent. In the atmosphere the Reynolds number is nearly always high enough that turbulence is fully developed. In the Kolmogorov model for turbulence, the kinetic energy associated with large-scale turbulent motions is transferred to smaller and smaller scale sizes of turbulence until it is finally dissipated into heat by viscous friction. If the turbulence is fully developed and isotropic, then the two-dimensional power spectrum of the phase fluctuations or the refractive index varies as $q_s^{-11/3}$, where q_s (cycles per meter) is the spatial frequency. (q_s, the conjugate variable of d, is analogous to q, the conjugate variable of θ.) The structure function for the refractive index $\mathscr{D}_n(d)$ is defined in a similar fashion to the structure function of phase in Eq. (13.77), that is, $\mathscr{D}_n(d)$ is the mean square deviation of the difference in the refractive index at two points a distance d apart, or $\mathscr{D}_n(d) = \langle [n(x) - n(x - d)]^2 \rangle$. For the conditions stated above, $\mathscr{D}_n$ can be shown to be given by the equation

$$\mathscr{D}_n(d) = C_n^2 d^{2/3}, \qquad L_i \ll d \ll L_o, \tag{13.96}$$

where L_i and L_o are called the inner and outer scales of turbulence, which may be less than a centimeter and a few kilometers, respectively. The parameter C_n^2 characterizes the strength of the turbulence. Note that water vapor, which is the dominant cause of fluctuation in the index of refraction, is poorly mixed in the troposphere and therefore may be only an approximate tracer of the mechanical turbulence.

The structure function of phase for an atmosphere where C_n^2 varies with height from the surface to an overall height L is given by [Tatarski 1961, Eq. (6.65)]

$$\mathscr{D}_\phi(d) = 2.91\left(\frac{2\pi}{\lambda}\right)^2 d^{5/3} \int_0^L C_n^2(h)\, dh, \qquad (13.97)$$

which is valid in the range $\sqrt{L\lambda} < d < L_o$. Note that the factor 2.91 is a dimensionless constant and C_n^2 has units of length$^{-2/3}$. The lower limit on d is equivalent to the requirement that diffraction effects be negligible. If C_n^2 is constant with height, then Eq. (13.97) reduces to

$$\mathscr{D}_\phi(d) = 2.91\left(\frac{2\pi}{\lambda}\right)^2 C_n^2 L\, d^{5/3}. \qquad (13.98)$$

Thus from Eq. (13.78) the rms phase deviation is

$$\sigma_\phi = 1.71\left(\frac{2\pi}{\lambda}\right)\sqrt{L C_n^2}\, d^{5/6}. \qquad (13.99)$$

We can calculate d_0, the baseline length for which $\sigma_\phi = 1$ rad, by setting $\mathscr{D}_\phi$ in Eq. (13.97) equal to 1 rad^2. The expression for d_0 is then

$$d_0 = 0.058\lambda^{6/5}\left[\int_0^L C_n^2(h)\, dh\right]^{-3/5}. \qquad (13.100)$$

The separation of two points along the wavefront for which the rms deviation in the difference phase is 2.62 radians is known as the Fried length (Fried 1965, 1967). Since σ_ϕ is proportional to $d^{5/6}$, the Fried length is 3.2 times d_0. Equation (13.100) shows that d_0 is proportional to $\lambda^{6/5}$, and thus the angular resoluion or seeing limit ($\simeq \lambda/d_0$) is proportional to $\lambda^{-1/5}$ [see Fig. 13.9 and Eq. (13.91)]. This relationship may hold over broad wavelength ranges when C_n^2 is constant. In the optical range C_n^2 is related to temperature fluctuations, whereas in the radio range C_n^2 is dominated by turbulence in the water vapor.

The two-dimensional power spectrum of phase, $\mathscr{S}_2(q_x, q_y)$, is the Fourier transform of the two-dimensional autocorrelation function of phase, $R_\phi(d_x, d_y)$. If R_ϕ is only a function of d, where $d^2 = d_x^2 + d_y^2$, then $\mathscr{S}_2$ is a function of q_s where $q_s^2 = q_x^2 + q_y^2$, and $\mathscr{S}_2(q_s)$ and $R_\phi(d)$ form a Hankel transform pair. Since $\mathscr{D}_\phi(d) = 2[R_\phi(0) - R_\phi(d)]$, we can write

$$\mathscr{D}_\phi(d) = 4\pi \int_0^\infty \left[1 - J_0(2\pi q_s d)\right] \mathscr{S}_2(q_s) q_s \, dq_s, \qquad (13.101)$$

where J_0 is the Bessel function of order zero. When $\mathscr{D}_\phi(d)$ is given by Eq. (13.98), $\mathscr{S}_2(q_s)$ is

$$\mathscr{S}_2(q_s) = 0.0097 \left(\frac{2\pi}{\lambda}\right)^2 C_n^2 L q_s^{-11/3}. \qquad (13.102)$$

The one-dimensional temporal spectrum of the phase fluctuations $\mathscr{S}_\phi'(f)$ (the two-sided spectrum) can be calculated from $\mathscr{S}_2(q_s)$ if we assume that the turbulent structure is frozen in the troposphere as it flows over the observer at velocity v_s. In this case,

$$\mathscr{S}_\phi'(f) = \frac{1}{v_s} \int_{-\infty}^\infty \mathscr{S}_2\left(q_x = \frac{f}{v_s}, q_y\right) dq_y. \qquad (13.103)$$

Substitution of Eq. (13.102) into Eq. (13.103) yields

$$\mathscr{S}_\phi'(f) = 0.016 \left(\frac{2\pi}{\lambda}\right)^2 C_n^2 L v_s^{5/3} f^{-8/3} \quad (\text{rad}^2 \, \text{Hz}^{-1}), \qquad (13.104)$$

where v_s is in meters per second. Examples of the temporal spectra of water vapor fluctuations can be found in Hogg, Guiraud, and Sweezy (1981).

The Allan variance $\sigma_y^2(\tau)$, or fractional frequency stability for time interval τ, associated with $\mathscr{S}_\phi'(f)$ has been defined in Section 9.4. It can be calculated by substituting Eq. (9.82) into Eq. (9.94), which gives

$$\sigma_y^2(\tau) = \left(\frac{2}{\pi v_0 \tau}\right)^2 \int_0^\infty \mathscr{S}_\phi'(f) \sin^4(\pi \tau f) \, df. \qquad (13.105)$$

By substituting Eq. (13.104) into Eq. (13.105), and noting that $\int_0^\infty [\sin^4(\pi x)]/x^{8/3} \, dx = 4.61$, we obtain

$$\sigma_y^2(\tau) = 1.3 \times 10^{-17} C_n^2 L v_s^{5/3} \tau^{-1/3}. \qquad (13.106)$$

Armstrong and Sramek (1982) give general expressions for the relations among $\mathscr{S}_2, \mathscr{S}_\phi', \mathscr{D}_\phi$, and σ_y for an arbitrary power-law index. If $\mathscr{S}_2 \propto q^{-\alpha}$, then $\mathscr{D}_\phi(d) \propto d^{\alpha-2}$, $\mathscr{S}_\phi' \propto f^{-\alpha+1}$, and $\sigma_y^2 \propto \tau^{\alpha-4}$. The exponent α equals 11/3 or 3.67 when the turbulence is isotropic and fully developed. For scale sizes larger than the tropospheric thickness, the turbulence becomes effectively two-dimensional, in which case $\alpha = 2.67$. Thus $\mathscr{D}_\phi(d) \propto d^{0.67}$ for d greater than a few

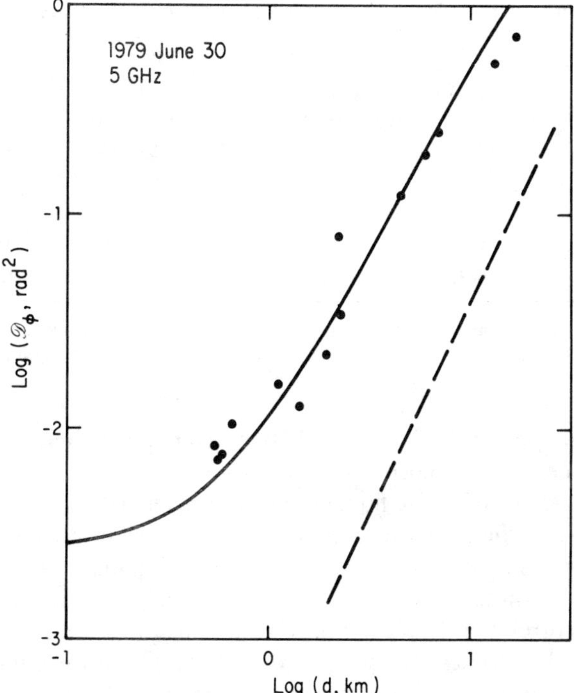

Figure 13.10   Phase structure function at 5 GHz measured with the VLA on June 30, 1979. The dashed line is the variation expected for source position errors (0.04 arcsec). The solid line is a fit to the data of a power-law structure function plus a baseline-independent contribution from the interferometer electronics of 3°. The structure function is $\mathscr{D}_\phi(d) = 5.3 \times 10^{-8} d^{1.74}$. From Armstrong and Sramek (1982).

kilometers. As d increases, one expects $\mathscr{D}_\phi$ to have an initial power-law coefficient of about 1.67, which decreases to 0.67 when d exceeds the tropospheric thickness, and which further decreases to zero when d exceeds L_o.

Armstrong and Sramek (1982) have collected much of the relevant data on phase fluctuations induced by the troposphere and shown that the Kolmogorov turbulence model fits the data reasonably well. They have presented data from the VLA at 6-cm wavelength that show that $\mathscr{D}_\phi(d)$ can be modeled by a power law whose exponent is generally slightly less than 1.67. An example of one of their measurements of $\mathscr{D}_\phi(d)$ is shown in Fig. 13.10. Measurements of $\mathscr{D}_\phi(d)$ have been made at 3-mm wavelength with the Hat Creek interferometer (Bieging et al. 1984) and at 6-cm wavelength with the Cambridge One-Mile interferometer (Hinder 1970, Hinder and Ryle 1971). The ranges of values of $C_n^2 L$ derivable from these sets of data are listed in Table 13.2. The outer scale length is not well defined (Hargrave and Shaw 1978, Hamaker 1978), but in the domain of very long baselines, as encountered in VLBI, $\mathscr{D}_\phi(d)$ should be independent of d. Rogers and Moran (1981) and Rogers et al. (1984) discuss

TABLE 13.2   Values of Turbulence Parameter $C_n^2 L$ Observed With Various Interferometers[†]

Location	Wavelength (cm)	d (km)	$C_n^2 L$ ($m^{1/3}$)	Reference
Green Bank	11	2.5	5×10^{-13}–5×10^{-11}	Baars (1967)
VLA	6	35	2×10^{-13}–2×10^{-9}	Armstrong and Sramek (1982)
Cambridge (UK)	6	1.5	8×10^{-12}–2×10^{-10}	Hinder (1970)
Hat Creek	0.35	0.18	8×10^{-13}–6×10^{-11}	Bieging et al. (1984)

[†] The seeing angle θ_s can be obtained from Eqs. (13.79a), (13.91), and (13.99) when $\beta = 5/6$, and is given by $\theta_s \simeq 1.08 \times 10^6 (C_n^2 L)^{3/5} \lambda^{-1/2}$ (arcsec). For the values of $C_n^2 L$ given in this table, θ_s ranges from $0''.05$ to $10''$.

the effects of the troposphere on VLBI measurements, and their plot of the Allan variance of the atmosphere is shown in Fig. 9.14.

The strength of the phase fluctuations, characterized by the parameter $C_n^2 L$, is difficult to predict. Measurements at the VLA show that $C_n^2 L$ is not well correlated with surface absolute humidity. The dominant correlation is probably with solar-induced convection. Baars (1967) and Hinder (1972) have studied the correlation between phase variations of radio interferometers and meteorological variables. Hinder found that the phase variations have strong positive correlation with temperature and duration of sunshine, and negative correlation with cloud cover.

Water Vapor Radiometry

The excess propagation path in a particular direction due to water vapor can be accurately determined from measurements of the brightness temperature in the same direction at frequencies near the 22.2-GHz water vapor resonance. This method was first investigated by Westwater (1967) and Schaper, Staelin, and Waters (1970). To appreciate the degree of correlation between wet path length and brightness temperature, we need to examine the dependence of these quantities on pressure, water vapor density, and temperature. The absorption coefficient given by Eq. (13.42) is complicated, but at line center it is given approximately by

$$\alpha_m \propto \frac{\rho_V}{PT^{1.875}} e^{-644/T}, \tag{13.107}$$

where T is in kelvins, and we have neglected the rightmost term in Eq. (13.42) and all terms of order ρ_V^2. We assume that the opacity given by Eq. (13.47) is small, so that the brightness temperature defined by Eq. (13.45) can be written

$$T_B = \int_0^\infty \frac{\rho_V}{PT^{0.875}} e^{-644/T} \, dh, \tag{13.108}$$

when we neglect the background temperature T_{B0}.

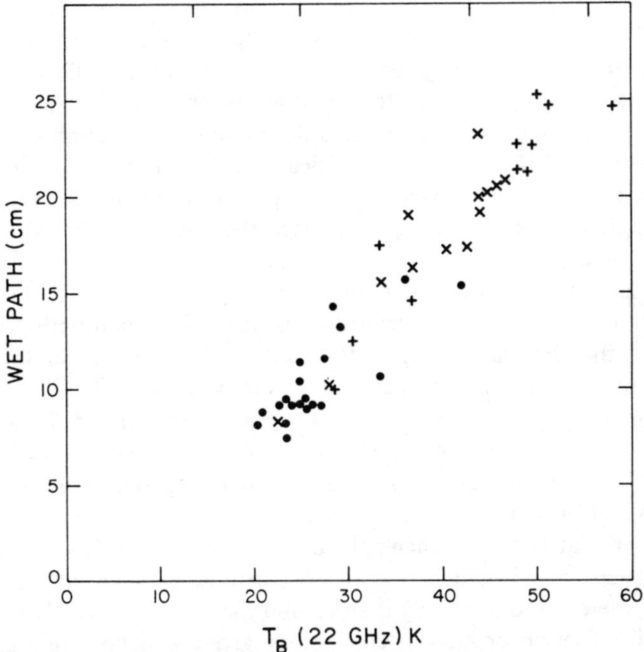

Figure 13.11   Wet path $\mathscr{L}_V$ estimated from radiosonde data versus the brightness temperature at 22 GHz. The scatter is due primarily to noise in the radiosonde sensors and the microwave radiometer. The different symbols represent different cloud-cover conditions. From Moran and Rosen (1981).

Recall that Eq. (13.16) shows that

$$\mathscr{L}_V \propto \int_0^\infty \frac{\rho_V}{T}\, dh.$$

(13.109)

Thus, if P and T were constant with height, then T_B and $\mathscr{L}_V$ would be proportional to each other, at least to the degree of approximation necessary to derive Eq. (13.108). The brightness temperature at 22.2 GHz has been found empirically to be related to the wet path length by the equation (e.g., Moran and Rosen 1981)

$$T_B(22.2\,\text{GHz}) \;\;(\text{K}) \simeq 2.1\mathscr{L}_V \;\;(\text{cm}).$$

(13.110)

Data from which this relation was derived are shown in Fig. 13.11.

Recall that ρ_V is approximately exponentially distributed with a scale height of 2 km. The temperature, on average, decreases by about 2% per kilometer. This change affects the proportionality between T_B and $\mathscr{L}_V$ only through the exponential factor in Eq. (13.108) and the slight difference in the power law for temperature. Thus, temperature has a small effect. The pressure decreases by 10% per kilometer so that water vapor at higher altitudes contributes more heavily to T_B than is desirable for estimation by radiometry. The sensitivity of T_B to pressure is decreased by moving off the resonance frequency to a frequency near the half-power point of the transition. The reason for this is that as pressure increases, the line profile broadens while the integrated line profile is constant. Therefore, the absorption at line center decreases and the absorption in the line wings increases. Westwater (1967) showed that at 20.6 GHz the absorption is nearly invariant with pressure. The opacity at this frequency is less than at the line center, so the nonlinear relationship between T_B and opacity is less important.

The foregoing discussion assumes that measurements of T_B are made in clear sky conditions. The water droplet content in clouds causes substantial absorption but small change in the index of refraction compared to that of water vapor. Fortunately, the effect of clouds can be eliminated by combining measurements at two frequencies. In nonprecipitating clouds, the sizes of the water droplets are generally less than 100 μm, and, at wavelengths greater than a few millimeters, the scattering is small and the attenuation is due primarily to absorption. The absorption coefficient is given by the empirical formula (Staelin 1966)

$$\alpha_{\text{clouds}} = \frac{\rho_L 10^{0.0122(291-T)}}{\lambda^2} \quad (\text{m}^{-1}), \tag{13.111}$$

where ρ_L is the density of liquid water droplets in grams per cubic meter, λ is the wavelength in meters, and T is in kelvins. A very wet cumulus cloud with a water density of 1 g m^{-3} and a size of 1 km will have an absorption coefficient of 7×10^{-5} m^{-1}, and will therefore have a brightness temperature of about 20 K at 22 GHz. The index of refraction of liquid water is about 5 at 22 GHz for $T = 280$ K (Goldstein 1951). The actual excess propagation path through the cloud due to liquid water would be about 4 mm, but the predicted excess path from Eq. (13.110) is 10 cm. Thus, the brightness temperature at a single frequency cannot be used to estimate reliably the excess path length when clouds are present. In order to eliminate the brightness temperature contribution of clouds, measurements must be made at two frequencies, ν_1 and ν_2, one near the water line and one well off the water line, respectively. The brightness temperature is

$$T_{Bi} = T_{BVi} + T_{BCi}, \tag{13.112}$$

where T_{BVi} and T_{BCi} are the brightness temperatures due to water vapor and clouds at frequency i. Here we neglect the effects of atmospheric O_2. Since, from Eq. (13.111), $T_{BC} \propto \nu^2$, we can form the observable

$$T_{B1} - T_{B2}\frac{\nu_1^2}{\nu_2^2} = T_{BV1} - T_{BV2}\frac{\nu_1^2}{\nu_2^2}, \qquad (13.113)$$

which eliminates the effect of clouds. The correlation between $T_{BV1} - T_{BV2}$ $\times \nu_1^2/\nu_2^2$ and $\mathscr{L}_V$ can be estimated from model calculations based on Eqs. (13.45) and (13.16). The off-resonance frequency ν_2 is generally chosen to be about 31 GHz. The problem of finding the best two frequencies and the appropriate correlation coefficients to use in predicting $\mathscr{L}_V$ has been widely discussed (Westwater 1978, Wu 1979, Westwater and Guiraud 1980). The liquid content of clouds can also be measured by dual-frequency techniques (Snider, Burdick, and Hogg 1980).

The application of dual-frequency microwave radiometry to the calibration of wet path length has been described by Guiraud, Howard, and Hogg (1979), Moran and Rosen (1981), Elgered, Rönnäng, and Askne (1982), and Resch (1984). The results show that $\mathscr{L}_V$ can be estimated to an accuracy better than 1 cm. This accuracy is useful for VLBI measurements of delay. Measurements of T_B at the antennas of short baseline interferometers can be useful in correcting the interferometer phase (Resch, Hogg, and Napier 1984). More accurate predictions of $\mathscr{L}_V$ or interferometer phase could be obtained by including measurements at other frequencies. For example, measurements of the wings of the terrestrial oxygen line near 50 GHz can be used to probe the vertical temperature structure of the troposphere (e.g., Miner, Thornton, and Welch 1972, Snider 1972). The additional accuracy may not justify the further complication.

13.2 IONOSPHERE

The ionosphere has been studied extensively since the pioneering experiments of Appleton and Barnett (1925) and Breit and Tuve (1926). The literature on the subject is vast. Magneto-ionic propagation theory relevant to the ionosphere is treated in depth by Ratcliffe (1962) and Budden (1961); the morphology of the ionosphere is described by Rawer (1956); and an excellent general treatment of ionospheric propagation is given by Davies (1965). Reviews of particular relevance to radio astronomy can be found in Evans and Hagfors (1968) and Hagfors (1976). Beynon (1975) gives interesting historical anecdotes on the early development of ionospheric research. In this section, we treat only those aspects of the ionosphere that have a deleterious effect on interferometric observations. Table 13.3 gives the magnitude of various propagation effects for the daytime and nighttime ionosphere. Most of these effects scale as ν^{-2}, and they can be minimized by observing at higher frequencies. The magnitude of the ionospheric excess path typically equals that of the

TABLE 13.3   Maximum Likely Values of Ionospheric Effects at 100 MHz for a Zenith Angle of 60°

Effect	Maximum[a] (Daytime)	Minimum[b] (Night)	Frequency Dependence
Faraday rotation	15 rotations	1.5 rotations	ν^{-2}
Group delay	12 μsec	1.2 μsec	ν^{-2}
Excess (phase) path length	3500 m	350 m	ν^{-2}
Phase change	7500 rad	750 rad	ν^{-1}
Phase stability (peak to peak)	± 150 rad	± 15 rad	ν^{-1}
Frequency stability (rms)	± 0.04 Hz	$\pm .004$ Hz	ν^{-1}
Absorption (in D and F regions)	0.1 dB[c]	0.01 dB	ν^{-2}
Refraction (ambient)	0.05°	0.005°	ν^{-2}

Source: Adapted from Evans and Hagfors (1968). For values of parameters at the zenith, divide numbers (except refraction) by sec z_i, which is approximately 1.7 [see Eq. (13.130)]. For typical (rather than maximum) parameters, divide numbers by 2.

[a] Total electron content = 5×10^{17} m^{-2}.

[b] Total electron content = 5×10^{16} m^{-2}.

[c] 1 dB = 0.230 Np.

neutral atmosphere at approximately 2 GHz, but the frequency of this equality can vary from about 1 to 5 GHz. Thus, at 20 GHz the ionospheric excess path length is typically only 1% of the tropospheric excess path length.

Basic Physics

The ionization of the upper atmosphere is caused by the ultraviolet radiation from the sun. Typical daytime and nighttime vertical profiles of the electron density are shown in Fig. 13.12. The electron distribution and the total electron content vary also with geomagnetic latitude, time of year, and sunspot cycle. There are also substantial winds, traveling disturbances, and irregularities in the ionosphere. The ionosphere is permeated by the quasi-dipole magnetic field of the earth. Propagation is governed by the theory of waves in a magnetized plasma with collisions.

We derive some of the fundamental properties of the ionosphere related to the propagation of electromagnetic waves by considering elementary cases. First, consider a plane monochromatic wave of linear polarization that propagates through a uniform plasma of electron density $\mathcal{N}_e$, where the magnetic field and collisions between particles can be neglected. The electrons oscillate with the electric field, but the protons, because of their greater mass, remain relatively unperturbed. The index of refraction can be found by calculating

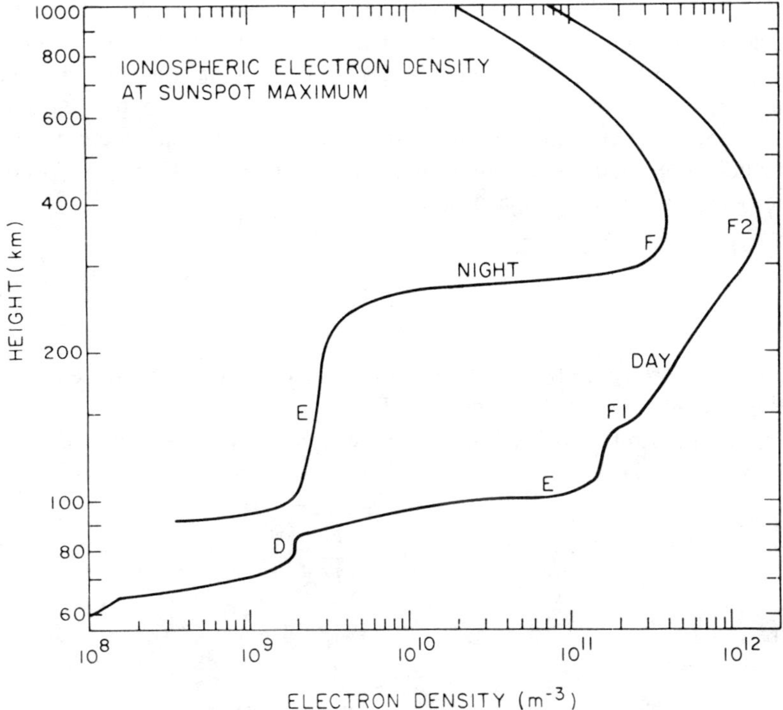

Figure 13.12 Idealized electron density distribution in the earth's ionosphere. The curves drawn indicate the densities to be expected at sunspot maximum in temperate latitudes. From Evans and Hagfors (1968).

either the induced current or the dipole moment. Either method yields the same result. We use the latter method as we did when considering the index of refraction of water vapor using the bound oscillator model in Section 13.1. The equation of motion of a free electron in the plasma is

$$m\ddot{\mathbf{x}} = -e\mathbf{E}_0 e^{-j2\pi\nu t}, \tag{13.114}$$

where m, e, and $\mathbf{x}$ are the mass, charge *magnitude*, and displacement of the electron, and $\mathbf{E}_0$ and ν are the amplitude and frequency of the electric field $\mathbf{E}$ of the incident wave. The magnetic field of the plane wave has negligible influence on the electrons as long as the electron velocity is much less than c, and the electric field has negligible influence on the motion of the protons. The steady-state solution to Eq. (13.114) is

$$\mathbf{x} = \frac{e}{(2\pi\nu)^2 m}\mathbf{E}_0 e^{-j2\pi\nu t}. \tag{13.115}$$

Note that the induced current density is $\mathbf{i} = \mathcal{N}_e e \dot{\mathbf{x}}$, where $\dot{\mathbf{x}}$, the velocity of the particle, is 90° out of phase with the driving electric field. Thus, the work done by the wave on the particles, which is $\langle \mathbf{i} \cdot \mathbf{E} \rangle$, is zero, and the wave propagates without loss, as expected, since Eq. (13.114) has no dissipative terms. The dipole moment per unit volume $\mathbf{P}$ is equal to $-\mathcal{N}_e e \mathbf{x}_0$, where $\mathbf{x}_0$ is the amplitude of oscillation. The dielectric constant ε is $1 + (\mathbf{P}/\mathbf{E}_0)/\epsilon_0$, where ϵ_0 is the permittivity of free space, so that

$$\varepsilon = 1 - \frac{\mathcal{N}_e e^2}{4\pi^2 \nu^2 \epsilon_0 m}. \tag{13.116}$$

The dielectric constant is real and less than unity because the induced dipole is 180° out of phase with the driving field.

The index of refraction n is equal to the square root of ε, and in this case is real, so

$$n = \sqrt{1 - \frac{\nu_p^2}{\nu^2}}, \tag{13.117}$$

where

$$\nu_p = \frac{e}{2\pi} \sqrt{\frac{\mathcal{N}_e}{\epsilon_0 m}} \simeq 9\sqrt{\mathcal{N}_e} \quad (\text{Hz}), \tag{13.118}$$

and $\mathcal{N}_e$ is in meters^{-3}. ν_p is known as the plasma frequency, which is also the natural frequency of mechanical oscillations in the plasma (e.g., Holt and Haskell 1965). The plasma frequency of the ionosphere (see Fig. 13.12) is usually less than 12 MHz. Waves normally incident on a plasma with frequencies below ν_p are perfectly reflected. The phase velocity of a wave with $\nu > \nu_p$ in the plasma is c/n, which is greater than c, and the group velocity of a wave packet is cn, which is less than c.

Now consider a plasma with a static magnetic field $\mathbf{B}$ in the direction of propagation of the plane wave. The equation of motion of an electron is

$$m\dot{\mathbf{v}} = -e[\mathbf{E} + \mathbf{v} \times \mathbf{B}]. \tag{13.119}$$

Let the incident field be a circularly polarized wave. If $\mathbf{B}$ is zero, the particle will follow the tip of the electric field vector and move in a circular orbit. If $\mathbf{B}$ is nonzero, the sum of the $\mathbf{v} \times \mathbf{B}$ force term, which will be in the radial direction, and the electric force term must be balanced by centripetal acceleration. Thus, there is a basic anisotropy in the plasma depending on whether the wave is right or left circularly polarized, since the sign of the $\mathbf{v} \times \mathbf{B}$ term changes between the two cases. The radius R_e of the circular orbit of the electron is derived from the balance-of-forces equation $eE_0 \pm evB = mv^2/R_e$, where $v = 2\pi\nu R_e$, B is the magnitude of the magnetic field, and the upper and lower signs refer to left and right circular polarization, respectively. Thus we obtain

$$R_e = \frac{eE_0}{4\pi^2 m\nu^2 \mp 2\pi\nu eB}. \tag{13.120}$$

Following the same procedure as the one below Eq. (13.115), we find that the index of refraction is given by the equation

$$n^2 = 1 - \frac{\nu_p^2}{\nu(\nu \mp \nu_B)}, \tag{13.121}$$

where ν_B is the gyro, or cyclotron, frequency given by

$$\nu_B = \frac{eB}{2\pi m}. \tag{13.122}$$

The gyrofrequency is the frequency at which an electron would spiral around a magnetic field line in the absence of any electromagnetic radiation. In the absence of damping, R_e would go to infinity if the applied electric field frequency were ν_B. The gyrofrequency of the earth's magnetic field in the ionosphere ($\simeq 0.5 \times 10^{-4}$ tesla) is about 1.4 MHz.

Equation (13.121) gives the index of refraction for the case of a longitudinal magnetic field, that is, where the field is parallel to the direction of wave propagation. The solution for the transverse case is different. The solution for the quasi-longitudinal case is obtained by replacing B with $B \cos\theta$, where θ is the angle between the propagation vector and the direction of the magnetic field. The quasi-longitudinal solution is applicable when the angle θ is less than that specified by the inequality (Ratcliffe 1962)

$$\tfrac{1}{2}\sin\theta \, \tan\theta < \frac{\nu^2 - \nu_p^2}{\nu\nu_B}. \tag{13.123}$$

When $\nu > 100$ MHz, $\nu_p \sim 10$ MHz and $\nu_B \sim 1.4$ MHz, the quasi-longitudinal solution is valid for $|\theta| < 89°$, or virtually all cases of interest. Therefore, to a high accuracy, when $\nu \gg \nu_p, \nu_B$ we can expand Eq. (13.121) to obtain

$$n \simeq 1 - \frac{1}{2}\frac{\nu_p^2}{\nu^2} \mp \frac{1}{2}\frac{\nu_p^2\nu_B}{\nu^3}\cos\theta, \tag{13.124}$$

where we neglect terms in ν^4 and higher order. For propagation in the direction of **B**, the index of refraction is lower for a left circularly polarized wave than for a right circularly polarized wave.

The difference in the index of refraction for right and left circularly polarized waves leads to the important phenomenon of Faraday rotation, whereby a linearly polarized wave has its plane of polarization rotated as it propagates through the plasma. A linearly polarized wave with position angle ψ can be decomposed into right and left circularly polarized waves of equal amplitude and phase difference 2ψ. The phase of the two circular waves as they propagate in the z direction through a plasma is $2\pi\nu n_R z/c$ and $2\pi\nu n_L z/c$, where n_R and n_L are the indices of refraction for the right circular and left circular modes, respectively. The phase difference between the waves is $2\pi\nu(n_R - n_L)z/c$. From Eq. (13.124), $n_R - n_L = \nu_p^2 \nu_B \nu^{-3}\cos\theta$, so it is clear that the plane of polarization is rotated by the angle

$$\Delta\psi = \frac{\pi}{c\nu^2} \int \nu_p^2 \nu_B \cos\theta \, dz, \tag{13.125}$$

where ν_p, ν_B, and θ may be functions of z.

For constant magnetic field and electron density, Eq. (13.125) can be written

$$\Delta\psi = 2.6 \times 10^{-13} \mathcal{N}_e B\lambda^2 L \cos\theta, \tag{13.126}$$

where $\Delta\psi$ is in radians, $\mathcal{N}_e$ is in meters^{-3}, B is in tesla and positive when the field is pointed toward the observer, λ is the wavelength in meters, and L is the path length in meters. A magnetic field pointed toward the observer causes the position angle to increase (i.e., a counterclockwise rotation of the plane of polarization of incident radiation as viewed from the surface of the earth).

Refraction and Propagation Delay

Refraction in the ionosphere decreases the zenith angle of signals arriving from outside the earth's atmosphere. This bending is caused by the curvature of the ionosphere. If the ionosphere were a plane-parallel layered structure, then from Eq. (13.26), the bending angle would be zero. In a well-known approximation (Bailey 1948), the ionosphere is assumed to consist of a layer of thickness Δh within which there is a parabolic distribution of electron density having a maximum at height h_i. The bending angle in this case is

$$\Delta z = \frac{2\Delta h \sin z}{3r_0}\left(\frac{\nu_p}{\nu}\right)^2\left(1 + \frac{h_i}{r_0}\right)\left(\cos^2 z + \frac{2h_i}{r_0}\right)^{-3/2}, \tag{13.127}$$

where ν_p is the plasma frequency at h_i. If the constant $\Delta h\nu_p^2$ is chosen properly, Δz is accurate to better than 5% for all values of z.

The excess path length in the zenith direction can be calculated using Eqs. (13.5), (13.118), and (13.124) with the assumption that $\nu \gg \nu_p, \nu_B$. The result is

$$\mathcal{L}_0 \simeq -\frac{1}{2}\int_0^\infty \left[\frac{\nu_p(h)}{\nu}\right]^2 dh \simeq -\frac{40.3}{\nu^2}\int_0^\infty \mathcal{N}_e(h)\,dh, \qquad (13.128)$$

where ν is in Hz and $\mathcal{N}_e(h)$ and $\nu_p(h)$ are the electron distribution (m^{-3}) and plasma frequency as a function of height. The integral of electron density over height in Eq. (13.128) is called the *total electron content* or *column density*. The excess path corresponds to a phase delay and is negative for the ionosphere. If

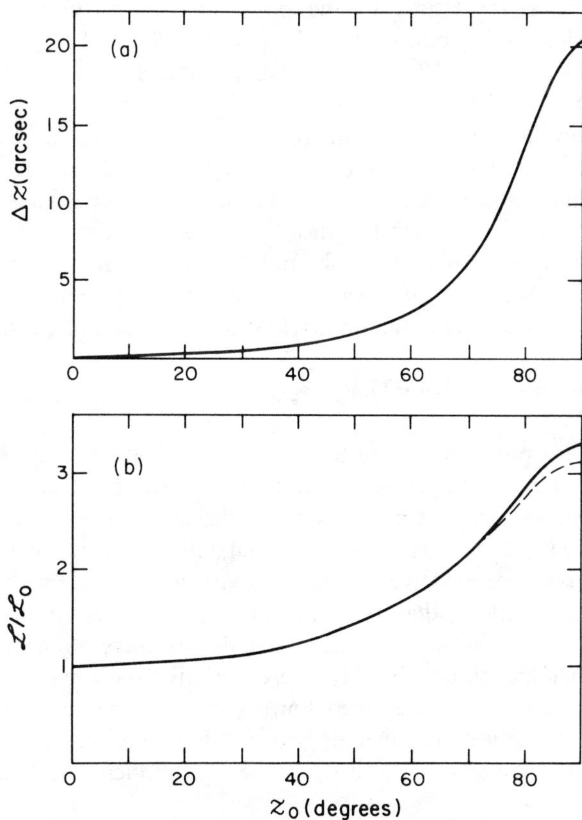

Figure 13.13   (a) Ionospheric bending angle versus zenith angle at 1000 MHz from a ray-tracing calculation for the daytime electron density profile in Fig. 13.12. The bending predicted by Eq. (13.127), with parameters $\nu_p = 12$ MHz, $h_i = 350$ km, $\Delta h = 225$ km, and $r_0 = 6370$ km, differs from the curve shown by no more than 5%. (b) Normalized ionospheric excess path length versus zenith angle for the same electron density profile from a ray-tracing calculation (solid curve) and from Eq. (13.129) (dashed curve). The total electron content is 6.03×10^{17} m^{-2}, and the excess path length at the zenith is 24.3 m. The bending and excess path length scale as ν^{-2}.

we approximate the ionosphere by a thin layer at height h_i, then the excess path length will vary as the secant of the zenith angle of the ray as it passes through the layer. Thus,

$$\mathcal{L} = \mathcal{L}_0 \sec z_i, \tag{13.129}$$

where z_i (see Fig. 13.4) is given by

$$z_i = \sin^{-1}\left[\left(\frac{r_0}{r_0 + h_i}\right)\sin z\right]. \tag{13.130}$$

When $z = 90°$, $\sec z_i$ is only $\simeq 3$ if $h_i = 400$ km. The secant law provides a reasonable model for estimating the excess ionospheric path length. A more complex model can be found in Spoelstra (1983). Plots of Δz and $\mathcal{L}$ obtained from Eqs. (13.127) and (13.129) as well as from actual ray-tracing calculations are shown in Fig. 13.13.

In some applications, it is required to correct the measurements of fringe frequency for the effects of ionospheric delay (see Section 12.2). The ionospherically induced frequency shift at an antenna is $(v/c)\,d\mathcal{L}/dt$. The time rate of change in excess path length $d\mathcal{L}/dt$ has two components: one caused by the time rate of change of zenith angle dz/dt, and the other caused by the time rate of change of $\mathcal{L}_0$, $d\mathcal{L}_0/dt$. At many times, especially near sunrise and sunset, the latter term may dominate (Mathur, Grossi, and Pearlman 1970).

Calibration of Ionospheric Delay

The excess ionospheric path length must be calibrated as accurately as possible in experiments involving precise determination of source positions or baselines. Three approaches are possible. Models of the ionosphere can be constructed that depend on parameters such as geomagnetic latitude, solar time, season, and solar activity. Alternatively, measurements of total electron content can be obtained from stations that routinely obtain such measurements via iono-sondes, Faraday rotation of satellite signals, incoherent backscatter (Evans 1969), or other techniques. Finally, the differential path length effects can be virtually eliminated for unresolved sources by making astronomical observations simultaneously at two widely separated frequencies, v_1 and v_2. If the interferometer phases are ϕ_1 and ϕ_2 at the two frequencies, then the quantity

$$\phi_c = \phi_2 - \left(\frac{v_1}{v_2}\right)\phi_1 \tag{13.131}$$

will be substantially free of ionospheric delay effects. A small residual error remains because of higher-order frequency terms in the index of refraction and because the rays at the two frequencies traverse slightly different paths through the ionosphere. Dual-frequency observations are widely used in astrometric radio interferometry where source structure can be neglected (e.g. Fomalont and Sramek 1975, Kaplan et al. 1982, Shapiro 1976). Note that the difference in total electron content along the ray paths to the interferometer elements can be estimated from measurement of $\phi_2 - (\nu_2/\nu_1)\phi_1$.

Absorption

Absorption in the ionosphere is caused by collisions of electrons with ions and neutral particles. At frequencies much greater than ν_p, the power absorption coefficient is

$$\alpha = 2.68 \times 10^{-7}\frac{\mathcal{N}_e\nu_c}{\nu^2} \quad (\mathrm{m}^{-1}), \tag{13.132}$$

where ν_c is the collision frequency and $\mathcal{N}_e$ is in meters^{-3}. The collision frequency in Hz is approximately

$$\nu_c \simeq 6.1 \times 10^{-9}\left(\frac{T}{300}\right)^{-3/2}\mathcal{N}_i + 1.8 \times 10^{-14}\left(\frac{T}{300}\right)^{1/2}\mathcal{N}_a, \tag{13.133}$$

where $\mathcal{N}_i$ is the ion density and $\mathcal{N}_a$ is the neutral particle density, both in meters^{-3} (Evans and Hagfors 1968). Numerical values of absorption are listed in Table 13.3.

Small- and Large-Scale Irregularities

The small-scale irregularities in the electron density distribution introduce random changes in the wavefront of a passing electromagnetic wave. As a consequence, fluctuations in fringe amplitude and phase can be readily observed with an interferometer at frequencies below a few hundred megahertz. In the early days of radio astronomy, signals from Cygnus A and other compact sources were observed to fluctuate on time scales of 0.1–1 min. At first these fluctuations were thought to be intrinsic to the sources (Hey, Parsons, and Phillips 1946), but later observations with spaced receivers showed that the fluctuations were uncorrelated for receiver separations of more than a few kilometers (Smith, Little, and Lovell 1950). This result led to the conclusion that irregularities in the ionosphere were perturbing the cosmic signals. The predominant scale sizes in the ionization irregularities (blobs) were found to be a few kilometers or less. The time scale of the fluctuations indicates that ionospheric wind speeds are in the range of 50–300 m sec^{-1}. The effects of these fluctuations have been studied extensively at frequencies between about 20 and 200 MHz and have been observed at frequencies as high as 7 GHz (Aarons et al. 1983). An example of the fluctuations seen in interferometer measurements is given in Fig. 13.14. Hewish (1952), Booker

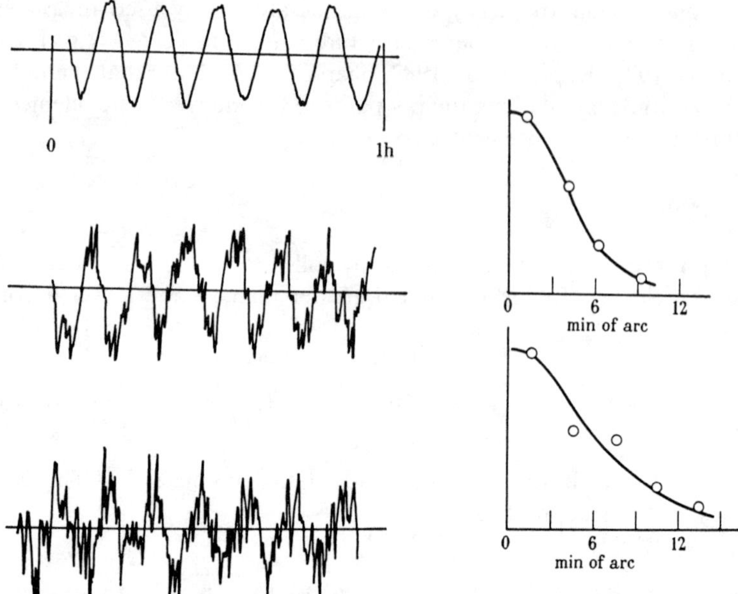

Figure 13.14   (Left). Typical records of the correlator output on three occasions from a phase-switching interferometer at Cambridge, England, having a 1-km baseline and operating at a wavelength of 8 m. The irregular responses are caused by disturbances in the ionosphere. (Right). Probability distributions of the angle of arrival deduced from the zero crossings of the correlator response. From Hewish (1952).

(1958), and Lawrence, Little, and Chivers (1964) have reviewed the early results and techniques. A comprehensive review of theory and observations of ionospheric fluctuations can be found in Crane (1977), Fejer and Kelley (1980), and Yeh and Liu (1982), and a summary of global morphology can be found in Aarons (1982). The effects of ionospheric scintillations on a synthesis telescope have been described by Spoelstra and Kelder (1984). In Section 13.3 we discuss a theory of scintillation, which can be applied to the ionosphere as well as to the interplanetary and interstellar media.

Large-scale variations in the electron density integrated along the line of sight are caused by traveling ionospheric disturbances (TIDs). TIDs, which are manifestations of acoustic–gravity waves in the upper atmosphere, are quasi-periodic large-scale perturbations in electron density. The atmosphere has a natural buoyancy so that a parcel of gas displaced vertically and released will oscillate at a frequency known as the Brunt–Väisälä or buoyancy frequency. This frequency is about 0.5–2 mHz (periods of 10–20 min) at ionospheric heights. For waves with frequencies above the buoyancy frequency, the restoring force is pressure (acoustic wave), and for waves with frequencies below the buoyancy frequency, the restoring force is gravity (gravity wave). Yeh and Liu (1974) and Hunsucker (1982) have reviewed the literature on acoustic–gravity waves. There are many potential sources of TIDs including auroral heating, severe weather fronts, earthquakes, and volcanic eruptions. Medium scale TIDs have scale lengths of 100–200 km, time scales of 10–20 min, and cause a variation in total electron content of 0.5–5%. Such TIDs are present for a substantial fraction of the time. Large-scale TIDs, which are relatively uncommon, have scale lengths of 1000 km, time scales of hours, and can cause variations in total electron content of up to 8%. One such disturbance, excited by a volcano, was observed by VLBI (Roberts et al. 1982). The effects of TIDs on radio interferometry, which are primarily slow phase variations, are described by Hinder and Ryle (1971). The effects on satellite tracking are described by Evans, Holt, and Wand (1983).

13.3   SCINTILLATION CAUSED BY PLASMA IRREGULARITIES

Understanding the propagation of radiation in a random medium is an important problem in many fields. The signals from cosmic radio sources propagate through several random media including the ionized interstellar gas of our galaxy, the solar wind, and the ionosphere as well as the neutral gas of the troposphere. We discuss here the diffractive effects caused by a random ionized medium. These effects, due to small-scale turbulence, cannot be corrected for, unlike the larger-scale refractive effects, which can be more easily calibrated.

Gaussian Screen Model

We begin the discussion by considering a simple model that serves to illustrate many features of the problem. This model was first developed by Booker, Ratcliffe, and Shinn (1950) to explain ionospheric scintillation and was refined by Ratcliffe (1956). Scheuer (1968) applied it to pulsar observations. The model assumes that the irregular medium is confined to a thin screen, and the irregularities have one characteristic scale size. Diffraction effects are neglected within the irregular medium; only the phase change imposed by the medium is considered. Diffraction is taken into account in the free-space region between the irregular medium and the receivers.

The geometrical situation is shown in Fig. 13.15. The thin screen assumption is not particularly restrictive. However, the assumption that the screen is filled with plasma blobs having one characteristic size a is restrictive, and distinguishes this model from the power-law model where a range of scale sizes is present. From Eqs. (13.118) and (13.124), the index of refraction of the plasma can be written

$$n \simeq 1 - \frac{r_e \mathcal{N}_e \lambda^2}{2\pi}, \tag{13.134}$$

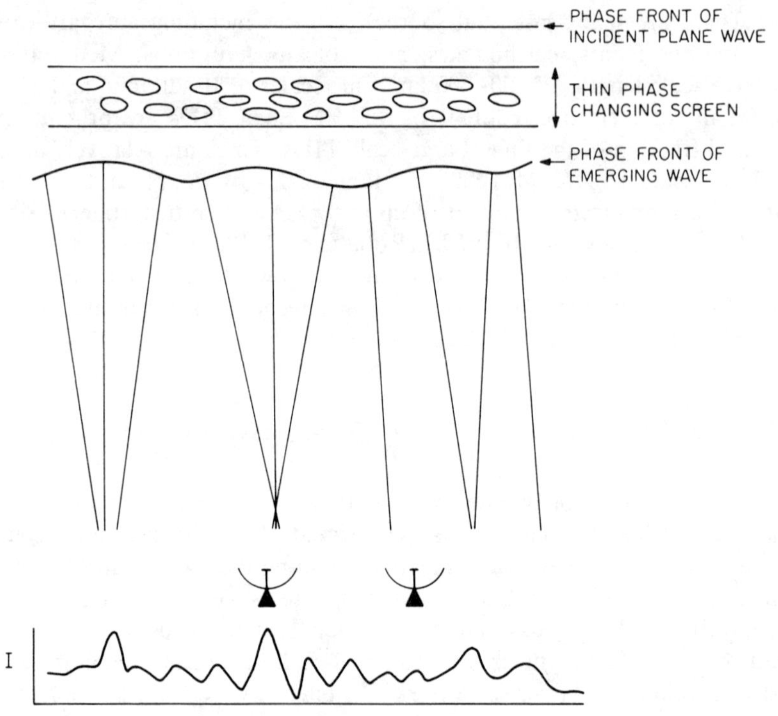

Figure 13.15   Geometry of thin screen scintillation model. An initially plane wave is incident on a thin phase changing screen. The emerging wavefront is irregular. As the wave propagates to the observer, amplitude fluctuations develop as suggested by the crossing rays. Below the antennas is a plot of intensity versus position along the wavefront. If there is motion between the screen and the observer, the spatial fluctuations will be observed as temporal fluctuations in the power received or the fringe visibility.

where r_e is the classical electron radius, equal to $e^2/4\pi\epsilon_0 mc^2$ or 2.82×10^{-15} m, and the term in ν_B is neglected. Thus, the excess phase shift across one blob is

$$\Delta\phi_1 = r_e \lambda a \Delta\mathcal{N}_e, \tag{13.135}$$

where $\Delta\mathcal{N}_e$ is the excess electron density in the blob over the ambient level. If the thickness of the screen is L, then the wave will encounter about L/a blobs, and the rms phase deviation $\Delta\phi$ will be $\Delta\phi_1\sqrt{L/a}$ or

$$\Delta\phi = r_e \lambda \Delta\mathcal{N}_e \sqrt{La}. \tag{13.136}$$

The wave emerging from the screen is crinkled, that is, the amplitude is unchanged, but the phase is no longer constant and has random fluctuations with rms deviation $\Delta\phi$. The wave therefore can be decomposed into an angular spectrum of waves propagating with a variety of angles. The full width of the

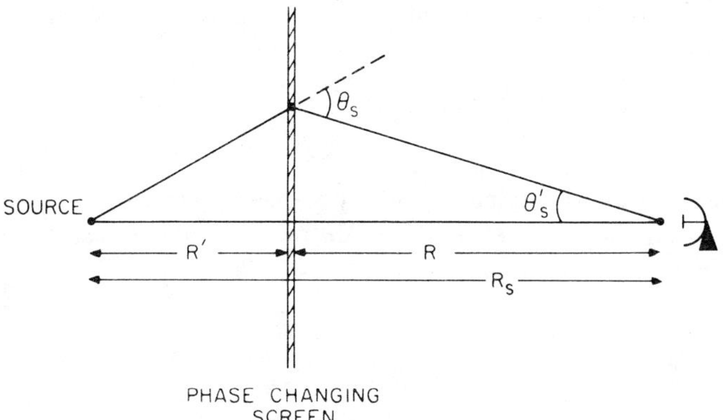

Figure 13.16 Path of a refracted ray in the thin phase screen model. θ_s is the rms scattering angle given by Eq. (13.137).

angular spectrum, θ_s, can be estimated by imagining that the random medium consists of refracting wedges that tilt the wavefront by the amount $\pm \Delta\phi\lambda/2\pi$ over a distance a. Thus,

$$\theta_s = \frac{1}{\pi} r_e \lambda^2 \Delta\mathcal{N}_e \sqrt{\frac{L}{a}} . \tag{13.137}$$

If the source is not infinitely distant, then the incident wave will not be plane. In that case the observed scattering angle θ'_s depends on the location of the screen with respect to the source and observer. Since θ_s and θ'_s are small angles, it follows from the geometry in Fig. 13.16 that

$$\theta'_s = \frac{R'}{R + R'} \theta_s. \tag{13.138}$$

Therefore, the effectiveness of the scattering screen is diminished if the screen is moved toward the source.

Amplitude fluctuations build up as the wave propagates away from the screen. If the phase fluctuations are large, $\Delta\phi > 1$, then significant amplitude fluctuations occur when rays cross (see Fig. 13.15). The critical distance beyond which large-amplitude fluctuations are observed is

$$R_f \simeq \frac{a}{\theta'_s} . \tag{13.139}$$

Note that if $\Delta\phi = 2\pi$, R_f is the distance for which the size of a blob is equal to the size of the first Fresnel zone. The random electric field distribution at the earth, in the plane perpendicular to the propagation direction, is called the diffraction pattern. It has a characteristic correlation length d_c given by

$$d_c \simeq \frac{\lambda}{\theta_s'}.$$
(13.140)

If the screen moves with relative velocity v_s in the direction perpendicular to the propagation direction, so that the diffraction pattern sweeps across the observer, then the time scale of variability is

$$\tau_d \simeq \frac{d_c}{v_s} \frac{R'}{R + R'} \simeq \frac{\lambda}{\theta_s v_s}.$$
(13.141)

The signal reaching the observer by traveling along the scattered ray path is delayed by an amount

$$\tau_c \simeq \frac{RR'\theta_s^2}{2c(R + R')}$$
(13.142)

with respect to the unscattered signal. The phase of the scattered wave is $2\pi\nu\tau_c$ with respect to the direct (unscattered) wave, and interference between these two waves causes scintillation. The bandwidth over which the relative phase changes by 2π is called the correlation bandwidth $\Delta\nu_c$. The correlation bandwidth is the reciprocal of τ_c, and for the case $R = R'$ is

$$\Delta\nu_c \simeq \frac{8c}{R_s\theta_s^2},$$
(13.143)

where R_s is the distance between the source and the observer. If observations are made with a receiver with a bandwidth greater than $\Delta\nu_c$, the amplitude fluctuations will be greatly reduced. Note from Eq. (13.143) and Eq. (13.137) that $\Delta\nu_c$ varies as λ^{-4}. Finally, if the source has two equal components separated by distance ℓ, then each component will produce the same diffraction pattern, but these patterns will be displaced at the earth by distance $\ell R/R'$. If this distance is greater than d_c, then the diffraction pattern will be smoothed and the amplitude fluctuations reduced. Thus, if the source size is greater than a critical size θ_c, amplitude fluctuations will be sharply reduced because the diffraction patterns from the component parts overlap and are smoothed out. θ_c is given by

$$\theta_c \simeq \frac{\lambda}{R\theta_s}.$$
(13.144)

A useful quantity is the average fringe visibility $\mathcal{V}_m$ measured by an interferometer in the presence of scintillation. Assume that the phases ϕ_1 and

ϕ_2 at two points along the phase screen separated by distance d are random variables with a joint Gaussian distribution with variance $\Delta\phi^2$ and normalized correlation $\rho(d)$. $\rho(d)$ is the correlation function of the phase or of the variable component of the index of refraction. The visibility can be calculated by the same method used in Appendix 3.1 to prove the Van Cittert–Zernike theorem using Huygens' principle, except that the phase changing screen is added. From the relation

$$\langle e^{j(\phi_1-\phi_2)}\rangle = e^{-\Delta\phi^2[1-\rho(d)]}, \tag{13.145}$$

one can show (Ratcliffe 1956, Mercier 1962) that

$$\langle \mathcal{V}_m\rangle = \mathcal{V}e^{-\Delta\phi^2[1-\rho(d)]}, \tag{13.146}$$

where $\Delta\phi$ is given by Eq. (13.136), $\mathcal{V}$ is the visibility, and $\langle \mathcal{V}_m\rangle$ is the expectation of the measured visibility. In much of the early radio astronomical literature, $\rho(d)$ is assumed to be a Gaussian function

$$\rho(d) = e^{-d^2/2a^2}, \tag{13.147}$$

where the characteristic scale length a corresponds to the blob size in the above discussion. This model, called the Gaussian screen model, is probably unrealistically restrictive because there are undoubtedly many scale sizes present. In the case where $\Delta\phi \gg 1$, $\mathcal{V}_m$ decreases rapidly as d increases, and we need only consider the case of $d \ll a$. Then substitution of Eq. (13.147) into Eq. (13.146) yields

$$\langle \mathcal{V}_m\rangle \simeq \mathcal{V}e^{-\Delta\phi^2 d^2/2a^2}. \tag{13.148}$$

Thus, the brightness distribution of a point source observed through a Gaussian screen is a Gaussian distribution with a diameter (full width at half-maximum) of [see, e.g., Eqs. (13.82)–(13.85)]

$$\theta_s \simeq \sqrt{2\ln 2}\,\frac{\Delta\phi\,\lambda}{\pi a} = \frac{\sqrt{2\ln 2}}{\pi}r_e\lambda^2\Delta\mathcal{N}_e\sqrt{\frac{L}{a}}. \tag{13.149}$$

This formula for θ_s is essentially equivalent to the one given in Eq. (13.137). In the case where $\Delta\phi \ll 1$, the normalized visibility function drops from unity to $e^{-\Delta\phi^2}$ when $d \gg a$. Therefore, the resulting brightness distribution for a point source is an unresolved core surrounded by a halo. The ratio of the flux density in the halo to the flux density in the core is $1 - e^{-\Delta\phi^2}$. The effects of a turbulent medium on the unaveraged visibility are discussed by Cronyn (1972).

Power-Law Model

A more realistic treatment involves modeling the spectrum of phase fluctuations as a power-law function similar to the Kolmogorov model for tropospheric turbulence. We assume that the power-law exponent is the same for plasma turbulence as it is for Kolmogorov turbulence in a neutral medium. With this assumption, which is consistent with observations, much of the theory of turbulence in the neutral atmosphere can be applied to the problem of turbulence in a plasma. The parameter C_n^2 is not particularly useful in characterizing plasma turbulence because it is a function of wavelength (since the index of refraction varies as λ^2). Thus we define a structure function of the electron density, similar to Eq. (13.96), given by

$$\mathscr{D}_{\mathcal{N}}(d) = C_{\mathcal{N}}^2 d^{2/3}, \tag{13.150}$$

where $C_{\mathcal{N}}^2$ characterizes the strength of the turbulence in the electron density distribution. From Eq. (13.134) the relationship between C_n^2 and $C_{\mathcal{N}}^2$ is

$$C_n^2 = \frac{r_e^2 \lambda^4}{4\pi^2} C_{\mathcal{N}}^2. \tag{13.151}$$

The units of $C_{\mathcal{N}}^2$ are meters$^{-20/3}$. There is some variation in the literature in the definition of $C_{\mathcal{N}}^2$. The structure function of phase, defined in Eq. (13.77), can be written in terms of the normalized correlation function $\rho(d)$ as

$$\mathscr{D}_{\phi}(d) = 2\Delta\phi^2 \left[1 - \rho(d)\right]. \tag{13.152}$$

Thus, by substituting Eq. (13.152) into Eq. (13.146), we obtain

$$\langle \mathscr{V}_m \rangle = \mathscr{V} e^{-\mathscr{D}_\phi/2}. \tag{13.153}$$

Using Eq. (13.98) for $\mathscr{D}_\phi$ and Eq. (13.151) for C_n^2, Eq. (13.153) becomes

$$\langle \mathscr{V}_m \rangle = \mathscr{V} \exp\left[-1.45 r_e^2 \lambda^2 C_{\mathcal{N}}^2 L d^{5/3}\right], \tag{13.154}$$

for baseline lengths between the inner and outer scales of turbulence. Equations (13.153) and (13.154) have rather general validity as shown by Lee and Jokipii (1975). For example, the phase of the wavefront need not be a Gaussian random process as assumed in Eq. (13.145). If the turbulent medium is not confined to a screen but distributed between the source and observer, then the term $C_{\mathcal{N}}^2 L$ in Eq. (13.154) must be replaced by a suitable integral of $C_{\mathcal{N}}^2$ along the line of sight that describes the scattering effectiveness at different distances (see Fig. 13.16 and Cordes, Weisberg, and Boriakoff 1985). If the turbulence is uniform between the source and observer, then $C_{\mathcal{N}}^2 L$ should be replaced by $C_{\mathcal{N}}^2 L/6$ in Eq. (13.154). The observed brightness distribution, the

Fourier transform of Eq. (13.154), differs slightly from a Gaussian distribution as can be seen in Fig. 13.9b. The scattering angle (full width at half-maximum) can be obtained from Eqs. (13.88) and (13.91) with $\beta = 5/6$ and $q = d/\lambda$,

$$\theta_s \simeq 4.1 \times 10^{-13}\left(C_{\mathcal{N}}^2 L\right)^{3/5}\lambda^{11/5} \quad \text{(arcsec)}, \tag{13.155}$$

where $C_{\mathcal{N}}^2 L$ and λ are in units of meters$^{-17/3}$ and meters, respectively. Thus, a difference between the power-law model and the Gaussian screen model is that θ_s, measured by Fourier transformation of visibility data over a range of baselines, is proportional to $\lambda^{2.2}$ in the former model and to λ^2 in the latter. Note that if $\langle \mathcal{V}_m \rangle$ were measured on a single baseline, that is, with d fixed, and if θ_s were estimated from the comparison of the measured visibility with the visibility expected for a Gaussian brightness distribution, then θ_s would appear to vary as λ^2 in both models.

13.4   INTERPLANETARY MEDIUM

Refraction

Radio waves passing near the sun are bent by the ionization of the solar corona and the solar wind. Calculation of the refraction in the extended solar atmosphere is important for the understanding of solar radio emission at low frequencies where the bending angles are large (Kundu 1965), and for tests of the general relativistic bending of electromagnetic radiation passing near the sun (Shapiro 1967, Fomalont and Sramek 1977). The accuracy of the measurement of relativistic bending in the radio region is limited by the accuracy with which the bending by the ionized media can be accounted for. We now discuss the refraction in the case relevant to relativistic bending experiments, that is, the microwave region where the bending is small.

The electron density as a function of distance from the sun can be measured in a variety of ways. Optical observations of Thomson scattering during solar eclipses have been analyzed to give an electron density model

$$\mathcal{N}_e = \left[1.55r^{-6} + 2.99r^{-16}\right] \times 10^{14} \quad (\text{m}^{-3}), \tag{13.156}$$

where r, the radial distance from the sun in units of the solar radius, is less than about 4. Equation (13.156) is the well-known Allen–Baumbach formula (Allen 1947). Eclipse data have also been interpreted by a model of the form

$$\mathcal{N}_e = a_1 r^{-6} + a_2 r^{-2.33}, \tag{13.157}$$

for $r > 3$ (see references in Muhleman, Ekers, and Fomalont 1970). The coefficients a_1 and a_2 depend on solar activity and can vary by a factor of 5 during an 11-year cycle. Scintillation measurements of the occultation of the

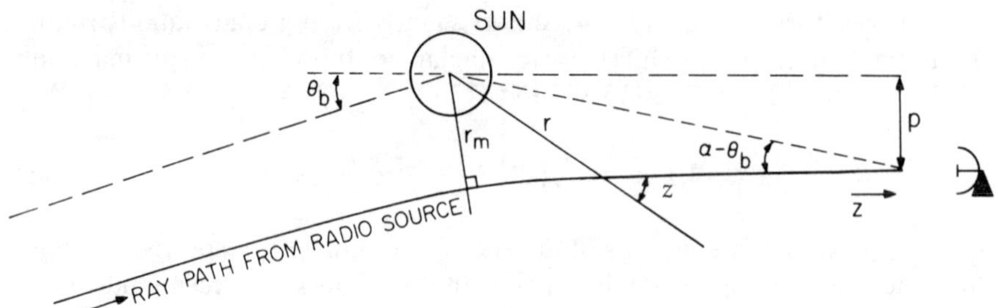

Figure 13.17 Path of a ray passing through the ionized gas surrounding the sun. p is the impact parameter and α is the angle between the sun and the source in the absence of solar bending.

Crab Nebula at 26 MHz can be represented reasonably well by the model

$$\mathcal{N}_e = 5 \times 10^{11} r^{-2} \quad (\mathrm{m}^{-3}), \tag{13.158}$$

for $4 < r < 20$ (Erickson 1964, Evans and Hagfors 1968). Dispersion measurements of pulsars during solar occultation give about the same result as Eq. (13.158) (Counselman and Rankin 1972, Counselman et al. 1974). The angle of refraction of a ray passing near the sun can be calculated readily for the case where this angle is small. A ray obeys Snell's law in spherical coordinates, $nr \sin z = \mathrm{constant}$ (Smart 1962), where z is the angle between the ray and a line from the center of the sun, as shown in Fig. 13.17. From this relation, the bending angle is found to be

$$\theta_b = \pi - 2 \int_{r_m}^{\infty} \frac{dr}{r\sqrt{(nr/p)^2 - 1}}, \tag{13.159}$$

where r_m is the distance of closest approach of the ray to the sun and p is the impact parameter (see Fig. 13.17). Assume that the electron density has a power-law distribution given by

$$\mathcal{N}_e = \mathcal{N}_0 r^{-\beta}, \tag{13.160}$$

where $\mathcal{N}_0$ is the electron density in meters^{-3} at one solar radius and β is a constant. For a fully ionized solar wind characterized by a constant mass loss rate and velocity, β is equal to 2. This case is applicable for $r \geq 10$.

The index of refraction is obtained by the substitution of Eqs. (13.160) and (13.118) into Eq. (13.124) and neglecting the term in ν_B. Graphical solutions of Eq. (13.159) for large bending angles are given by Jaeger and Westfold (1950). For small bending angles, an approximate solution to Eq. (13.159) can be obtained by the use of the substitution, $nr/p = \sec\theta$,

$$\theta_b \simeq 80.6\sqrt{\pi}\,\frac{\mathcal{N}_0}{v^2}\,\frac{\Gamma\left(\dfrac{\beta+1}{2}\right)}{\Gamma\left(\dfrac{\beta}{2}\right)}\,p^{-\beta}, \tag{13.161}$$

where p is in units of solar radii and Γ is the Gamma function. Note that the rays are bent away from the sun. The bending angle associated with the model in Eq. (13.158) is

$$\theta_b \simeq 2.4\lambda^2 p^{-2} \quad \text{(arcmin)}, \tag{13.162}$$

where λ is the wavelength in meters. For a multiple power-law model of electron density such as given in Eqs. (13.156) and (13.157), the bending angles for each component can be summed when the bending angles are small.

The *general relativistic bending* can be described classically by an effective index of refraction given by $1 + 2GM_\odot/rc^2$, where G is the gravitational constant, and $M_\odot$ is the mass of the sun. The bending angle, for small values of p, is (Weinberg 1972)

$$\theta_{GR} \simeq -1\rlap{.}''75\,p^{-1}. \tag{13.163}$$

The relativistic effect bends the rays toward the sun. In experiments to measure the relativistic bending, a model of the interferometer phase is formulated using Eq. (13.161) and a solar model with power-law components given by Eq. (13.160) with a coefficient $\mathcal{N}_{0i}$ for each component. The relativistic bending and the coefficients can be estimated simultaneously from the interferometer data. If, however, the electron density distribution has a component with $\beta = 1$, the relativistic effect would be masked if measurements were made at only one frequency. The solar wind is highly variable and attempts to characterize it by a power-law model with constant coefficients may not be adequate for high-precision experiments.

At large angular distances from the sun, a more appropriate approximation for the general relativistic bending is (Misner, Thorne, and Wheeler 1973)

$$\theta_{GR} \simeq -0\rlap{.}''00407\sqrt{\frac{1 + \cos\alpha}{1 - \cos\alpha}}, \tag{13.164}$$

where α is the solar elongation, that is, the angle between the sun and the source. This bending, which is about 4.1 milliarcsec at $\alpha = 90°$, can be detected at almost all values of solar elongation with VLBI measurements (Robertson and Carter 1984). Correction for the effect of relativistic bending must be made for many interferometric observations. Eq. (13.164) is appropriate for a source at infinity. This equation must be modified if the source under investigation is within the solar system (Shapiro 1967).

The excess phase path for a ray passing through the corona, for the case where the effect of ray bending can be neglected, is, from Eq. (13.128),

$$\mathcal{L} \simeq -\frac{40.3}{\nu^2} \int_{-\infty}^{\infty} \mathcal{N}_e \, dz. \tag{13.165}$$

For a power-law model given by Eq. (13.160), the excess path is

$$\mathcal{L} \simeq -\frac{40.3\,\mathcal{N}_0}{\nu^2} \int_{-\infty}^{\infty} \frac{dz}{\left(p^2 + z^2\right)^{\beta/2}}, \tag{13.166}$$

which can be integrated to give

$$\mathcal{L} \simeq -\frac{40.3\sqrt{\pi}}{\nu^2} \frac{\Gamma\left(\dfrac{\beta-1}{2}\right)}{\Gamma\left(\dfrac{\beta}{2}\right)} \mathcal{N}_0 p^{1-\beta}. \tag{13.167}$$

Note that the change in $\mathcal{L}$ with p describes the tilting of the wavefront and is the bending angle; hence $\theta_b \simeq d\mathcal{L}/dp$ (Bracewell, Eshleman, and Hollweg 1969). Differentiation of Eq. (13.167) gives Eq. (13.161).

Interplanetary Scintillation

The earliest investigation of scintillation of extragalactic radio sources due to irregularities in the solar wind was carried out by Hewish, Scott, and Wills (1964). Interplanetary scintillation is readily distinguishable from ionospheric scintillation since the time scale [Eq. (13.141)] and critical source size [Eq. (13.144)] are approximately 1 sec and 0.5 arcsec for interplanetary scintillation and 30 sec and 10 arcmin for ionospheric scintillation. Further observations of interplanetary scintillation by Cohen et al. (1967) showed that the angular size of the radio source 3C273B is smaller than $0\rlap{.}''02$, based on the application of Eq. (13.144). This result and the long-baseline interferometric results stimulated the development of VLBI. A comprehensive discussion of the interpretation of interplanetary scintillations can be found in Salpeter (1967), Young (1971), and Scott, Coles, and Bourgois (1983). For rough calculations, the scattering angle due to the interplanetary medium may be approximated by

(Erickson 1964)

$$\theta_s \simeq 50\left(\frac{\lambda}{p}\right)^2 \quad \text{(arcmin)}, \tag{13.168}$$

where λ is in meters and p, the impact parameter, is in solar radii.

The concept that extended sources do not scintillate as much as point sources [see Eq. (13.144)] can be generalized to obtain more information about source structure. We assume that the scintillation is caused by a screen at a distance R from the earth as shown in Fig. 13.16, where $R \ll R_s$, and that the intensity at the earth is $I(x, y)$, where x and y are coordinates in a plane parallel to the screen (see Fig. 13.15). The function $\Delta I(x, y) = I(x, y) - \langle I(x, y)\rangle$, where $\langle I(x, y)\rangle$ is the mean intensity, has a power spectrum $\mathcal{S}_{I0}(q_x, q_y)$ for a point source and $\mathcal{S}_I(q_x, q_y)$ for an extended source where q_x and q_y are the spatial frequencies (cycles per meter). If the visibility of the source is $\mathcal{V}(q_x R, q_y R)$, then it can be shown that (Cohen 1969)

$$\mathcal{S}_I(q_x, q_y) = \mathcal{S}_{I0}(q_x, q_y)\left|\mathcal{V}(q_x R, q_y R)\right|^2, \tag{13.169}$$

where $q_x R$ and $q_y R$ are the projected baseline coordinates u and v. The scintillation index of the source m_s is defined by

$$m_s^2 = \frac{\langle \Delta I(x, y)^2\rangle}{\langle I(x, y)\rangle^2} = \frac{1}{\langle I(x, y)\rangle^2}\int_{-\infty}^{\infty}\int_{-\infty}^{\infty}\mathcal{S}_I(q_x, q_y)\, dq_x\, dq_y. \tag{13.170}$$

In principle, $\mathcal{S}_I(q_x, q_y)$ could be computed from the simultaneous measurements of $\Delta I(x, y)$ with a large number of spaced receivers. In practice, the motion of the solar wind sweeps the diffraction pattern across a single telescope so that, from measurements of $\Delta I(t)$, the temporal power spectrum $\mathcal{S}(f)$ can be calculated. If the diffraction pattern moves with velocity v_s in the x direction, then $\mathcal{S}(f)$ can be related to the spatial spectrum since $q_x = f/v_s$:

$$\mathcal{S}(f) = \frac{1}{v_s}\int \mathcal{S}_I(q_x, q_y)\, dq_y. \tag{13.171}$$

In principle, $|\mathcal{V}|^2$ can be recovered from Eq. (13.169) by observing a source over a range of different orientations with respect to the solar wind vector. The situation is entirely analogous to that of lunar occultation observations (Chapter 15) except that with lunar occultation observations the visibility phase can also be obtained. An estimate of the source diameter can be deduced from the width of the temporal power spectrum (Cohen, Gundermann, and Harris 1967) or from the scintillation index [Eq. (13.170)] (Little and Hewish 1966).

13.5 INTERSTELLAR MEDIUM

Dispersion and Faraday Rotation

The smooth, ionized component of the interstellar medium of our galaxy affects propagation by introducing delay and Faraday rotation. The time of arrival of a pulse of radiation, such as that from a pulsar, is

$$t_p = \int_0^L \frac{dz}{v_g},$$
(13.172)

where L is the propagation path, v_g is the group velocity (cn), and n is given by Eq. (13.117). Differentiation of Eq. (13.172) gives

$$\frac{dt_p}{dv} \simeq -\frac{e^2}{4\pi\epsilon_0 mcv^3}\int_0^L \mathcal{N}_e \, dz.$$
(13.173)

The integral of $\mathcal{N}_e$ over the path length is called the *dispersion measure*,

$$D_m = \int_0^L \mathcal{N}_e \, dz,$$
(13.174)

which is the same quantity as the total electron content. dt_p/dv can be estimated by measuring the time of arrival of pulsar pulses at different frequencies, and the dispersion measure can then be found from Eq. (13.173). If the distance to the pulsar is known, then the average electron density can be calculated. A typical value of $\langle \mathcal{N}_e \rangle$ in the plane of our galaxy is 0.03 cm^{-3} (Weisberg, Rankin, and Boriakoff 1980). Alternatively, if a pulsar's distance is unknown, it can be estimated from Eq. (13.174) using an estimated average value of $\mathcal{N}_e$.

The magnetic field of the galaxy causes Faraday rotation of the polarization plane of radiation from extragalactic radio sources. Equation (13.125) can be rewritten

$$\Delta\psi = \lambda^2 R_m,$$
(13.175)

where R_m is the *rotation measure* given by

$$R_m = 8.1 \times 10^5 \int \mathcal{N}_e B_\parallel \, dz.$$
(13.176)

Here R_m is in radians per square meter, λ is in meters, $B_\parallel$ is the longitudinal component of magnetic field in gauss (1 gauss = 10^{-4} tesla), $\mathcal{N}_e$ is in centimeters^{-3}, and dz is in parsecs (pc) (1 pc = 3.1×10^{16} m). The interstellar magnetic field can be estimated by dividing the rotation measure by the dispersion measure. Typical values of the magnetic field obtained in this way are 2 μG (Heiles 1976). This procedure underestimates the magnetic field if the field reverses direction along the line of sight. A formula for roughly estimating the rotation measure due to the galactic magnetic field is (Spitzer 1978)

$$R_m = -18|\cot b|\cos(l - 94°), \tag{13.177}$$

where l and b are the galactic longitude and latitude. Extensive measurements of rotation measure as a function of direction can be found in Simard-Normandin and Kronberg (1980).

Faraday rotation that occurs within a radio source depolarizes the emergent radiation. This depolarization happens because radiation emitted from different depths in the source suffers different amounts of Faraday rotation. Such a source might be a relativistic gas emitting polarized synchrotron radiation immersed in a thermal plasma that causes the Faraday rotation. The degree of polarization of the observed radiation can be succinctly described in a Fourier transform relationship when self-absorption is negligible. We first introduce the function M, the complex degree of linear polarization, defined by

$$M = m_\ell e^{j2\psi} = \frac{Q + jU}{I}, \tag{13.178}$$

where m_ℓ is the degree of linear polarization, ψ is the position angle of the electric field, and Q, U, and I are the Stokes parameters as defined in Section 4.9. If z is the linear distance into the source, $\psi(z)$ is the intrinsic position angle of the radiation at depth z, $j_\nu(z)$ is the volume emissivity of the source, and $\lambda^2\beta(z)$ is the Faraday rotation suffered by radiation emitted at depth z, then the degree of polarization of the observed radiation can be written

$$M(\lambda^2) = \frac{\int_0^\infty m_\ell(z) j_\nu(z) e^{j2[\psi(z)+\lambda^2\beta(z)]}\, dz}{\int_0^\infty j_\nu(z)\, dz}. \tag{13.179}$$

The denominator in Eq. (13.179) is the total brightness. $\beta(z)$ is the Faraday depth, which increases monotonically into the source as long as the sign of the longitudinal magnetic field direction does not change. In any case, we can superpose all the radiation from the same Faraday depth and write the integrals in Eq. (13.179) as a function of β instead of z, yielding

$$M(\lambda^2) = \int_{-\infty}^\infty F(\beta) e^{j2\lambda^2\beta}\, d\beta, \tag{13.180}$$

where

$$F(\beta) = \frac{m_\ell(z)\,j_\nu(z)\,e^{j2\psi(z)}}{\displaystyle\int_0^\infty j_\nu(z)\,dz}.$$

(13.181)

Thus $M(\lambda^2)$ and $F(\beta)$ form a Fourier transform pair. $F(\beta)$ is sometimes called the Faraday dispersion function. Unfortunately, $F(\beta)$, in general, cannot be found since M cannot be measured for negative values of λ^2. Because of this difficulty with the Fourier transform, $F(\beta)$ is usually estimated by model fitting. However, if $\psi(z)$ is constant, then $M(-\lambda^2) = M^*(\lambda^2)$, and $F(\beta)$ can be obtained by Fourier transformation.

Consider the result for a simple source model for which ψ and j_ν are constant. From Eq. (13.179) or Eq. (13.180),

$$M(\lambda^2) = M(0)\left[\frac{\sin \lambda^2 R_m}{\lambda^2 R_m}\right]e^{j\lambda^2 R_m},$$

(13.182)

where R_m is the Faraday rotation measure through the whole source. If the Faraday rotation originates in front of the radiation source, the complex degree of polarization is

$$M(\lambda^2) = M(0)\,e^{j2\lambda^2 R_m}.$$

(13.183)

In this case there is no depolarization, and the Faraday rotation is twice that of Eq. (13.182) in which the source is uniformly distributed throughout the rotation medium. For detailed treatment of intrinsic Faraday rotation, see Burn (1966) and Gardner and Whiteoak (1966).

Interstellar Scintillation

Interstellar scintillation has been extensively investigated by observation of pulsars and compact extragalactic radio sources. For pulsars the temporal broadening of the pulses [Eq. (13.142)], the decorrelation bandwidth [Eq. (13.143)], and the angular broadening [Eq. (13.138)] can be measured. Interpretation of the measurements in terms of a thin screen model suggests that $\Delta\mathcal{N}_e/\mathcal{N}_e \simeq 10^{-3}$ and that the scale size responsible for the scintillation is on the order of 10^{11} cm. The temporal variations or fading of the signal from a pulsar are caused by the motions of the observer and the pulsar relative to the quasi-stationary interstellar medium. A measurement of the decorrelation bandwidth can be used to estimate the scattering angle [Eq. (13.143)]. This estimate of the scattering angle and the measurement of the time scale of fading (10^2–10^3 sec at 408 MHz) can be used to estimate the relative velocity of the scattering screen by Eq. (13.141). From the relative velocity of the screen, the transverse velocity of the pulsar can be found. Velocities, and thus proper motions, of pulsars estimated in this way agree well with those measured directly with interferometers (Lyne and Smith 1982). The transverse component of the orbital velocity of a binary pulsar has also been measured (Lyne 1984).

More sophisticated analyses of interstellar scintillation (Rickett 1977; Cordes, Weisberg, and Boriakoff 1985) show that the fluctuations in electron density can be described by a power-law spectrum with a power-law exponent of about 3.7 ± 0.3, which is similar to the value of $11/3$ for Kolmogorov turbulence in the neutral atmosphere. The power-law spectrum appears to extend over a range of scale sizes from less than 10^{10} cm to greater than 10^{15} cm. Irregularities on the largest scales are thought to be responsible for the slow fading of signals from pulsars and the low-frequency variability of some extragalactic sources, which have time scales of days to months (Rickett, Coles, and Bourgois 1984)

Extensive measurements of the angular sizes of extragalactic radio sources have been used to derive an approximate formula for θ_s based on the Gaussian screen model, by Harris, Zeissig, and Lovelace (1970), Readhead and Hewish (1972), Cohen and Cronyn (1974), Duffett-Smith and Readhead (1976), and others. This formula is

$$\theta_s \simeq \frac{15}{\sqrt{|\sin b|}} \lambda^2 \quad \text{(milliarcsec)}, \qquad |b| > 15° \qquad (13.184)$$

where b is the galactic latitude and λ is the wavelength in meters.

The pulsar data have been interpreted by Cordes (1984) in terms of the power-law model to arrive at approximate formulas for θ_s:

$$\theta_s \simeq 7.5\lambda^{11/5} \qquad \text{(arcsec)}, \qquad |b| \leq 0°\!.6$$

$$\simeq 0.5|\sin b|^{-3/5}\lambda^{11/5} \quad \text{(arcsec)}, \qquad 0°\!.6 < |b| < 3°\text{–}5°$$

$$\simeq 13|\sin b|^{-3/5}\lambda^{11/5} \quad \text{(milliarcsec)}, \qquad |b| \geq 3°\text{–}5°. \qquad (13.185)$$

The accuracy of the representations in Eq. (13.185) decreases with decreasing $|b|$. In particular, the scattering angle at low latitudes, $|b| < 1°$, can take on a large range of values (Cordes, Ananthakrishnan, and Dennison 1984).

An example of a compact radio source that suffers a large degree of interstellar scattering is the one known as Sagittarius A in the center of our galaxy. This source has an angular size of about 15 milliarcsec at a wavelength of 3.6 cm [compared with 7.7 milliarcsec predicted by Eq. (13.185)]. The angular size varies approximately as the wavelength squared over the entire measuring range 1.3–30 cm (Lo et al. 1985).

Interstellar scattering probably places the ultimate limit on the angular size measurements that can be made with an interferometer. The apparent sizes of interstellar masers, which are mostly found in the galaxy at low galactic latitudes, are sometimes set by interstellar scattering (Little 1973). In some cases, it may be possible to determine source sizes smaller than the scattering limit by using bandwidths and integration times smaller than the corresponding characteristics of the scintillation (Cohen and Cronyn 1974).

REFERENCES

Aarons, J., Global Morphology of Ionospheric Scintillations, *Proc. IEEE*, **70**, 360–378, 1982.

Aarons, J., J. A. Klobuchar, H. E. Whitney, J. Austen, A. L. Johnson, and C. L. Rino, Gigahertz Scintillations Associated with Equatorial Patches, *Radio Sci.*, **18**, 421–434, 1983.

Allen, C. W., Interpretation of Electron Densities from Corona Brightness, *Mon. Not. R. Astron. Soc.*, **107**, 426–432, 1947.

Allen, C. W., *Astrophysical Quantities*, 2nd ed., University of London, London, 1964, p. 119.

Appleton, E. V. and M. A. F. Barnett, On Some Direct Evidence for Downward Atmospheric Reflection of Electric Rays, *Proc. R. Soc. Lond. A*, **109**, 621–641, 1925.

Armstrong, J. W. and R. A. Sramek, Observations of Tropospheric Phase Scintillations at 5 GHz on Vertical Paths, *Radio Sci.*, **17**, 1579–1586, 1982.

Baars, J. W. M., Meteorological Influences on Radio Interferometer Phase Fluctuations, *IEEE Trans. Antennas Propag.*, **AP-15**, 582–584, 1967.

Bailey, D. K., On a New Method of Exploring the Upper Atmosphere, *J. Terr. Mag. Atmos. Elec.*, **53**, 41–50, 1948.

Bean, B. R., B. A. Cahoon, C. A. Samson, and G. D. Thayer, *A World Atlas of Atmospheric Refractivity*, U.S. Government Printing Office, Washington, DC, 1966.

Bean, B. R. and E. J. Dutton, *Radio Meteorology*, National Bureau of Standards Monograph 92, U.S. Government Printing Office, Washington, DC, 1966.

Beynon, W. J. G., Marconi, Radio Waves, and the Ionosphere, *Radio Sci.*, **10**, 657–664, 1975.

Bieging, J. H., J. Morgan, W. J. Welch, S. N. Vogel, and M. C. H. Wright, Interferometer Measurements of Atmospheric Phase Noise at 86 GHz, *Radio Sci.*, **19**, 1505–1509, 1984.

Bohlander, R. A., R. W. McMillan, and J. J. Gallagher, Atmospheric Effects on Near-Millimeter-Wave Propagation, *Proc. IEEE*, **73**, 49–60, 1985.

Booker, H. G., The Use of Radio Stars to Study Irregular Refraction of Radio Waves in the Ionosphere, *Proc. IRE*, **46**, 298–314, 1958.

Booker, H. G., J. A. Ratcliffe, and D. H. Shinn, Diffraction from an Irregular Screen with Applications to Ionospheric Problems, *Philos. Tran. R. Soc. Lond. A*, **242**, 579–607, 1950.

Bracewell, R. N., *The Fourier Transform and Its Applications*, McGraw-Hill, New York, 1978.

Bracewell, R. N., V. R. Eshleman, and J. V. Hollweg, The Occulting Disk of the Sun at Radio Wavelengths, *Astrophys. J.*, **155**, 367–368, 1969.

Breit, G. and M. A. Tuve, A Test of the Existence of the Conducting Layer, *Phys. Rev.*, **28**, 554–575, 1926.

Budden, K. G., *Radio Waves in the Ionosphere*, Cambridge University Press, Cambridge, England, 1961.

Burn, B. J., On the Depolarization of Discrete Radio Sources by Faraday Dispersion, *Mon. Not. R. Astron. Soc.*, **133**, 67–83, 1966.

COESA, *U.S. Standard Atmosphere, 1976*, NOAA-S/T 76-1562, U.S. Government Printing Office, Washington, DC, 1976.

Cohen, M. H., High-Resolution Observations of Radio Sources, *Ann. Rev. Astron. Astrophys.*, **7**, 619–664, 1969.

Cohen, M. H. and W. M. Cronyn, Scintillation and Apparent Angular Diameter, *Astrophys. J.*, **192**, 193–197, 1974.

Cohen, M. H., E. J. Gundermann, H. E. Hardebeck, and L. E. Sharp, Interplanetary Scintillations. II. Observations, *Astrophys. J.*, **147**, 449–466, 1967.

Cohen, M. H., E. J. Gundermann, and D. E. Harris, New Limits on the Diameters of Radio Sources, *Astrophys. J.*, **150**, 767–782, 1967.

Cordes, J. M., Interstellar Scattering, in *VLBI and Compact Radio Sources, IAU Symposium 110*, R. Fanti, K. Kellermann, and G. Setti, Eds., D. Reidel, Dordrecht, Holland, 1984, pp. 303–307.

Cordes, J. M., S. Ananthakrishnan, and B. Dennison, Radio Wave Scattering in the Galactic Disk, *Nature*, **309**, 689–691, 1984.

Cordes, J. M., J. M. Weisberg, and V. Boriakoff, Small Scale Electron Density Turbulence in the Interstellar Medium, *Astrophys. J.*, **288**, 221–247, 1985.

Coulman, C. E., Fundamental and Applied Aspects of Astronomical "Seeing", *Ann. Rev. Astron. Astrophys.*, **23**, 19–57, 1985.

Counselman, C. C., III, S. M. Kent, C. A. Knight, I. I. Shapiro, T. A. Clark, H. F. Hinteregger, A. E. E. Rogers, and A. R. Whitney, Solar Gravitational Deflection of Radio Waves Measured by Very-Long-Baseline Interferometry, *Phys. Rev. Lett.*, **33**, 1621–1623, 1974.

Counselman, C. C., III and J. M. Rankin, Density of the Solar Corona from Occultations of NP0532, *Astrophys. J.*, **175**, 843–856, 1972.

Crane, R. K., Refraction Effects in the Neutral Atmosphere, in *Methods of Experimental Physics*, Vol. 12B, M. L. Meeks, Ed., Academic Press, New York, 1976, pp. 186–200.

Crane, R. K., Ionospheric Scintillation, *Proc. IEEE*, **65**, 180–199, 1977.

Crane, R. K., Fundamental Limitations Caused by RF Propagation, *Proc. IEEE*, **69**, 196–209, 1981.

Cronyn, W. M., Interferometer Visibility Scintillation, *Astrophys. J.*, **174**, 181–200, 1972.

Davies, K., *Ionospheric Radio Propagation*, National Bureau of Standards Monograph 80, U.S. Government Printing Office, Washington, DC, 1965.

Davis, J. L., T. A. Herring, I. I. Shapiro, A. E. E. Rogers, and G. Elgered, Geodesy by Radio Interferometry: Effects of Atmospheric Modeling Errors on Estimates of Baseline Length, *Radio Sci.*, **20**, 1593–1607, 1985.

Debye, P., *Polar Molecules*, Dover, New York, 1929.

Duffett-Smith, P. J. and A. C. S. Readhead, The Angular Broadening of Radio Sources by Scattering in the Interstellar Medium, *Mon. Not. R. Astron. Soc.*, **174**, 7–17, 1976.

Elgered, G., B. O. Rönnäng, and J. I. H. Askne, Measurements of Atmospheric Water Vapor with Microwave Radiometry, *Radio Sci.*, **17**, 1258–1264, 1982.

Erickson, W. C., The Radio-Wave Scattering Properties of the Solar Corona, *Astrophys. J.*, **139**, 1290–1311, 1964.

Evans, J. V., Theory and Practice of Ionospheric Study by Thomson Scatter Radar, *Proc. IEEE*, **57**, 496–530, 1969.

Evans, J. V. and T. Hagfors, *Radar Astronomy*, McGraw-Hill, New York, 1968.

Evans, J. V., J. M. Holt, and R. H. Wand, A Differential-Doppler Study of Traveling Ionospheric Disturbances from Millstone Hill, *Radio Sci.*, **18**, 435–451, 1983.

Fejer, B. G. and M. C. Kelley, Ionospheric Irregularities, *Rev. Geophys. Space Sci.*, **18**, 401–454, 1980.

Fomalont, E. B. and R. A. Sramek, A Confirmation of Einstein's General Theory of Relativity by Measuring the Bending of Microwave Radiation in the Gravitational Field of the Sun, *Astrophys. J.*, **199**, 749–755, 1975.

Fomalont, E. B. and R. A. Sramek, The Deflection of Radio Waves by the Sun, *Comments Astrophys.*, **7**, 19–33, 1977.

Fried, D. L., Statistics of a Geometric Representation of Wavefront Distortion, *J. Opt. Soc. Am.*, **55**, 1427–1435, 1965.

Fried, D. L., Optical Heterodyne Detection of an Atmospherically Distorted Signal Wave Front, *Proc. IEEE*, **55**, 57–67, 1967.

Gardner, F. F. and J. B. Whiteoak, The Polarization of Cosmic Radio Waves, *Ann. Rev. Astron. Astrophys.*, **4**, 245–292, 1966.

Goldstein, H., Attenuation by Condensed Water, in *Propagation of Short Radio Waves*, M.I.T. Radiation Laboratory Series, Vol. 13, D. E. Kerr, Ed., McGraw-Hill, New York, 1951, p. 671.

Guiraud, F. O., J. Howard, and D. C. Hogg, A Dual-Channel Microwave Radiometer for Measurement of Precipitable Water Vapor and Liquid, *IEEE Trans. Geosci. Electron.*, **GE-17**, 129-136, 1979.

Hagfors, T., The Ionosphere, in *Methods of Experimental Physics*, Vol. 12B, M. L. Meeks, Ed., Academic Press, New York, 1976, pp. 119–135.

Hamaker, J. P., Atmospheric Delay Fluctuations with Scale Sizes Greater than One Kilometer, Observed with a Radio Interferometer Array, *Radio Sci.*, **13**, 873–891, 1978.

Hargrave, P. J. and L. J. Shaw, Large-Scale Tropospheric Irregularities and Their Effect on Radio Astronomical Seeing, *Mon. Not. R. Astron. Soc.*, **182**, 233–239, 1978.

Harris, D E., G. A. Zeissig, and R. V. Lovelace, The Minimum Observable Diameter of Radio Sources, *Astron. Astrophys.*, **8**, 98–104, 1970.

Heiles, C., The Interstellar Magnetic Field, *Ann. Rev. Astron. Astrophys.*, **14**, 1–22, 1976.

Hess, S. L., *Introduction to Theoretical Meteorology*, Holt, Rinehart and Winston, New York, 1959.

Hewish, A., The Diffraction of Galactic Radio Waves as a Method of Investigating the Irregular Structure of the Ionosphere, *Proc. R. Soc. Lond. A*, **214**, 494–514, 1952.

Hewish, A., P. F. Scott, and D. Wills, Interplanetary Scintillation of Small Diameter Radio Sources, *Nature*, **203**, 1214–1217, 1964.

Hey, J. S., S. J. Parsons, and J. W. Phillips, Fluctuations in Cosmic Radiation at Radio Frequencies, *Nature*, **158**, 234, 1946.

Hill, R. J. and S. F. Clifford, Contribution of Water Vapor Monomer Resonances to Fluctuations of Refraction and Absorption for Submillimeter through Centimeter Wavelengths, *Radio Sci.*, **16**, 77–82, 1981.

Hill, R. J., R. S. Lawrence, and J. T. Priestley, Theoretical and Calculational Aspects of the Radio Refractive Index of Water Vapor, *Radio Sci.*, **17**, 1251–1257, 1982.

Hinder, R. A., Observations of Atmospheric Turbulence with a Radio Telescope at 5 GHz, *Nature*, **225**, 614–617, 1970.

Hinder, R. A., Fluctuations of Water Vapour Content in the Troposphere as Derived from Interferometric Observations of Celestial Radio Sources, *J. Atmos. Terr. Phys.*, **34**, 1171–1186, 1972.

Hinder, R. A. and M. Ryle, Atmospheric Limitations to the Angular Resolution of Aperture Synthesis Radio Telescopes, *Mon. Not. R. Astron. Soc.*, **154**, 229–253, 1971.

Hogg, D. C., F. O. Guiraud, and W. B. Sweezy, The Short-Term Temporal Spectrum of Precipitable Water Vapor, *Science*, **213**, 1112–1113, 1981.

Holt, E. H. and R. E. Haskell, *Foundations of Plasma Dynamics*, MacMillan, New York, 1965, p. 254.

Humphreys, W. J., *Physics of the Air*, 3rd ed., McGraw-Hill, New York, 1940.

Hunsucker, R. D., Atmospheric Gravity Waves Generated in the High-Latitude Ionosphere: A Review, *Rev. Geophys. Space Phys.*, **20**, 293–315, 1982.

Jackson, J. D., *Classical Electrodynamics*, 2nd ed., Wiley, New York, 1975, pp. 811–821.

Jaeger, J. C. and K. C. Westfold, Equivalent Path and Absorption for Electromagnetic Radiation in the Solar Corona, *Aust. J. Phys.*, **3**, 376–386, 1950.

Kaplan, G. H., F. J. Josties, P. E. Angerhofer, K. J. Johnston, and J. H. Spencer, Precise Radio Source Positions from Interferometric Observations, *Astron. J.*, **87**, 570–576, 1982.

Kundu, M. R., *Solar Radio Astronomy*, Wiley-Interscience, New York, 1965, p. 104.

Lawrence, R. S., C. G. Little, and H. J. A. Chivers, A Survey of Ionospheric Effects Upon Earth-Space Radio Propagation, *Proc. IEEE*, **52**, 4–27, 1964.

Lee, L. C. and J. R. Jokipii, Strong Scintillations in Astrophysics. I. The Markov Approximation, Its Validity and Application to Angular Broadening, *Astrophys. J.*, **196**, 695–707, 1975.

Liebe, H. J., Calculated Tropospheric Dispersion and Absorption Due to the 22-GHz Water Vapor Line, *IEEE Trans. Antennas Propag.*, **AP-17**, 621–627, 1969.

Liebe, H. J., Modeling Attenuation and Phase of Radio Waves in Air at Frequencies below 1000 GHz, *Radio Sci.*, **16**, 1183–1199, 1981.

Liebe, H. J., An Updated Model for Millimeter Wave Propagation in Moist Air, *Radio Sci.*, **20**, 1069–1089, 1985.

Little, L. T., Interstellar Scattering and the Angular Diameters of OH Components, *Astrophys. Lett.*, **13**, 115–118, 1973.

Little, L. T. and A. Hewish, Interplanetary Scintillation and Its Relation to the Angular Structure of Radio Sources, *Mon. Not. R. Astron. Soc.*, **134**, 221–237, 1966.

Lo, K. Y., D. C. Backer, R. D. Ekers, K. I. Kellermann, M. J. Reid, and J. M. Moran, On the Size of the Galactic Centre Compact Radio Source: Diameter < 20 AU, *Nature*, **315**, 124–126, 1985.

Loudon, R., *The Quantum Theory of Light*, Oxford University Press, London, 1973.

Lyne, A. G., Orbital Inclination and Mass of the Binary Pulsar PSR0655 + 64, *Nature*, **310**, 300–302, 1984.

Lyne, A. G. and F. G. Smith, Interstellar Scintillation and Pulsar Velocities, *Nature*, **298**, 825–827, 1982.

Marini, J. W., Correction of Satellite Tracking Data for an Arbitrary Tropospheric Profile, *Radio Sci.*, **7**, 223–231, 1972.

Mathur, N. C., M. D. Grossi, and M. R. Pearlman, Atmospheric Effects in Very Long Baseline Interferometry, *Radio Sci.*, **5**, 1253–1261, 1970.

Mercier, R. P., Diffraction by a Screen Causing Large Random Phase Fluctuations, *Proc. R. Soc. Lond. A*, **58**, 382–400, 1962.

Miner, G. F., D. D. Thornton, and W. J. Welch, The Inference of Atmospheric Temperature Profiles from Ground-Based Measurements of Microwave Emission from Atmospheric Oxygen, *J. Geophys. Res.*, **77**, 975–991, 1972.

Misner, C. W., K. S. Thorne, and J. A. Wheeler, *Gravitation*, W. H. Freedman, San Francisco, 1973, Section 40.3.

Moran, J. M. and B. R. Rosen, Estimation of the Propagation Delay through the Troposphere from Microwave Radiometer Data, *Radio Sci.*, **16**, 235–244, 1981.

Muhleman, D. O., R. D. Ekers, and E. B. Fomalont, Radio Interferometric Test of the General Relativistic Light Bending Near the Sun, *Phys. Rev. Lett.*, **24**, 1377–1380, 1970.

Owens, J. C., Optical Refractive Index of Air: Dependence on Pressure, Temperature, and Composition, *Appl. Opt.*, **6**, 51–58, 1967.

Ratcliffe, J. A., Some Aspects of Diffraction Theory and Their Application to the Ionosphere, *Rep. Prog. Phys.*, **19**, 188–267, 1956.

Ratcliffe, J. A., *The Magneto-Ionic Theory and Its Application to the Ionosphere*, Cambridge University Press, Cambridge, England, 1962.

Rawer, K., *The Ionosphere*, Ungar, New York, 1956.

Readhead, A. C. S. and A. Hewish, Galactic Structure and the Apparent Size of Radio Sources, *Nature*, **236**, 440–443, 1972.

Reber, E. E. and J. R. Swope, On the Correlation of Total Precipitable Water in a Vertical Column and Absolute Humidity, *J. Appl. Meteorol.*, **11**, 1322–1325, 1972.

Resch, G. M., Water Vapor Radiometry in Geodetic Applications, in *Geodetic Aspects of Electromagnetic Wave Propagation Through the Atmosphere*, F. K. Brunner, Ed., Springer-Verlag, Berlin, 1984.

Resch, G. M., D. E. Hogg, and P. J. Napier, Radiometric Correction of Atmospheric Path Length Fluctuations in Interferometric Experiments, *Radio Sci.*, **19**, 411–422, 1984.

Rickett, B. J., Interstellar Scattering and Scintillation of Radio Waves, *Ann. Rev. Astron. Astrophys.*, **15**, 479–504, 1977.

Rickett, B. J., W. A. Coles, and G. Bourgois, Slow Scintillation in the Interstellar Medium, *Astron. Astrophys.*, **134**, 390–395, 1984.

Roberts, D. H., A. E. E. Rogers, B. R. Allen, C. L. Bennett, B. F. Burke, P. E. Greenfield, C. R. Lawrence, and T. A. Clark, Radio Interferometric Detection of a Traveling Ionospheric Disturbance Excited by the Explosion of Mt. St. Helens, *J. Geophys. Res.*, **87**, 6302–6306, 1982.

Robertson, D. S. and W. E. Carter, Relativistic Deflection of Radio Signals in the Solar Gravitational Field Measured with VLBI, *Nature*, **310**, 572–574, 1984.

Roddier, F., The Effects of Atmospheric Turbulence in Optical Astronomy, in *Progress in Optics XIX*, E. Wolf, Ed., North-Holland, Amsterdam, 1981, pp. 281–376.

Rogers, A. E. E., A. T. Moffet, D. C. Backer, and J. M. Moran, Coherence Limits in VLBI Observations at 3-Millimeter Wavelength, *Radio Sci.*, **19**, 1552–1560, 1984.

Rogers, A. E. E., and J. M. Moran, Coherence Limits for Very Long Baseline Interferometry, *IEEE Trans. Inst. Meas.*, **IM-30**, 283–286, 1981.

Rybicki, G. B. and A. P. Lightman, *Radiative Processes in Astrophysics*, Wiley-Interscience, New York, 1979.

Saastamoinen, J., Introduction to Practical Computation of Astronomical Refraction, *Bull. Geodesique*, **106**, 383–397, 1972a.

Saastamoinen, J., Atmospheric Correction for the Troposphere and Stratosphere in Radio Ranging of Satellites, in *The Use of Artificial Satellites for Geodesy*, Geophysical Monograph 15, American Geophysical Union, Washington, DC, 1972b, pp. 247–251.

Salpeter, E. E., Interplanetary Scintillations. I. Theory, *Astrophys. J.*, **147**, 433–448, 1967.

Schaper, L. W., Jr., D. H. Staelin, and J. W. Waters, The Estimation of Tropospheric Electrical Path Length by Microwave Radiometry, *Proc. IEEE*, **58**, 272–273, 1970.

Scheuer, P. A. G., Amplitude Variations in Pulsed Radio Sources, *Nature*, **218**, 920–922, 1968.

Scott, S. L., W. A. Coles, and G. Bourgois, Solar Wind Observations near the Sun using Interplanetary Scintillation, *Astron. Astrophys.*, **123**, 207–215, 1983.

Shapiro, I. I., New Method for the Detection of Light Deflection by Solar Gravity, *Science*, **157**, 806–808, 1967.

Shapiro, I. I., Estimation of Astrometric and Geodetic Parameters, in *Methods of Experimental Physics*, Vol. 12C, M. L. Meeks, Ed., Academic Press, New York, 1976, pp. 261–276.

Simard-Normandin, M. and P. P. Kronberg, Rotation Measures and the Galactic Magnetic Field, *Astrophys. J.*, **242**, 74–94, 1980.

Smart, W. M., *Textbook on Spherical Astronomy*, 5th ed., Cambridge University Press, Cambridge, England, 1962.

Smith, E. K., Jr. and S. Weintraub, The Constants in the Equation for Atmospheric Refractive Index at Radio Frequencies, *Proc. IRE*, **41**, 1035–1037, 1953.

Smith, F. G., C. G. Little, and A. C. B. Lovell, Origin of the Fluctuations in the Intensity of Radio Waves from Galactic Sources, *Nature*, **165**, 422–424, 1950.

Snider, J. B., Ground-Based Sensing of Temperature Profiles from Angular and Multi-Spectral Microwave Emission Measurements, *J. Appl. Meteorol.*, **11**, 958–967, 1972.

Snider, J. B., H. M. Burdick, and D. C. Hogg, Cloud Liquid Measurement with a Ground-Based Microwave Instrument, *Radio Sci.*, **15**, 683–693, 1980.

Spitzer, L., *Physical Processes in the Interstellar Medium*, Wiley-Interscience, New York, 1978, p. 65.

Spoelstra, T. A. T., The Influence of Ionospheric Refraction on Radio Astronomy Interferometry, *Astron. Astrophys.*, **120**, 313–321, 1983.

Spoelstra, T. A. T. and H. Kelder, Effects Produced by the Ionosphere on Radio Interferometry, *Radio Sci.*, **19**, 779–788, 1984.

Staelin, D. H., Measurements and Interpretation of the Microwave Spectrum of the Terrestrial Atmosphere near 1-Centimeter Wavelength, *J. Geophys. Res.*, **71**, 2875–2881, 1966.

Sutton, E. C., S. Subramanian, and C. H. Townes, Interferometric Measurements of Stellar Positions in the Infrared, *Astron. Astrophys.*, **110**, 324–331, 1982.

Tatarski, V. I., *Wave Propagation in a Turbulent Medium*, Dover, New York, 1961.

Tatarski, V. I., *The Effects of the Turbulent Atmosphere on Wave Propagation*, National Technical Information Service, Springfield, Virginia, 1971.

Thayer, G. D., An Improved Equation for the Radio Refractive Index of Air, *Radio Sci.*, **9**, 803–807, 1974.

Van Vleck, J. H., E. M. Purcell, and H. Goldstein, Atmospheric Attenuation, in *Propagation of Short Radio Waves*, M.I.T. Radiation Laboratory Series, Vol. 13, D. E. Kerr, Ed., McGraw-Hill, New York, 1951, pp. 641–692.

Waters, J. W., Absorption and Emission by Atmospheric Gases, in *Methods of Experimental Physics*, Vol. 12B, M. L. Meeks, Ed., Academic Press, New York, 1976, pp. 142–176.

Weinberg, S., *Gravitation and Cosmology: Principles and Applications of the General Theory of Relativity*, Wiley, New York, 1972, p. 188.

Weisberg, J. M., J. Rankin, and V. Boriakoff, HI Absorption Measurements of Seven Low Latitude Pulsars, *Astron. Astrophys.*, **88**, 84–93, 1980.

Westwater, E. R., *An Analysis of the Correction of Range Errors due to Atmospheric Refraction by Microwave Radiometric Techniques*, ESSA Tech. Rpt. IER 30-ITSA 30, Inst. for Telecommunication Sciences and Aeronomy, Boulder, Colorado, 1967.

Westwater, E. R., The Accuracy of Water Vapor and Cloud Liquid Determination by Dual-Frequency Ground-Based Microwave Radiometry, *Radio Sci.*, **13**, 677–685, 1978.

Westwater, E. R. and F. O. Guiraud, Ground-Based Microwave Radiometric Retrieval of Precipitable Water Vapor in the Presence of Clouds with High Liquid Content, *Radio Sci.*, **15**, 947–957, 1980.

Woolf, N. J., High Resolution Imaging from the Ground, *Ann. Rev. Astron. Astrophys.*, **20**, 367–398, 1982.

Wu, S. C., Optimum Frequencies of a Passive Microwave Radiometer for Tropospheric Path-Length Correction, *IEEE Trans. Antennas Propag.*, **AP-27**, 233–239, 1979.

Yeh, K. C. and C. H. Liu, Acoustic–Gravity Waves in the Upper Atmosphere, *Rev. Geophys. Space Phys.*, **12**, 193–216, 1974.

Yeh, K. C. and C. H. Liu, Radio Wave Scintillations in the Ionosphere, *Proc. IEEE*, **70**, 324–360, 1982.

Young, A. T., Interpretation of Interplanetary Scintillations, *Astrophys. J.*, **168**, 543–562, 1971.

14

RADIO INTERFERENCE

With the increasing use of the radio spectrum for communications, navigation, and other services, the avoidance of unwanted signals is an essential practical concern in radio astronomy. Interference poses particular problems to the radio astronomer because the signal levels from cosmic sources are much lower than the operating levels in active (transmitting) services, and wide bandwidths are required for adequate sensitivity. Although certain frequency bands are allocated solely to radio astronomy and passive sensing, those at meter and centimeter wavelengths are mostly too narrow to allow the desired sensitivity to be obtained within them. Also, cosmic spectral-line frequencies often fall outside the radio astronomy bands. Thus it is often necessary for radio astronomers to observe within bands that are allocated to other services. Interference can then best be avoided by placing radio telescopes in locations remote from centers of industrial and similar activity and by taking advantage of shielding from transmitters by terrain features. A basic parameter in site selection and coordination with other spectrum users is the threshold of harmful interference, that is, the flux density above which an interfering signal falling within the passband of the radio telescope causes degradation of astronomical observations. The harmful threshold is a function of the type and operating parameters of the radio telescope, and this dependence is the subject of the present chapter. A description of the system of allocation and protection of frequencies is given by Pankonin and Price (1981).

14.1 GENERAL CONSIDERATIONS

The ultimate limit on the sensitivity of a radio telescope is set by the system noise, and an interfering signal can generally be tolerated if its contribution to the output is small compared with the noise fluctuations. A response to

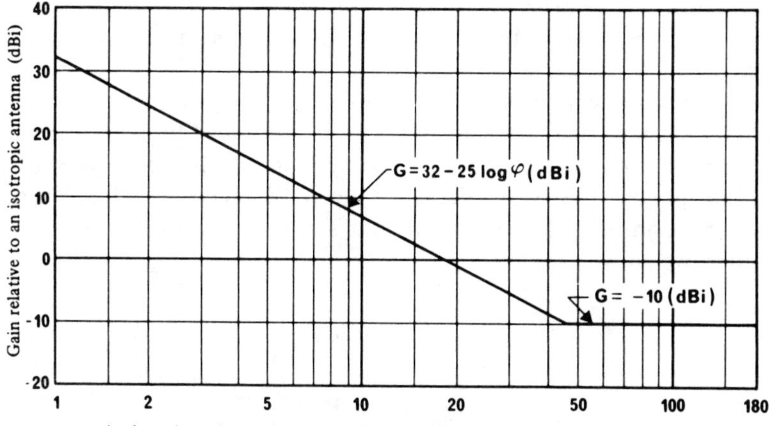

Figure 14.1 Empirical sidelobe-envelope model for reflector antennas of diameter greater than 100 wavelengths. Measurements on actual antennas show that 90% of sidelobe peaks lie below the curve. Sidelobe levels can be reduced by 3 dB or more in designs in which aperture blockage by feed structure is avoided, but the curve shown is representative of large antennas of the type used in radio astronomy. From CCIR (1982a).

interference of 1/10 of the rms level of the noise in the measurements is a useful criterion in interference-threshold calculations. The corresponding flux density of such a signal can be calculated if the effective collecting area of the antenna is known. Radio astronomy antennas usually have narrow beams and the probability of the interfering signal being received in the main beam or nearby sidelobes is low, especially if the interfering transmitter is ground based. Thus it can be assumed that interference usually enters the far sidelobes of the antenna. Figure 14.1 shows an empirical-model curve for the maximum sidelobe gain as a function of angle from the main-beam axis. This curve is derived from the measured response patterns of a number of large reflector antennas. For the present calculations it is convenient to use a gain of 0 dBi (i.e., 0 dB with respect to an isotropic radiator) which occurs at about 19° from the main beam. 0 dBi is also the mean gain of an antenna over 4π steradians, and the effective collecting area for this gain is equal to $\lambda^2/4\pi$, λ being the wavelength. If F_h (Wm^{-2}) is the flux density of an interfering signal within the receiver passband, the interference-to-noise power ratio in the receiver is

$$\frac{F_h\lambda^2}{4\pi kT_S\Delta\nu},\tag{14.1}$$

where k is Boltzmann's constant, T_S is the system noise temperature, and $\Delta\nu$ is the receiver bandwidth. In this expression we have assumed that the polarization of the interfering signal matches that of the antenna, that is, the worst-case condition. In practice the interfering signal level is likely to fluctuate with time because of propagation effects and the tracking motion of the radio telescope, which sweeps the sidelobe pattern across the direction of the transmitter.

For comparison with correlator systems we first consider the simpler case of a receiver that measures the total power at the output of a single antenna. The signal-to-noise ratio of the output, after averaging for a time τ, is Expression (14.1) multiplied by $\sqrt{\Delta\nu\,\tau}$. This result is derived from considerations similar to those in Section 6.2. Then for an output signal-to-noise ratio of 0.1 for the interfering signal we obtain

$$F_h = \frac{0.4\pi k T_S \nu^2 \sqrt{\Delta\nu}}{c^2 \sqrt{\tau}}. \tag{14.2}$$

Note that the harmful threshold increases with frequency as ν^2. With increasing frequency the system temperature and the usable bandwidth also tend to increase.

The total-power type of radio telescope is the most sensitive to interference. Thus, the result in Eq. (14.2) provides a worst-case specification for the harmful limits for radio astronomy. For broadband interference the corre-

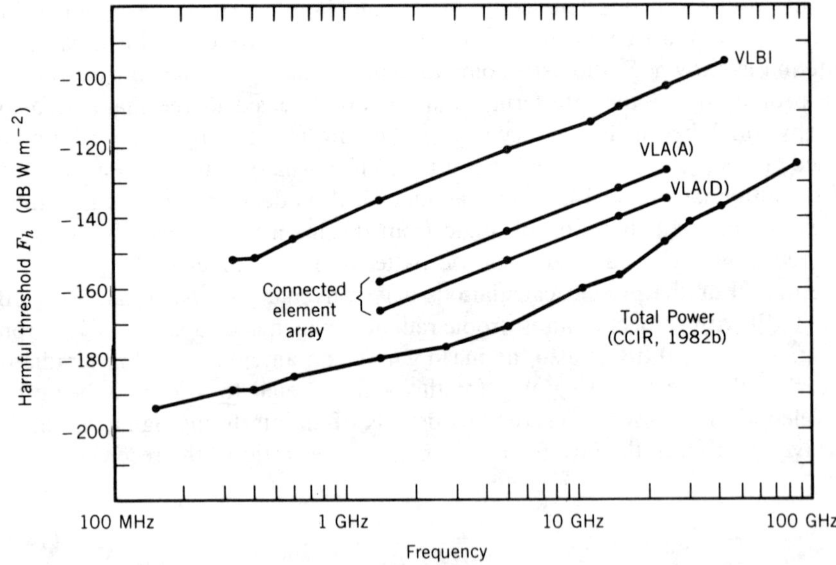

Figure 14.2   Curves of the harmful threshold of interference F_h for radio telescopes computed using typical instrumental characteristics for each frequency band and type of instrument. The curve for total-power radiometers is based on Eq. (14.2) with values from CCIR (1982b), that for connected-element arrays on Eq. (14.12) with parameters for the VLA, and that for VLBI systems on Eq. (14.22). The increase in F_h for total-power systems above 15 GHz results from the increase in the widths of radio astronomy bands. The ordinate scale is decibels relative to 1 W m^{-2}.

sponding threshold level is S_h (Wm^{-2} Hz^{-1}) $= F_h \Delta \nu^{-1}$. Values of F_h and S_h computed for a total-power system using typical parameters for the various radio astronomy bands are given in CCIR (1982b), as general limits for the radio astronomy service. These values of F_h are plotted as the lowest curve in Fig. 14.2.

14.2 CONNECTED-ELEMENT ARRAYS

We now consider the interference response of a correlator array with antenna spacings up to a few tens of kilometers. Two effects reduce the response to interference compared with that of a total-power system. First, the source of interference does not move across the sky with the sidereal motion of the object under observation, and thus it produces fringe oscillations of a different frequency from those of the wanted signal. Second, the instrumental delays are adjusted to equalize the signal paths for radiation incident from the direction under observation, and signals from another direction, if they are broadband, are to some extent decorrelated. In examining these effects we follow the analysis by Thompson (1982).

Fringe-Frequency Averaging

Consider first the fringe-frequency effect. Suppose that instrumental phase shifts are introduced as described in Section 6.1 to slow the fringe oscillations of the wanted signal to zero frequency. The removal of the fringe-frequency phase shifts from the cosmic signals introduces corresponding shifts into the interfering signals. If the source of interference is stationary with respect to the antennas, the interference at the correlator output has the form of oscillations at the natural fringe frequency for the source under observation, which from Eq. (4.23) (omitting the sign of dw/dt) is

$$\nu_f = \omega_e u \cos \delta. \tag{14.3}$$

Here ω_e is the angular rotation velocity of the earth, u is a component of antenna spacing (see Section 4.2), and δ is the declination of the source under observation. Averaging of such a fringe-frequency waveform for a period τ is equivalent to convolution with a rectangular function of width τ. The amplitude is thus decreased by a factor that follows from the Fourier transform of the convolving function:

$$F_1 = \frac{\sin(\pi \nu_f \tau)}{\pi \nu_f \tau}. \tag{14.4}$$

In order to derive a harmful threshold of interference we compute the ratio of the rms level of interference to the rms level of noise in a radio map and, as

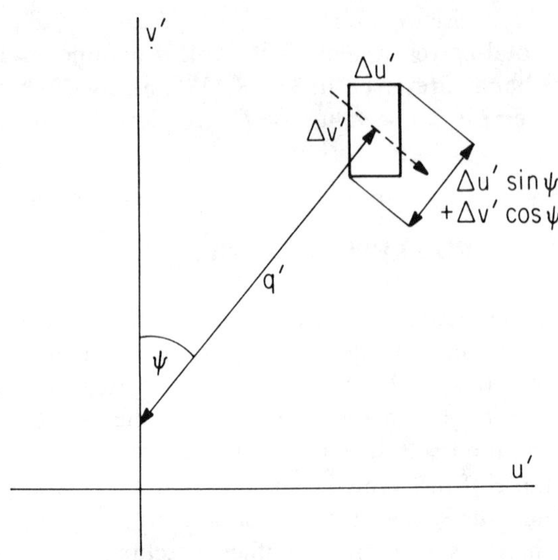

Figure 14.3   Derivation of the mean cell crossing time for the spatial-frequency locus indicated by the broken line. The velocity of the spatial-frequency vector in the (u', v') plane is $\omega_e q'$. The mean path length through the cell in the direction of the broken line is the cell area $\Delta u' \Delta v'$ divided by the cell width projected normal to that direction.

before, equate the result to 0.1. The first step is to determine the mean squared value of the modulus of the interference component in the visibility data. Figure 6.6b, which depicts the spectral components at the correlator output, shows that the output from the correlated signal component, in this case the interference, is represented by a delta function. Assuming, as before, that the interference enters sidelobes of gain 0 dBi, and that the polarization is matched, we substitute in the magnitude of the delta function $kT_A \Delta v = F_h c^2 / 4\pi v^2$. Thus, the sum of the squared modulus of the interference over n_r grid points in the (u, v) plane is

$$\sum_{n_r} \langle |r_i|^2 \rangle = \left(\frac{H_0^2 F_h c^2}{4\pi v^2} \right)^2 n_r \langle F_1^2 \rangle. \tag{14.5}$$

Here r_i is the correlator response to the interference, H_0 is a voltage gain factor, and $\langle F_1^2 \rangle$ is the mean squared value of F_1 as given in Eq. (14.4), which represents the effect of the visibility averaging on the fringe-frequency oscillations. To determine the mean squared value of F_1 consider the variation of this factor as the spacing vector for an antenna pair sweeps out a circular locus in the (u', v') plane, as described in Section 4.3. Also, suppose that cell averaging (see Section 4.10) is used to determine visibility points on a rectangular grid. The averaging time τ is equal to the cell crossing time as shown in Fig. 14.3. Note from Eq. (14.3) that the fringe frequency goes through zero at the v' axis, and F_1 is then unity. For small values of ψ, as defined in Fig. 14.3, the path length through a cell is closely equal to Δu, the cell crossing time is $\tau = \Delta u / \omega_e q'$, and $v_f \tau = \Delta u \sin \psi \cos \delta$. Now Δu is equal to the reciprocal of the width of the synthesized field, which, except at long wavelengths, is unlikely to be more than $0.5°$. We therefore assume that Δu is of order 100 or greater, which permits the following simplification. For $\Delta u = 100$ and $\delta < 70°$, F_1^2 goes from 1 to 10^{-3} as ψ goes from 0 to $< 17°$. Thus, most of the contribution to F_1^2 occurs for small ψ, and we can substitute $v_f \tau = \psi \Delta u \cos \delta$ in Eq. (14.4) and obtain

$$\langle F_1^2 \rangle = \frac{2}{\pi} \int_0^{\pi/2} \frac{\sin^2(\pi \psi \Delta u \cos \delta)}{(\pi \psi \Delta u \cos \delta)^2} d\psi \simeq \frac{1}{\pi \Delta u \cos \delta}. \tag{14.6}$$

Since Δu is large we have used an upper limit of ∞ in evaluating the integral.

For the noise we again refer to Fig. 6.6b. The power spectral density of the noise near zero frequency is $H_0^4 k^2 T_S^2 \Delta v$, and an equivalent bandwidth τ^{-1}, including negative frequencies, is passed by the averaging process: see Eq. (6.38). The mean squared component of the noise over the n_r grid points is thus

$$\sum_{n_r} \langle |r_n| \rangle^2 = H_0^4 k^2 T_S^2 \Delta v n_r \langle \tau^{-1} \rangle, \tag{14.7}$$

where $\langle \tau^{-1} \rangle$ is the mean value of τ^{-1}. From Fig. 14.3 the mean cell crossing time is

$$\tau = \frac{\Delta u |\csc \delta|}{q' \omega_e (|\sin \psi| + |\csc \delta||\cos \psi|)}, \tag{14.8}$$

where $q' = \sqrt{X_\lambda^2 + Y_\lambda^2}$, X_λ and Y_λ being the components of antenna spacing projected onto the equatorial plane, as defined in Section 4.2. We have assumed that $\Delta u' = \Delta v' \sin \delta$ (i.e., $\Delta u = \Delta v$), and that for all but a negligibly small number of cells the path of the spatial-frequency locus can be approximated by a straight line. The mean value of τ^{-1} around a (u', v') locus is, from Eq. (14.8),

$$\frac{2}{\pi} \int_0^{\pi/2} \tau^{-1} d\psi = \frac{2 \omega_e q'}{\pi \Delta u} (1 + |\sin \delta|), \tag{14.9}$$

and the mean for the n_r points in the (u, v) plane is

$$\langle \tau^{-1} \rangle = \frac{2 \omega_e}{\pi \Delta u} (1 + |\sin \delta|) \frac{1}{n_r} \sum_{n_r} q'. \tag{14.10}$$

From Eqs. (14.5), (14.6), (14.7), and (14.10)

$$\frac{(|r_i|)_{\text{rms}}}{(|r_n|)_{\text{rms}}} = \frac{F_h c^2}{4\pi k T_S \nu^2 \sqrt{2\Delta\nu\omega_e}\cos\delta(1+|\sin\delta|)} \frac{1}{\sqrt{\dfrac{1}{n_r}\displaystyle\sum_{n_r} q'}}. \tag{14.11}$$

By Parseval's theorem the ratio of the rms values of the interference and noise in the map is equal to Eq. (14.11), so to evaluate F_h we equate the right-hand side to 0.1 and obtain

$$F_h = \frac{0.4\pi k T_S \nu^2 \sqrt{2\,\Delta\nu\,\omega_e}}{c^2}\sqrt{\frac{1}{n_r}\sum_{n_r} q'}. \tag{14.12}$$

The factor $\sqrt{\cos\delta(1+|\sin\delta|)}$ has been replaced by unity, the resulting error being less than 1 dB for $0 < |\delta| < 71°$, and 2.3 dB for $\delta = 80°$. Note that with fixed antenna positions q' is proportional to ν, so F_h is proportional to $\nu^{2.5}$. The number of points in the (u', v') plane to which an antenna pair contributes is proportional to q', so in evaluating Eq. (14.12) it is convenient to write

$$\frac{1}{n_r}\sum_{n_r} q' = \frac{\displaystyle\sum_{n_p} q'^2}{\displaystyle\sum_{n_p} q'}, \tag{14.13}$$

where n_p is the number of correlated antenna pairs in the array.

Values of F_h from Eq. (14.12) are given in Fig. 14.2 using parameters for the VLA. The range of 7.7 dB between the curves corresponds to the four scaled configurations of the VLA (Thompson et al. 1980) in which the maximum distance of an antenna from the array center (see Fig. 5.17b) varies from 0.59 to 21 km.

Since the averaging is ineffective in reducing the interference when u goes through zero, visibility values containing the greatest contributions from interference cluster around the v axis. Some degree of randomness in the occurrence of high values is to be expected, as a result of the varying sidelobe levels. Because of the (u, v) distribution, the interference in the (ξ, η) domain takes the form of quasi-random structure that is elongated in the east–west direction: for an example see Thompson (1982). The clustering also suggests the possibility of reducing the interference response by deleting visibility data near the v axis, particularly those data with unusually high values. Although this may add sidelobes into the synthesized beam, they can be largely removed through CLEAN and related processes discussed in Chapter 11.

The above discussion applies to cases where the observation is of long enough duration that the (u, v) plane is well sampled, and the strength of the interfering signal remains approximately constant during this time. If only a fraction α of the (u, v) loci cross the v axis, then a factor of $\sqrt{\alpha}$ should be introduced into the denominator of Eq. (14.12). Strong, sporadic interference, of course, can produce different responses from the one that we have considered.

Decorrelation of Broadband Signals

Since interfering signals are usually incident from directions other than that of the desired radiation, their time delays to the correlator inputs are generally not equal. Broadband interfering signals are thereby decorrelated, which further reduces their response. The reduction is not amenable to a general-case analysis like that resulting from averaging of the fringe frequency, but it can be computed for each particular antenna configuration and position of the interfering source. For this reason, and the fact that only broadband signals are reduced, the effect has not been included in the threshold equation (14.12) or in Fig. 14.2.

At any instant during an observation let θ_s be the angle between a plane normal to the baseline for a pair of antennas and the direction of the source under observation. θ_s defines a small circle on the celestial sphere for which the delays are equalized. Similarly, let θ_i be the corresponding angle for the source of interference. The delay difference for the interfering signals at the correlator is

$$\tau_d = \frac{D|\sin\theta_s - \sin\theta_i|}{c}.\tag{14.14}$$

Expressions for θ_s and θ_i can be derived from Eq. (4.17), since $\sin\theta_s = w/D_\lambda$. Suppose that the received interfering signal has an effectively rectangular spectrum of width $\Delta\nu$ and center frequency ν_0, defined either by the signal itself or by the receiving passband. By the Weiner–Khinchin relation the autocorrelation function of the signal is equal to

$$\frac{\sin(\pi\Delta\nu\,\tau)}{\pi\Delta\nu\,\tau}\cos(2\pi\nu_0\tau).\tag{14.15}$$

Expression (14.15) represents the real output of a complex correlator as a function of the differential delay τ. The imaginary output is represented by a similar expression in which the cosine function is replaced by a sine. Thus, the decorrelation of the modulus of the complex output for a delay τ_d is given by the factor

$$F_2 = \frac{\sin(\pi\Delta\nu\,\tau_d)}{\pi\Delta\nu\,\tau_d}.\tag{14.16}$$

For a fixed transmitter location, θ_i remains constant, but θ_s varies as the antennas track. Thus τ_d may go through zero, causing F_2 to peak, but unlike F_1, a peak in F_2 can occur at any point on the (u, v) plane. Those antenna pairs for which the F_1 and F_2 peaks overlap contribute most strongly to the interference in the map, and those for which the peaks are well separated contribute little. Therefore, for broadband signals, the fringe-frequency and decorrelation effects should be considered in combination. In sample calculations for the response of the VLA to a geostationary satellite on the meridian, a factor

$$\sqrt{\frac{\Sigma q' F_1^2 F_2^2}{\Sigma q' F_1^2}} \tag{14.17}$$

was computed which represents the additional decrease in the rms interference resulting from decorrelation. The summations in Expression (14.17) were taken over all antenna pairs for equal increments in hour angle, and the q' factors were inserted to compensate for the uneven density of points in the (u, v) plane resulting from this method of sampling. The results showed that over the range of antenna spacings, wavelengths, and bandwidths considered, the suppression of broadband interference resulting from decorrelation can vary from 4 to 34 dB, with strong dependence on the observing declination.

14.3   VERY-LONG-BASELINE SYSTEMS

In VLBI arrays in which the antenna spacings are hundreds or thousands of kilometers, the output resulting from correlated components of an interfering signal at the correlator inputs is usually negligible. This is because the natural fringe frequencies are higher than those in connected-element arrays, and the delay inequalities for signals that do not come from the direction of observation are also greater. Furthermore, unless the interfering signal originates in a satellite or spacecraft, it is unlikely to be present at two widely separated locations.

Consider an interfering signal entering one antenna of a correlated pair. The interference reduces the measured correlation, and the overall effect is similar to an increase in the system noise for the antenna. In Fig. 14.4, $x(t)$ and $y(t)$ represent the signals plus system noise from two antennas in the absence of interference, and $z(t)$ represents an interfering signal at one antenna. The three waveforms have zero means and the standard deviations are σ for x and y and σ_i for z. In the absence of interference, the measured correlation coefficient is

$$\rho_1 = \frac{\langle xy \rangle}{\sqrt{\langle x^2 \rangle \langle y^2 \rangle}} = \frac{\langle xy \rangle}{\sigma^2}. \tag{14.18}$$

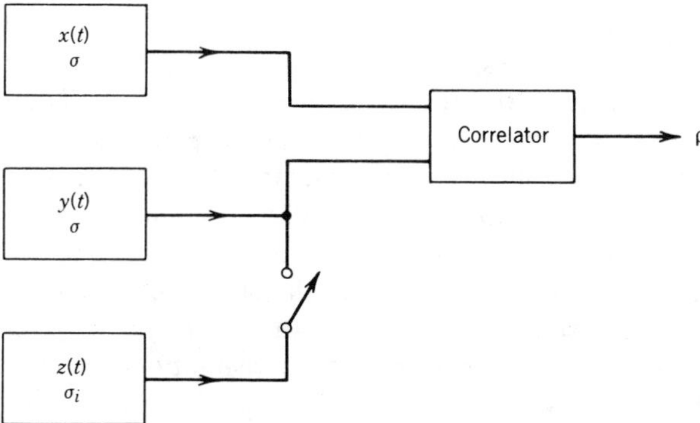

Figure 14.4   Components of the correlator input signals used in the discussion of the effects of interference on VLBI observations.

When the interference is present the correlation becomes

$$\rho_2 = \frac{\langle xy \rangle + \langle xz \rangle}{\sqrt{\langle x^2 \rangle (\langle y^2 \rangle + 2\langle yz \rangle + \langle z^2 \rangle)}}.$$
(14.19)

The interference is uncorrelated with x and y, so $\langle xz \rangle = \langle yz \rangle = 0$. Also, at the harmful threshold, $\sigma_i^2 \ll \sigma^2$. Thus, from Eqs. (14.18) and (14.19) we obtain

$$\rho_2 \simeq \rho_1 \left[1 - \frac{1}{2} \left(\frac{\sigma_i}{\sigma} \right)^2 \right].$$
(14.20)

The interference reduces the measured correlation, and one can envisage it as causing an increase in the system temperature and a corresponding decrease in the effective system gain. Thus, the error introduced in the measurements is multiplicative, rather than additive, as occurs in the cases discussed earlier in which the detector or correlator responds directly to the interference. The different effects of these two types of errors have been discussed in Section 10.3. In principle, the changes in the effective gain can be monitored by using a calibration signal as discussed in Section 7.6. However, this calibration could be difficult if the strength of the interference varies rapidly. The harmful threshold should therefore be specified so that it is just small enough that the errors introduced do not significantly increase the level of uncertainty in the measurements resulting from other factors. For the general case this level can be specified only roughly, and we use a value of 1%. If we include the possibility of simultaneous but uncorrelated interference in both antennas, the resulting condition is

$$\left(\frac{\sigma_i}{\sigma} \right)^2 \leq 0.01.$$
(14.21)

It follows from Parseval's theorem that a 1% rms error in the visibility introduces into the brightness an error of which the rms over the map is 1% of the corresponding rms of the true brightness distribution. The effect on the dynamic range of brightness within the map depends on the form of the brightness distribution and of the error distribution. For a map of a single point source the rms brightness error would be about $10^{-2}\sqrt{f/n_r}$ times the peak brightness, where f is the fraction of the n_r gridded visibility data that contains interference. Here it is assumed that the fluctuations in the received interfering signal are sufficiently fast that the values of the interference level are essentially independent for each gridded visibility point. If this is not the case the resulting error will be greater.

In terms of the criterion in Eq. (14.21) the ratio of interference to system noise powers as given by Expression (14.1) must not exceed 0.01, and

$$F_h = \frac{0.04\pi k T_S \nu^2 \Delta\nu}{c^2}.$$

(14.22)

The curve for VLBI in Fig. 14.2 shows values of F_h from Eq. (14.22). Typical practical values are used for T_S and $\Delta\nu$: for example, at 0.6, 5, and 24 GHz, T_S is 55, 35, and 45 K and $\Delta\nu$ is 6, 50, and 100 MHz, respectively. The harmful thresholds are approximately 40 dB greater than those for total-power systems.

14.4   INTERFERENCE FROM AIRBORNE AND SPACE TRANSMITTERS

In application of the F_h values derived above it should be remembered that they are relevant to interference from stationary, ground-based transmitters. It is generally possible to make observations at high enough angles of elevation to point no closer than about 19° to any such transmitter: 19° is the angle from the main beam at which most sidelobes fall below the isotropic level in the model in Fig. 14.1. Airborne and satellite transmitters present a special problem. Motion of the transmitter across the sky is most likely to increase the frequency at the correlator outputs of a synthesis array and thereby reduce the response, but the signals may be received in high-level sidelobes near the main beam. Transmitters on geostationary satellites represent the greatest hazard to radio astronomy because of their fixed locations at high elevation angles near the celestial equator. Interfering signals from a series of such satellites distributed along the geostationary orbit would result in a band of sky centered on the orbit in which high-sensitivity observations would be precluded. Thus, tolerable levels for interference from geostationary satellites should be based on sidelobe levels higher than 0 dBi to reduce the area of sky in which the threshold is exceeded.

REFERENCES

CCIR, Generalized Space Research Earth Station Antenna Radiation Pattern for Use in Interference Calculations, Including Coordination Procedures, Recommendation 509, in *Recommendations and Reports of the International Radio Consultative Committee*, Vol. II, International Telecommunication Union, Geneva, 1982a.

CCIR, Interference Protection Criteria for the Radioastronomy Service, Report 224-5, in *Recommendations and Reports of the International Radio Consultative Committee*, Vol. II, International Telecommunication Union, Geneva, 1982b.

Pankonin, V. and R. M. Price, Radio Astronomy and Spectrum Management: The Impact of WARC-79, *IEEE Trans. Electromagn. Compat.*, **EMC-23**, 308–317, 1981.

Thompson, A. R., The Response of a Radio-Astronomy Synthesis Array to Interfering Signals, *IEEE Trans. Antennas Propag.*, **AP-30**, 450–456, 1982.

Thompson, A. R., B. G. Clark, C. M. Wade, and P. J. Napier, The Very Large Array, *Astrophys. J. Suppl.*, **44**, 151–167, 1980.

15

RELATED TECHNIQUES

Concepts and techniques similar to those used in radio interferometry and synthesis mapping occur in various areas of astronomy. Here we introduce a few of them, including optical techniques, to leave the reader with a broader view. All of these subjects are described in detail elsewhere, so here the aim is mainly to outline the principles involved.

15.1 INTENSITY INTERFEROMETER

In long-baseline interferometry the intensity interferometer offers some technical simplifications that were mainly of importance in radio astronomy during the early development of the subject. As mentioned in Chapter 1, its practical applications in radio astronomy have been limited (Jennison and Das Gupta 1956, Carr et al. 1970, Dulk 1970), because in comparison with a conventional interferometer it requires a much larger signal-to-noise ratio in the receiving system, and only the modulus of the visibility function is measured. The intensity interferometer was devised by Hanbury Brown who has described its development and application (Hanbury Brown 1974).

In the intensity interferometer, the signals from the antennas are amplified and then passed through square-law (power-linear) detectors before being applied to a correlator, as shown in Fig. 15.1. As a result, the rms signal voltages at the correlator inputs are proportional to the intensity (i.e., the power) of the received signals. No fringes are formed because the phase of the radiofrequency signals is lost in the detection, but the correlator output indicates the degree of correlation of the detected waveforms. Let the voltages at the detector inputs be V_1 and V_2. Then the correlator output is proportional to $\langle V_1^2 V_2^2 \rangle$. From the fourth-order moment relation for Gaussian processes in

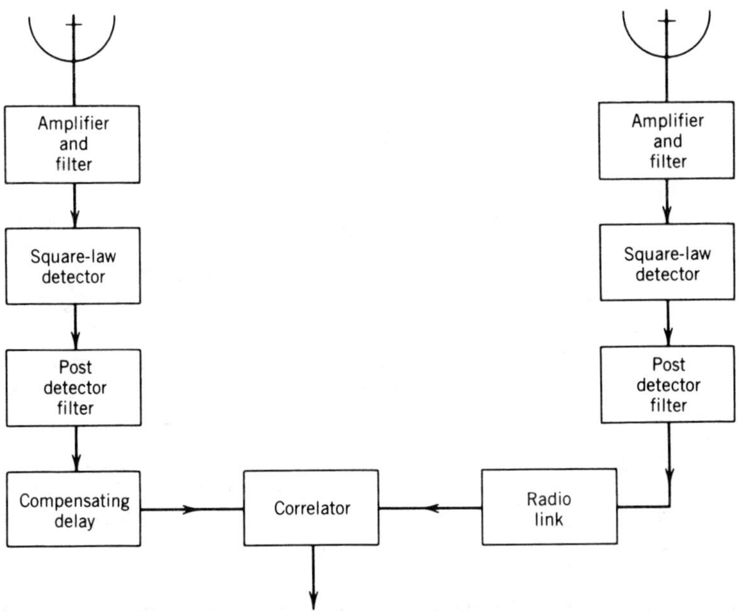

Figure 15.1 Intensity interferometer. The amplifier and filter block may also incorporate a local oscillator and mixer. The compensating delay equalizes the time delays of signals from the source to the correlator inputs. The post detector filters remove dc and RF components.

Eq. (6.30) we obtain

$$\langle V_1^2 V_2^2 \rangle = \langle V_1^2 \rangle \langle V_2^2 \rangle + 2\langle V_1 V_2 \rangle^2. \tag{15.1}$$

The terms $\langle V_1 \rangle^2$ and $\langle V_2 \rangle^2$ represent the dc parts of the detector outputs and these are filtered out and do not reach the correlator inputs. Thus the correlator output is proportional to $\langle V_1 V_2 \rangle^2$, which is the square of the correlator output for a conventional interferometer. The response, therefore, is proportional to the squared modulus of the complex visibility.

We now give an alternative derivation of the response which provides a physical picture of how the signals from different parts of the source combine within the instrument. The source is represented as a one-dimensional brightness distribution in Fig. 15.2. We suppose that it can be considered as a linear distribution of many small regions, each of which is large enough to emit a signal with the characteristics of stationary random noise, but of angular width small compared with $1/u$ which defines the angular resolution of the interferometer. The source is assumed to be incoherent so the signals from different regions are uncorrelated. Consider two regions of the source, k and ℓ, at angular positions θ_k and θ_ℓ and subtending angles $d\theta_k$ and $d\theta_\ell$ as in Fig. 15.2. Each radiates a broad spectrum, but we first consider only the output resulting from a Fourier component at frequency ν_k from region k and similarly a component at ν_ℓ from region ℓ. Let $A_1(\theta)$ be the power reception pattern of

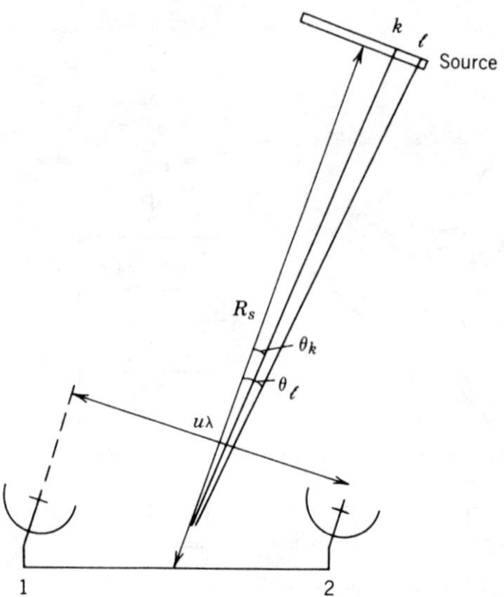

Figure 15.2   Distances and angles used in the discussion of the intensity interferometer.

the two antennas and $B_1(\theta)$ the brightness distribution of the source, these two functions being one-dimensional representations. Then the detector output of the first receiver is equal to

$$\left[V_k \cos 2\pi \nu_k t + V_\ell \cos(2\pi \nu_\ell t + \phi_1)\right]^2, \tag{15.2}$$

where ϕ_1 is a phase term resulting from path-length differences, and the signal voltages V_k and V_ℓ are given by

$$V_k^2 = A_1(\theta_k) B_1(\theta_k) d\theta_k d\nu_k \tag{15.3}$$

and

$$V_\ell^2 = A_1(\theta_\ell) B_1(\theta_\ell) d\theta_\ell d\nu_\ell. \tag{15.4}$$

After expanding Expression (15.2) and removing the dc and RF terms we obtain for the detector output from receiver 1

$$V_k V_\ell \cos\left[2\pi(\nu_k - \nu_\ell)t - \phi_1\right]. \tag{15.5}$$

Similarly, the detector output from receiver 2 is

$$V_k V_\ell \cos\left[2\pi(\nu_k - \nu_\ell)t - \phi_2\right]. \tag{15.6}$$

The correlator output is proportional to the time-averaged product of Expressions (15.5) and (15.6), that is, to

$$A_1(\theta_k) A_1(\theta_\ell) B_1(\theta_k) B_1(\theta_\ell) d\theta_k d\theta_\ell d\nu_k d\nu_\ell \cos(\phi_1 - \phi_2). \tag{15.7}$$

Note that Expression (15.7) is independent of the frequencies ν_k and ν_ℓ, so

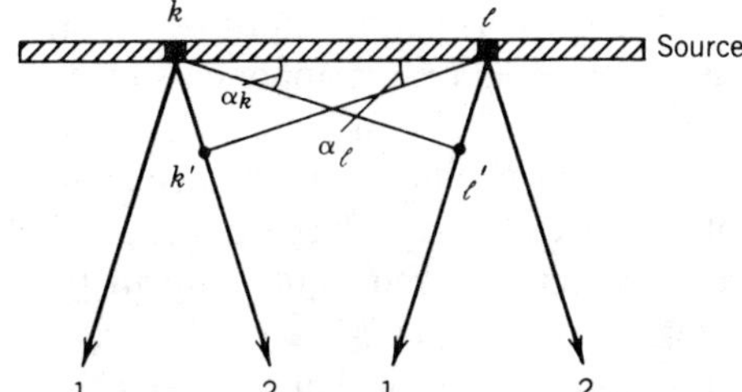

Figure 15.3 Relative delay paths kk' and $\ell\ell'$ from regions k and ℓ of the source for rays traveling in the directions of antennas 1 and 2.

that if we integrate it with respect to ν_k and ν_ℓ over a rectangular receiving passband of width $\Delta\nu$, $d\nu_k d\nu_\ell$ is replaced by $\Delta\nu^2$.

The phase angles ϕ_1 and ϕ_2 result from the path differences kk' and $\ell\ell'$ shown in Fig. 15.3. Note that ϕ_1 and ϕ_2 have opposite signs since the excess path length to antenna 1 is from point ℓ and that to antenna 2 is from point k. If R_s is the distance of the sources from the antennas, the distance $k\ell$ in the source is approximately equal to $R_s(\theta_k - \theta_\ell)$. The angle $\alpha_k + \alpha_\ell$ is approximately equal to $u\lambda/R_s$ since u represents the antenna spacing projected normal to the source and measured in wavelengths. The above approximations are accurate if α_k, α_ℓ, and the angle subtended by the source are all small. Thus, the difference of the phase angles is

$$\phi_1 - \phi_2 = 2\pi R_s(\theta_k - \theta_\ell)\frac{(\sin\alpha_k + \sin\alpha_\ell)}{\lambda}$$

$$\simeq 2\pi u(\theta_k - \theta_\ell). \tag{15.8}$$

From Expression (15.7) the output of the correlator now becomes

$$A_1(\theta_k)A_1(\theta_\ell)B_1(\theta_k)B_1(\theta_\ell)\Delta\nu^2\cos\left[2\pi u(\theta_k - \theta_\ell)\right]d\theta_k d\theta_\ell. \tag{15.9}$$

To obtain the output from all pairs of regions within the source Expression (15.9) can be integrated with respect to θ_k and θ_ℓ over the source, giving

$$\left[\Delta\nu\int A_1(\theta_k)B_1(\theta_k)\cos(2\pi u\theta_k)d\theta_k\right]\left[\Delta\nu\int A_1(\theta_\ell)B_1(\theta_\ell)\cos(2\pi u\theta_\ell)d\theta_\ell\right]$$

$$+\left[\Delta\nu\int A_1(\theta_k)B_1(\theta_k)\sin(2\pi u\theta_k)d\theta_k\right]\left[\Delta\nu\int A_1(\theta_\ell)B_1(\theta_\ell)\sin(2\pi u\theta_\ell)d\theta_\ell\right]$$

$$= A_0^2\Delta\nu^2\left[\mathscr{V}_R^2 + \mathscr{V}_I^2\right] = A_0^2\Delta\nu^2|\mathscr{V}|^2, \tag{15.10}$$

where we assume that $A_1(\theta)$ has a constant value A_0 over the source, and the subscripts R and I denote the real and imaginary parts of the visibility. The above result follows from the definition of visibility which is given for a two-dimensional source in Section 2.4. Thus, the correlator output is proportional to the square of the modulus of the complex visibility.

In the above discussion we have integrated the correlator output over both frequency and angle, and in so doing have considered only components of the two detector outputs that arise from the same pairs of signal frequencies and the same pairs of regions in the source. This is justified for an incoherent source since only such combinations of postdetector signal components are correlated. Other combinations have zero expectation but contribute to the noise in the correlator output. The assumption of a square-law characteristic for the detectors is also important, since no combinations of signal components higher than second order need then be considered. Thus, the response of the correlator to signals from the source is a linear combination of terms having the form of Expression (15.7). For a more detailed discussion following the same approach see Hanbury Brown and Twiss (1954). An analysis based on the mutual coherence of the radiation field is given by Bracewell (1958).

Some characteristics of the intensity interferometer offer advantages over the conventional interferometer, which is the main subject of this book. The intensity interferometer is much less sensitive to atmospheric phase fluctuations because each signal component at the correlator input is generated as the difference between two radiofrequency components that have followed almost the same path through the atmosphere. The phase fluctuations in the difference-frequency components at the detectors are less than those in the radiofrequency signals by the ratio of the difference frequency to the radiofrequency, which may be of order 10^{-5}. In the conventional interferometer such phase fluctuations can make the amplitude as well as the phase of the visibility very difficult to measure. Similarly, fluctuations in the phases of the local oscillators in the two receivers do not contribute to the phases of the difference-frequency components. Thus, it is not necessary to synchronize the local oscillators or even to use high-stability frequency standards as in VLBI. These advantages were helpful, though by no means essential, in the early radio implementation of the intensity interferometer. Had the diameters of the sources under investigation then been of order of arcseconds, rather than arcminutes, the characteristics of the intensity interferometer would have played a more essential role. The serious disadvantage of the intensity interferometer is its relative lack of sensitivity. Because of the action of the detectors in the receivers, the ratio of the signal power to the noise power at the correlator inputs is proportional to the square of the corresponding ratio in the RF (predetector) stages [see Eq. (9.73)], the exact value being dependent on the bandwidths of these and the postdetector stages (Hanbury Brown and Twiss 1954). In a conventional interferometer, it is possible to detect signals that are ~ 60 dB below the noise at the correlator inputs. In the intensity interferometer, a similar signal-to-noise ratio at the correlator output would require signal-to-noise ratios greater by ~ 30 dB in the RF stages. This effect, together with the lack of sensitivity to the visibility phase, has greatly restricted the radio usage of the intensity interferometer. The most important results have come from the optical application of intensity interferometry discussed in Section 15.4.

15.2 LUNAR OCCULTATION OBSERVATIONS

Observations of the occultation of a radio source by the moon, in which a single antenna is used to measure the total received power, provide estimates of the position and one-dimensional brightness distribution with an angular resolution of the order of 1 arcsecond. Such measurements contributed to the development of interferometry in the 1960s by providing accurate positions of calibration sources. Two decades later they are still of some importance, but mainly at low frequencies (meter wavelengths). At centimeter wavelengths the high thermal flux density from the moon complicates the observations, and high resolution can be obtained more easily by interferometry. Reviews of the occultation technique are given by Cohen (1969) and Hazard (1976). Lunar occultations are also used at optical and infrared wavelengths and provide measurements of stellar diameters with angular resolution in the milliarcsecond range (e.g., Nather and McCants 1970, Ridgway et al. 1980).

The lunar occultation response can be described in terms of the sensitivity to spatial frequencies, and from this viewpoint it is interesting to compare it with the response of an array. Figure 15.4 shows the geometrical situation and the form of an occultation record. The departure of the moon's limb from a straight edge, as a result of curvature and roughness, is small compared with the size of the first Fresnel zone at radiofrequencies. Thus the point-source response is the well-known diffraction pattern of a straight edge, which is derived in most texts on physical optics. The main change in the received power in Fig. 15.4b corresponds to the covering or uncovering of the first Fresnel zone by the moon, and the oscillations result from higher-order zones. The appearance of the curve suggests that the angular resolution of the observation is less than would be obtained if geometrical optics prevailed, in which case the point-source occultation curve would be a Heaviside step function. However, examination of the Fourier transforms of a step function and of the curve in Fig. 15.4b shows that their spectral sensitivity functions are identical in amplitude and differ only in the phases of the frequency components. This result was obtained by Scheuer (1962) who solved the problem of deriving the one-dimensional brightness distribution B_1 from the occultation curve. In the case of the hypothetical geometrical-optics occultation, the observed curve would be the integral of B_1 as a function θ, the angle between the source and the moon's limb measured as in Fig. 15.4a. B_1 could then be obtained by a differentiation. In the actual case the observed occultation curve $\mathscr{G}(\theta)$ is equal to convolution of $B_1(\theta)$ with the point-source diffraction pattern of the moon's limb $\mathscr{P}(\theta)$. This convolution is $B_1(\theta) * \mathscr{P}(\theta)$. Differentiation with respect to θ yields

$$\mathscr{G}'(\theta) = B_1(\theta) * \mathscr{P}'(\theta), \tag{15.11}$$

where the primes indicate derivatives. Fourier transformation of the two sides

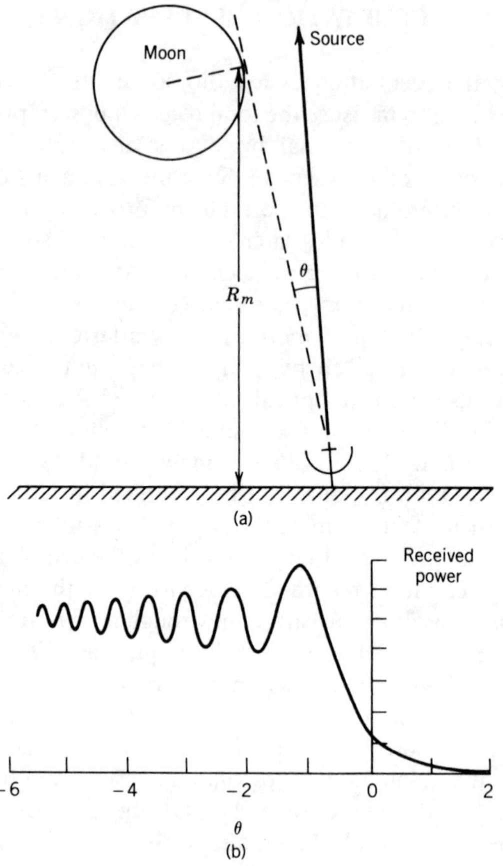

Figure 15.4   Occultation of a radio source by the moon. (a) The geometrical situation. (b) The occultation curve for a point source in which the units of θ on the abscissa are equal to $\sqrt{\lambda/2R_m}$, λ being the wavelength and R_m the moon's distance.

of Eq. (15.11) gives

$$\bar{\mathcal{G}}'(u) = \bar{B}_1(u)\bar{\mathcal{P}}'(u), \qquad (15.12)$$

where the bar indicates the Fourier transform and the prime indicates a derivative in the θ domain. Now in the geometrical-optics case $\mathcal{P}'(\theta)$ would be a delta function, for which the Fourier transform is a constant. For the diffraction-limited case the function $\bar{\mathcal{P}}(u)$ is given by

$$\bar{\mathcal{P}}(u) = \frac{j}{u}\exp\!\left[j2\pi\theta_F^2 u^2 R_m^2 \,\text{sgn}\, u\right], \qquad (15.13)$$

where θ_F is the angular size of the first Fresnel zone, which is equal to

$\sqrt{\lambda/2R_m}$, R_m is the distance of the moon, and sgn is the sign function which takes values ± 1 to indicate the sign of u. It follows from the derivative theorem of Fourier transforms that $\overline{\mathscr{P}}'(u)$ has a constant amplitude with no zeros and can be divided out from Eq. (15.12). Thus $B_1(\theta)$ is equal to $\mathscr{G}'(\theta)$ convolved with a function whose Fourier transform is $1/\overline{\mathscr{P}}'(u)$. Scheuer (1962) shows that this last function is proportional to $\mathscr{P}'(-\theta)$, which can be used as a restoring function as follows:

$$B_1(\theta) = \mathscr{G}'(\theta) * \mathscr{P}'(-\theta)$$

$$= \mathscr{G}(\theta) * \mathscr{P}''(-\theta). \tag{15.14}$$

The second form of the right-hand side is more useful since it avoids the practical difficulty of differentiating a noisy occultation curve. Now in principle the above restoration provides B_1 without limit on the angular resolution, in contrast to the performance of an array. Remember, however, that the amplitude of the spatial-frequency sensitivity of the occultation curve, which is given by Eq. (15.13), is proportional to $1/u$. Thus in the restoration in Eq. (15.14) the amplitudes of the Fourier components, which include also the noise, are increased in proportion to u. The increase of the noise sets a limit to the useful resolution. This limit can be conveniently introduced by replacing $\mathscr{P}''(\theta)$ in (15.14) by $\mathscr{P}''(\theta)$ convolved with a Gaussian function of θ. One then derives B_1 as it would be observed with a beam of the same Gaussian shape. In practice, the introduction of the Gaussian function is essential to the method since it ensures the convergence of the convolution integral in Eq. (15.14). The discussion above follows the classical approach to reduction of moon-occultation observations, which developed from the geometrical-optics analogy. One can envisage the reduction more succinctly as taking the Fourier transform of the occultation curve, dividing by $\overline{\mathscr{P}}(u)$ (with suitable weighting to control the increase of the noise), and retransforming to the θ domain. This process is mathematically equivalent to that in Eq. (15.14).

An approximate estimate of the noise-imposed limit on the angular resolution can be obtained as follows. Consider the region of an occultation curve (see Fig. 15.4b) in which the main change in the received power occurs, and let τ be a time interval in which the change in the record level is equal to the rms noise. Then if v_m is the rate of angular motion of the moon's limb over the radio source, the obtainable angular resolution is approximately

$$\Delta\theta = v_m\tau. \tag{15.15}$$

During the interval τ the flux density at the antenna changes by ΔS. Let θ_s be the width of the main structure of the source in a direction normal to the moon's limb, and let S be the total flux density of the source. Then for simple source structures the average brightness is approximately S/θ_s^2, and the change in solid angle of the covered part of the source in time τ is $\theta_s \Delta\theta$. Thus we have:

$$\frac{\Delta\theta}{\theta_s} \simeq \frac{\Delta S}{S} . \tag{15.16}$$

The signal-to-noise ratio at the receiver output for a component of flux density ΔS is

$$\mathcal{R}_{sn} = \frac{A\,\Delta S\sqrt{\Delta\nu\tau}}{2kT_S} , \tag{15.17}$$

where A is the collecting area of the antenna, $\Delta\nu$ and T_S are the bandwidth and system temperature of the receiving system, and k is Boltzmann's constant. The conditions that we are considering correspond to $\mathcal{R}_{sn} \simeq 1$, and from Eqs. (15.15), (15.16), and (15.17) we obtain

$$\Delta\theta = \left(\frac{2kT_S\theta_s}{AS}\right)^{2/3}\left(\frac{v_m}{\Delta\nu}\right)^{1/3} . \tag{15.18}$$

As an example consider an observation at a frequency in the 100–300 MHz range for which we use $A = 2000$ m^2, $T_S = 200$ K, and $\Delta\nu = 2$ MHz. For a fairly weak radio source we take $S = 10^{-26}$ W m^{-2} Hz^{-1} (1 Jy) and $\theta_s = 5$ arcsec. v_m varies with the position on the moon's limb but is typically 0.3 arcsec sec^{-1}. With these values, Eq. (15.18) gives $\Delta\theta = 0.7$ arcsec. Although Eq. (15.18) is derived using a geometrical-optics approach, this does not limit its applicability. For an observed occultation curve, the equivalent curve for the geometrical-optics case can be obtained by adjustment of the phases of the Fourier components, which does not affect the signal-to-noise ratio.

The bandwidth of the receiving system has the effect of smearing out angular detail in an occultation observation in a similar manner to that for an array. Thus, since the signal-to-noise ratio increases with bandwidth, for any observation there exists a bandwidth with which the sensitivity to fine angular structure is maximized. Further discussion of such details and of the practical implementation of Scheuer's restoration technique is given by von Hoerner (1964), Cohen (1969), and Hazard (1976). Note that a source may undergo a number of occultations within a period of a few months, with the moon's limb traversing the source at various position angles. The one-dimensional brightness distributions can be combined to obtain a two-dimensional image of the source.

15.3   MEASUREMENTS ON ANTENNAS

Measurement of the electric field distribution over the aperture of an antenna is an important step in optimizing the performance, especially in the case of large reflector antennas for which such results indicate the accuracy of the surface adjustment. If x and y are axes in the aperture plane, the field distribution $\mathscr{E}(x_\lambda, y_\lambda)$ is the Fourier transform of the far-field voltage radiation (reception) pattern $G(\xi, \eta)$, where ξ and η are here the direction cosines measured with respect to the x and y axes and the subscript λ indicates measurement in wavelengths:

$$\mathscr{E}(x_\lambda, y_\lambda) \rightleftharpoons G(\xi, \eta). \tag{15.19}$$

This relationship has also been encountered in Section 3.2. Direct measurement of $\mathscr{E}$ is difficult and involves introducing a probe in the aperture without disturbing the field, whereas measurement of G is relatively easy, especially for an antenna on a fully steerable mount. It is necessary to measure both the amplitude and phase of $G(\xi, \eta)$ in order to perform the Fourier transform for $\mathscr{E}(x_\lambda, y_\lambda)$. To accomplish this the beam of the antenna under test may be scanned over the direction of a distant transmitter and a second, fixed antenna used to receive a phase reference signal. The function $G(\xi, \eta)$ can be obtained from the product of the signals from the two antennas. This technique resembles the use of a reference beam in optical holography, and antenna measurements of this type have been described as holographic (Napier and Bates 1973, Bennett et al. 1976). See Morris (1985) for a phaseless method.

The same principle applies to observations using interferometers and synthesis arrays, in which the correlator-type receiving system is ideally suited for measurement of $G(\xi, \eta)$. Consider such an instrument with tracking antennas observing an unresolved cosmic source at the phase reference position. If the instrumental parameters (baselines, etc.) and the source position are accurately known, the calibrated visibility values will have a real part corresponding to the flux density of the source and an imaginary part equal to zero (except for the noise). Then if one antenna of a correlated pair is scanned in a raster pattern over the source, while the other antenna continues to track the source, the corresponding visibility values will be proportional to the amplitude and phase of $G(\xi, \eta)$ for the scanning antenna. The sampling theorem of Fourier transforms indicates that for an antenna of width x_λ wavelengths the scanning should be performed in angular steps no greater than x_λ^{-1}. A discrete Fourier transformation then provides an array of relative values of $\mathscr{E}(x_\lambda, y_\lambda)$ in amplitude and phase. The spacing between the measurements in x_λ and y_λ is inversely proportional to the dimensions of the transformation array and, thus, to the total angular range of the scan of the antenna. The phase values provide a measure of the surface accuracy. If it is necessary to optimize only the focus position of the feed, a few measurements across the aperture will suffice (Godwin, Anderson, and Bennett 1978).

Measurement of synthesis-array antennas as outlined above was first described by Scott and Ryle (1977), who also discuss the signal-to-noise ratio required. An example of measured amplitude and phase distributions is shown in Fig. 15.5.

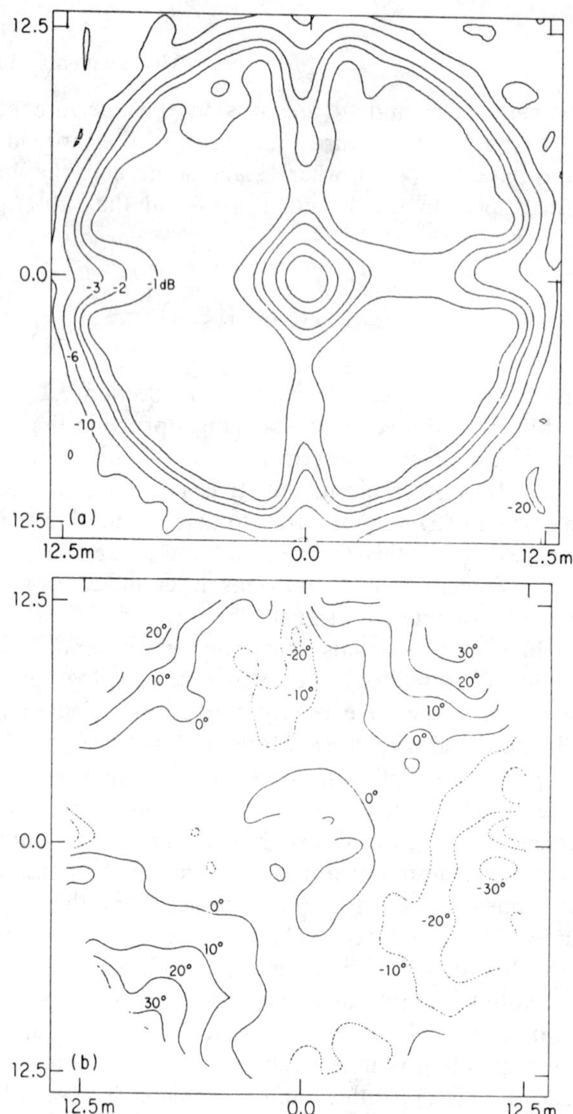

Figure 15.5 (a) Contours of power and (b) contours of phase obtained for a VLA antenna (diameter 25 m) at a frequency of 4.86 GHz. The resolution of the measurements in the aperture plane is approximately 1.4 m. The contours in (a) show the effect of blockage by the subreflector support structure and of the off-axis position of the feed. The phase contours in (b) reflect the effect of small pointing and focusing errors. From Napier, Thompson, and Ekers (1983); © 1983 IEEE.

15.4   OPTICAL INTERFEROMETRY

The principles of optical interferometry are essentially identical to those at radiofrequencies, but accurate measurements are more difficult to make. One difficulty arises because irregularities in the atmosphere introduce variations in the effective path length that are large compared with the wavelength, and thus cause the phase to vary irregularly by many rotations. Also, obtaining the mechanical stability of an instrument required to obtain fringes at a wavelength of order 500 nm presents a formidable problem. Thus it is difficult or impossible to calibrate the instrumental phase response, and in many cases only the visibility amplitude is measured. The resulting data can be interpreted in terms of models, as discussed in Section 1.3, or the autocorrelation of the brightness distribution as explained in Section 11.2. More importantly, the techniques for two-dimensional reconstruction without phase data (e.g., Bates 1984) are applicable in optical interferometry. Numerous techniques have been devised to measure and interpret visibility amplitudes and the literature is extensive: see reviews by Worden (1977), Labeyrie (1978), Dainty (1981), and Bates (1982). Before discussing three types of instruments we introduce briefly some relevant atmospheric parameters.

The irregularities in the atmosphere give rise to random variations in the refractive index over a large range of linear scales. For any particular wavelength there exists a scale size over which portions of a wavefront remain substantially plane compared with the wavelength, that is, atmospheric phase variations are small compared with 2π. This scale size is represented by a parameter, the Fried length r_0 (Fried 1965), which is equal to $3.2d_0$: see Eq. (13.100). Regions of such uniform characteristics are sometimes referred to as seeing cells. The scale size r_0 and the height at which the dominant irregularities occur define an isoplanatic angle (or isoplanatic patch size), that is, an angular range of the sky within which the incoming wavefronts from different points encounter similar phase shifts. Within an isoplanatic patch the point spread function remains constant, so the convolution relationship between source and image holds. Typical figures for the 50 percentile value of r_0, which scales as $\lambda^{6/5}$ [see Eq. (13.100)], and the isoplanatic angle, are given in Table 15.1 from Woolf (1982). Also included for comparison are the corresponding values of the diffraction-limited resolution of a telescope of 1-m aperture.

TABLE 15.1   Atmospheric and Instrumental Parameters at Visible and Infrared Wavelengths

Wavelength (nm)	r_0 (m)	Isoplanatic Angle at Zenith	Resolution of 1-m Diameter Aperture
0.5 (visible)	0.14	5.5″	0.13″
2.2 (infrared)	0.83	33″	0.55″
20 (infrared)	11.7	8′	5.0″

Modern Michelson Interferometer

The original Michelson instrument was briefly discussed in Section 1.3, and it was pointed out that the instability of the fringes was a limiting factor in estimation of the visibility. The time scale of the atmospheric fluctuations is of the order of 10 msec, so if the source is bright enough it is possible to make a measurement on this time scale, and to average many such measurements using electronic instrumentation. The results are then much more accurate than can be obtained visually. A modern implementation of a Michelson-type interferometer is shown in Fig. 15.6. The mirrors M_1 and M_2 track the object under investigation and the corner mirrors C_1 and C_2 are adjusted to keep the path lengths constant. The rest of the equipment remains fixed and is thus more stable than a telescope-mounted system. In either case the mechanical bearings must be highly accurate to avoid unwanted motions. The apertures of the interferometer determined by M_1 and M_2 are made no larger than the scale size r_0 so the effect of the irregularities is mainly to vary the phase and not the amplitude of the fringes. This limitation of the aperture size also limits the sensitivity of the Michelson interferometer. If the two detectors D_1 and D_2 are made to respond to points on the fringe pattern spaced by one-quarter of a fringe cycle, it is possible to combine their outputs to measure the instantaneous amplitude and phase of the fringes. This method is described, for example, by Rogstad (1968) who has also pointed out that with a multielement system the phase information can be utilized by means of closure relationships as discussed in Section 11.2. Optical interferometers can be built with very wide bandwidths, that is, $\Delta\lambda/\lambda$ approaching unity, so that the central or white fringe is readily identifiable. If such a system is made to operate at two such wide wavelength bands simultaneously, the effects of the atmosphere, which is slightly dispersive, can be removed. Ground-based optical astrometry with

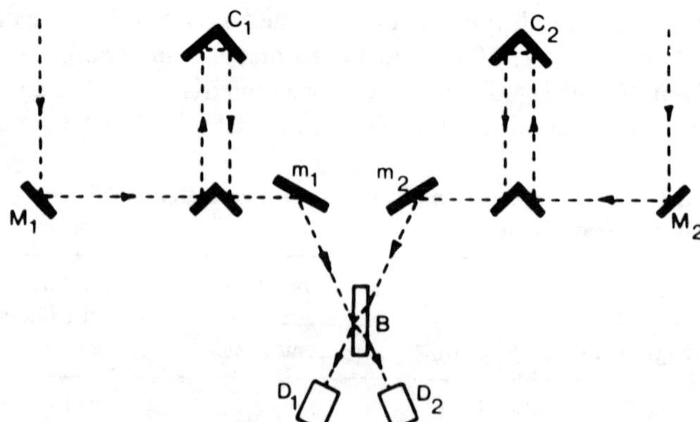

Figure 15.6 Simplified diagram of a modern implementation of the Michelson interferometer. From Davis (1976).

dual-wavelength phase-tracking interferometers can yield accurate positions of stars (Shao and Staelin 1980). For further discussion of practical systems, see Currie, Knapp, and Liewer (1974), Labeyrie (1975), and Davis (1976).

An alternative to the optical system of combining the light from the two apertures as in Fig. 15.6 is a heterodyne system in which the light from each aperture is mixed with coherent light from a central laser to produce an intermediate frequency. The IF waveforms are then amplified and correlated in an electronic system essentially identical to that used in radio interferometry. In comparison with a fully optical system, the sensitivity is greatly limited by the bandwidth that can be handled by the electronic amplifiers, unless the mixer outputs are split into many frequency channels, each of which is again converted to the IF range. A large bandwidth can then be processed using a large number of amplifiers and correlators. The bandwidth division also has the effect of increasing the path length difference over which the signals remain coherent. The heterodyne technique has been used in infrared interferometry by Johnson, Betz, and Townes (1974), and application to large multielement telescopes with multiband processing in the infrared and visible ranges has been discussed by Swenson, Gardner, and Bates (1986).

Factors that determine the sensitivity of optical systems such as losses due to scattering, partial reflection, and absorption are different from corresponding effects at radio wavelengths. The most important difference is the role of quantum effects. The energy of optical photons is five or more orders of magnitude greater than that of radio photons, and quantum effects are usually negligible in the radio domain.

At infrared and optical frequencies, the differences in noise performance between a direct detection system and a heterodyne system are significant. The number of photons received from a source of intensity I is

$$N = \frac{I\Omega_s A \, \Delta\nu}{h\nu} \quad (\text{photons/sec}), \qquad (15.20)$$

where Ω_s is the solid angle of the source (with no atmospheric blurring), A is the collecting area of the telescope, $\Delta\nu$ is the bandwidth, ν is the frequency, and h is Planck's constant. If the source is a black body at temperature T, then Planck's law gives

$$I = \frac{2h\nu^3}{c^2} \frac{1}{e^{h\nu/kT} - 1}, \qquad (15.21)$$

and therefore

$$N = \frac{2\Omega_s A \, \Delta\nu}{\lambda^2} \frac{1}{e^{h\nu/kT} - 1} \quad (\text{photons/sec}). \qquad (15.22)$$

The received power is

$$P = h\nu N. \tag{15.23}$$

The fluctuations in the power, ΔP_D, are caused by photon shot noise and therefore are proportional to $\sqrt{N}$. Thus,

$$\Delta P_D = h\nu\sqrt{N}. \tag{15.24}$$

ΔP_D is known as the noise equivalent power (NEP). The signal-to-noise ratio in one second is $P/\Delta P_D$, and therefore for an integration time τ the signal-to-noise ratio is

$$\mathscr{R}_{\mathrm{snD}} = \left[\left(\frac{2\Omega_s A}{\lambda^2}\right)\frac{\Delta\nu\tau}{e^{h\nu/kT}-1}\right]^{1/2}. \tag{15.25}$$

Note that $\mathscr{R}_{\mathrm{snD}}$ is proportional to $\sqrt{A}$, because of the shot noise, rather than to A as in the radio case. In a heterodyne system, the noise is determined by the uncertainty principle since the mixer is a linear device that preserves phase (see Section 1.3 under *Very-Long-Baseline Interferometry*). The minimum noise is one photon per mode (one photon per Hz per second). This is equivalent to saying that the system temperature is $h\nu/k$ (see Heffner 1962; Caves 1982). Hence, in a period of one second the uncertainty in power is

$$\Delta P_H = h\nu\sqrt{\Delta\nu}. \tag{15.26}$$

The heterodyne detector responds only to the component of radiation to which its polarization is matched, and the total received power is half that in Eq. (15.23). The signal-to-noise ratio is therefore

$$\mathscr{R}_{\mathrm{snH}} = \frac{\Omega_s A}{\lambda^2}\sqrt{\Delta\nu\,\tau}\,\frac{1}{e^{h\nu/kT}-1}. \tag{15.27}$$

Note that Eq. (15.27) reduces to the usual expression for the radio case when $h\nu/kT \ll 1$, and T_S is set equal to $h\nu/k$. The ratio of signal-to-noise ratios for the two optical systems when the bandwidth and integration time are the same is

$$\frac{\mathscr{R}_{\mathrm{snH}}}{\mathscr{R}_{\mathrm{snD}}} = \sqrt{\frac{\Omega_s A}{2\lambda^2}\frac{1}{e^{h\nu/kT}-1}}. \tag{15.28}$$

In the weak signal case, the heterodyne system is generally much less sensitive than the direct detection system. However, there are conditions that favor the use of heterodyne systems (Townes and Sutton 1981; de Graauw and van de Stadt 1981).

Optical Intensity Interferometer

The use of the intensity interferometer for optical measurements on stars was demonstrated by Hanbury Brown and Twiss (1956) shortly after the success of the radio intensity interferometer described in Chapter 1. In the optical case, a photomultiplier tube at the focus of each telescope mirror replaces the RF and IF stages and the detectors of the radio instrument. The photomultiplier outputs are amplified and fed to the inputs of the correlator. The optical intensity interferometer is largely insensitive to atmospheric phase fluctuations for the same reasons explained in Section 15.1. The size of the light-gathering apertures, therefore, is unrestricted by the scale size of the irregularities. Also, it is not necessary that the reflecting mirrors produce a diffraction-limited image, and their accuracy need be sufficient only to deliver all the light to the photomultiplier cathodes. This is fortunate since the low sensitivity mentioned earlier for the radio case necessitates the use of large light-gathering areas. Hanbury Brown (1974) has given an analysis of the response of the optical instrument and shown that it is proportional to the square of the visibility modulus as in the radio case. Either a correlator or a photon coincidence counter can be used to combine the photomultiplier outputs.

The intensity interferometer constructed at Narrabri, Australia (Hanbury Brown, Davis, and Allen 1967; Hanbury Brown 1974), used two 6.5-m diameter reflectors and a bandwidth of 60 MHz for the signals at the correlator inputs. The resulting limiting magnitude of $+2.5$ enabled measurements of 32 stars to be made. Davis (1976) has discussed the relative merits of the intensity interferometer and modern implementations of the Michelson interferometer for development of a more sensitive instrument.

Speckle Interferometry

The image of an unresolved point source observed with a large-aperture telescope, that is, one in which the width of the aperture is large compared with r_0, depends on the exposure time over which the image is averaged. An exposure no longer than 10 msec shows a group of bright speckles, each of which is of the approximate size of the Airy disk (i.e., the diffraction-limited point-source image) of the telescope. If the exposure is much longer the pattern is blurred into a single patch (the "seeing" disk) of typical diameter 1 arcsec determined by the atmosphere. The characteristic fluctuation time of 10 msec in the optical range corresponds to the time taken for a cell of size r_0 to move past any point in the telescope aperture at a typical wind speed of 10–20 m sec^{-1}. The use of sequences of short-exposure images to obtain information at the diffraction limit of a large telescope originated with Labeyrie (1970) and is known as speckle interferometry. Speckle patterns reflect the random distribution of atmospheric irregularities over the aperture and differ from one exposure to the next on the 10-msec time scale. Reduction of the many exposures required to observe faint objects by this technique involves computer image processing.

For the theory of the speckle response see, for example, Dainty (1973) or the review by Bates (1982). Here we note that the high-resolution image represented by a single speckle can be understood if one considers each speckle as resulting from several seeing cells of the wavefront, located at points distributed across the telescope aperture. These cells are the ones that present approximately equal phase shifts in the ray paths from the wave front to the speckle image (Worden 1977). Then, by analogy with an array of antennas, the resolution corresponds to the maximum spacing of the cells, that is, it is of the order λ/D, where D is the telescope aperture. Aberrations in the reflector do not significantly degrade the speckle pattern as long as the dominant phase irregularities are those of the atmosphere. The area of the image over which the speckles are spread corresponds to λ/r_0 on the sky and becomes the seeing disk in a long exposure. The seeing cells can be regarded as subapertures within the main telescope aperture, the responses of which combine with random phases in the image. The number of speckles is of the order of the number of subapertures, that is, $(D/r_0)^2$. With a large telescope ($D \sim 1$ m) this number is of the order of 100 at optical wavelengths. As shown in Table 15.1 the size of the seeing cells increases with wavelength, and in the infrared only a few speckles appear in the image. Examples of speckle images can be found in the review papers cited above.

The modulus of the visibility can be determined from the speckle images. This procedure can be understood from the following simplified discussion. On a single image of short exposure, a number of approximately diffraction-limited speckles appear at random locations within the seeing disk. The speckle image $B_s(\xi, \eta)$ can be described as the convolution of the actual brightness distribution $B(\xi, \eta)$ with the speckle point spread function $\mathscr{P}(\xi, \eta)$. Thus

$$B_s(\xi, \eta) = B(\xi, \eta) * * \mathscr{P}(\xi, \eta). \tag{15.29}$$

The function $\mathscr{P}(\xi, \eta)$ is a random function that cannot be specified exactly. As a first approximation, we will assume that $\mathscr{P}(\xi, \eta)$ is the point spread function of the telescope in the absence of atmospheric effects, $b_0(\xi, \eta)$, replicated at the position of each speckle. Thus we can write

$$\mathscr{P}(\xi, \eta) = \sum b_0(\xi - \xi_i, \eta - \eta_i), \tag{15.30}$$

where ξ_i and η_i are the locations of the speckles, all of which are assumed to have the same intensity. From Eqs. (15.29) and (15.30), we obtain

$$B_s(\xi, \eta) = \sum B(\xi, \eta) * * b_0(\xi - \xi_i, \eta - \eta_i). \tag{15.31}$$

If the Fourier transform of $b_0(\xi, \eta)$ is $\bar{b}_0(u, v)$, then the Fourier transform of $b_0(\xi - \xi_i, \eta - \eta_i)$ is $\bar{b}_0(u, v)\exp[j2\pi(u\xi_i + v\eta_i)]$. Hence, the Fourier transform of Eq. (15.31) can be written

$$\overline{B}_s(u, v) = \sum \mathcal{V}(u, v)\overline{b}_0(u, v)e^{j2\pi(u\xi_i + v\eta_i)}, \tag{15.32}$$

where $\mathcal{V}$ and $\overline{B}_s$ are the Fourier transforms of B and B_s, respectively. The speckle transforms $\overline{B}_s$ cannot be summed directly because of the random phase factors in Eq. (15.32). To eliminate these phase factors, we calculate $|\overline{B}_s|^2$ (that is, $\overline{B}_s\overline{B}_s^*$), which is

$$|\overline{B}_s(u, v)|^2 = \sum_i \sum_k |\mathcal{V}(u, v)|^2|\overline{b}_0(u, v)|^2 e^{j2\pi[u(\xi_i - \xi_k) + v(\eta_i - \eta_k)]}, \tag{15.33}$$

whence

$$|\overline{B}_s(u, v)|^2 = |\mathcal{V}(u, v)|^2|\overline{b}_0(u, v)|^2\left[N + \sum_{i \neq k} e^{j2\pi[u(\xi_i - \xi_k) + v(\eta_i - \eta_k)]}\right], \tag{15.34}$$

where N is the number of speckles. Since the expectation of the summation term in Eq. (15.34) is zero, the expectation of Eq. (15.34) is

$$\langle|\overline{B}_s(u, v)|^2\rangle = N_0|\mathcal{V}(u, v)|^2|\overline{b}_0(u, v)|^2, \tag{15.35}$$

where N_0 is the average number of speckles. Hence, the average of a series of measurements of $|B_s(u, v)|^2$, estimated from short exposures, is proportional to the squared modulus of $\mathcal{V}(u, v)$ times the squared modulus of $\overline{b}_0(u, v)$. Since $b_0(u, v)$ is nonzero for $|u|, |v| < D/\lambda$, the function $|\mathcal{V}(u, v)|^2$ can be determined over the same range of u and v, if $\overline{b}_0(u, v)$ is known.

In practice the speckles cannot be accurately modeled by Eq. (15.30). However, we can write

$$\langle|\overline{B}_s(u, v)|^2\rangle = |\mathcal{V}(u, v)|^2\langle|\overline{\mathcal{P}}(u, v)|^2\rangle, \tag{15.36}$$

where $\overline{\mathcal{P}}(u, v)$ is the Fourier transform of $\mathcal{P}(\xi, \eta)$. $\langle|\overline{\mathcal{P}}(u, v)|^2\rangle$ is proportional to $|\overline{b}_0(u, v)|^2$ and is called the renormalization function. It is obtained by observing a point source under the same seeing conditions as the source of interest. In the method used in the first speckle observations (Gezari, Labeyrie, and Stachnik 1972) the speckle images are recorded on film and the Fourier transform obtained optically. An alternative technique involves digitizing the speckle images and computing the two-dimensional autocorrelation function. The Fourier transform of the autocorrelation is $|\overline{B}_s(u, v)|^2$. In either case the procedure is repeated using an unresolved star to determine $|\overline{b}_0(u, v)|^2$. Methods of image restoration using only the visibility modulus, as discussed in Chapter 11, can then be applied.

Reduction techniques also exist in which information from the visibility phase is retained. Examination of individual speckles as diffraction-limited images with some atmospheric distortion is an approach in which the asymmetry of the source, and thus its visibility phase, can be preserved: see, for example, Lynds, Worden, and Harvey (1976). For brightness distributions that include a point source within the same isoplanatic patch, holographic methods are possible (Bates, Gough, and Napier 1973). Knox and Thompson (1974) originated an algorithm which in effect measures the derivative of the visibility with respect to u and v. In this technique the individual images are sampled digitally and their Fourier transforms calculated at the Nyquist interval corresponding to a field size somewhat larger (say, a factor of 2) than the seeing size. Such an interval guarantees that the phase change due to speckle phase (the term in the exponential in Eq. (15.32)) from one cell to the next will be less than $\pi/2$. For each transform, the difference in phase between adjacent cells in the u and in the v direction is calculated, generating the derivative of phase with respect to u and v. The fringe amplitudes and phase differences obtained from all the images are averaged. From Eq. (15.32) we see that the part of phase difference due to speckle motion averages to zero, leaving only the part due to source structure. Finally, the phase differences in the u and v directions can be integrated to obtain the visibility phase. Note that because the phase differences are in two dimensions, the phase can be recovered in such a way that errors do not build up as they would in a one-dimensional integration (Hunt 1979). Examples of brightness distributions recovered by this technique can be found in Papaliolios, Nisenson, and Ebstein (1985).

REFERENCES

Bates, R. H. T., Astronomical Speckle Imaging, *Phys. Rep.*, **90**, 203–297, 1982.

Bates, R. H. T., Uniqueness of Solutions to Two-Dimensional Fourier Phase Problems for Localized and Positive Images, *Comp. Vision, Graphics Image Process.*, **25**, 205–217, 1984.

Bates, R. H. T., P. T. Gough, and P. J. Napier, Speckle Interferometry Gives Holograms of Multiple Star Images, *Astron. Astrophys.*, **22**, 319–320, 1973.

Bennett, J. C., A. P. Anderson, and P. A. McInnes, Microwave Holographic Metrology of Large Reflector Antennas, *IEEE Trans. Antennas Propag.*, **AP-24**, 295–303, 1976.

Bracewell, R. N., Radio Interferometry of Discrete Sources, *Proc. IRE*, **46**, 97–105, 1958.

Carr, T. D., M. A. Lynch, M. P. Paul, G. W. Brown, J. May, N. F. Six, V. M. Robinson, and W. F. Block, Very Long Baseline Interferometry of Jupiter at 18 MHz, *Radio Sci.* **5**, 1223–1226, 1970.

Caves, C. M., Quantum Limits on Noise in Linear Amplifiers, *Phys. Rev.*, **26D**, 1817–1839, 1982.

Cohen, M. H., High Resolution Observations of Radio Sources, *Ann. Rev. Astron. Astrophys.*, **7**, 619–664, 1969.

Currie, D. G., S. L. Knapp, and K. M. Liewer, Four Stellar-Diameter Measurements by a New Technique: Amplitude Interferometry, *Astrophys. J.*, **187**, 131–134, 1974.

Dainty, J. C., Diffraction-Limited Imaging of Stellar Objects using Telescopes of Low Optical Quality, *Opt. Comm.*, **7**, 129–134, 1973.

Dainty, J. C., Speckle Interferometry in Astronomy, in *Symposium on Recent Advances in Observational Astronomy*, *Ensenada, Mexico*, 1979, H. L. Johnson and C. Allen, Eds, Universidad Nacional Autonóma de México, Ciudad Universitaria, Mexico, 1981, pp. 95–111.

Davis, J., High Angular Resolution Stellar Interferometry, *Proc. Astron. Soc. Aust.*, **3**, 26–32, 1976.

de Graauw, T. and H. van de Stadt, Coherent Versus Incoherent Detection for Interferometry at Infrared Wavelengths, *Proc. of ESO Conference on Scientific Importance of High Angular Resolution at Infrared and Optical Wavelengths*, M. H. Ulrich and K. Kjär, Eds., European Southern Observatory, Garching, 1981.

Dulk, G. A., Characteristics of Jupiter's Decametric Radio Source Measured with Arc-Second Resolution, *Astrophys. J.*, **159**, 671–684, 1970.

Fried, D. L., Statistics of a Geometric Representation of Wavefront Distortion, *J. Opt. Soc. Am.*, **55**, 1427–1435, 1965.

Gezari, D. Y., A. Labeyrie, and R. V. Stachnik, Speckle Interferometry: Diffraction-Limited Measurements of Nine Stars with the 200-Inch Telescope, *Astrophys. J.*, **173**, L1–L5, 1972.

Godwin, M. P., A. P. Anderson, and J. C. Bennett, Optimisation of Feed Position and Improved Profile Mapping of a Reflector Antenna from Microwave Holographic Measurements, *Electron. Lett.*, **14**, 134–136, 1978.

Hanbury Brown, R., *The Intensity Interferometer*, Taylor and Françis, London, 1974.

Hanbury Brown, R., J. Davis, and L. R. Allen, The Stellar Interferometer at Narrabri Observatory-I, *Mon. Not. R. Astron. Soc.*, **137**, 375–392, 1967.

Hanbury Brown, R. and R. Q. Twiss, A New Type of Interferometer for Use in Radio Astronomy, *Philos. Mag.*, ser. 7, **45**, 663–682, 1954.

Hanbury Brown, R. and R. Q. Twiss, A Test of a New Type of Stellar Interferometer on Sirius, *Nature*, **178**, 1046–1048, 1956.

Hazard, C., Lunar Occultation Measurements, in *Methods of Experimental Physics*, Vol. 12C, M. L. Meeks, Ed., Academic Press, New York, 1976.

Heffner, H., The Fundamental Noise Limit of Linear Amplifiers, *Proc. IRE*, **50**, 1604–1608, 1962.

Hunt, B. R., Matrix Formulation of the Reconstruction of Phase Values from Phase Differences, *J. Opt. Soc. Amer.*, **69**, 393–399, 1979.

Jennison, R. C. and M. K. Das Gupta, The Measurement of the Angular Diameter of Two Intense Radio Sources, *Philos. Mag.*, ser. 8, **1**, 55–75, 1956.

Johnson, M. A., A. L. Betz, and C. H. Townes, 10-μm Heterodyne Stellar Interferometer, *Phys. Rev. Lett.*, **33**, 1617–1620, 1974.

Knox, K. T. and B. J. Thompson, Recovery of Images from Atmospherically Degraded Short-Exposure Photographs, *Astrophys. J.*, **193**, L45–L48, 1974.

Labeyrie, A., Attainment of Diffraction Limited Resolution in Large Telescopes by Fourier Analysing Speckle Patterns in Star Images, *Astron. Astrophys.*, **6**, 85–87, 1970.

Labeyrie, A., Interference Fringes Obtained on Vega with Two Optical Telescopes, *Astrophys. J.*, **196**, 671–675, 1975.

Labeyrie, A., Stellar Interferometry Methods, *Ann. Rev. Astron. Astrophys.*, **16**, 77–102, 1978.

Lynds, C. R., S. P. Worden, and J. W. Harvey, Digital Image Reconstruction Applied to Alpha Orionis, *Astrophys. J.*, **207**, 174–180, 1976.

Morris, D., Phase Retrieval in the Radio Holography of Reflector Antennas and Radio Telescopes, *IEEE Trans. Antennas Propag.*, **AP-33**, 749–755, 1985.

Napier, P. J. and R. H. T. Bates, Antenna-Aperture Distributions from Holographic Type of Radiation-Pattern Measurements, *Proc. IEE*, **120**, 30–34, 1973.

Napier, P. J., A. R. Thompson, and R. D. Ekers, The Very Large Array: Design and Performance of a Modern Synthesis Radio Telescope, *Proc. IEEE*, **71**, 1295–1320, 1983.

Nather, R. E. and M. M. McCants, Photoelectric Measurements of Lunar Occultations. IV. Data Analysis, *Astron. J.*, **75**, 963–968, 1970.

Papaliolios, C., P. Nisenson, and S. Ebstein, Speckle Imaging with the PAPA Detector, *Applied Optics*, **24**, 287–292, 1985.

Ridgway, S. T., G. H. Jacoby, R. R. Joyce, and D. C. Wells, Angular Diameters by the Lunar Occultation Technique. III., *Astron. J.*, **85**, 1496–1504, 1980.

Rogstad, D. H., A Technique for Measuring Visibility Phase With an Optical Interferometer in the Presence of Atmospheric Seeing, *Appl. Opt.*, **7**, 585–588, 1968.

Scheuer, P. A. G., On the Use of Lunar Occultations for Investigating the Angular Structure of Radio Sources, *Aust. J. Phys.*, **15**, 333–343, 1962.

Scott, P. F. and M. Ryle, A Rapid Method for Measuring the Figure of a Radio Telescope Reflector, *Mon. Not. R. Astron. Soc.*, **178**, 539–545, 1977.

Shao, M. and D. H. Staelin, First Fringe Measurements with a Phase-Tracking Stellar Interferometer, *Appl. Opt.*, **19**, 1519–1522, 1980.

Swenson, G. W., Jr., C. S. Gardner, and R. H. T. Bates, Optical Synthesis Telescopes, *Proc. S.P.I.E.*, **643**, 129–140, 1986.

Townes, C. H. and E. C. Sutton, Multiple Telescope Infrared Interferometry, *Proc. of ESO Conference on Scientific Importance of High Angular Resolution at Infrared and Optical Wavelengths*, M. H. Ulrich and K. Kjär, Eds., European Southern Observatory, Garching, 1981.

von Hoerner, S., Lunar Occultations of Radio Sources, *Astrophys. J.*, **140**, 65–79, 1964.

Woolf, N. J., High Resolution Imaging from the Ground, *Ann. Rev. Astron. Astrophys.*, **20**, 367–398, 1982.

Worden, S. P., Astronomical Image Reconstruction, *Vistas in Astron.*, **20**, 301–318, 1977.

PRINCIPAL SYMBOLS

Listed below are the principal symbols used throughout the book. Subscripted variations and locally defined symbols with restricted usage are selectively included.

a	model dimension, scale size
α	scale size of ionospheric irregularities (blobs), (Chapter 13)
A	antenna collecting area (reception pattern)
A_1	one-dimensional reception pattern
A_0	antenna collecting area on axis
A_N	normalized reception pattern
$\mathscr{A}$	mirror-image reception pattern, azimuth
b_0	synthesized beam pattern, point-source response
b_N	normalized synthesized beam pattern
$\mathfrak{b}$	galactic latitude (Chapter 13)
B	brightness, magnetic field magnitude
$\mathbf{B}$	magnetic field vector
B_1	one-dimensional brightness
B_a	brightness distribution with visibility averaging (Chapter 6)
B_b	brightness distribution with bandwidth smearing (Chapter 6)
B_0	peak brightness of point-source map, derived (synthesized) brightness distribution
$\mathscr{B}$	bandwidth pattern (Chapter 3)
c	speed (velocity) of light
C	coherence function (Chapter 9), convolving function (Chapter 10)
$C_n^2,\, C_{\mathcal{N}}^2$	turbulence strength parameters (Chapter 13)

d	distance, baseline declination, projected baseline (Chapter 13)
d_0	distance over which rms phase deviation $= 1$ rad (Chapter 13)
D	baseline (antenna spacing), antenna diameter
$\mathbf{D}$	baseline vector
$D_\lambda, \mathbf{D}_\lambda$	baseline measured in wavelengths
$D_a, \mathbf{D}_a$	inter-axis distance of antenna mount (Chapter 4)
D_E	equatorial component of baseline
D_m	dispersion measure (Chapter 13)
D_R	delay resolution function (Chapter 9)
$\mathscr{D}$	sensitivity degradation factor (Chapter 7)
$\mathscr{D}_\phi$	structure function of phase (Chapter 13)
$\mathscr{D}_n$	structure function of refractive index (Chapter 13)
e	spatial spectrum of electric field (Chapter 3), electronic charge (Chapter 13)
$E, \mathbf{E}$	electric field, spectral components of electric field
E_x, E_y	components of electric field
$\mathscr{E}$	complex amplitude of electric field, elevation.
f	frequency of Fourier components of power spectrum (Chapter 9)
f_i	oscillator strength (Chapter 13)
F	flux density (W m^{-2}), fringe function
F_h	threshold of harmful interference (W m^{-2}) (Chapter 14)
$\mathscr{F}$	quantized fringe-rotation function (Chapter 9)
g	aperture illumination function of an antenna (Chapter 3), voltage gain constant for a single antenna, gravitational acceleration (Chapter 13)
G	one-dimensional voltage reception pattern of an antenna (Chapter 3)
G_{mn}	gain factor for a correlated antenna pair
G_N	normalized voltage reception pattern
G_0	gain factor (Chapter 7)
$\mathscr{G}$	occultation response function (Chapter 15)
h	Planck's constant, hour angle, height
H	hour angle, voltage frequency response
H_0	gain constant
i	electric current
I	Stokes parameter, standard deviation of fractional frequency deviation (Chapter 9)
$\mathbf{I}$	unit vector in direction of polar or azimuth axes
I_0	modified Bessel function of zero order

I_1	modified Bessel function of first order
j	$\sqrt{-1}$
j_ν	volume emissivity (Chapter 13)
J_0	Bessel function of first kind and zero order
J_1	Bessel function of first kind and first order
k	Boltzmann's constant, propagation constant (Chapter 13)
$\mathbf{k}$	propagation vector with magnitude $2\pi/\lambda$ (Chapter 9)
ℓ	length, direction cosine
ℓ_1	unit spacing in a grating array (Chapter 5)
l	galactic longitude (Chapter 13)
L	antenna dimension, length of a transmission line, path length, likelihood function (Chapter 12)
$\mathbf{L}$	probability function (Chapter 8)
L_1	loss in a transmission line (Chapter 7)
$\mathscr{L}$	latitude, *excess* path length (Chapter 13)
m	direction cosine, electron mass (Chapter 13)
m_ℓ, m_c, m_t	degree of linear, circular, and total polarization
m	measured quantity (Appendix 12-1)
M	complex degree of linear polarization (Chapter 13)
$\mathscr{M}$	molecular weight (Chapter 13)
n	weighting factor in quantization (Chapter 8), noise component, index of refraction (Chapter 13)
n_a	number of antennas
n_p	number of antenna pairs
n_r	number of points in a rectangular array
N	refractivity (Chapter 13)
$\mathscr{N}$	density of gas molecules (Chapter 13)
$\mathscr{N}_e, \mathscr{N}_i, \mathscr{N}_a$	electron, ion, and neutral particle density (Chapter 13)
p	probability density or probability distribution [i.e. $p(x)dx$ is the probability that the random variable lies between x and $x + dx$], bivariate normal probability function (Chapter 8), partial pressure (Chapter 13), impact parameter (Chapter 13)
p_D	partial pressure of dry air (Chapter 13)
p_V	partial pressure of water vapor (Chapter 13)
P	cumulative probability, interferometer response function (Chapter 3), total atmospheric pressure (Chapter 13)
$\mathbf{P}$	dipole moment per unit volume
P_{mnp}	instrumental polarization factor
$\mathscr{P}$	interferometer response function (Chapter 3), point-source response (point spread function) (Chapter 15)

q	distance in (u, v) plane
q'	distance in (u', v') plane
q_s	radial distance in the (q_x, q_y) spatial frequency plane (cycles per meter)
Q	Stokes parameter, number of quantization levels, quality factor for a line or cavity (Chapter 9)
r	correlator output, distance in the (ξ, η) plane, radial distance
$\mathbf{r}$	position vector of antenna relative to center of earth
r'	distance in the (ξ', η') plane
r_e	classical electron radius (Chapter 13)
r_0	distance from the center of the earth (Chapter 13), Fried length (Chapter 15)
R	autocorrelation function, correlator output, distance, gas constant (Chapter 13)
R_a	response with visibility averaging (Chapter 6)
R_b	response with finite bandwidth (Chapter 6)
R_m	rotation measure (Chapter 13), distance of the moon's limb (Chapter 15)
R_y	autocorrelation function of fractional frequency deviation (Chapter 9)
R_ϕ	autocorrelation function of phase
$\mathscr{R}_{sn}$	signal-to-noise ratio
s	signal component
$\mathbf{s}$	unit position vector
$\mathbf{s}_0$	unit position vector of field center
S	(spectral) flux density $(\mathrm{W\,m^{-2}\,Hz^{-1}})$
S_c	flux density of a calibrator
S_h	threshold of harmful interference $(\mathrm{W\,m^{-2}\,Hz^{-1}})$ (Chapter 14)
$\mathscr{S}$	power spectrum
$\mathscr{S}_y, \mathscr{S}_y'$	single-sided and double-sided power spectra of fractional frequency deviation
$\mathscr{S}_\phi, \mathscr{S}_\phi'$	single-sided and double-sided power spectra of phase fluctuations (single-sided power spectra, which contain only positive frequencies, are used only in Section 9.4)
t	time
T	temperature, time interval
T_A	component of antenna temperature resulting from wanted signal
T_A'	total antenna temperature
T_B	brightness temperature
T_c	noise temperature of calibration signal

T_c	period of the earth's rotation (Chapter 12)
T_g	gas temperature
T_R	receiver temperature
T_S	system temperature
u	antenna spacing coordinate in units of wavelength (spatial frequency)
u'	u coordinate measured in the equatorial plane
U	Stokes parameter
v	antenna spacing coordinate in units of wavelength (spatial frequency), velocity
v'	v coordinate measured in the equatorial plane
v_g	group velocity (Chapter 13)
v_p	phase velocity (Chapter 13)
v_s	velocity of scintillation screen
v_0	quantization level (Chapter 8), particle velocity (Chapter 9)
V	voltage, Stokes parameter
$\mathcal{V}, \boldsymbol{\mathcal{V}}$	complex visibility, vector visibility
$\mathcal{V}_m$	measured complex visibility
$\mathcal{V}_M$	Michelson's fringe visibility
$\mathcal{V}_N$	normalized complex visibility
w	antenna spacing coordinate in units of wavelength (spatial frequency), weighting function, column height of precipitable water (Chapter 13)
w'	w coordinate measured in the polar direction
w_a	atmospheric weighting function (Chapter 13)
w_t	visibility tapering function
w_u	function that adjusts visibility magnitude for effective uniform weighting
W	spectral sensitivity function, transfer function
x	general position coordinate, coordinate in antenna aperture, signal voltage
x_λ	x coordinate measured in wavelengths
X	coordinate of antenna spacing, coordinate of point on a source (Appendix 3.1), signal waveform measured in units of rms amplitude (Chapter 8)
X_λ	X coordinate measured in wavelengths
y	general position coordinate, coordinate in antenna aperture, signal voltage, fractional frequency deviation (Chapter 9)
y_λ	y coordinate measured in wavelengths
Y	coordinate of antenna spacing, coordinate of point on a source (Appendix 3.1), signal waveform measured in units of rms

	amplitude (Chapter 8)
Y_λ	Y coordinate measured in wavelengths
z	general position coordinate, signal voltage
z_λ	z coordinate measured in wavelengths
$\varkappa$	zenith angle (Chapter 13)
Z	coordinate of antenna spacing, signal plus noise in correlator output, compressibility factor (Chapter 13)
$\mathbf{Z}$	signal-plus-noise vector
Z_λ	Z coordinate measured in wavelengths
α	right ascension, power absorption coefficient, solar elongation (Chapter 13)
β	oversampling factor
γ	source coherence function (Chapter 3)
Γ	mutual coherence function (Chapter 3), damping constant (Chapter 13), gamma function
Γ_{12}	mutual coherence function
δ	declination, increment prefix, (Dirac) delta function
$^2\delta$	delta function in two dimensions
Δ	small length, increment prefix
$\Delta\nu$	bandwidth
$\Delta\nu_{IF}$	intermediate frequency bandwidth
$\Delta\nu_{LF}$	low frequency bandwidth
$\Delta\nu_{LO}$	frequency difference of local oscillators
$\Delta\tau$	delay error
$\Delta u, \Delta v$	increments in (u, v) plane
$\Delta\xi, \Delta\eta$	increments in (ξ, η) plane
ε	residual, error component, noise voltage, dielectric constant (Chapter 13)
$\boldsymbol{\varepsilon}$	noise vector
ϵ	permittivity (Chapter 13)
ϵ_0	permittivity of free space (Chapter 13)
ζ	direction cosine with respect to w axis
ζ'	direction cosine with respect to w' axis
η	direction cosine with respect to v axis, loss factor
η'	direction cosine with respect to v' axis
η_D	discrete delay step loss factor
η_Q	efficiency (loss) factor for Q-level quantization
η_R	fringe rotation loss factor
η_S	fringe sideband rejection loss factor
θ	general angle, angle measured from a plane normal to the

	baseline, instrumental phase angle
θ_0	angular position of source or field center
θ_b	width of synthesized beam, bending angle (Chapter 13)
θ_f	width of synthesized field
Θ	variation in earth-rotation angle (UT1 $-$ UTC) (Chapter 12)
λ	wavelength
μ	power-law exponent in Allan variance (Chapter 9)
ν	frequency
ν'	frequency measured with respect to center frequency or local oscillator frequency
ν_b	bit rate
ν_B	gyro frequency (Chapter 13)
ν_d	intermediate frequency at which delay is inserted
ν_{ds}	delay step frequency (Chapter 9)
ν_f	fringe frequency
ν_{LO}	local oscillator frequency
ν_0	center frequency of an IF or RF band
ν_p	plasma frequency (Chapter 13)
ξ	direction cosine with respect to u axis
ξ'	direction cosine with respect to u' axis
Π	unit rectangle function
ρ	autocorrelation function, cross correlation function, reflection coefficient (Chapter 7), gas density (Chapter 13)
ρ_σ	area density in the (u, v) plane (Chapter 10)
σ	standard deviation, rms noise level
$\boldsymbol{\sigma}$	unit position vector
σ_y	Allan standard deviation (square root of Allan variance)
Σ	summation
τ	time interval
τ_a	averaging time, atmospheric delay error
τ_c	coherent integration time (Chapter 9)
τ_e	clock error
τ_g	geometrical delay
τ_i	instrumental delay
τ_0	unit increment of instrumental delay, duration of an observation (Chapter 6), zenith optical depth (opacity) of the atmosphere (Chapter 13)
τ_s	sampling interval (Chapter 8)
τ_ν	optical depth (opacity) (Chapter 13)
ϕ	phase angle

ϕ_v	visibility phase
ϕ_G	instrumental phase for correlated antenna pair
Φ	probability integral (Chapter 8)
χ	arctangent of axial ratio of polarization ellipse
ψ	position angle
ψ_p	parallactic angle
ω_e	angular rotation velocity of the earth
Ω	solid angle
Ω_0	equivalent solid angle of main lobe of synthesized beam

FREQUENTLY USED SUBSCRIPTS

$1, 2$	antenna designation
$2, 3, 4, \infty$	quantization levels (Chapter 8)
A	antenna
d	delay, double sideband
D	dry component (Chapter 13)
I	imaginary part
ℓ	lower sideband
L	left circular polarization
LO	local oscillator
0	center of frequency band or angular field, earth's surface (Chapter 13)
m, n	antenna designation
N	normalized
R	real part
R	right circular polarization
S	system
T	truncated
u	upper sideband
V	water vapor (Chapter 13)
λ	measured in wavelengths

SUPERSCRIPTS

$\hat{}$	function of frequency (Chapter 3), quantized variable (Chapter 8)
$-$	Fourier transform

SPECIAL SYMBOLS

Ш	shah function in one dimension
$^2\text{Ш}$	shah function in two dimensions
$\rightleftharpoons$	"is the Fourier transform of"
$*$	convolution in one dimension
$**$	convolution in two dimensions
$\star$	cross correlation in one dimension
$\star\star$	cross correlation in two dimensions
$\langle\ \ \rangle$	expectation, in practice often approximated by a finite average
$\dot{\ }$	first derivative with respect to time (e.g., $\dot{x}$)
$\ddot{\ }$	second derivative with respect to time (e.g., $\ddot{x}$)
$\cdot$	scalar product of vectors

FOURIER PAIRS

Denoted by: upper and lower case letters (used only as noted in particular cases)
circumflex ($\hat{\ }$) (functions of frequency and time in Chapter 3)
superscript bar

SPECIAL FUNCTIONS

For definitions see: Abramowitz, M., and I. A. Stegun, *Handbook of Mathematical Functions*, National Bureau of Standards, Washington, D. C., 1964, reprinted by Dover, New York, 1965.

erf	error function
J_0	Bessel function of first kind and zero order
J_1	Bessel function of first kind and first order
I_0	Modified Bessel function of zero order
I_1	Modified Bessel function of first order
Γ	gamma function

AUTHOR INDEX

Aarons, J., 447, 448
Ables, J. G., 349, 350
Abramowitz, M., 211, 511
Alder, B., 33
Allan, D. W., 272, 291
Allen, C. W., 428, 455
Allen, L. R., 183, 497
Ananthakrishnan, S., 463
Anderson, A. P., 491
Anderson, B., 23, 193, 364
Anderson, J. A., 12
Apostol, T. M., 80
Appleton, E. V., 14, 439
Archer, J. W., 183, 208
Armstrong, J. W., 434, 435, 436
Arsac, J., 127
Ashby, N., 291
Askne, J. I. H., 439
Audoin, C., 272

Baade, W., 17
Baars, J. W. M., 28, 128, 180, 436
Bailey, D. K., 444
Bajaja, E., 338
Baldwin, J. E., 354, 355, 356, 359
Ball, J. A., 243
Bare, C., 30, 294
Barnes, J. A., 269, 270, 276, 282
Barnett, M. A. F., 439

Bartel, N., 371, 377
Bates, R. H. T., 353, 354, 491, 493, 495, 498, 500
Batty, M. J., 181
Bean, B. R., 409, 426
Beauchamp, K. G., 203
Becker, R. H., 321
Bennett, A. S., 22
Bennett, J. C., 491
Beran, M. J., 57, 59, 60
Berkner, L. V., 33
Bernier, L. G., 280
Betz, A. L., 495
Bevington, P. R., 388
Beynon, W. J. G., 439
Bieging, J. H., 435, 436
Biraud, F., 33, 310
Blackman, R. B., 240
Blair, B. E., 269, 292
Block, W. F., 30
Blum, E. J., 28, 159
Blythe, J. H., 120, 121
Bohlander, R. A., 406
Boischot, A., 28
Bolton, J. G., 16, 17, 21
Booker, H. G., 64, 447, 449
Booth, R. S., 384
Boriakoff, V., 454, 460, 463
Born, M., 55, 59, 63, 73, 98

Bourgois, G., 458, 463
Bowers, F. K., 217, 225, 226, 230
Bowyer, S., 398
Braccesi, A., 23
Bracewell, R. N., 28, 33, 34, 48, 49, 55, 59,
    84, 86, 107, 109, 110, 115, 117, 127,
    129, 130, 131, 170, 180, 213, 235, 321,
    322, 323, 325, 326, 327, 328, 344, 431,
    458, 486
Brady, F. B., 310
Bregman, J. D., 240
Breit, G., 439
Bridle, A. H., 9
Brigham, E. O., 106
Brosche, P., 373
Broten, N. W., 30, 126, 294
Brouw, W. N., 84, 128, 330, 332
Brown, G. W., 30
Bruck, Y. M., 354
Bryan, R. K., 352
Budden, K. G., 439
Burdick, H. M., 439
Burge, R. E., 353
Burke, B. F., 30, 32, 138
Burn, B. J., 462
Burns, W. R., 217, 221

Calame, O., 401
Caloccia, E. M., 183
Cannon, W. H., 259
Carr, T. D., 30, 482
Carter, A. W. L., 358
Carter, W. E., 382, 383, 458
Carver, K. R., 98
Casse, J. L., 182
Caves, C. M., 496
CCIR, 471, 472, 473
Champeney, D. C., 49, 76
Chandler, S. C., 380
Chesalin, L. S., 294
Chi, A. R., 310
Chicada, Y., 243
Chie, C. M., 290
Chivers, H. J. A., 448
Chow, Y. L., 132
Christiansen, W. N., 28, 33, 131, 326
Clark, B. G., 26, 86, 205, 206, 269, 294, 298,
    348, 349
Clark, T. A., 32, 285, 292, 294, 384
Clemence, G. M., 379, 380
Clemmow, P. C., 64
Clifford, S. F., 427
Cochran, W. T., 106
COESA, 414, 417

Cohen, M. H., 30, 258, 305, 377, 458, 459,
    463, 464, 487, 490
Cole, T., 217, 327
Coles, W. A., 458, 463
Collin, R. E., 114
Colvin, R. S., 26, 153, 159
Conway, R. G., 4, 104, 105, 106
Cooley, J. W., 106
Cooper, B. F. C., 217, 221, 225, 226
Cordes, J. M., 454, 463
Cornwell, T. J., 346, 348, 361, 364
Cotton, W. D., 359
Coulman, C. E., 432
Counselman, C. C., III, 377, 384, 456
Covington, A. E., 126
Cowan, J. J., 29, 363
Crane, P. C., 333
Crane, R. K., 406, 410, 448
Cronyn, W. M., 453, 463, 464
Currie, D. G., 495
Cutler, L. S., 289

D'Addario, L. R., 195, 200, 232, 233, 293,
    295, 339, 350, 351
Dainty, J. C., 493, 498
Daniell, G. J., 350, 351
Das Gupta, M. K., 18, 19, 482
Davies, J. G., 23, 193, 364
Davies, K., 439
Davis, J., 494, 495, 497
Davis, J. L., 417
Davis, M. M., 282, 292
Davis, W. F., 230
Debye, P., 426
de Graauw, T., 496
de Jager, J. T., 153
Dennison, B., 463
Diamond, P. J., 384
Dicke, R. H., 10
Douglas, J. N., 72
Drane, C. J., 57, 60, 66
Dreher, J. W., 29, 363
Dreyer, J. L. E., 9
Duffet-Smith, P. J., 463
Dulk, G. A., 482
Dutta, P., 276
Dutton, E. J., 426

Ebstein, S., 500
Edge, D. O., 9, 22
Edson, W. A., 269, 288
Ekers, R. D., 132, 133, 134, 180, 181, 234,
    335, 346, 362, 455, 492
Elgaroy, O., 23

Elgered, G., 439
Elsmore, B., 17, 27, 28, 180, 370, 371
Emerson, D. T., 230
Erb, K., 181
Erickson, W. C., 181, 456, 459
Eshleman, V. R., 458
Evans, J. V., 439, 440, 441, 446, 447, 449, 456

Fanselow, J. L., 377, 398, 401
Fanti, R., 310
Farley, D. T., 217, 221, 222, 225, 227
Fejer, B. G., 448
Fenstermacher, D. L., 182
Fernbach, S., 33
Feshbach, H., 353
Fienup, J. R., 354
Findlay, J. W., 33
Fisher, D. O., 292
Flinn, A. E., 32
Fomalont, E. B., 23, 24, 320, 371, 447, 455
Foreman, P., 285
Fort, D. N., 358
Frank, R. L., 292
Frater, R. H., 135, 183, 326
Fried, D. L., 433, 493
Frieden, B. R., 349, 350

Gallagher, J. J., 406
Garcia-Barreto, J. A., 384
Gardner, C. S., 495
Gardner, F. F., 462
Gardner, F. M., 193
Gauss, K. F., 390
Genzel, R., 32, 384, 386
Gezari, D. Y., 499
Gilbert, S. W., 291
Ginat, M., 28
Giuffrida, T. S., 387
Godwin, M. P., 491
Gold, B., 107
Gold, T., 32
Goldenberg, H. M., 285, 287, 288, 289
Goldstein, H., 422, 438
Goldstein, R. M., 211
Goldstein, S. J., Jr., 72
Gough, P. T., 500
Gower, J. F. R., 22
Granlund, J., 191, 192, 205, 206, 208
Grossi, M. D., 446
Guinot, B., 380
Guiraud, F. O., 434, 439
Gull, S. F., 350, 351
Gundermann, E. J., 459
Gwinn, C. R., 379

Haddock, F. T., 33
Hagan, J. B., 217, 221, 222, 225, 227
Hagfors, T., 56, 439, 440, 441, 447, 456
Hall, T., 390
Hamaker, J. P., 335, 435
Hamilton, W. C., 388, 395
Hanbury Brown, R., 12, 18, 23, 482, 486, 497
Hargrave, P. J., 27, 318, 435
Harmuth, H. F., 203
Harris, D. E., 459, 463
Harris, F. J., 240
Harris, R. W., 182
Harvey, J. W., 500
Haskell, R. E., 442
Hatanaka, T., 28
Hazard, C., 22, 371, 487, 490
Heffner, H., 496
Heiles, C., 232, 233, 461
Hellwig, H., 283
Herring, T. A., 296, 301, 377, 379, 384, 398,
    401
Hess, S. L., 408, 410
Hewish, A., 21, 22, 121, 140, 447, 448, 458,
    459, 463
Hey, J. S., 17, 447
Hill, E. R., 22
Hill, R. J., 426, 427
Hinder, R. A., 435, 436, 449
Hinteregger, H. F., 377
Hjellming, R. M., 25, 339, 373
Högbom, J. A., 33, 128, 344, 345, 350, 351
Hogg, D. C., 434, 439
Hogg, D. E., 28, 439
Hollweg, J. V., 458
Holt, E. H., 442
Holt, J. M., 449
Hooghoudt, B. G., 128
Horn, P. M., 276
Howard, J., 439
Hudson, J. A., 86·
Hughes, M. P., 26
Humphreys, W. J., 408
Hunsucker, R. D., 448
Hunt, B. R., 500

IAU, 5, 9, 100
IEEE, 6, 100
Ishiguro, M., 127

Jackson, J. D., 422
Jacobs, I. M., 156
Jaeger, J. C., 456
Jansky, K. G., 14
Jasik, H., 114

516   Author Index

Jaynes, E. T., 350
Jennison, R. C., 18, 19, 357, 482
Johnson, D. R., 5
Johnson, M. A., 495
Johnson, R. C., 114
Johnston, K. J., 371, 382
Jokipii, J. R., 454

Kahlman, H. C., 339
Kalachev, P. D., 23
Kaplan, G. H., 377, 447
Kardashev, N. S., 30
Kartashoff, P., 282
Kelder, H., 448
Kellermann, K. I., 4, 136, 138, 310
Kelley, M. C., 448
Kenderdine, S., 180, 325
Keshner, M. S., 276
Kesteven, M. J. L., 9
Kikuchi, R., 350
Klemperer, W. K., 249
Klepczynski, W. J., 291
Kleppner, D., 285, 287, 288, 289
Klingler, R. J., 217, 225, 226
Knapp, S. L., 495
Knowles, S. H., 30
Knox, K. T., 500
Ko, H. C., 99
Kock, W. E., 71
Kogan, L. R., 294
König, A., 81
Kraus, J. D., 33, 98, 181
Krishnan, T., 126
Kronberg, P. P., 104, 105, 106, 319, 461
Kroupa, V. F., 310
Kulkarni, S. R., 232, 233
Kundu, M. R., 455

Labeyrie, A., 493, 495, 497, 499
Labrum, N. R., 126
Lambeck, K., 378
Lampton, M., 398
Latham, V., 18
Lawrence, R. S., 426, 448
Lawson, J. L., 156
Lee, L. C., 454
Leech, J., 127
Legg, T. H., 186
Lesage, P., 272
Levine, M. W., 289
Lewis, P. A. W., 106
Liebe, H. J., 418, 424, 425
Lieske, J. H., 379
Liewer, K. M., 495

Lightman, A. P., 4, 5, 6, 419
Lim, J. S., 352
Lindsey, W. C., 290
Linfield, R., 32
Little, A. G., 118, 192, 358
Little, C. G., 447, 448
Little, L. T., 459, 464
Liu, C. H., 448
Lo, K. Y., 463
Lo, W. F., 235, 238
Long, R. J., 4
Louden, R., 32, 423
Lovas, F. J., 5
Love, A. W., 114
Lovelace, R. V., 463
Lovell, A. C. B., 447
Lynds, C. R., 500
Lyne, A. G., 463

McCants, M. M., 487
McCarthy, D. D., 382, 401
McCready, L. L., 15
MacDonald, G. J. F., 378
MacKay, J. R., 383
McKenzie, A. A., 291
McMillan, R. W., 406
MacPhie, R. H., 66, 71
Mahoney, M. J., 181
Maltby, P., 23, 320
Mandel, L., 57
Marcaide, J. M., 377
Margon, B., 398
Marini, J. W., 416
Markowitz, W., 380
Mathewson, D. S., 131
Mathur, N. C., 57, 71, 132, 133, 135, 446
Matsakis, D. N., 384
Matveenko, L. I., 30
Mauzy, R. E., 208
Meeks, M. L., 33
Meinel, A. B., 308
Melchior, P., 383
Mercier, R. P., 453
Michelson, A. A., 11
Middleton, D., 156, 218
Milligan, T. A., 114
Mills, B. Y., 17, 21, 22, 118
Miner, G. F., 439
Minkowski, R., 17
Misner, C. W., 457
Moffet, A. T., 23, 127, 128, 320, 321
Moran, J. M., 5, 30, 56, 166, 243, 258, 259, 277, 282, 294, 384, 386, 435, 437, 439
Morimoto, M., 192

Morison, I., 23, 193, 364
Morris, D., 23, 102, 491
Morris, D., 310
Morrison, N., 398
Morse, P. M., 353
Muhleman, D. O., 455
Mullaly, R. F., 28
Munk, W. H., 378

Napier, P. J., 132, 133, 134, 180, 181, 234,
    333, 346, 353, 362, 439, 491, 492, 500
Narayan, R., 350, 351
NASA, 310, 401
Nather, R. E., 487
Neville, A. C., 27, 28
Nisenson, P., 500
Nityananda, R., 350, 351
Norris, R. P., 384
NRAO, 132, 191, 192
Nyquist, H., 10, 213

O'Brien, P. A., 23
Oliver, B. M., 33
Oppenheim, A. V., 107, 213, 352
Oster, L., 5
Owens, J. C., 427

Palmer, H. P., 23
Pankonin, V., 470
Papaliolios, C., 500
Papoulis, A., 107, 165, 260
Parkinson, B. W., 291
Parrent, G. B., Jr., 57, 59, 60, 61, 62, 66, 70
Parsons, S. J., 17, 447
Pauliny-Toth, I. I. K., 4
Pawsey, J. L., 15, 22, 34
Payne-Scott, R., 15
Pearlman, M. R., 446
Pearson, T. J., 31, 32, 352
Pease, F. G., 11, 12
Perley, R. A., 29, 363, 377
Petley, B. W., 370
Phillips, J. W., 17, 447
Phillips, T. G., 182
Picken, J. S., 126
Pierce, J. A., 291
Pilkington, J. D. H., 22, 382, 401
Pollak, H. O., 332
Ponsonby, J. E. B., 349
Potts, C. E., 292
Press, W. H., 276
Preston, R. A., 138, 139
Price, R., 222
Price, R. M., 470

Priestly, J. T., 426
Purcell, E. M., 422

Rabiner, L. R., 107
Radhakrishnan, V., 26, 102
Raimond, E., 104
Ramsey, N. F., 285, 287, 288, 289
Rankin, J. M., 456, 460
Ratcliffe, J. A., 439, 443, 449, 453
Rawer, K., 439
Read, R. B., 23, 154, 180
Readhead, A. C. S., 352, 358, 359, 463
Reber, E. E., 410
Reber, G., 14
Reid, M. J., 5, 258, 384
Reid, M. S., 182
Resch, G. M., 439
Rhodes, D. R., 332
Rice, S. O., 165
Rickett, B. J., 463
Riddle, A. C., 326
Ridgway, S. T., 487
Roberts, D. H., 449
Roberts, J. A., 48, 110, 323, 344, 366
Robertson, D. S., 383, 458
Robinson, B. J., 153
Roddier, F., 423
Rogers, A. E. E., 32, 277, 282, 294, 305, 306,
    308, 358, 359, 435
Rogstad, D. H., 494
Rönnäng, B. O., 439
Rosen, B. R., 437, 439
Rotenberg, M., 33
Rowson, B., 23, 28, 88
Rutman, J., 269, 270
Rybicki, G. B., 4, 5, 6, 419
Ryle, M., 14, 15, 17, 19, 20, 21, 22, 26, 27,
    28, 29, 121, 127, 140, 180, 318, 370,
    371, 435, 449, 491

Saastamoinen, J., 414, 416
Salpeter, E. E., 458
Schafer, R. W., 107, 213
Schaper, L. W., Jr., 436
Scheuer, P. A. G., 449, 487, 489
Schwab, F. R., 328, 332, 361, 363, 364
Schwarz, U. J., 346, 347
Scott, P. F., 22, 458, 491
Scott, S. L., 458
Searle, C. L., 289
Seidelmann, P. K., 371
Seielstad, G. A., 102, 136
Serna, R., 183
Setti, G., 310

Shaffer, D. B., 258, 377
Shakeshaft, J. R., 22, 121, 140
Shannon, C. E., 213
Shao, M., 495
Shapiro, I. I., 259, 377, 379, 383, 384, 447,
    455, 458
Shapiro, L. D., 292
Shaw, L. J., 435
Shimoda, K., 285
Shinn, D. H., 449
Shklovsky, I. S., 34
Sholomitskii, G. B., 30
Siegman, A. E., 288
Silver, S., 64
Simard-Normandin, M., 461
Skellern, D. J., 326
Skilling, J., 352
Slee, O. B., 17, 21, 22
Slepian, D., 332
Smart, W. M., 413, 414, 456
Smith, E. K., Jr., 427
Smith, F. G., 17, 18, 358, 370, 447, 463
Smith, H. M., 380
Snider, J. B., 439
Snyder, L. E., 5
Sodin, L. G., 354
Soffer, B. H., 350
Southworth, G. C., 14
Sovers, O. J., 401
Spitzer, L., 461
Spoelstra, T. A. T., 446, 448
Sramek, R. A., 434, 435, 436, 447, 455
Stachnik, R. V., 499
Staelin, D. H., 418, 436, 438, 495
Stanier, H. M., 23
Stanley, G. J., 16, 17
Stegun, I. A., 211, 511
Stone, J. L., 71
Strange, W. E., 382
Subrahmanya, C. R., 350
Subramanian, S., 432
Sullivan, W. T., III, 34
Sutton, E. C., 7, 432, 496
Swarup, G., 28, 126, 131, 135, 185, 186
Sweezy, W. B., 434
Swenson, G. W., Jr., 57, 71, 135, 136, 495
Swope, J. R., 410

Taff, L. G., 379
Tatarski, V. I., 432, 433
Taylor, J. H., 377
Têtu, M., 280
Thayer, G. D., 426
Thomas, J. B., 296

Thompson, A. R., 4, 23, 26, 28, 109, 126,
    129, 130, 132, 133, 134, 138, 180, 181,
    193, 195, 200, 205, 206, 234, 322, 323,
    328, 339, 346, 362, 473, 476, 492
Thompson, B. J., 500
Thompson, M. C., 185
Thorne, K. S., 457
Thornton, D. D., 338, 439
Tiuri, M. E., 10, 159, 164
Townes, C. H., 285, 432, 495, 496
Tukey, J. W., 106, 240
Tuve, M. A., 439
Twiss, R. Q., 18, 358, 486, 497

Uhlenbeck, G. E., 156
U.S.A.F., 184
U.S. Naval Observatory, 291

van Albada, G. D., 338
Vander Vorst, A. S., 153
van de Stadt, H., 496
Vanier, J., 280
van Schooneveld, C., 366
Van Vleck, J. H., 218, 422
Vessot, R. F. C., 276, 280, 281, 284, 285,
    286, 289
Vinokur, M., 165
Visser, J. J., 182
Vitkevich, V. V., 23
Vonberg, D. D., 14
von Hoerner, S., 490

Wade, C. M., 95, 371, 373
Wahr, J. M., 379
Walker, R. C., 137, 138, 384, 387, 388
Walsh, D., 22
Wand, R. H., 449
Wang, T. C., 285
Warburton, J. A., 28, 326
Wardle, J. F. C., 319
Warner, P. J., 354, 355, 356, 359
Waters, J. W., 418, 419, 421, 422, 436
Webber, J. C., 136
Weiler, K. W., 102, 104
Weimer, R., 298
Weinberg, S., 457
Weinreb, S., 182, 183, 211, 217, 298
Weintraub, S., 427
Weisberg, J. M., 454, 460, 463
Welch, P. D., 106
Welch, W. J., 180, 338, 439
Wernecke, S. J., 326, 350, 351
Westfold, K. C., 456
Westwater, E. R., 436, 438, 439

Wheeler, J. A., 457
White, R. L., 321
Whiteoak, J. B., 462
Whitney, A. R., 32, 257, 291
Wieder, B., 292
Wild, J. P., 34, 135
Wilkinson, P. N., 359, 361, 364
Willis, A. G., 240, 339
Wills, D., 22, 458
Wilson, R. W., 26
Wineland, D. J., 283
Woestenburg, E. E. M., 182
Wolf, E., 55, 57, 59, 63, 73, 98
Woodward, R. H., 291
Woody, D. P., 182

Woolard, E. W., 379, 380
Woolf, N. J., 432, 493
Worden, S. P., 493, 498, 500
Wozencraft, J. M., 156
Wright, M. C. H., 153, 180, 320
Wu, S. C., 439

Yang, K. S., 185, 186
Yao, S. S., 217, 221
Yee, H. K. C., 358
Yeh, K. C., 448
Yen, J. L., 30, 34, 247
Young, A. T., 458

Zeissig, G. A., 463

SUBJECT INDEX

Aberration, diurnal, 9, 259, 379
Absorption:
  bound oscillator model, 422–425
  chopper wheel calibration, 421
  in clouds, 438–439
  coefficient, 406, 419, 424
  ionospheric, 440, 447
  spectra, 25
  tipping scan calibration, 420
  tropospheric, 417–422
  water vapor, 418, 436, 438
Adaptive calibration, 199, 363, 430. *See also*
     Hybrid mapping; Self-calibration
  comparison of methods, 363–364
  limitations of, 365–366
Agassiz station, Fort Davis, Texas, 136
Airy disk, 497
Algol, 371
Aliasing, 78, 85, 165, 212, 339
  definition, 110–111
  suppression in maps, 328–334
Allan variance:
  atmosphere, 434–436
  frequency standards, 272–283, 288–289
Allen-Baumbach model, 455
Altazimuth mount, 94–96, 104, 319
Amplifier, *see* Receiver
Amplitude closure, 357–358

Analog processing hardware, 326
  comparison with digital, 210
Analytic signal, 59
Antenna, 63, 65, 114
  angular resolution, 1
  aperture efficiency, 9
  aperture illumination (excitation), 64, 68,
     115, 490
  axis offsets, 94–95, 315, 398
  collecting area, 9
  cost scaling, 308
  design, 114
  mounts, 94–96, 105, 398
  output power, 48–49, 54
  polarization model, 99–101
  reception pattern, 48, 49, 115, 322
  sidelobe levels, 471
  surface measurements, 490–492
  temperature, 10
  tracking effect on fringe frequency, 123
Antennas:
  circularly polarized, 104–105
  cross polarized, 103–105, 168
  identically polarized, 102–103
  minimum number, 136, 360
  oppositely polarized, *see* Antennas, cross
     polarized
  in space, 138–139

Antenna-spacing coordinates, 86–90
Aperture efficiency, 9
Aperture synthesis, 139–140
Area density, 322
Arrays:
  circular (ring), 135, 192
  correlator, 121–123
  cross, 132. *See also* Mills cross
  effective collecting area of, 164
  grating, 28, 126
  minimum redundancy, 127, 131
  mixed, 97
  nontracking, 115, 118–121
  one-dimensional, 125–131
  phased, *see* Phased array
  tracking, 123–125
  transfer function of, 125
  T-shaped, 119–120, 132
  two-dimensional, 131–138, 338
  VLBI, 135–138
  Y-shaped, 132
Astrometry, 2, 259, 369–404, 447
  accuracy, 377
Astronomical Almanac (Ephemeris), 379
  Explanatory Supplement, 9, 112
Atmosphere, neutral, 405–439
  absorption, 315, 417–421
  brightness temperature, 436–439
  effects:
    on astrometry, 377, 386
    of clouds, 436, 438–439
    on fringe frequency and delay, 376
    on mapping, 335, 336, 338, 356–357,
      366
    on visibility, 428–432
    on VLBI, 429–432, 435–436
  excess path length, 315, 414–417
  Fried length in, 433, 493
  oxygen, 417, 418, 439
  phase fluctuations, 428–436
  refraction, 411–414
  water vapor, *see* Water vapor
Attenuation in cable, 190–191
Australia telescope, 135
Autocorrelation function, 45, 60, 115, 116, 156,
    213, 270, 354, 477
  brightness distribution, 352, 354–356, 493
  definition, 46, 116
  map, 354–356
  of speckles, 499
Autocorrelator, 211, 230, 237
Automatic level control (ALC), 206, 231, 315
Automatically correcting system, 191–192
Azimuth, 89, 112

Bandwidth:
  correlation, 452
  effect in maps, 169–174, 196
  Gaussian, 46–47
  output (postcorrelator), 158
  pattern, 44–46, 67, 71, 76, 147, 372
  rectangular, 47, 196
  rms, 306, 376, 394
  synthesis, 305–308, 375–377, 394
Bartlett weighting, 239
Baseband response, 183, 212–213, 305
Baseline:
  calibration, 93–94, 372–377, 398
  coordinates, 86–90
  definition, 14, 41
  equatorial component, 374
  non-coplanar, 81–83
  reference point, 95
  retarded, 258–259
  surveying, 3, 89, 370
Beam:
  clean, 344–346
  fan, 21, 52, 126, 129, 325–326
  fringe frequency, 387
  pencil, 22, 129
  synthesized (dirty), 169, 170, 322–325, 334,
      338, 342–349, 400, 476
Beamshape, 96–97
Beamwidth, 96–97, 322
Besselian year, 379
Bias:
  in polarization measurements (Rice
      distribution), 319
  in variance of mean, 390
  in visibility measurements, 262, 263, 265
Bivariate Gaussian probability distribution,
      211–212, 218, 396–397
Blackbody, *see* Planck's law
Blackman weighting, 239
Bologna, Italy, 23
Bose-Einstein statistics, 350
Brewster angle, 97
Brightness, 6–8, 51, 54, 70
  derivation, 320
  interpretation, 334–335
  scale, 334–335, 359
  temperature, 8
Brunt-Väisälä frequency, 448
Bureau International de l'Heure (BIH), 381,
      383

C_n^2, 433
C_N^2, 454

Cables, 183, 291
  attenuation, 190–191
  reflections, 187, 190, 196
  velocity dispersion, 187
Calibration, 314–319
  adaptive, 363, 365–366
  bandpass, 317–318, 366
  of cables, 291
  gain, 207
  interference errors, 479
  ionospheric delay, 446–447
  phase, 290
  polarization, 318–319
  sources, 251, 314, 316–317, 356
  spectral-line, 257, 317–318
Cambridge (England):
  Five-Kilometer radio telescope, 27, 28, 325
  fourth survey, 22, 23
  One-Mile radio telescope, 26, 27, 28, 127,
      128, 435–436
  radio instruments, 17, 21–22, 29, 118,
      120–121, 448
  third survey, 9, 22
Canadian VLBI array, 138
Cassiopeia A, 15, 17–18, 28
Cavity pulling, 288, 289
CCIR, 471, 472, 473
Celestial:
  coordinates, 88, 112
  equator, 87, 88, 131
  sphere, 81, 87
Cell:
  averaging, 108, 109, 164, 201, 329, 333–334
  crossing time, 475
Centaurus A, 17
Chandler wobble, 380
Chopper wheel calibration, 421
Circular polarization:
  degree, 98
  IEEE definition, 100
Circulator, 186
Classical electron radius, 450
Clausius-Clapeyron equation, 410
CLEAN algorithm, 131, 343–349, 351
  application, 365
  Clark's algorithm, 349
  comparison with MEM, 350
  extended sources, 345, 348, 349, 351
  hybrid mapping, 358–359
  interference reduction, 476
  loop gain, 345, 348
  self-calibration, 361
  smoothness-stabilized, 348
  spectral line data, 365

Clipping, 217. See also Quantization
Closure relationships, 19, 32, 136, 357–360,
      372, 494
  accuracy of, 364
Clouds, atmospheric:
  absorption, 438–439
  effect on phase, 436
  index of refraction of, 438
COESA (Committee on the Extension to the
      Standard Atmosphere), 414, 417
Coherence:
  complex degree of, 61
  function, source, 60–63, 66
  function, temporal, 277–281. See also
      Autocorrelation function
  of hydrogen maser, 281
  mutual, 61, 74–76, 486
  of oscillator, 277–281
  self, 62
  time, 251, 262, 267, 277–279, 365–366,
      375
Coherency matrix (polarization), 99
Coherent source, 70
Collecting area, 9
Comb spectrum, 193–194, 290
Compensating delay, see Delay, instrumental
Complex visibility, 14
  definition, 55
Compound interferometer, 126
Compton loss, 249
Confusion of radio sources, 22
Connected-element array (definition),
      32
Constrained optimization, 349–352
Continuum:
  mapping, 169
  radiation, 4
Conventional International Origin (CIO), 380.
      See also Polar motion
Conversion:
  frequency, see Frequency conversion
  serial-to-parallel, 235, 244
Convolution, 48–50, 76, 164, 173, 238
  theorem, 50
Convolving function, 326, 327–334. See also
      Smoothing functions
  Gaussian, 329
  Gaussian-sinc, 331, 333
  rectangular, 329
  spheroidal, 332
Coordinate:
  conversion, 112
  systems, 78
Correlation, 76–77

524 Subject Index

Correlator, 20, 41, 44, 64, 65, 211
  analog, 183
  complex, 121, 150–151, 161, 167
  digital, 236
  hybrid, 243
  recirculating, 242–243
  simple (single-multiplier), 150, 161, 213
  subsystem, 183
  voltage offset, 202, 230, 336, 339
Cosmic background, 419
Crab Nebula, 17, 456
Cross, *see* Mills cross
Cross correlation, 45, 76
  coefficient, 214
Cross power spectrum, 77, 237, 238, 253–258, 300, 305, 393. *See also* Spectral line
Cross-talk, 336
Cryogenic cooling, 153, 182
Crystal oscillator, 193, 269, 280, 283, 287
Culgoora array (Australia), 135
Culmination, 17
Cyclotron:
  frequency, 443
  radiation, 97
Cygnus A, 4, 8, 9, 15, 16, 17, 18, 19, 27, 28, 29, 32, 363, 447

dBi, 471
Declination, 8
  coordinate conversion, 112
  measurement of, 370, 372–375, 398
Deconvolution, *see* CLEAN algorithm; Maximum entropy method (MEM)
Delay:
  adjustment, 46, 200–205
  analog, 183
  circuits, digital, 234–235
  compensating, *see* Delay, instrumental
  errors, 200
  geometrical, 42, 253, 370, 411–412, 414
  group, 251, 256–257, 305, 375, 377, 382, 384, 440
  instrumental, 44–46, 91, 145–146, 149, 151, 200, 296
  measurement error, 305–306
  reference, 149, 201
  subsystem, 183–184
  tracking, 149
Delay pattern, *see* Bandwidth, pattern
Delay resolution function, 307, 308. *See also* Bandwidth, synthesis
Delay-setting tolerances, 151, 200–201

Delta function:
  CLEAN components, 345–347
  Fourier transform of series of, 129
  LO frequency, 290
  point source, 92, 125, 175, 371
  Price's theorem, 222, 227
  Shah function, 110, 327
  visibility sampling, 51, 162
Depolarization, 461–462
Detector:
  power-linear (square law), 19, 482, 486
  synchronous, 19, 185, 203, 421
Diameter, stellar, 12
Dielectric constant, 422
  of plasma, 442
Diffraction pattern:
  lunar occultation, 487
  scintillation, 452–453, 459
Digital processing, 210–246
  multiplier, 236
  sampling, 212–234
  spectral measurements, 236
Diode, 143–144, 185, 193, 288, 290
Direct detection, optical, 495
Directional coupler, 186
Direction cosine, 42, 80–81
Dirty beam, *see* Synthesized beam (dirty beam)
Dirty map, 345
Discrete Fourier transform, 106–109, 327–328
Dispersion, classical theory, 422–425
Dispersion measure, 460
Doppler effect, 41, 242, 284–287, 289, 384
Double sideband system, 150–155, 160, 167, 168
Dynamic range, 327, 343, 364, 480

Earth:
  equatorial bulge, 378
  magnetic field, 443
  radius vector, 258, 414
  tectonic plate motion, 3, 383
  tides, 383
  see also Geodetic measurements; Polar motion
Earth rotation:
  scanning, 121
  synthesis, 26–28, 81, 326
  see also Universal Time
East-west array, 83–85, 125–131, 322
East-west baseline, 52, 176
Ecliptic, 378
Editing of data, 314, 335, 365
  for interference, 476

Efficiency:
 aperture, 9
 quantization, 160, 229, 304
Electron density:
 galaxy, 460
 interplanetary medium, 455–457
 ionosphere, 441
Electronics:
 historical development, 180–181
 subsystems, 180–184
Elevation, 89, 112
Entropy, 349–351
Equatorial mount, 94–96
Equinox, 371, 378
Ergodic waveform, 4
Error function, 165, 173, 511
Errors:
 (ξ, η) origin, 339
 additive, 336, 479
 clock (VLBI), 253–254, 257
 in maps, 335–338
 multiplicative, 336, 479
 phase, 338
 pointing, 337
Excess path length, 407
 interplanetary medium, 458
 ionosphere, 439–440, 445–446
 troposphere, 414–417, 437–439
Explanatory Supplement to Astronomical
   Ephemeris, 9, 112
Extended sources, 48–50
 cleaning, 345, 348, 349, 351
 self-calibration, 360
 signal-to-noise ratio, 164, 335, 338

Faraday depth, 461
Faraday dispersion function, 462
Faraday rotation, 4, 319, 444
 interstellar, 460–461
 intrinsic, 461–462
 ionospheric, 440, 446
Far-field assumption, 55–56, 76
Fast Fourier transform (FFT), 106, 243, 327
 rearrangement of data, 108–109
 spectral line processors, 243
Fast Hartley transform, 107
Field:
 far, requirement, 76
 incident, 57–60
 near, observations in, 56
Field of view:
 bandwidth smearing, 169–174
 fringe-frequency mapping, 386–387

restrictions, 76, 82–83
 visibility averaging, 173–178
Filter, 64, 65, 144, 183
 baseband, 183
 Butterworth, 305
 effect on signal-to-noise ratio, 196
 narrow-band, 195
 number of poles, 194, 196
 phase-locked oscillator, use of, 195
 phase stability, 194–195
 Q-factor, 194
 spatial frequency, see Spatial frequency
 spectral-line, 183, 242, 243
Fleurs, Australia, 21, 28
Flux density, 6
Fort Davis, Texas, see Agassiz station
Fourier transform:
 analog hardware, 326–327
 derivative property, 271, 321, 489
 direct, 321–327
 discrete, 106–109, 327–334
 fast, 106, 243, 327
 integral theorem, 157
 linearity property, 325
 relationships, 59, 115
 sign of exponent, 55
 similarity theorem, 170
 spectrometer, 243
 three-dimensional, 85
Fourth-order moment relation, 156, 215,
   482
Fractional bit shift loss, see VLBI
Fractional frequency deviation, 270
Fraunhofer diffraction, 75, 116. See also Field,
   far, requirement
Frequency channels, 237
Frequency conversion, 143–149
 multiple, 149, 152
 optical, 495
Frequency multiplication, 193, 290
Frequency-passband shaping, 72
Frequency protection, 470
Frequency response, 195–201
 optimum, 195–196
 tolerances, 196–200
Frequency standards:
 cesium beam, 269, 285
 crystal oscillator, 193, 280, 283, 287
 hydrogen maser, 267, 285–289
 phase noise processes, 274–277
 rubidium vapor, 267, 283–285
Fresnel zone, 451, 487, 488
Fried length, 433, 493

Fringe:
  envelope, 46
  function (pattern), 14, 42, 43, 68
  rotation (stopping), 149, 155, 167, 182, 205,
    297–299
  visibility, 12
  washing function, 46
  white light, 250, 494
Fringe frequency, 41, 91–92, 148, 149, 150
  in astrometry, 374–376
  baseline solution, 374–376
  beam, 387
  definition, 91–92
  interference suppression effect, 473–477
  ionospheric effect, 446
  mapping, 386–387
  measurement accuracy, 392–393
  natural, 148, 149, 205, 473
  spectrum, 268, 387, 393
  tracking and nontracking interferometers,
    123–124
  in VLBI, 256, 262–267
Fringe rate, *see* Fringe frequency
Fringes:
  first radio record, 15
  interference, 12
Front end, 182. *See also* Receiver

Gain:
  calibration, 206–207
  errors, 198–200, 232
  factor, 147, 357
Gamma function, 457
Gaussian convolving function, 329–330
Gaussian random variable, 3, 156, 259–262,
    278, 389, 429. *See also* Bivariate
    Gaussian probability distribution
Gaussian-sinc function, 331
Gaussian taper, 119, 132, 324–325, 346
Geodetic measurements, 32, 259, 369, 383–384
Geometrical delay, 42, 253, 370, 411–412, 414
Gibbs phenomenon, 240
Goldstone, California, 136
Granlund system, 191–192
Grating array, 28, 126
Gravitational deflection, *see* Relativity
Green Bank, W. Virginia, 28, 136, 267, 436
Greenwich meridian, 86, 259, 374, 380, 382
Gridding, 327, 338. *See also* Cell, averaging
Group delay, 251, 256–257, 305, 375, 377,
    382, 384, 440
Group velocity, 442, 460
Gyro frequency, 443

Hadamard matrices, 204
Half-order derivative, 130
Hamming weighting, 239
Hankel transform, 431, 433–434
Hanning weighting, 239
Hartley transform, fast, 107
Hat Creek Observatory, California, 29, 136,
    435–436
Haystack Observatory, Westford, Massachusetts,
    30, 136, 267
Heaviside step function, 487
Heterodyne conversion, *see* Frequency
    conversion
Hilbert transform, 59, 150, 161, 235, 238, 353,
    422
Holography, 491, 500
Hour angle, 41, 86–88, 112
Huygens' principle, 453
Hybrid junction, 121
Hybrid mapping, 32, 356, 358–363
Hydrogen line, 5, 285, 286
Hydrostatic equilibrium, 408, 410

IAU:
  polarization standard, 100
  radio lines, 5
  radio-source nomenclature, 9
IEEE:
  committee on frequency stability, 269
  polarization standard, 100
  power flux density, 6
Illumination, aperture, *see* Antenna, aperture
    illumination (excitation)
Image defects:
  correlator offset, 339
  distortion, 337
  visibility errors, 335–336
  see also Scintillation; Seeing
Incoherence assumption, 56, 62, 73, 139, 483,
    486
Incoherent source, response to, 67–70
Index of refraction, *see* Refraction, Refractivity
    index of;
Instrumental (compensating) delay, *see* Delay,
    instrumental
Instrumental polarization, 105–106, 201–202,
    318
Integrator, 44
Intensity, 6. *See also* Brightness
Intensity interferometer, 18, 352,
    482–486
  optical, 497
  sensitivity of, 269, 486

Interference, 470–480
connected-element arrays, 473–478
decorrelation effect, 477–478
fringe-frequency averaging, 473–477
geostationary satellites, 480
harmful thresholds, 470–480
single antenna, 472–473
solar, 336
(u, v) plane distribution, 476
VLBI, 472, 478–480
Interferometer:
adding, 19
basic features, 40–43, 63–64
broadband, 44–48
compound, 126
correlator, 19–21
infrared, 495
intensity, 18, 269, 352, 482–486, 497
Michelson, 11–13, 494–497
nontracking, 118–121
optical, 11–13, 352, 493–500
sea, 14–17, 21
sensitivity advantage of, 20, 25
spectral-line, 25
Intermediate frequency (IF), 144
amplifier, 64
subsystem, 182–183
Interplanetary medium:
electron density, 455–457
excess path length, 458
refraction, 455–458
scintillation, 458–459
Interpolation, 107, 110–111, 328. *See also*
Gridding
Interstellar medium, 460–464
dispersion measure, 460
Faraday rotation, 460–462
pulsar signals, effects on, 460
scintillation, 462–464
Invisible distribution, 344
Ionosphere, 439–449
absorption, 447
acoustic-gravity waves, 448–449
effects of irregularities, 447–449
electron density distribution, 441, 444
Faraday rotation, 440, 444, 446
index of refraction, 442–444
phase stability, 284, 335, 356, 440
propagation delay, 444–446
refraction, 440, 444–446
scintillation, 447–449
total electron content, 440, 445, 446, 447,
449

travelling ionospheric disturbances (TIDs),
448–449
Isoplanatic:
angle, 493
condition, 364

Jansky (unit), 6
Jodrell Bank, England, 18, 23, 28, 29, 247
J^2 synthesis (or J-squared synthesis), 135
Julian year, 379
Jupiter, burst radiation, 30, 247

Kolmogorov turbulence, 431, 432, 435, 454,
463
Kramers-Kronig relation, 422, 425

Leakage, 110
Leap second, 381
Least-mean-squares analysis, 373–374, 388–400
accuracy, 397
CLEAN, 347
correlated measurements, 395
design matrix, 397
error ellipse, 396–397, 400
estimation of delay, 393–394
estimation of fringe frequency, 392–393
matrix formulation, 395
nonlinear case, 397–398
normal equation matrix, 395, 397, 399
partial derivative matrix, 395
precision, 397
self-calibration, 361
source position errors, 398–400
variance matrix, 395
weighted, 389
Light, speed of, 370
Likelihood function, 389
Line of nodes, 378. *See also* Equinox
Lines, radio, *see* Spectral line
Linked-element array, 32
Lloyd's mirror, 14
Local oscillator, 143–149, 182. *See also*
Frequency standards
independent, 18, 29. *See also* VLBI
laser, 495
phase switching of, 205–206
signed-sum of frequencies, 149
synchronization of, 184–194
Long baseline interferometer, 23, 29, 247, 482
Loran, 291–292, 381
Lorentzian profile, 287, 424, 425, 429, 431
Low-noise input stage, 153, 181–182
Lunar occultation, 25, 371, 459, 487–490

Magnetic fields:
  in frequency standards, 286, 287, 289
  interstellar, 460–461
  terrestrial, 443
Magnetic tape recording, 18, 30, 292–295
Mapping:
  synthesis, definition, *vii*
  two-dimensional, 52–56, 78–86
  wide field, 85–87, 174
Maryland Point, Maryland, 136
Maser frequency standard, 285–289
Maser radio source, 5, 30, 247, 249, 267, 463
  mapping procedures, 384–388
Master oscillator, 181, 182, 184, 186, 187, 190
Maximum entropy method (MEM), 349–352
Maximum likelihood method, 308, 389–390, 400
Maxwell's relation, 423
Meridian, 86, 119
  Greenwich, 86, 259, 374, 380, 382
  local, 17, 87, 88, 370, 382
  plane, 86, 372
  transit (crossing), 17, 120, 381
MERLIN array, 23, 364
Meter, definition of, 369–370
Michelson interferometer, 11–13, 494–496
Microwave link, *see* Radio link
Millibar, 407
Millimeter wave instruments, 29, 153, 243, 421
Mills cross, 21–23, 28, 118–120
Minimum redundancy arrays, *see* Arrays,
    minimum redundancy; Bandwidth,
    synthesis
Mirror-image reception pattern, 50
Mixer, 143–145, 153, 182
  digital single sideband, 297–299
  single sideband (image rejecting), 183, 195,
    207–208
MKSA units, 422
Model:
  in adaptive calibration, 364
  circular disk, 12, 13, 349
  Cygnus A, 18, 19
  delta function (CLEAN), 345
  elliptical, 17
  fitting, 320–321, 349
  Gaussian, 13, 249, 320
  moments of, 320–321
  rectangular, 13
  without phase, 320
Modulated reflector, 185
Molonglo, Australia, 22
Moon, *see* Lunar occultation; Precession
Mosaic mapping, 85

Mullard Radio Astronomy Observatory, *see*
    Cambridge (England)
Multiplier, 44. *See also* Correlator
Mutual coherence function, 61, 74–76, 486

Nançay, France, 28
Narrabri, Australia, 497
National Aeronautics and Space Administration
    (NASA), 2, 32
National Geodetic Survey (NGS), 2
National Radio Astronomy Observatory
    (NRAO), 28, 29, 30, 132, 248, 267,
    363. *See also* Green Bank, W. Virginia;
    Very Large Array (VLA)
Natural weighting, 164, 325, 326
Naval Observatory, U.S. (USNO), 2, 285, 291,
    381
Naval Research Laboratory (NRL), 2, 267
NAVSTAR, Global Positioning System, 291, 381
Near field observations, 56
Negative frequencies, 46, 51
NGC 4486, 17, 18
NGC 5128, 17, 18
NGC 7027, 4, 5, 8, 9
Nobel Lecture, 29
Nobeyama, Japan, 29
Noise:
  amplitude and phase, 165–168
  available power, 10
  in complex and simple correlators, 160–162
  equivalent power (NEP), 496
  Gaussian, 3, 156, 165
  in map, 162–165
  photon shot noise, 284, 289, 496
  quantum limit, 33
  response to, 155–168
  visibility, 160–162
  VLBI, 259–262
  see also Signal-to-noise ratio
Non-coplanar baselines, 81–83
North Liberty, Iowa, 136
Nuffield Radio Astronomy Laboratories, 23. *See*
    also Jodrell Bank, England
Nutation, 2, 9, 378–379
Nyquist rate, 212–213

Observation, planning and reduction, 338–339
Occultation observations, *see* Lunar occultation
1548 + 115, 362
Ootacamund (India) array, 135
Opacity, 419
  measurement of, 420–421
Optical depth, *see* Opacity

Optical fibers, 183
Optical interferometry, 11–13, 352, 493–500
Orion Nebula, 5, 7
Oversampling, 213, 215–216, 221, 225–226, 228
Owens Valley Radio Observatory, California,
    23–24, 29, 136

Parallactic angle, 88, 97, 104, 105, 319
Parallax, 379
Parametric amplifier, 153
Parseval's theorem, 165, 244, 267, 343, 432,
    476, 480
Partial coherence, 61
Passband:
    Gaussian, 46–47, 173, 174
    rectangular, 46–47, 173, 196
    shaping, 72
Pencil beam, 22
Permittivity, 422
Phase closure, 18–19, 32, 249, 317, 357
Phased array, 121–123, 126, 135, 326
    correlator array, comparison with, 123
    randomly phased, 309
    as VLBI element, 308–310
Phase data:
    imaging without, 352–356
    incomplete, 352–364
    uncalibrated, 356–364
Phase-locked loop, 186, 192–194, 280
    natural frequency, 193
Phase noise:
    effects on maps, 386
    in frequency standards, 267, 269–280
    ionospheric, 447–449
    in multipliers, 290
    neutral atmospheric, 428–436, 486
    receiver noise, 165–167, 260–262
Phase reference:
    feature, 385
    position, 80, 89, 91
Phase stability:
    filters, 194–195
    frequency standards, 280–289
    local oscillator, 290
    reference signal distribution system,
        184–194
    theory of, 269–280
Phase switching, 19–21, 126, 202–206, 230,
    232, 336
Planck's law, 8, 419, 495
Planet, 349, 371, 378, 390. See also Jupiter,
    burst radiation
Planetary nebula, 5

Plasma:
    absorption in, 447
    frequency, 442
    index of refraction, 442–444
    oscillations, 97
    propagation in, 440–464
    RF discharge, 283, 286
    turbulence, 454–455
    see also Interplanetary medium; Interstellar
        medium; Ionosphere
Plateau de Burre, France, 29
Pointing correction, 315
Point-source response, 47, 116, 125, 344, 365,
    429. See also Beam, synthesized (dirty);
        Voltage reception pattern
Polarimetry, 97–106
Polarization, 97–106
    calibration, 105–106, 318–319
    circular, 98, 100–101, 102, 104, 318–319
    complex degree of, 461–462
    degree of, 98, 319
    ellipse, 99–100
    emission processes, 4
    instrumental, 105–106, 201–202, 318
    linear, 98, 104, 318
    measurement bias, 319
    mismatch tolerance, 201–202
    parallactic angle effect, 104, 105
    position angle, 98, 100, 319. See also
        Faraday rotation
Polar motion, 2, 378, 379–380
    measurement of, 381–383
Position measurements:
    early, 17
    methods, see Astrometry
Power (density) spectrum, 45
    atmospheric fluctuations, 433–434
    correlator output, 159
    interplanetary scintillation, 459
    phase and frequency fluctuations, 269–280
Power-law spacing, 133–134
Power reception pattern, see Antenna, reception
        pattern; Voltage reception pattern
Poynting vector, 6
Precession, 2, 9, 378–379
Price's theorem, 222, 227, 228
Principal response, 130, 323, 324, 342–343,
    349
Principal solution, see Principal response
Probability:
    cumulative, 262–263
    of error, 264
    of misidentification, 265–266

Probability distribution:
  of angle of arrival, 448
  bivariate Gaussian, 211–212, 218, 396–397
  of delay-setting error, 200
  Gaussian, 156, 259–262, 278, 389, 429
  Rayleigh, 166–167, 260–262
  Rice, 165–166, 260, 319
  uniform, 200, 260–261
Prolate spheroidal wave functions, 332
Propagation:
  constant, 258, 406–407
  interplanetary, 455–459
  interstellar, 460–464
  ionospheric, 439–449
  Loran, 291–292, 381
  neutral atmospheric, 405–439
Proper motion, 9, 371, 379
Pulsar, 249
  astrometry, 377
  coherence, 56
  determination of vernal equinox, 371
  dispersion measure, 456, 460
  proper motions, 463
  scintillation, 449, 462–463
  timing accuracy, 282, 292

Q-factor:
  cavity, 287–289
  filter, 194
Quadrature:
  network, 150, 231, 238
  phase shift, 161, 168, 203, 207, 235
Quadruple moment theorem, *see* Fourth-order
    moment relation
Quantization, 211, 216
  comparison of schemes, 228–230
  efficiency factor, 160, 229, 292, 304
  four-level, 221–227
  indecision region, 233
  noise, 211
  in phased arrays, 309
  sixteen-level (four-bit), 230
  three-level, 227–228
  thresholds, 221–222, 227, 231
  two-level, 217–221
  in VLBI systems, 248, 304–305, 309
Quantum noise, 33, 284, 289, 495–496
Quantum paradox, 32–33
Quasar, 4, 32, 249, 371, 377. *See also* Radio
    source

Radamacher functions, 203, 204
Radial smearing, *see* Bandwidth, effect in maps

Radiation, incident, 57–60
Radiative transfer, equation of, 419
Radio lines, 5, 6, 7
  CO, 6, 8
  frequencies of, 6
  H, 5, 6, 25, 285
  H_2O, 5, 6, 267, 384
  OH, 5, 6, 30, 247, 384
  SiO, 5, 6
Radio link, 17, 18, 23, 32, 183, 185, 230
Radiosonde, 437
Radio source:
  Cassiopeia A, 15, 17, 18, 28
  Centaurus A, 17
  Crab Nebula, 17, 456
  Cygnus A, 4, 8, 9, 15, 16, 17, 18, 19, 27,
      28, 29, 32, 363, 447
  Jupiter, 30, 247
  NGC 4486, 17, 18
  NGC 5128, 17, 18
  NGC 7027, 4, 5, 8, 9
  1548 + 115, 362
  Orion Nebula, 5, 7
  Sagittarius A, 463
  Sun, 14, 15, 23, 28, 131, 135, 160, 326. *See
      also* Sun
  Taurus A, 17
  3C10, 346
  3C33.1, 24
  3C48, 4, 8, 9
  3C273, 31, 371, 458
  Venus, 211
  Virgo A, 17
  W3(OH), 267
  W49, 384
Radio source nomenclature, 8–9
Raised cosine weighting, 239–240
Rayleigh distribution, 166–167, 260–262
Rayleigh-Jeans approximation, 5, 8, 419. *See
    also* Planck's law
Receiver, 9
  electronics, 153, 180–184
  phase switching, 126
  temperature, 10, 181
Reception pattern, *see* Antenna, reception
    pattern; Voltage reception pattern
Redundancy factor, 127
Reflections in cable, 187, 190, 196
Refraction:
  index of, 406–407, 423–428, 442
  interplanetary, 455–458
  ionospheric, 444–446
  origin of, 421–425

plane parallel atmosphere, 412
  spherically symmetric, 413, 455–458
  tropospheric, 291–292, 411–417
Refractivity, 407.
  optical, 428
  Smith-Weintraub equation, 426–428
  See also Refraction, index of
Relativity:
  general relativistic bending, 377, 455,
    457–458
  time transfer effects, 291
Resolution:
  atmospheric limitation of, 428–436
  MEM, 351
  of source, 14, 249
Restoration from samples, *see* Sampling theorem
Retarded baseline, 258–259
Reynolds number, 432
Rice distribution, 165–166, 260, 319
Right Ascension, 8
  measurement of, 371, 372–375, 398
  origin of, 371, 378
Ringlobes, 128–131, 171, 339
rms bandwidth, 306, 376, 394
Rotation measure, 460–462
Round-trip phase, 184–191, 315

Sagittarius A, 463
Sampling, 212–234
  of bandpass spectrum, 213
  digital, errors in, 230–234
  threshold levels, 233
  unquantized, 213
Sampling theorem, 93, 128, 131, 212, 354,
    491. *See also* Nyquist rate
  derivation, 110–111
Satellite:
  data link, 30, 247, 250
  interference from, 478, 480
  laser ranging, 383
  signals, Faraday rotation, 446
  time transfer, 291
Scintillation:
  angular spectrum of, 450–451
  correlation bandwidth, 452
  correlation length, 452
  critical source size, 452
  Gaussian screen model, 449–453, 463
  index, 459
  interplanetary, 458–459
  interstellar, 138, 462–464
  ionospheric, 447–449
  neutral atmosphere, 432–436
  plasma irregularities, 449–455

power-law model, 454–455, 463
  scattering angle, 451, 455, 457, 459, 463
  thin screen, 449
Sea interferometer, 14–17, 21
Second, definition of, 285. *See also* Time
Seeing, 405. *See also* Scintillation
  angle, 430–436
  cell, 493, 497, 498
  disk, 497
Self-absorption, 5
Self-calibration, 360–364
Serial-to-parallel conversion, 235
Serpukhov, Soviet Union, 23
Shadowing, 315, 336
Shah function, 110
  two-dimensional, 327
Short-spacing data, 131, 138–139, 337–338, 365
Shot noise, photon, 284, 289, 496
Sideband, 144–145
  double, 150–154, 167, 168, 171
  fringe-frequency dependence, 92
  separation, 154–155, 168
  single (upper, lower), 145–148, 153–154, 168
Sidelobe:
  bandwidth smearing of, 171–172
  model, 471
  see also Ringlobes; Synthesized beam
    (dirty beam)
Sidereal rate, 91
Signals:
  cosmic, 3–8
  ergodic, 4
  spurious, 202–203
Signal-to-noise ratio:
  aliasing effect, 332–334
  basic derivations, 155–168
  coherent averaging, 281
  extended source, 335, 338
  frequency response, effect of, 195–198
  fringe-frequency mapping, 387
  incoherent averaging, 266–269
  intensity interferometer, 269, 486
  in interference calculations, 471
  loss factors, VLBI, 296–305
  lunar occultation, 490
  in map, 162–165
  nontracking array, 121
  optical, 496
  phased array, 308–310
  quantized signals, 219–228
  quantum limit, 32–33, 496
  radiometer, 10
  unquantized signals, 215
  see also Noise

Signal transmission, 182–183
Simeiz, Soviet Union, 30
Sinc function, 43
Slave oscillator, 182
Smearing:
    circumferential, *see* Visibility, averaging
    radial, *see* Bandwidth, effect in maps
Smith-Weintraub equation, 407, 425, 426–428
Smoothing functions, 239
Snapshot, 134, 338
Snell's law, 411–413
    spherical coordinates, 414, 456
Soil temperature, 184
Solar system studies, 56, 131, 458
Source:
    calibration, 251, 314, 316–317
    coherence function, 60–63, 66
    coherent, 70
    extended, *see* Extended sources
    incoherence requirement, 56, 62, 73, 139,
        483, 486
    incoherent, 67
    model, *see* Model
    radio, *see* Radio source
    subtraction, 338–339. *See also* CLEAN
        algorithm
Space-frequency equivalence, 71–73, 174
Space VLBI, 138–139, 250
Spatial frequency, 48, 53, 115
    filter, 115, 123–125
    spectrum, 51, 58
Spatial incoherence, 56, 62, 73, 139, 483,
        486
Speckle interferometry, 497–500
Spectral (power) flux density, 5–6
Spectral line:
    absorption, 25–26
    adaptive calibration, 366
    analog correlator, 183
    atmospheric absorption, 405, 417–421
    bandpass calibration, 240–241, 317–318
    baseline ripple, 241
    calibration procedures, 317–318
    CLEAN procedures, 365
    digital correlator system, 236–244, 250,
        384
    double sideband observation, 154
    frequencies, 6, 470
    mapping procedures, 365, 384–388
    radiation, *see* Maser radio source; Radio lines
    systems, 25, 69, 174, 183, 210
    VLBI procedures, 257–258, 295, 302,
        384–388

Spectral sensitivity, 115–118, 321
    of aperture antenna, 337
    of correlator array, 116
    support of, 117
Spectrum:
    spatial frequency, 58. *See also* van Cittert-
        Zernike theorem
    temporal frequency, 58–60. *See also* Wiener-
        Khinchin relation
Spheroidal wave functions, 332
Stanford, California, 28
Stars:
    observation of, 11, 12, 25–26, 381, 384,
        497
    proper motion, 379
    visibility modelling, 321
Stokes parameters, 98–106, 318
Structure function:
    electron density, 454
    phase, 278, 428, 433–435, 454
    refractive index, 432
Sun:
    coronal refraction, 455–458
    gravitational deflection, 457–458
    gravitational effects, 378–379
    interference from, 336
    ionization of ionosphere, 440
    observation of, 14, 15, 23, 28, 131, 135,
        160, 326
    solar time, 381
    solar wind, 455–457
Superluminal motion, 31–32
Support of a function, 117
Survey instruments, 21–23
Swarup and Yang system, 185–186
Synchrotron radiation, 4, 97, 249, 461
Synthesis mapping:
    definition, *vii*
    simplified discussion, 51–53
Synthesized beam (dirty beam), 169, 170,
        322–325, 334, 338, 342–349, 400,
        476
System temperature, 10, 157–168, 315
    for cascaded components, 181
    correction for atmospheric absorption, 420
    measurement of, 207

Tangent plane, 81, 85
Taper, *see* Gaussian taper; Weighting
T-array, 22, 28, 119–120, 132
Taurus A, 17
Tectonic plates, 3, 383
Telephone signal transmission, 18, 250

Temperature:
  antenna, 10
  receiver, 10
  system, *see* System temperature
Temperature coefficient of length, 184
Thomson scattering (incoherent backscatter),
    446, 455
3C10, 346
3C33.1, 24
3C48, 4, 8, 9
3C273, 31, 371, 458
Time:
  averaging, 174–178
  definition of second, 285
  International Atomic (IAT), 285, 381
  multiplexing, 193
  solar, 381
  transfer methods, 291–292
  UT, 291, 381
Timing accuracy, 94, 291–292
Tipping scan calibration, 420
Tolerances:
  bandpass (frequency response), 196–200, 201
  delay-setting, 200–201
  polarization, 201–202
  three-level sampling, 231–234
Tomography, 326
Total electron content, 440, 445, 446, 447, 449.
    See also Dispersion measure
Toyokawa, Japan, 28
Transfer function, 123–125, 140, 169, 170, 321
  space VLBI, 138–139
  VLA, 134
  VLB Array, 137
Transmission lines, *see* Cables; Local oscillator,
    synchronization; Optical fibers; Radio
    link; Waveguide
Travelling Ionospheric Disturbances (TIDs),
    448–449
Troposphere, *see* Atmosphere, neutral
Turbulence:
  Allan variance, 434
  inner and outer scales of, 433, 454
  neutral atmosphere, 432–436
  spectrum of phase fluctuations, 434
  structure function of phase, 433–435
Two-dimensional array, 52–53, 131–138

(u, v) plane, 80, 81
  CLEAN algorithm, 349
  coordinates, 48, 79–81
  coverage, 125–139, 343. *See also* Transfer
    function

holes in coverage, 139, 337–338
interference susceptibility, 476
interpolation in, 107–109, 327–334
loci, 86–90
(u, v, w) components, 79, 80, 83, 86–88, 120,
    326
(u', v') plane, 83–85, 90–93
  fringe-frequency averaging, 474–476
  visibility averaging, 175–178
Uncertainty principle, 33, 282, 496
Undersampling, 213
Uniform weighting, 239
Unit rectangle function, 197, 329
Universal Time, 380–383

van Cittert-Zernike theorem, 55, 62–63, 73–76,
    453
VanVleck relationship, 218, 223
VanVleck-Weisskopf profile, 419
Varactor diode, 193, 288
Venus, 211
Vermilion River Observatory, Illinois, 136
Very Large Array (VLA):
  antenna configuration, 131–135
  antenna illumination measurement, 492
  atmospheric phase noise, 435, 436
  CLEAN algorithm, 346, 349
  delay increments, 200
  images, 29, 346, 362, 363
  interference thresholds, 472, 476, 478
  phased-array mode, 308
  phase switching, 205, 206
  practical considerations, 335
  self-calibration, 362
  snapshot mode, 134, 338
  (u, v) plane tracks (transfer function), 134
  VLBI element, 308
Very Long Baseline Array (VLBA), 137, 138
Very long baseline interferometry, *see* VLBI
Video spectrum, 213
Virgo A, 17
Visibility:
  averaging, 174–178
  complex, 14, 52
  complex, defined, 55
  frequencies, 92–93
  fringe, 12
  model fitting, 320–321
  most likely value, 262
  reduction due to phase noise, 428–432, 453,
    454
  stellar envelope model, 321
  Taylor expansion of, 320–321

Visibility-brightness relationship, 54–56
VLA, *see* Very Large Array (VLA)
VLBI, 29–33, 135–139, 247–313
   antenna polarization angle, 104
   antennas:
      nonidentical, 97
      in space, 138–139
   arrays, 135–138
   astrometry, 374–377, 398
   atmospheric limitations, 342, 429–432,
      435–436
   bandwidth synthesis, 305–308
   calibration sources, 251, 317
   clock errors, 253–255, 257
   closure phase, 358
   coherence time, 248, 277–280
   coherent and incoherent averaging, 266–269
   data encoding, 292–295
   development, 3, 29–33, 247–249, 458
   discrete delay step loss, 301–304
   double sideband system, 155, 168
   field of view, 169
   fractional bit shift loss, 301–304
   frequency standards, precise, 280–289
   fringe rotation, 296–299, 304
   fringe sideband rejection loss, 299–301, 304
   geodesy, 32, 383–384
   group delay, 256, 305
   hybrid mapping, 356–364, 366
   interference sensitivity, 472, 478–480
   local oscillator stability, 290
   Mark I system, 248, 267, 294–295
   Mark II and III systems, 294–295
   networks, 135–136, 250
   phase calibration system, 290–291
   phased-array elements, 308–310
   phase noise, 167, 342
   phase stability, 269–280
   polar motion observations, 382–383
   probability distributions, 259–266
   quantization loss, 248, 304, 310
   recording systems, 292–295
   relativistic bending measurements, 457–458
   retarded baseline, 258–259
   satellite link, 30, 247, 250
   sideband separation, 155, 168
   signal-to-noise ratio, 268–269, 296–304
   spectral line, 241, 257–258, 294, 295, 300,
      302
   TID, observation of, 449
   time synchronization, 291–292
   water vapor radiometry, 439
Voltage reception pattern, 63–64, 115
   measurement of, 490–492

W3(OH), 267
W49, 384
Walsh functions, 203–205
Water vapor:
   22-GHz line, 425
   absorption, 417–419
   compressibility factor, 427
   maser, 267, 384–388
   refractivity, 407–409, 426–428
   resonance model, 421–426
   turbulence, 433
   worldwide distribution, 409
Water vapor radiometry, 315, 436–439
Waveguide, 183
w component, 82, 91, 365, 370, 477
Weighting:
   antenna excitation, 119
   functions:
      atmospheric, 430
      spatial, 132, 321–326, 343, 346
      spectral, 239–240
   natural, 164, 325, 326
Westerbork synthesis radio telescope, 28, 104,
      128, 129, 308, 330, 335
Westford, Massachusetts, *see* Haystack
      Observatory
White light fringe, 250, 494
Wide field mapping, 85–86, 174
Wiener-Khinchin relation, 45, 63, 76–77, 116,
      157, 213, 477
WWV, 291, 381

Young's slit interferometer, 32
Y-shaped array, 132

Zeeman effect, 97, 287, 289
Zero spacing problem, *see* Short-spacing data